About the Editor

Roger S. Gottlieb is professor of
philosophy at Worcester Polytechnic
Institute. He is the author or editor
of fourteen books and more than fifty
articles on political philosophy, religious
life, the Holocaust, environmentalism,
and disability, including *A Greener Faith:
Religious Environmentalism and Our
Planet's Future* (OUP 2006), *This Sacred
Earth: Religion, Nature, Environment*
(second edition, 2003), and *Joining
Hands: Politics and Religion Together
for Social Change* (2002). He writes a
column for the national magazine *Tikkun*
and serves on the editorial boards of four
scholarly journals.

THE OXFORD HANDBOOK OF

RELIGION AND ECOLOGY

THE OXFORD HANDBOOK OF

RELIGION
AND ECOLOGY

Edited by

ROGER S. GOTTLIEB

OXFORD
UNIVERSITY PRESS
2006

OXFORD
UNIVERSITY PRESS

Oxford University Press, Inc., publishes works that further
Oxford University's objective of excellence
in research, scholarship, and education.

Oxford   New York
Auckland   Cape Town   Dar es Salaam   Hong Kong   Karachi
Kuala Lumpur   Madrid   Melbourne   Mexico City   Nairobi
New Delhi   Shanghai   Taipei   Toronto

With offices in
Argentina   Austria   Brazil   Chile   Czech Republic   France   Greece
Guatemala   Hungary   Italy   Japan   Poland   Portugal   Singapore
South Korea   Switzerland   Thailand   Turkey   Ukraine   Vietnam

Copyright © 2006 by Oxford University Press, Inc.

Published by Oxford University Press, Inc.
198 Madison Avenue, New York, New York 10016

www.oup.com

Oxford is a registered trademark of Oxford University Press

Library of Congress Cataloging in Publication Data
The Oxford handbook of religion and ecology / edited by Roger S. Gottlieb.
p. cm.
Includes bibliographical references and index.
ISBN-13 978-0-19-517872-2
ISBN 0-19-517872-6
1. Ecology—Religious aspects.   I. Gottlieb, Roger S.
BL65.E36O94 2006
201'.77—dc22   2006004401

1 3 5 7 9 8 6 4 2

Printed in the United States of America
on acid-free paper

To every human being who has the courage to love;

And to the trees and birds, swamps and mountains,

air, water, and earth—who sustain us all.

ACKNOWLEDGMENTS

Once again thanks to the supportive and extremely competent staff at Oxford University Press, especially its editor for this book, Theo Calderara.

Roger S. Gottlieb, "Religious Environmentalism in Action," first appeared in *A Greener Faith: Religious Environmentalism and Our Planet's Future* (New York: Oxford University Press, 2006).

Contents

PART II: RELIGION AND ECOLOGY:
CONFLICTS AND CONNECTIONS

ABBREVIATIONS

AAEC   African Association of Earthkeeping Churches
AIC   African Initiated Churches
AZTREC   Association of Zimbabwean Traditional Ecologists
COEJL   Coalition on the Environment and Jewish Life
DEFRA   Department for Environment, Food, and Rural Affairs
EEN   Evangelical Environmental Network
ERN   *The Encyclopedia of Religion and Nature.* Edited by Bron
                Raymond Taylor. 2 vols. London: Continuum, 2005.
IEEN   International Evangelical Environmental Network
NCC   National Council of Churches
PG   Patrologia graeca
PL   Patrologia latina
UCC   United Church of Christ
WCC   World Council of Churches
ZIRRCON   Zimbabwean Institute of Religious Research
                and Ecological Conservation

Contributors

DAVID LANDIS BARNHILL is director of environmental studies and professor of English at the University of Wisconsin–Oshkosh. He has published translations of the Japanese nature poet Bashō (*Bashō's Haiku* and *Bashō's Journey*), edited an anthology of nature essays (*At Home on the Earth*), coedited the anthology of essays *Deep Ecology and World Religions*, and written articles on Gary Snyder. He is currently working on a book analyzing radical politics in American nature writing.

JOHN BERTHRONG, educated in Sinology at the University of Chicago, has been the associate dean for academic and administrative affairs and associate professor of comparative theology at the Boston University School of Theology since 1989. Active in interfaith dialogue projects and programs, his teaching and research interests are in the areas of interreligious dialogue, Chinese religions, and comparative philosophy and theology. His publications include *All under Heaven: Transforming Paradigms in Confucian-Christian Dialogue*; *The Transformations of the Confucian Way*; and *Concerning Creativity: A Comparison of Chu Hsi, Whitehead, and Neville*. He is coeditor of a volume on Confucianism and ecology. In 1999 he published *The Divine Deli*, a study of religious pluralism and multiple religious participation in North America. Most recently he collaborated with Evelyn Nagai Berthrong on *Confucianism: A Short Introduction*.

CHRISTOPHER KEY CHAPPLE is professor of theological studies and associate academic vice president at Loyola Marymount University. He has published several books, including *Karma and Creativity*; *Nonviolence to Animals, Earth, and Self in Asian Traditions*; and *Reconciling Yogas*. He has three books on the topic of religion and ecology: *Ecological Prospects: Religious, Scientific, and Aesthetic Perspectives*; *Hinduism and Ecology: The Intersection of Earth, Sky, and Water* (coeditor); and *Jainism and Ecology: Nonviolence in the Web of Life*. He serves on the advisory boards for the Green Yoga Association, the Forum on Religion and Ecology, the Global Ethics and Religion Forum, the Ahimsa Center, and the Yadunandan Center for India Studies.

JOHN CHRYSSAVGIS cofounded St. Andrew's Theological College in Sydney (1985), where he was lecturer in the divinity school (1986–90) and the school of studies in religion (1990–95) at the University of Sydney. In 1995 he was appointed professor of theology at Holy Cross School of Theology, directing the religious studies program at Hellenic College until 2002 and serving as academic dean of both schools. He established the environment office at the same school in 2001. His recent publications include *Soul Mending: The Art of Spiritual Direction*; *In the Heart of the Desert: The Spirituality of the Desert Fathers and Mothers*; *Letters from the Desert: A Selection from Barsanuphius and John*; *Light through Darkness*; *John Climacus: From the Egyptian Desert to the Sinaite Mountain*; and *Beyond the Shattered Image*. He also edited the official volume on the ecological initiatives of the Ecumenical Patriarch Bartholomew, *Cosmic Grace, Humble Prayer*, and serves as theological advisor to the Ecumenical Patriarch on environmental issues.

JOHN B. COBB JR. received his graduate education at the University of Chicago Divinity School. He is emeritus professor of theology at the Claremont School of Theology and founding codirector of the Center for Process Studies. Among his writings are *Is It Too Late? A Theology of Ecology*; *The Liberation of Life: From the Cell to the Community* (coauthor); *For the Common Good: Redirecting the Economy toward Community, the Environment, and a Sustainable Future*; *Sustainability*; and *The Earthist Challenge to Economism*.

MARTHINUS L. DANEEL is emeritus professor of missiology at the University of South Africa; currently part-time professor of missions at Boston University School of Theology; codirector of the Center for Global Christianity and Mission; and director of the Ecumenical Foundation of Zimbabwe for ongoing research and development work in Zimbabwe. He studied and served African Initiated Churches in Zimbabwe for forty years. His numerous publications on AICs and African traditional religion include several rated as classics: *God of the Matopo Hills*; *Quest for Belonging: Introduction to a Study of African Independent Churches*; and *Guerrilla Snuff* (under the pseudonym Mafuranhunzi Gumbo)—the latter recently selected as one of the seventy-five classics of Zimbabwean literature of the twentieth century.

CALVIN B. DEWITT is professor of environmental studies in the Gaylord Nelson Institute for Environmental Studies in the University of Wisconsin–Madison; president emeritus of Au Sable Institute of Environmental Studies, Mancelona, Michigan, and Coupeville, Washington; cofounder of the Evangelical Environmental Network; and the president of the Academy of Evangelical Scientists and Ethicists.

O. P. DWIVEDI, Order of Canada, PhD, LLD (hon.), FRS (Canada), is university professor emeritus in the department of political science at the University of Guelph, Canada. He has published thirty-two books on various subjects, including several on environmental issues such as *Environmental Crisis and Hindu Religion* (coauthor); *World Religions and the Environment* (edited); *Environmental*

Ethics: Our Dharma to the Environment; India's Environmental Policies, Programmes, and Stewardship; and *Sustainable Development and Canada* (coauthor).

RICHARD C. FOLTZ is associate professor of religion at Concordia University. He has published numerous books and articles on issues dealing with the Muslim world, particularly Iran.

ROGER S. GOTTLIEB is professor of philosophy at Worcester Polytechnic Institute. He is the author or editor of fourteen books and more than seventy articles on political philosophy, religious life, the Holocaust, environmentalism, and disability. He is editor of five academic book series; book review editor of *Social Theory and Practice* and *Capitalism, Nature, Socialism;* on the editorial board of *Worldviews;* and has a column in the national magazine *Tikkun.* His earlier books include *This Sacred Earth: Religion, Nature, Environment* (known internationally as the first comprehensive collection on the topic); *A Spirituality of Resistance: Finding a Peaceful Heart and Protecting the Earth;* and *Joining Hands: Politics and Religion Together for Social Change.* Most recently he has published *A Greener Faith: Religious Environmentalism and Our Planet's Future.*

JOHN A. GRIM is a visiting scholar at Yale University and series coeditor of World Religions and Ecology, from Harvard Divinity School's Center for the Study of World Religions. Until recently he was professor of religion at Bucknell University, where he taught courses in Native American and indigenous religions and religion and ecology. His published works include *The Shaman: Patterns of Religious Healing among the Ojibway Indians* and coedited volumes entitled *Worldviews and Ecology: Religion, Philosophy, and the Environment* and *Indigenous Traditions and Ecology: The Interbeing of Cosmology and Community.* He is currently president of the American Teilhard Association.

JOHN HART is professor of Christian ethics at Boston University School of Theology. His books include *The Spirit of the Earth* and *Sacramental Commons.* He was a delegate of the International Indian Treaty Council to the U.N. International Human Rights Commission and involved with the *Earth Charter.* He has ghost-written several church documents and lectured on social ethics and theology-ecology relationships on four continents: in seven countries and twenty-seven U.S. states.

STEPHANIE KAZA is professor of environmental studies at the University of Vermont, where she teaches religion and ecology, ecofeminism, and unlearning consumerism. Kaza is a longtime practitioner of Soto Zen Buddhism, affiliated with Green Gulch Zen Center, California; she has also studied with Thich Nhat Hanh and Joanna Macy. She is the author of *The Attentive Heart: Conversations with Trees* (meditative essays on deep ecological relations with trees) and coeditor of *Dharma Rain: Sources of Buddhist Environmentalism.* Her latest book is *Hooked! Buddhist Writings on Greed, Desire, and the Urge to Consume.*

SALVADOR LEAVITT-ALCANTARA is a graduate student of systematic theology at the Graduate Theological Union in Berkeley, California. He received a Master of Arts in theology from Harvard University and is a recipient of the Hispanic Theological Initiative scholarship. He is studying the contributions of Latin American liberation theologies to global concepts of violence, justice, democracy, and environmental issues. He was born and raised in El Salvador.

ANDREW LINZEY is a member of the faculty of theology in the University of Oxford and holds the world's first post in theology and animal welfare: the Bede Jarrett Senior Research Fellowship at Blackfriars Hall, Oxford. He is also honorary professor of theology in the University of Birmingham and special professor at Saint Xavier University, Chicago. He has written or edited twenty books, including *Animal Theology; Animal Gospel;* and *Animal Rites: Liturgies of Animal Care;* and is coeditor of *Animal Rights: A Historical Anthology.*

LOIS ANN LORENTZEN is professor of social ethics, associate director of the Center for Latino Studies in the Americas, and principal investigator for the Religion and Immigration Project at the University of San Francisco. She has authored or edited seven books and has written numerous articles in the fields of environmental ethics, religion and immigration, and gender and the environment. She is a former wilderness guide and misses it desperately.

DANIEL C. MAGUIRE is a professor of ethics at Marquette University and past president of the Society of Christian Ethics. He is the author or editor of thirteen books and some two hundred articles and president of the Religious Consultation on Population, Reproductive Health, and Ethics, an international collegium of eighty scholars from all the world religions. *Ms Magazine* in their tenth anniversary issue listed Maguire as "one of the forty male heroes of the past decade, men who took chances and made a difference."

JAMES MILLER is assistant professor of east Asian traditions and coordinator of the graduate program in religion and modernity at Queen's University, Canada. He is the author of *Daoism: A Short Introduction;* editor of *Chinese Religions in Contemporary Societies;* and coeditor of *Daoism and Ecology.* His current research focuses on the intersection of religion, nature, and modernity in contemporary China.

JACOB OLUPONA is professor and director of African American and African studies program at the University of California, Davis. His research interest includes religion and Immigration, religion of traditional and modern Africa, and African religions in the Americas. He is the author of *Kinship, Religion and Rituals in a Nigerian Community: a Phenomenological Study of the Ondo Yoruba Festival*

(1983), editor and co-editor of several works including *African Spirituality* (2003) and *Beyond Primitivism: Indigenous Religious Traditions and Modernity* (2004). He is the co-editor of a University of Wisconsin book series on African and African Diaspora religions and he recently served as one of the associate editors of the recently published second edition of the *Encyclopedia of Religions*. Olupona is completing a book titled *The City of 201 Gods: Ile-Ife in Time, Space and the Imagination*. With a grant from the Ford Foundation, Olupona has been pioneering a new research project on African Immigrant Religious Communities in the United States. He received the John Simon Guggenheim Fellowship in 1997 and was the Davidson Distinguished Visiting Professor in the Humanities at the International University of Florida, Miami, in 1999.

HOLMES ROLSTON III is university distinguished professor and professor of philosophy at Colorado State University. He has written seven books, most recently *Genes, Genesis, and God*; *Science and Religion: A Critical Survey*; *Philosophy Gone Wild*; *Environmental Ethics*; and *Conserving Natural Value*. He gave the Gifford Lectures at the University of Edinburgh in 1997–98. Rolston has spoken as distinguished lecturer on seven continents. He is featured in *Fifty Key Thinkers on the Environment* (edited by Joy A. Palmer). He received the Templeton Prize in Religion in 2003, in amount about $1.3 million, greater than a Nobel Prize, and the largest such award in the world. The award was given by Prince Philip in Buckingham Palace.

ROSEMARY RADFORD RUETHER is the Carpenter Professor Emeritus of Feminist Theology at the Graduate Theological Union in Berkeley, California. She is currently teaching as a visiting scholar in feminist theology at the Claremont Graduate University and Claremont School of Theology. For twenty-eight years she was the Georgia Harkness Professor of Applied Theology at the Garrett Theological Seminary and Northwestern University in Evanston, Illinois. She is author or editor of twenty-eight books and twelve book collections in the areas of feminist and liberation theologies.

H. PAUL SANTMIRE is the author of *Brother Earth: Nature, God, and Ecology in a Time of Crisis*; *The Travail of Nature: The Ambiguous Ecological Promise of Christian Theology*; and *Nature Reborn: The Ecological and Cosmic Promise of Christian Theology*. His Harvard doctoral dissertation (1966) was on Karl Barth's theology of nature. He also served as a coauthor of statements on the environment by the Lutheran Church in America (1972), the Evangelical Lutheran Church in America (1993), and the National Council of Churches of Christ (2005).

THOMAS A. SHANNON is professor emeritus in the department of humanities and arts at Worcester Polytechnic Institute. He is the author or coauthor of numerous articles and books on bioethics and social justice, including *The New Genetic Medicine*; *Catholic Perspectives on Peace and War*; and *Introduction to Contemporary*

Bioethics. He is the editor of a series of readers in bioethics published by Sheed and Ward: *Reproductive Technologies; Death and Dying; Health Care Policy*; and *Genetics: Science, Ethics, and Public Policy*. He is an associate editor of *Modern Catholic Social Teaching: Commentary and Interpretation.*

LISA H. SIDERIS teaches religious ethics and environmental ethics in the department of religious studies at Indiana University in Bloomington. She has particular interests in the intersection of science, religion, and the environment and in Darwinism generally and has published *Environmental Ethics, Ecological Theology, and Natural Selection* on this subject. She is currently coediting a volume of interdisciplinary essays on the life and work of Rachel Carson.

BRON TAYLOR is the Samuel S. Hill Eminent Scholar at the University of Florida, where he has led the development of its graduate program in religion and nature. A winner of numerous teaching and research awards, he is editor-in-chief of the *Encyclopedia of Religion and Nature*. His edited volume *Ecological Resistance Movements: The Global Emergence of Radical and Popular Environmentalism* was one of the first volumes to critically examine such social movements around the world, and he has written many articles about such movements and their political, religious, and ethical dimensions. A gateway to the encyclopedia, the graduate program in religion and nature, his own work, the society, and the *Journal for the Study of Religion, Nature, and Culture* can be found at www.religionandnature.com.

HAVA TIROSH-SAMUELSON is professor of history at Arizona State University in Tempe. She holds a PhD in Jewish philosophy and kabbalah from the Hebrew University of Jerusalem (1978) and a BA in Religious Studies from SUNY in Stony Brook (1974). Prior to joining the faculty of Arizona State University, she taught at Columbia University, Emory University, and Indiana University. Her research focuses on medieval and early-modern Jewish intellectual history, feminism and Jewish philosophy, and Judaism and ecology. In addition to many articles and book chapters, she is the author of *Between Worlds: The Life and Work of Rabbi David ben Judah Messer Leon*, which received the award of the Hebrew University of Jerusalem for the best work in Jewish history for 1991, and *Happiness in Premodern Judaism: Virtue, Knowledge, and Well-Being in Pre-modern Judaism*. She is also the editor of *Judaism and Ecology: Created World and Revealed World* and *Women and Gender in Jewish Philosophy*. She is currently at work on a book tentatively titled *Judaism and Nature* and edits *Judaism and the Phenomenon of Life: The Legacy of Hans Jonas.*

MARY EVELYN TUCKER is a visiting scholoar at Yale University. She jointly organized a series of ten conferences on World Religions and Ecology at Harvard Divinity School's Center for the Study of World Religions and is series coeditor for the ten volumes from the conferences. She is the author of *Worldly Wonder:*

Religions Enter Their Ecological Phase and *Moral and Spiritual Cultivation in Japanese Neo-Confucianism.* She coedited *Worldviews and Ecology; Buddhism and Ecology; Confucianism and Ecology; Hinduism and Ecology; When Worlds Converge;* and *Confucian Spirituality.*

RELIGION AND ECOLOGY

INTRODUCTION

Religion and Ecology—What Is the Connection and Why Does It Matter?

ROGER S. GOTTLIEB

FROM NATURE TO ENVIRONMENT

FOR as long as human beings have practiced them, the complex and multifaceted beliefs, rituals, and moral teachings known as religion have told us how to think about and relate to everything on earth that we did not make ourselves. Whether as "nature," "creation," the "ten thousand things," or "all our relations," humanity's surroundings were both a gift and a problem. Because they were the source of our sustenance, a source for which we clearly were not responsible, they were a gift. Because we had to think about what they meant morally and spiritually, and because while we had to use them to meet our needs, it often seemed intuitively clear that in some sense these other beings had their own integrity, purposefulness, and value, they were a problem.

In the Western monotheistic religions, there was (until recently) a general consensus that we should never forget that nature was God's creation, and ultimately God's property, not our own. However, once divine ownership was acknowledged, we had the full right to use that property to our own advantage. Because of our distinctive spiritual nature we had been given special privileges among all other inhabitants of the earth. As long as we did not waste what we used (and there was rather wide latitude about what legitimate use included) we could do as we wished.[1] On the other hand, both individual injunctions concerning our treatment of nature (e.g., not to muzzle the ox while it threshes grain, to leave spontaneous growth in the fields for "wild animals" during the sabbatical year; see Deut. 20:19–20; 22:6–7; 25:4) and local customs (bringing animals to church for blessing, desert monks being instructed in the gospel by animals) seemed to contradict—or at least limit—the ontological gulf which much of scripture emphasized.

In other religious traditions the distinction between humans and nature was not nearly so clear. Indigenous traditions for the most part saw the natural world as "peopled" by beings with whom it was necessary to cultivate mutually respectful relationships. Daoism viewed humans as an essential part of nature, eschewing as well any fixed distinction between the mind or soul and the body. In Hinduism the entire universe is God, and for Buddhism reincarnation as an animal in a future life is fully compatible with being a human in this one. And in any case the goal of a realized Buddhist (at least in the Mahayana tradition) was to ease the suffering of "all sentient beings," not just of people. Often, however, this more encouraging metaphysical attitude was unaccompanied by actual care for the natural world—and in any case Eastern religions did not have much of a prophetic tradition with which to galvanize adherents to socially critical responses to injustice to people or nature. And with the advent of modernity (or perhaps much earlier) indigenous traditions were marginalized by modern states.

Now all this has changed. Varying perspectives which seemed more or less adequate to the first fifteen thousand or so years of human history have been rendered, if not irrelevant, then clearly insufficient by the environmental crisis. This crisis has at least eight major dimensions, each of which by itself would be a critical problem, but all of which together make for perhaps the most significant challenge human beings have ever faced:[2]

1. *Global climate change* has already damaged, and will damage at an increasing rate, agriculture, wild lands, and animals; raise the ocean level and precipitate more intense storms and worse draughts; expand the range of tropical insects and diseases and kill coral; and in all likelihood have effects that we cannot foresee.
2. A staggering accumulation of *chemical, heavy metal, biological, and nuclear wastes* is found in every region, no matter how remote, and leads to a plague of environmentally caused diseases—most obviously the dramatic increase in cancer, immune-system problems, and birth defects.
3. From overuse of chemical agriculture and the destruction of forests, the *loss of topsoil* threatens the production of food throughout the developing nations and leads to erosion and desertification everywhere. Massive erosion can also destroy ecosystem balance in rivers and coastal fishing areas.

4. In what some call a *crisis of biodiversity*, the decimation of habitats through expanding human settlements, logging, mining, agriculture, and pollution and the killing of animals for sport, use, or food have raised rates of extinction to the highest they have been for sixty-five million years. Potential medicines vanish, ecosystems are destabilized, water supplies threatened, and irreplaceable natural beauties are lost forever. As we witness the harm we are doing we also lose ethical confidence in humanity's own worth.

5. *Loss of wilderness* is seen in the increasing rarity of ecosystems that are free to develop without human interference or intrusion. Besides the dwindling of biodiversity that this entails, human beings face a paradoxical loneliness. People are everywhere; yet we are haunted by a deep loneliness for those natural others who have been our companions for biological ages.

6. The last examples of human communities integrated into nonhuman nature are giving way to *devastation of indigenous peoples*. As their environments are poisoned, native peoples lose their land and culture and too often their lives.[3]

7. *Unsustainable patterns and quantities of consumption* deplete natural resources and contribute to global warming and the accumulation of waste. In the underdeveloped world, overpopulation relative to existing technological resources and political organization decimates the landscape.

8. *Genetic engineering* menaces us with the dismal prospects of engineered life-forms and the potentially catastrophic invention of insufficiently tested organisms. Given our track record with nuclear wastes and toxic chemicals and our political and economic elites' pronounced tendency to short-sightedness and greed, it seems highly doubtful that we are ready to create new life-forms in a cautious and sensible way.

The sheer scope of this crisis means that *nature*—however it was thought of before this time—has been transformed into something new: the *environment*, that is, a nonhuman world whose life and death, current shape and future prospects, are in large measure determined by human beings. If the rest of the universe is beyond our reach, the earth—or at least the earth's atmosphere, surface, waters, and ecosystems—plainly is not. In a sense modern industry, development, land use, and technology means that if a clear-cut distinction between nature and people was ever possible, it is so no longer. Human beings and the environment now form a dialectical totality, each side affecting, and being affected by, the other. If we still depend on nature for food and water, air and minerals, every wild ecosystem depends on some political arrangement for protection, and every living thing is affected by human-made climate change, importation of exotic species, habitat loss, and pollution.

While environmental devastation is not in and of itself a new thing, the scope of the current crisis makes for a totally new life situation. Millennia ago overirrigation may have destroyed the fertility of much of Babylonia, and Native Americans may have extinguished several species of megafauna before they developed their nature-honoring spiritual traditions, but humans simply did not have the power to transform climate, initiate mass extinctions, or make sunlight more dangerous. Above all, earlier ecological problems were *local*—confined to a region, a community, even an empire. Our plight today is *global*: there simply is no escape from it on this planet.

It should be stressed that the environmental crisis is not just a problem "out there." It has decisively changed people as well, inscribing itself in our bloodstreams, our breasts and prostates, our very mothers' milk, all of which carry unhealthy amounts of toxins. It also taints our sense of what is to come, as we realize, perhaps only subliminally, that the future is likely to be worse than the past. The environmental crisis includes not only devastating particular events (Chernobyl, Bhopal, a thousand-square-mile dead zone in the Gulf of Mexico) but a way of life which appears as a slow, seemingly unstoppable, deterioration in the quality of air, food, water, land, and life.[4] At the worst, the crisis provokes a deep fear that the earth will cease to be a healthy setting for human life.[5]

A CHANGED WORLD, A CHANGED FAITH

Thus the subject of this book—religion and ecology—reflects religions' historical concern with the natural world and their response to the current crisis. What have the world's faiths believed about the human relation to *nature*? And how must beliefs (and actions) change as we face the *environment*? These two questions form the heart of the study of religion and ecology.

Ecology matters to religion for a number of reasons. First, any dispassionate view of the past indicates that religion is partly responsible for the environmental crisis. Religions have been, at turns, deeply anthropocentric, otherworldly, ignorant of the facts, or blindly supportive of "progress" defined as more science, more technology, and much more "development." The first critical questions about humanity's modern relation to nature, wilderness, and industry were not raised by prominent theologians or religious leaders, but by freelance spiritual types, anticommunist Western Marxists, secular philosophers, or nature lovers.[6]

Thankfully, however, this is no longer the case. As the following essays show beyond doubt, world religion has entered into an "ecological phase"[7] in which environmental concern takes its place alongside more traditional religious focus on sexual morality, ritual, helping the poor, and preaching the word of God. In order to make this change religions have had to engage in several arduous and problematic tasks to discover their own distinctive ecological vocation.

To begin, theologians have had to *reevaluate their traditions*. Classic texts have been read, and interpreted, anew. Marginalized elements which support an ecological ethic have been recovered and stressed, and some previously unchallengeable teachings have been rejected.

This reevaluation, as necessary as it is, is not enough. Not unlike the changes which religions have had to face in coming to terms with women's equality, preserving the faith has required that it transform itself in fundamental ways. The essays in this book show that this process is underway. Theologies have been created which stress the spiritual value of nature, our kinship with the nonhuman, and our ethical responsibilities to the earth. New concepts of the divine, holiness, spiritual life, and sin are being forged. Innovative liturgies and rituals are being practiced, and a unique sense of moral responsibility that stresses the interdependence of our treatment of nature and our treatment of other people has emerged as the strikingly new concept of "ecojustice."[8] These developments are embraced by theologians, intellectuals, and laypeople, to be sure, but can also be seen at the highest levels of institutional authority. Popes and bishops, leading rabbis and mediation teachers, nationwide groups which include presidents of religious universities, heads of their most important affiliated institutions, and famous public figures have all said in no uncertain terms that what we have been doing is wrong and that it is time to change our ways.[9]

As well, and perhaps most importantly, coming to grips with the environmental crisis has meant that religious people have had to become *political and ecological activists*. It is clear to most religious environmentalists that pious words about "caring for God's creation" or "having compassion on all sentient beings" will not come to much unless there are dramatic changes in the way we produce and consume, grow food and get from place to place, build houses and use energy. Yet when environmentalists try to help create the needed changes, they frequently come up against the dominant social structures of industrialized society: profit-oriented corporations and a political elite more interested in preserving power than the environment. Consequently, religious environmentalists are mounting a widespread challenge to the prerogatives of private property and the complicity of do-nothing (or do-too-little) governments. Surprisingly—and ironically—one finds deeper and more sweeping criticisms of the environmental consequences of globalization from the National Council of Churches than from the AFL-CIO, more awareness of the pernicious consequences of an ever higher gross national product among religious people oriented to ecojustice than among some secular progressives who concentrate on who gets the wealth more than on the environmental consequences of how we produce it.

To take but one of hundreds of remarkable particular examples, consider the following statement by Catholic bishops from Alberta, Canada, a statement which reveals how *religious* authorities, once awakened to environmental issues, are led to progressive, one might even say quite radical, *political* views:

> Over the years, Albertans have lived as if the abundant forests, minerals, oil, gas and coal deposits, fertile prairie topsoil and clean air and water extended without limit.... However, times have changed. Our stewardship of this abundance is now being questioned. Our *economic model of maximizing profit* in an increasingly *global market* is unsustainable.... The issue of global climate change being pushed by rising fossil fuel consumption and deforestation goes to the heart of Alberta's *economic priorities.*
>
> The move to *large-scale corporate agriculture* in search of greater *economic efficiencies* runs the risk of destroying the agricultural foundations of fertile topsoil, clean air and water as well as the social ecology of vibrant rural human communities.
>
> The rapid, *widespread harvesting* of the boreal forest is testing the limits of ecosystem integrity and risks the future of what should be a renewable resource for future generations.[10]

In short, as religions become ecologically oriented they are at once theologically revitalized and political energized. Committed religious groups have challenged the World Bank's development programs, engaged in nonviolent civil disobedience at the Department of Energy in defense of the Arctic National Wildlife Refuge, confronted auto manufacturers on automobile fuel efficiency, and worked together with (gasp!) *scientists* to demand an ecologically responsible social order (see the discussion of these issues in part III).

OBJECTIONS TO ECOLOGICAL RELIGION

For my part, not surprisingly, all this is a good thing. The unique cultural status, financial and organizational resources, and moral experience of religions turned to saving the planet—what could be wrong with that? However, not everyone would agree. So we must at this point consider three objections that might be made to my optimistic assessment of religion's contribution to solving the environmental crisis.

To begin, some might point out that religion's proclamations about the sacredness of nature or our necessity to care for God's creation are no more important than anything else religion preaches—which is to say not particularly important at all. In the Christian-dominated United States, for example, one sees very little of the Golden Rule, not to mention "loving one's enemies" or eschewing wealth in order to follow Jesus. In supposedly Buddhist Thailand and Hindu

India, religious teachings on the importance of nonviolence are violated without much fuss being raised. [In short, what a particular religion says and what that religion's (self-proclaimed) followers actually do are two very different things.]

There is much truth in this objection. But in the end it amounts to little more than saying that most people, most of the time, go along with whatever is being done by everyone else and do so with little moral concern beyond their family, neighborhood, or village. It is only rarely—during a civil-rights movement, a revolutionary war, an active struggle for economic justice or human rights—that large numbers of people actually live out the highest aspirations of their moral code, whether that code is religious or secular, anthropocentric or environmental. So if people do not always, or even for the most part, get inspired by religion to act in moral ways, at least they do sometimes. Of course, religious support for environmental sanity cannot *guarantee* victory to the environmental movement—but then again, what can? At the very least having large, wealthy, and highly respected institutions throw some of their weight in that direction can only increase our chances of a modicum of success. If we are to make the necessary but extraordinarily difficult changes in the way we live, we will certainly benefit from every voice which can help motivate us.

A deeper complaint is that the *last* thing a democratic society needs to help solve its problems is the participation of religion in political life. The Iranian mullahs' repression of women and the American religious right's attacks on gay rights, religious incitement of ethnic violence in the Middle East and opposition to teaching evolution in the United States—all these and more show that we would be better off if religion, like sex, were practiced only by consenting adults in private.[11] The more religion in public life the more intolerance, bigotry, and rejection of the gains of human rights, women's equality, and social justice that have been won by the secular left over the last two hundred years. Religion is in fact the enemy of those movements, which have helped make society more just and humane and which are environmentalism's necessary allies in the attempt to make society sustainable as well.

I agree that there is much to lament in the religious presence in modern society. But despite the fact that conservative and fundamentalist forces in religion have been at the center of public attention for some time, a blanket claim about religion's backward social role is dreadfully sloppy. Thinking that all of religion is conservative obviously ignores those many instances in which religious political action furthered, rather than hindered, the expansion of democracy, human rights, and simple justice. In countless instances leaders such as Martin Luther King Jr., Desmond Tutu, or Ang San Suu Kyi saw themselves (and, more importantly, were seen by their followers) as embodying *religious* values as they struggled for values which are the hallmarks of *liberal, democratic* modernity. There have been powerful religious voices in the fall of communism and the challenge to American imperialism, in the antiglobalization movement and in support of feminism and even gay rights. These voices find no incompatibility between devoted faith and democracy, no tension between serious commitment to their own faith and accepting that the faith of others deserves respect as well.[12]

Further, any unbiased look at the last century of political life shows that society is endangered as much by fanaticism of the secular variety as it is by that of the faithful. For every Al-Qaeda there is a Sadaam Hussein. In fact, it could well be argued that the most effective forms of antidemocratic repression and totalitarianism have been, at least until recently, movements that were explicitly secular. It has even been suggested that central ideals of democracy such as human rights are actually rooted in religious ideas. What, ultimately, justifies the idea that "all men are created equal" if we are not all "created in the image of God"?[13] The upshot of these points is that identifying a person or a movement as either religious or secular tells us nothing about their political commitments. Violence and repression of difference are no more the province of one than the other.

The third caution against religious environmentalism comes from those who may wonder if a direct confrontation with society's politics, economics, energy policy, transportation, and agriculture is really religion's business. Perhaps the problem is not that activist religion is bad for society, but that it diminishes religion.[14] When the values of environmentalism are put into practice, after all, they involve complex changes of a nation's legal and industrial infrastructure. It is much more complicated than dismantling segregation or stopping a war. To accomplish such changes, environmentalists must engage in the political process: making alliances, promoting a partisan point of view, compromising certain principles in order to win on other issues. Above all, this kind of political activism is aimed at political *power*: to change laws, limit what corporations can do, prohibit certain kinds of production, support new technologies, and educate our children to be environmentalists rather than consumers. Yet—many religious thinkers argue—the pursuit of these kinds of political and social power is anathema to the religious goals of creating a community governed by values of love of God, discipleship of Christ, following the Mitzvot, or seeking enlightenment. Politics is grubby, and if we seek to be holy we should avoid it.

Perhaps unfortunately, as distasteful and morally complicated as the political process may be in today's world, seriously religious people are not free to refrain from it. Minimal reflections show why this is so. Obviously, any serious religious commitment includes an ethical one, and the simple fact is that in a technologically and politically globalized world, ethics requires politics. Are we to treat our neighbors as ourselves? Then our gasoline use—for shopping, commuting, or even going to Mass—had better not threaten the health and livelihood of other people. But given the relation between fossil-fuel use and global warming, that is just what is happening. To change U.S. energy policies, however—or to challenge the use of pesticides which spread throughout the world or acid rain caused by

Midwest smokestacks which can kill forests in Canada or northern Europe—requires precisely the political organization and clout which religious quietists so shun. And this is not even to mention that actively following specifically religious commands to respect God's creation or prevent needless pain to other sentient beings requires a wholesale alteration of current environmental practices. Thus as difficult as it is, religious people must engage in political life if they are to fulfill the minimum ethical requirements of their faiths—if, that is, the consequences of the way they live are not to make a mockery of what they claim are their values. (People do often make much of how hard this is, ignoring the fact that a committed religious life is always difficult: loving our enemies or giving up attachment to desire is, after all, no day at the beach.)

Spiritual Challenge, Spiritual Opportunity

Besides the fact that it has an obligation to help clean up the mess it helped make (or which it ignored), there are other reasons why religion must confront the environmental crisis. This crisis is, among other things, a spiritual problem, affecting both the passion and the intimacy of religious life.

It does so first by raising in a particularly compelling form the problem of evil. If one believes in a transcendent God we can ask—as the twentieth century has compelled us to do in increasingly urgent ways after its historically unprecedented world wars and genocides—where God is in a world filled with so much pain and loss. Of course, there is no purely logical reason why familiar solutions to the problem of evil—that suffering is produced by human freedom, that God is a mystery, that later on all will be made clear—cannot be applied in this context as well. Yet (as Hegel observed) sometimes a change in quantity leads to a change in quality. And in this case—irreversible damage permeating the fabric of the earth's life-forms—we have a scope of destruction which is so great that the problem of evil may threaten us anew.

In a way this spiritual quandary is less an issue of arguments about how God can coexist with evil than it is about our sense of God's own limits and vulnerability and about our own (in)ability to feel God's presence. If nature has ended, as Bill McKibben suggests, because the impact of human beings is everywhere, then this will surely, and tragically, diminish our sense of the sacred as found in nature.[15] We will look to the earth for comfort and find broken beer bottles on mountaintops or seabirds choking on plastic bags; we will find, that is, only ourselves. For those to whom creation embodies God's presence more than scripture does, or who do not directly hear God's voice, this is an irreparable loss. Or, at the very least, it is a challenge we have never faced before. If God is, as some say, everywhere, then she must be found in the toxic-waste dumps, the clear-cut forests, and your aunt dying of breast cancer as easily as in a majestic mountain peak or a meadow filled with wildflowers. As the Raji people, forest dwellers facing extreme deforestation on the Nepal-India border, put it: "Before, we knew where the gods were. They were in the trees. Now there are no more trees."[16]

This dilemma can arise in a number of religious settings, not only those based in transcendent monotheism. For example, the center of much traditional Buddhist meditation is using a mental focus on the breath to calm the mind or reveal the mind's deep currents of unregulated attachment, fear, or anger. Yet what is the meaning of a meditation teacher's injunction to "focus on your breathing" on a day when ground-level ozone readings have reached dangerous levels and the weather forecasts warns the old and the ill to "stay inside until the air" (the *air!*) "becomes less dangerous"?

From the most sophisticated theologian to someone who takes religion seriously but not intellectually, from the most devout monotheist to someone who is "spiritual but not religious," this problem requires a vital reorientation. As we face a decimated forest after a clear-cut or read about skyrocketing cancer rates after mining companies start to operate on native lands, we must deal with our shame and despair and struggle to retain a sense of God's presence and our belief that existence really has meaning. It remains a grave question as to how world religions will manage to do this.

What Can Religion Offer?

Once focused on the environmental crisis, the resources of religion have a distinct—and I would argue enormously valuable—role to play in trying to turn things around. It is not just a matter of tens of thousands of Catholics, Methodists, or Buddhists joining Greenpeace or demonstrating to demand stricter fuel-efficiency standards (not that those would not be good things!). If the environmental crisis means that religion has to change, it is also the case that over centuries religions have developed powerful resources to help us understand and respond to critical forms of suffering and injustice.

To begin with, we should remember that for hundreds of millions of people religion remains the arbiter and repository of life's deepest moral values. In this context, religions provide a rich resource to mobilize people for political action.

As one observer trenchantly stated: "Only our religious institutions, among the mainstream organizations of Western, Asian, and indigenous societies, can say with real conviction, and with any chance of an audience, that there is some point to life beyond accumulation."[17] More broadly we might say that it is part of the essential role of religion in social life to serve as a realm in which the pursuit of power, pleasure, and wealth is suspended in favor of attention to and conformity with humanity's "ultimate concern" (borrowing the phrase from Paul Tillich). Religion prompts us to pursue the most long-lasting and authentic values. If religious institutions themselves often reveal an all-too-common attachment to secular goals, our disappointment with them only underscores how such behavior violates our expectations.

Thus if religious leaders start to preach a green gospel, condemning human treatment of nature for its effects on the nonhuman as well as the human—it is likely to have more of an effect than statements by, say, a comparable number of college professors. To take but one example, consider the by-now well-known 1997 pronouncement by Bartholomew, spiritual leader of three hundred million orthodox Christians worldwide: "To commit a crime against the natural world is a sin . . . to cause species to become extinct and to destroy the biological diversity of God's creation . . . to degrade the integrity of the Earth by causing changes in its climate, stripping the Earth of its natural forests, or destroying its wetlands . . . to contaminate the Earth's waters, its land, its air, and its life with poisonous substances—these are sins."[18] Bartholomew went on to suggest that we should seek "ethical, legal recourse where possible, in matters of ecological crimes."[19]

It is important to note that Bartholomew did not simply say that bad environmental policies are wasteful, inefficient, or troublesome. In using the language of "sin," he asserted that these matters are of the highest importance, essential to the health not only of our rivers and lungs, but of our souls as well. I believe that many in the general public, and in the ostensibly secular environmental community as well, can resonate with the inherent solemnity of this language and the moral posture it expresses. With or without a literal belief in God or scripture, our society can make the requisite changes to keep us from drowning in our own wastes only if we take our relations to nature with grave seriousness, a seriousness which is still, despite everything, perhaps best articulated by religion.

Though often criticized as being an escape from real life, religion can also be of service in one of the central problems facing the human community: our tendency for avoidance and denial in the face of the looming environmental threat. Both as individuals and in our major institutions we expend a good deal of energy in simply pretending that these problems do not exist. Individually, we read the morning paper and skip over the latest environmental disaster story ("30% of world's coral dying from human causes"; "Burning of rainforest consumes 73,000 square miles"; "U.S. government gives $14 billion for road building project in Boston, $14 for public transportation"), immediately suppressing that slightly queasy mix of fear, anxiety, and guilt by focusing on something, anything, else.[20] On a global level many of the world's most powerful institutions—governments, the World Bank and World Trade Organization, major corporations from Exxon-Mobil to Ford—act as if business as usual can continue indefinitely. Their executives seem to assume that the oil will last forever and that using it does not change the climate; that large dam projects are not by and large destructive (even though their own reports tell them that they are); that the earth will continue to be fertile no matter how many chemicals we drench it with; or that it makes some kind of sense to spend ten times as much money on advertising their environmental responsibility than on exercising it. They seem to think that even though the climate changes, the land becomes less fertile, and clean water is increasingly hard to find, their own children will be exempt from the social turmoil and health dangers that will arise.

If religions themselves are often oriented to their own forms of escapism, they are also at times deeply immersed in realities which are frightening. They can provide a saving impulse to face life—and our own moral failings—as they actually are. Catholic confession can be trivialized or serve as a constant reminder to assess one's behavior. Through a special prayer intended to honor the memory of those we have lost, the reality of death is brought into every single Jewish worship service. Buddhist meditation can tell us to focus on our own mortality in distressingly concrete ways (what will your body look like, one is asked, in a hundred years). Thus religion attuned to the environmental crisis can help us, in Joanna Macy's phrase, "sustain the gaze,"[21] that is, to focus on what is actually happening long enough to see what we are doing to ourselves, our planet, and our future. In helping us maintain this focus, religion can thus enable us to take at least the first step toward collective change.

ENVIRONMENTALISM AND SPIRITUALITY

If the environmental crisis represents both a deep obligation for religious response and an important opportunity for a specifically religious contribution, it is also the case that environmental movements are by their very nature hospitable to religion. This is because environmentalism (though, of course, not without some very unpleasant exceptions)[22] tends to have a spiritual dimension which other liberal or leftist political movements lack. Compared to often partial and partisan struggles for democracy, in support of rights for workers, women, or racial minorities, against colonialism, or for more economic justice, environmentalism bears remarkable and crucially important affinities with religion. These affinities make the emerging alliance between secular environmental organizations and institutional religion particularly appropriate; and they mean that at times it is quite difficult to talk about the "relations" between religion and environmentalism since the two so shade together that it becomes hard to tell them apart.

In the contemporary environmental movement even those groups totally unconnected to religiously identified organizations are often inspired by a political ideology, or at least by a moral sensibility, with powerful religious overtones. This sensibility has been present in much environmentalism since its origins in the mid-nineteenth century and has evolved into a comprehensive worldview which in many respects is often undeniably spiritual in nature.

These claims are supported by the fact that much of early conservationism itself emerged from a religious sense of the earth as God's creation, a "temple" that we should not despoil. Thus in the initial years of conservation leading voices as disparate as Thoreau, John Muir, Robert Marshall, Sigurd Olson, and John Burroughs celebrated nature not only for its physical beauty and utility, but for its spiritual value as well. As historian Michael P. Nelson observes, it is quite common for people to argue for the preservation of wilderness as a "a site for spiritual, mystical, or religious encounters: places to experience mystery, moral regeneration, spiritual revival, meaning, oneness, unity, wonder, awe, inspiration, or a sense of harmony with the rest of creation—all essential religious experiences."[23]

A similar orientation can be found within all aspects of the contemporary environmental movement. From Greenpeace to Forest Service Employees for Environmental Ethics to the environmental justice-oriented Southwest Organizing Committee, care for the earth is motivated by two inescapable principles: because "human life depends on it, and because there is a widespread sense that the earth is kin to us."[24] The Principles of Environmental Justice adopted at the historic first meeting of people of color environmental activists begins with a remarkable statement whose tone simply would not be conceivable in the Socialist Party, Democratic National Committee, NAACP, or the National Organization of Women: "Environmental justice affirms the *sacredness* of Mother Earth, ecological unity and the interdependence of all species, and the right to be free from ecological destruction."[25]

Most contemporary environmental organizations repeatedly stress that their goal is not just to save wilderness, but to protect all of life.[26] At its best the religious spirit has a similarly inclusive goal. We are all, says the Bible, made in the image of God. We all, says Buddhism, suffer and deserve release from our pain. Each community, says the Qur'an, has its own purpose and value. Any violence against one of us, teach the Jains, can only hurt us all. Looked at in this light, then, the universal missions of truly compassionate religion and of truly global environmental politics naturally converge, at least in the attempt to forge the widest possible social and ecological ethic. Both believe that life deserves a reverence that cannot be reduced to dollar value, that self-examination and spiritual practice make the most important kind of sense, and that there is more to human well-being than money, power, and pleasure.

It should be obvious that no society can function without some comprehensive framework of values. Every time we apply (or fail to apply) the Endangered Species Act or choose between energy efficiency and more oil-drilling (no matter where or with what effects), we are expressing a sense of what is important to us, how we ought to live, and what we regard with reverence. The spiritual dimension of secular environmentalism and the new religious environmentalism are joining forces to offer us a fresh choice as to how we should answer those questions. Since spirituality has been a key part of the environmental movement from its inception to the present, it makes environmental politics particularly fertile ground for an alliance with religion. This alliance has been manifest in a host of particular circumstances, including common work between the Sierra Club and the National Council of Churches and the U.K.-based Alliance for Religion and Conservation and the World Wildlife Fund.[27]

ABOUT THIS BOOK

In the last two decades the connections between religion and ecology have been manifest by explosive growth in theological writings, scholarship, institutional commitment, and public action. Theologians from every religious tradition—along with dozens of nondenominational spiritual writers—have confronted religions' attitudes toward nature and complicity in the environmental crisis. This confrontation has given rise to vital new theologies based in the recovery of marginalized elements of tradition, profound criticisms of the past, and new visions of God, the sacred, the earth, and human beings. Religious morality has expanded to include our relations to other species and ecosystems, and religious practice has come to include rituals to help us express our grief and remorse and also to celebrate what is left. Further, dialogues on how traditional religions viewed nature and how these views should be reinterpreted or altered in light of the environmental crisis now join criticisms of economics, technology, energy policies, science, transportation, agriculture, taxation, and education.

This book reflects and furthers this fundamental shift toward ecological concern and commitment—as well as the academic study of this shift. In terms of scholarship, the last twenty years have witnessed the birth of what is virtually an entirely new field: the academic study of religion and ecology. Of the many developments which this birth has occasioned, we can list a few highlights.

Harvard Divinity School has sponsored a comprehensive series of conferences and subsequent publications on the connections between ecology and virtually all of the world's religious traditions.[28] Academic journals now focus on the subject.[29] A massive two-volume encyclopedia with more than one thousand entries and a rich online resource has recently been published.[30] The University of Florida now offers a PhD concentration in religion and ecology. And an academic society focusing on the topic—the Society for the Study of Religion, Nature, and Culture—has been formed. Even a partial bibliography of relevant books, articles, and websites would run to nearly a thousand entries. Twelve years ago, while I was editing the first comprehensive text on the topic,[31] I would receive many a quizzical query from academic colleagues: "Religion and ecology? What is the connection?" Now just about everyone knows.

The writers whose essays grace these pages are among the key voices that have helped make this change possible. Within their own faith traditions, in academia, and as members of society at large, they have been leaders in goading religions, scholars, and their fellow citizens to take nature, the environmental crisis, and the connections between God, holiness, ecology, and more traditional social-justice issues more seriously. They have analyzed texts, reported on institutional changes, and themselves drafted key documents for the leaders of their own faiths. Here are just a few of the many examples of how these authors not only study this movement but are part of it: John Hart, whose essay focuses on Catholicism, helped write major public environmental statements for the church, including the Columbia River Statement of American and Canadian bishops of the Columbia River region;[32] Protestant John Cobb was part of a group writing early—and highly influential material—for the World Council of Churches;[33] Mary Evelyn Tucker, whose essay surveys the academic field of religion and ecology, is cochair of one of that field's most important resources: the Forum on Religion and Ecology; Bron Taylor, who describes the role of spirituality in American environmental activism, edited the massive and groundbreaking *Encyclopedia of Religion and Nature*; and John Chryssavgis, who wrote the essay on Orthodox Christianity, advises that religion's highest authority on environmental matters. Most of these authors, including the editor, have been active in the American Academy of Religion's Religion and Ecology section, organizing and participating in panels which have spread the word to hundreds (if not thousands) of other scholars and teachers at the premier meeting of college-level teachers of religion in the United States.

The goal of the book as a whole is to make available in one place a comprehensive, organized, and high quality survey of all this field's essential concerns. It will be of use to the scholar or intellectual who is unacquainted with the subject; to the scholar knowledgeable about one area (say, Christianity and ecology) who desires to know more about related issues (ecotheology from other traditions, religious environmental activism, etc.); to the reasonably educated nonacademic environmentalist or religious professional who is interested in this vital connection; and to the undergraduate and graduate student researching religion and ecology. The essays are serious, readable, and important.

In terms of organization, the book's three sections reflect the several dimensions of this field. Each chapter in part I is shaped by the boundaries of a particular religious tradition. Theologians, leaders, clergy, and laypeople have had to ask how their own faiths—so essential in defining their understanding of the cosmos and their guidelines for living within it—have to change to face this new reality. These essays describe the resources with which each religion began and the varying contours of their responses.

Part II contains essays which explore some of the subject's complex and multifaceted connections and internal tensions. Separate papers on genetic engineering, animal rights, population, the contested ethical and religious meanings of ecology, and ecofeminism show that despite a large amount of agreement among religious environmentalists, many controversies remain. David Barnhill's treatment of the spiritual dimension of nature writing and Holmes Rolston's essay on religion and science show the significant ways in which any study of religion and ecology takes us beyond the limits of religion considered in isolation. Mary Evelyn Tucker surveys religion and ecology as a field of academic study.

Part III is rooted in what should be the obvious premise that the ultimate goal of the religious response to the environmental crisis is moral—and hence political—activism. It surveys concrete social practices of religious environmentalism throughout the world, showing that from the United States and Latin America to south Asia and Africa people of faith now make up a vital presence in the global environmental movement. People of faith, motivated in part *by* their faith, are demonstrating in Washington, planting trees in Zimbabwe, resisting deforestation in Latin America, and seeking sustainable development in Sri Lanka. As Bron Taylor argues, such activity resonates with an international environmental movement which often has its own spiritual dimension.

It is heartening to realize that, because of the sheer number of groups and activities involved, even the excellent essays included here can only scratch the surface of what is actually going on. What they do reveal is that to a greater extent than at any previous time in history religious people from around the world are active members of a progressively oriented global movement for social change. And as Calvin DeWitt's essay shows, this activism is not limited to the "usual suspects" of Reform Jews, liberal Protestants, and dissident Catholics—but includes some socially conservative evangelical Christians as well.

It is my firm belief that religion's response to the environmental crisis, as well as to the social forces of industrialization, globalization, militarization, and consumerism which give rise to the crisis, will be the single most important factor in determining whether religion will be a vital part of humanity's future or sink into increasing irrelevance. The essays in this book—and the realities they describe—show that the world's faiths are well on their way to meeting this challenge and thus that there is every reason to expect that if humanity can somehow learn to live without destroying other species and poisoning itself, religion will have been one of the forces teaching us how to do it and encouraging us to do so.

NOTES

1. In Jewish tradition the emphasis on not wasting could be stated quite seriously. For a summary of this position in the tradition, consider the following from a widely respected nineteenth-century German rabbi: " 'Do not destroy anything' is the first and most general call of God. . . . If you should now raise your hand to play a childish game, to indulge in senseless rage, wishing to destroy that which you should only use, wishing to exterminate that which you should only exploit, *if you should regard the beings beneath you as objects without rights*, not perceiving God Who created them, and therefore desire that they feel the might of your presumptuous mood, instead of *using them only as the means of wise human activity*—then God's call proclaims to you, 'Do not destroy anything! Be a mentsh [*responsible person*]! Only if you use the things around you for wise human purposes, sanctified by the word of My teaching, only then are you a mentsh and have the right over them which I have given you as a human. However, if you destroy, if you ruin, at that moment you are not a human but an animal and have no right to the things around you. I lent them to you for wise use only; never forget that I lent them to you. As soon as you use them unwisely, be it the greatest or the smallest, you commit treachery against My world, you commit murder and robbery against My property, you sin against Me!' " Samson Raphael Hirsch, *Horeb*, chap. 56 §398, citing Babylonian Talmud, tractate *Shabbat* 105b; quoted in "Bal Taschit" on the COEJL website: http://www.coejl.org/learn/je_tashchit.shtml. For an important earlier source, see Maimonides, *Mishneh Torah*, Laws of Kings 6.10.

2. There is no end of websites, books, articles, and so on, which detail the crisis. One excellent summary of the dismal facts is Frederic Buell, *From Apocalypse to Way of Life* (New York: Routledge, 2004).

3. "The discovery of anything which can be exploited is tantamount to the crack of doom for the Indians, who are pressured to abandon their lands or be slaughtered on them. And economic discoveries do not have to be exceptional for the Indians to be plundered." Brazilian anthropologist Darcy Ribeiro, quoted in Al Gedicks, *The New Resource Wars* (Boston: South End, 1992), 13.

4. Buell, *From Apocalypse to Way of Life*.

5. Joanna Macy, *World as Lover, World as Self* (Berkeley: Parallax, 1991), 114.

6. Social critics like Thoreau, visionary poets like William Blake or Walt Whitman, disaffected political or cultural radicals like Max Horkheimer, Theodor Adorno, and Martin Heidegger, or vaguely spiritual nature lovers like John Muir.

7. Mary Evelyn Tucker, *Worldly Wonder: Religions Enter Their Ecological Phase* (Chicago: Open Court, 2004).

8. For example, consider this quote from a leading Protestant ecotheologian: "The structural institutions and systemic forms separating the haves and the have-nots in our time, [we must] . . . name them for what they are: evil. They are the collective forms of 'our sin.' They are the institutions, laws, and international bodies of market capitalism." Sallie McFague, *Life Abundant: Rethinking Theology and Economy for a Planet in Peril* (Minneapolis: Fortress, 2001), 176.

9. For an overview see Roger S. Gottlieb, *A Greener Faith: Religious Environmentalism and Our Planet's Future* (New York: Oxford University Press, 2006), chap. 3.

10. "Celebrate Life: Care for Creation, the Alberta Bishops' Letter on Ecology for October 4, 1998" (emphasis added), on the *Western Catholic Reporter* website: http://www.wcr.ab.ca/bin/eco-lett.htm.

11. There is by now an extended debate on this position. My own treatment can be found in *Joining Hands: Politics and Religion Together for Social Change* (Cambridge, MA: Westview, 2002), chap. 2. See also John Rawls, *Political Liberalism* (New York: Columbia University Press, 1996); J. Judd Owen, *Religion and the Demise of Liberal Rationalism: The Foundational Crisis of the Separation of Church and State* (Chicago: University of Chicago Press, 2001); Robert Audi and Nicholas Wolterstorff, *Religion in the Public Square: The Place of Religious Convictions in Political Debate* (Lanham, MD: Rowman & Littlefield, 1997); and Kent Greenwalt, *Religious Convictions and Political Choice* (New York: Oxford University Press, 1988).

12. For a large collection of writings which confirm this claim, see *Liberating Faith: Religious Voices for Justice, Peace, and Ecological Wisdom* (ed. Roger S. Gottlieb; Lanham, MD: Rowman & Littlefield, 2003).

13. Michael J. Perry, *The Idea of Human Rights: Four Inquiries* (New York: Oxford University Press, 1998). Others argue that John Locke, whose political philosophy is an essential foundation to modern democracy, rooted his account of political equality in an essentially religious perspective. See Jeremy Waldron, *God, Locke, and Equality: Christian Foundations in Locke's Political Thought* (New York: Cambridge University Press, 2002).

14. This viewpoint permeates the position of Stanley Hauerwas. The most effective critique of Hauerwas is Jeffrey Stout, *Democracy and Tradition* (Princeton: Princeton University Press, 2004). I also provide an extended justification for religion's need of secular political insights in *Joining Hands*.

15. Bill McKibben, *The End of Nature* (10th anniversary ed.; New York: Anchor, 1997).

16. Eric Valli, "Golden Harvest of the Raji," *National Geographic* (June 1998).

17. Bill McKibben, "Introduction," *Daedalus* 130.4 (fall 2001): 1.

18. "Address of His Holiness Ecumenical Patriarch Bartholomew at the Environmental Symposium, Santa Barbara, CA, November 8, 1997," in *This Sacred Earth: Religion, Nature, Environment* (ed. Roger S. Gottlieb; 2nd ed.; New York: Routledge, 2004), 229–30.

19. This position reverberates throughout a good deal of Orthodox Christianity. Metropolitan John of Pergamon, a theologian and church leader, points the finger at Christians as well as humanity in general: "The ecological crisis is the most serious contemporary problem facing us. To some extent the Christian tradition bears responsibility for causing it." John Pergamon, "Orthodoxy and the Ecological Problems: A Theological Approach," on the Ecumenical Patriarchate of Constantinople website: http://www.patriarchate.org/.

20. See extended discussion of this in Roger S. Gottlieb, *A Spirituality of Resistance: Finding a Peaceful Heart and Protecting the Earth* (Lanham, MD: Rowman & Littlefield, 2003), chap. 2.

21. "Guardians of the Future" (interview with Joanna Macy), on the *In Context* website: http://www.context.org/ICLIB/IC28/Macy.htm.

22. For example, Michael Zimmerman describes the congruence of Nazism and environmentalism in "Ecofascism: A Threat to American Environmentalism?" in *The Ecological Community* (ed. Roger S. Gottlieb; New York: Routledge, 1997).

23. Michael P. Nelson, "An Amalgamation of Wilderness Preservation Arguments," in *The Great New Wilderness Debate* (ed. J. Baird Callicott and Michael P. Nelson; Athens: University of Georgia Press, 1998), 168.

24. Interview with Lisa Grob, publications manager at the National Resource Defense Council; 29 Oct. 2004.

25. Charles Lee, ed., *Proceedings of the First National People of Color Environmental Leadership Summit* (New York: United Church of Christ Commission for Racial Justice, 1992), xiii (emphasis added); repr. in *This Sacred Earth: Religion, Nature, Environment* (ed. Roger S. Gottlieb; 2nd ed.; New York: Routledge, 2004).

26. A summary of similar statements from a variety of groups, including Greenpeace, Friends of the Earth International, and National Resources Defense Council, can be found in Gottlieb, *Greener Faith*, chap. 5.

27. See ibid.

28. See the Harvard Forum on Religion and Ecology website: http://environment .harvard.edu/religion.

29. *Worldviews: Religion, Culture, Environment* (http://www.brill.nl) and *Journal for the Study of Religion, Nature, and Culture.*

30. Bron Taylor, ed., *Encyclopedia of Religion and Nature* (2 vols.; London: Continuum, 2005); see http://www.religionandnature.com.

31. Roger S. Gottlieb, ed., *This Sacred Earth: Religion, Nature, Environment* (2nd ed.; New York: Routledge, 2004 [orig. 1996]).

32. John Hart's contribution was to "The Columbia River Watershed: Caring for the Creation and the Common Good," online at http://www.columbiariver.org/main_pages/press.html.

33. John Cobb helped write "Liberating Life: A Report to the World Council of Churches," in *Liberating Life: Contemporary Approaches to Ecological Theology* (ed. Charles Birch, William Eakin, and Jay B. McDaniel; Maryknoll, NY: Orbis, 1990).

PART I

TRANSFORMING TRADITION

CHAPTER 1

JUDAISM

HAVA TIROSH-SAMUELSON

JUDAISM is the religious civilization of the Jewish people whose foundational document is the Bible. Believed to be divine instruction (i.e., Torah) revealed to God's chosen people, Israel, the Bible was canonized as sacred scripture during the Second Temple period (516 BCE–70 CE). Biblical law and ethics, however, reflect not only the conditions at the time when the texts were edited and accepted as canonical, but also the life of ancient Israel during the First Temple period (ca. 900–587 BCE) and earlier centuries, as well as the interaction of Israel with ancient Near Eastern civilizations through partial acceptance or rejection. While the Bible is the basis of Jewish beliefs and practices, Judaism as a comprehensive life of Torah is a postbiblical development, the creation of the rabbis after the destruction of the Second Temple (70 CE). Even after the redaction of its main literary products, rabbinic Judaism (70–600 CE) continued to evolve during the Middle Ages through interaction with Christianity and Islam. With the dawn of modernity in the eighteenth century, rabbinic Judaism and the traditional Jewish way of life experienced serious challenges, resulting in a variety of responses that made modern Judaism highly pluralistic.

Throughout Jewish history, attitudes toward the natural world reflected both changing historical conditions of the Jewish people and foundation beliefs of the Jewish religious worldview, namely, that God, Yahweh, is the sole creator of the universe; that God created humans in his own image; that God revealed his will to Israel in the form of law, the Torah; and that God will redeem Israel and the world from the imperfection of the present. These beliefs anchor the Jewish ethics of responsibility, including responsibility toward beings and nature created by God, but they also established an intrinsic tension between the created world of nature (i.e., what is) and the revealed word of God (i.e., what should be), that is, between nature and Torah. Precisely because nature is created, Judaism does not take nature to be inherently sacred or worthy of veneration. In fact, such worship is precisely what the Bible considers to be idolatry,[1] which monotheism is determined to eradicate. Instead, nature is viewed as imperfect, requiring human management and care: only human actions in accord with divine commands sanctify nature, making it holy.

JUDAISM AND ECOLOGY:
HISTORICAL OVERVIEW

The ancient Israelites were an agrarian society that occupied a particular land, the land of Israel, which the Israelites believed had been promised to them by God. In a semiarid land that depended on rainfall and dew for its fertility, ancient Israelites developed an agricultural society that had to pay attention to environmental factors in order to struggle with land degradation and desertification. It was most likely the ecological fragility of this area and the difficulties of agriculture that generated much of the ecological insights found in the Bible.[2] A theocracy governed by a hereditary priesthood, which was barred from owning land, the ancient Israelites were mostly farmers who lived in accord with the seasonal cycle of nature, and their annual pilgrim festivals celebrated the completion of agricultural activities. Thus Sukkot (Feast of Booths or Feast of Ingathering) celebrated the harvest of summer crops and the preparation of the fields for winter; Pessach (Passover or Feast of the Unleavened Bread) began with the new moon of the month just preceding the hardening of the barley; and Shavuot (Feast of Weeks or Pentecost) celebrated the barley harvest, when reaping started (Exod. 34:22; Deut. 16:10). These agrarian activities were given historical-religious meaning in the Bible, linking them to the exodus of Israel from Egypt, the wilderness experience, and the giving of the Torah at Sinai, respectively.

Whether ancient Israel actually emerged as the Bible describes is a hotly debated issue among modern historians. Most would accept that during the Late Bronze Age (1200–1000 BCE) ancient Israel was beginning to emerge as a distinct society on the fringes of Canaanite civilization of the hill country in the land of Canaan. About 1000 BCE this society transformed itself from a confederacy of tribes led by occasional charismatic leaders (known as "judges") into a monarchy that offered leadership alongside a hereditary priesthood that communicated between the people of Israel and God by means of an elaborate sacrificial cult. Whereas animals and agricultural produce were offered to God as expressions of gratitude, only animals, defined as ritually clean, were sacrificed to God to alleviate guilt and purify ritual contamination and sinfulness.[3] Through the sacrificial cult, Israel could maintain its ritual purity, enabling God to reside in its midst. Biblical law specified in great detail the proper treatment of the soil, animals, and vegetation of the land of Israel in order to maintain Israel's religious ritual purity and moral integrity. This linkage is the distinctive contribution of Judaism to religious ecology.

In 587 BCE the political sovereignty of Judea came to an end with the destruction of the Jerusalem temple by the Babylonians. Yahwism, however, did not come to an end but was preserved in the Babylonian exile through fasts, assemblies, prophetic preaching, and, most of all, the writing and editing of the sacred texts that would eventually make up the Bible. When King Cyrus of Persia allowed the Jews to return to the land of Israel and rebuild their temple (538 BCE) a new phase in Jewish history developed as Israel would communicate with God both through the sacrificial cult of the temple and through the study of scripture. The canonization of the Bible was a long process lasting throughout the Persian period (538–333 BCE) and the Hellenistic period (332–63 BCE), generating intense debates among the Jews about the scope and meaning of Torah and the administration of the Jerusalem temple. The debates generated factions and sects out of which emerged Judaism.

In the Second Temple period the land of Israel was increasingly urbanized, a process that intensified after the Romans gained control of Israel, ruling it first indirectly (63BCE–6 CE) and later directly (6–633 CE). While many Jews in Judea and the Galilee remained farmers, a growing conflict between the urban classes and the impoverished population of the countryside would become an important contributing factor to rebellion of the Jews in Judea against Rome in 66 CE, which ended with the destruction of the Second Temple (70 CE).[4] Once again, Judaism was not destroyed but was transformed under the guidance of a small scholarly elite—the rabbis—who perpetuated the religious orientation of the Pharisees, one of the sects in the Second Temple period. The rabbis elaborated and expanded many biblical laws, including laws concerning the land, its flora and fauna, claiming the status of oral Torah to their legal deliberations. Together the written Torah and the oral Torah constituted the ideal way of life that all Jews should follow.[5] By 600 CE the Judaism of the rabbis would become normative.

Pharisaic-rabbinic Judaism was a scriptural religion that enabled Jews to be holy as God is holy outside the precincts of the Jerusalem temple and even without the temple altogether. All aspects of life—space, time, the human body, and human relations—were sanctified by following a prescribed way of life that was all encompassing. These prescriptions, or commandments, capture the creative tension between nature and Torah. On the one hand, the sacred texts as interpreted by the rabbis specified normative behavior, ethical values, and social ideals that shaped all aspects of Jewish life, including attitudes toward the natural world. On the other hand, the veneration of and dedication to the Torah caused the distancing of religious Jews from the natural world. Since the commandment to study Torah was presented as the most important commandment, equivalent in worth to all other commandments combined, a rabbinic text declares that scripture regards the one who stops Torah study to appreciate the beauty of nature "as if he has forfeited his soul" (Mishnah, tractate *Avot* 3.7).[6] Precisely because rabbinic Judaism placed Torah at the center of Jewish life, rabbinic Jews would experience the natural world through the prism of Torah.

Rabbinic Judaism posed an elaborate program for the sanctification of nature through observance of divine commandments (*mitzvot*). In daily prayers, the Jewish worshiper sanctifies nature by expressing gratitude to the creator "who in his goodness creates each day." The prayers recognized the daily changes in the rhythm of nature—morning, evening, and night—and recognized the power of God to bring about changes. Similarly when Jews witness natural phenomena such as a storm or a tree blossoming they are obligated to say a blessing (*berakhah*) that bears witness to God's power in nature. The observant Jew blesses God for the natural functions of the human body and for the food that God provides to nourish the human body. Through such blessings, acts from which the worshiper derives either benefit or pleasure are consecrated to God. To act otherwise is considered a form of theft (Tosefta, tractate *Berakhot* 6.3). Yet, the more Jews lived in accordance to the religious prescriptions of the rabbinic tradition, the less they were interested in the natural world for its own sake. Thus the sacred text and its ongoing hermeneutics both sanctified the natural world and called on Jews to aspire to transcend nature and its demands on humans. In this sense rabbinic Judaism gave rise to the "unnatural Jew."[7]

An example of the sanctification of nature in rabbinic Judaism can be seen in the festival of Sukkot.[8] Originally celebrated at the end of the summer harvest and the preparation for the rainy season in the land of Israel, Sukkot was associated with the redemption of Israel from Egypt. In Leviticus 23:42 Israel was commanded to dwell in booths for seven days so "that your generations may know that I made the people of Israel dwell in booths when I brought them out of the land of Egypt." Removed from the protection of their regular dwelling, the temporary booth compelled the Israelites to experience the power of God in nature more directly and become even more grateful to God's power of deliverance. In addition to dwelling in a *sukkah*, the Israelites were commanded "to take the fruit of the goodly tree, palm branches, foliage of leafy trees, and willows of the brook and you shall rejoice before your God for seven days" (Lev. 23:40). In this manner nature became a means for Israel's fulfillment of the commandment to rejoice before God. After the destruction of the temple, the complex ritual of this pilgrimage festival could no longer be carried out in the temple.[9] Hence the rabbis elaborated the symbolic meaning of the *sukkah*, viewing it as a sacred home and the locus for the divine presence. They homiletically linked the four species to parts of the human body, types of people, the four patriarchs, the four matriarchs, and even to God. The festival of Sukkot was concluded by yet another festival, known as Shemini Atzert ("Eighth Day of Assembly"), which included prayers to God to deliver rain. Water indeed was an important theme in the celebration of Sukkoth when water libations were added to the daily morning offering and both wine and water were ritually poured along the sides of the altar. The most joyous event of the year was *Simchat bet ha-Shoevah*, the water-drawing festival. An additional feature was the ritual use of willow branches eleven cubits long (Babylonian Talmud, tractate *Sukkah* 45a) with which a special ritual was performed on the seventh day (*Hoshana Rabbah*).

The main literary sources of the rabbinic movement—Mishnah, Talmud, Tosefta, and some Midrashim—were redacted by the beginning of the Islamic conquest in the early seventh century. After the Islamic conquest of the Middle East, Jews in the lands of Islam underwent a major economic transformation. Heavy taxation (both land and poll taxes) imposed on *dhimmi* (i.e., the "protected people" who could remain infidels within the Islamic state) and the so-called Bourgeois Revolution of the tenth century led Jews to forsake agriculture and to enter domestic and international commerce, trade, finance, arts, and crafts.[10] This economic transformation deepened the growing alienation of Jews from the natural world characteristic of Jewish life in the Middle Ages, even though it was in medieval Islam that Jews began to reflect systematically about the laws of nature, stimulated by the flourishing of the sciences in Islamicate civilization from the ninth to the thirteenth centuries.

The involvement of Jews in commerce and trade made them particularly attractive to rulers in the northwest Europe, with the rise of towns in the tenth century. Beginning with the Carolingian ruler Louis the Pious (814–40), Jews were invited to settle in new areas to stimulate trade and commerce. While in the tenth and eleventh centuries Jews (especially in France and Spain) were still engaged in land cultivation, in general, commerce and trade were the basis of Jewish life rather than land cultivation. Jews did not take part in the feudal relations precisely because land was the basis of the personal bond between lord and vassal, which was sealed with a Christian oath. After the First Crusade (1096) Jews increasingly moved into moneylending in a rapidly growing urban society that was pressed for credit and ridden with debt. The church regarded moneylending at interest an odious activity forbidden to Christians, even though there were Christians (especially in Italy and in southern France) involved in this hateful profession.[11]

In the Late Middle Ages, European Jews had to rely on moneylending, pawnbroking, and commerce in secondhand goods for survival, which became increasingly more difficult due to conversionary pressures, discriminatory legislation, and frequent expulsions. Again, the harsh conditions did not stifle Jewish creativity: the rabbinic tradition continued to evolve and give rise to new modes of thought—rationalist philosophy, pietism, and kabbalah. Each of these intellectual movements had a distinctive theology of nature that reflected its conception of God. For the rationalist philosophers, nature was to be studied in order to fathom how its regularity and orderliness express the wisdom of God. For the pietists of Germany, the natural world was a mystery that encoded the hidden will of God, which the pietists were supposed to decode in their resourcefulness and ascetic life style. And for the kabbalists, the natural world was a metaphor of reality, an elaborate symbolic structure that mirrors the inner dynamic life of God as a unity within plurality of ten Sefirot. These diverse conceptions of nature illustrated the shared belief that for medieval Jews nature was experienced through the interpretation Torah, because in some way the Torah was the blueprint of the created world. In contrast to Christianity, medieval and early modern Jews did not use the two-books metaphor (i.e., the "book of nature" and the "book of Torah") but only one book—the Torah—which was believed to encompass corporeal and noncorporeal dimensions.[12] It was through the study of Torah that the Jew could communicate with God and bring about the salvation of the individual soul in the world-to-come and the redemption of the people in the messianic age.

By the beginning of the sixteenth century, western Europe was largely devoid of professing Jews due to expulsions. Jews moved eastward to the Ottoman Empire, Poland, and eastern Europe, and their life was largely a continuation of medieval patterns, albeit with some changes that signaled the dawn of modernity. With the spread of mercantilism (seventeenth and eighteenth centuries) and thereafter capitalism (nineteenth century) the Jewish preponderance toward finance, commerce, and trade was no longer viewed negatively. Jews and ex-conversos (namely, Jews who previously converted to Christianity but afterward returned to Judaism) would be in the forefront on commercial and financial life in western Europe and the New World. Locations from which Jews have been previously expelled would allow Jews to resettle for the expressed benefit of stimulating commerce. With the rise of the centralized modern state and the spread of democratic principles, the continued existence of a Jewish community as a separated legal entity, one governed by its own religious laws and by the laws imposed on Jews by rulers, became untenable. In western and central Europe Jews demanded civil rights and the end to centuries of exclusion, discrimination, and marginalization. The Jews in France were emancipated in 1790–91 as the logical outcome of the democratic principles of the French Revolution. By 1870, after a long and bitter struggle with many reversals, the Jews of central Europe were granted civil rights with the expectation that they were to be integrated into European culture and society and cease to exist as Jews.[13]

The emancipation opened new doors for Jews in professions and avenues of life previously closed to them. Many flocked to the universities, excelling in the natural sciences such as physics, chemistry, and biology and embracing science as a substitute to traditional Torah study. The awareness that traditional Judaism was lacking in comparison to European culture led to the emergence of a movement to institute religious reforms so as to modernize Judaism and make it fit into the style and ambiance of modern Europe. But the more that Jews integrated themselves into European life, benefiting from the breakdown of the ghetto walls, the more they were resented by Europeans who experienced the darker side of industrialization and modernization. Modern anti-Semitism based on the pseudo-scientific concept of race emerged in the 1870s as a response to the emancipation of the Jews. In racial theory the difference between Jews and non-Jews was now part of biology and not just religious belief and the refusal to accept Jesus as a savior. Nature itself now separated Jews and Gentiles.

The most original response to modern anti-Semitism was Zionism, a Jewish secular nationalist movement whose diverse goals included the reestablishment of Jewish political sovereignty in the land of Israel, the creation of modern Hebraic secular culture, and the establishment of a just society based on equality and transformation of Jewish occupational structure. Contrary to the anti-Semitic claim that the Jew is inherently unnatural or subnatural, the Zionist goal was to create a new "muscular," fearless, and agile Jew who was rooted in the land and able to respond to physical attacks.[14] The new Jew would be rooted in the soil rather than in the study of sacred texts and the performance of religious rituals. As much as the return to nature in the land of Israel was supposed to liberate the Jews from the negative character traits that they had acquired during the long exilic life, secular Hebrew culture was to highlight the agricultural basis of many Jewish festivals and celebrate the land without referring to God and without linking the abundance of the land to religious performance. Even though Zionism began in the 1880s, it failed to save the Jews from the catastrophe of the Holocaust. The Nazis succeeded in annihilating a third of the Jewish people, making the physical survival of the Jews the main priority of Jewish life in the second half of the twentieth century.

Ironically and tragically, the return to agriculture in the land of Israel did not improve the conditions of nature in the Holy Land. During the twentieth century, intensive agriculture, massive urbanization, rapid population growth, industrialization, and the perpetual state of war with Arab neighbors dictated overuse of preciously scarce natural resources, especially water.[15] Furthermore, the influx of Jews from the Arab world, which had not been exposed to Western modernization, reintroduced traditional Jewish life and values to the young state, including a certain indifference to the physical environment. The social agenda of these immigrants, as well as of the refugees from Europe after the Holocaust, has had little to do with protection of the land and its limited natural resources.

Environmentalism is thriving in Israel today, because the citizens of the Jewish state have finally recognized the environmental price that nature paid for the successes of Israel "to make the desert bloom." The vibrant environmental movement in Israel boasts about ten major nonprofit organizations that provide information about the environmental crisis, engage in legal action, create educational materials, and even sustain two political parties. However, a closer look at environmentalism in Israel sheds light on the uneasy relationship between Judaism and ecology. In Israel the concern for the physical landscape and action in regard to air pollution, water pollution, treatment of waste products, conservation of coastal lines, and legal action in parliament are most popular among secular Israelis. These concerns reflect the desire of Israelis to improve the quality of life and create a healthy lifestyle, and its justification does not come from Jewish religious sources. For secular Israelis, attention to environmental issues reflects contact with environmental movements in Europe and North America rather than with the Jewish tradition. Conversely, Israeli Jews who are anchored in the Jewish tradition tend to link their love of the land of Israel to a certain religious nationalist vision, the so-called Greater Israel vision, rather than preservation of nature. Even though the religious, nationalist parties now promote outdoor activities for their constituents, these activities were not grounded in the values and sensibilities of the environmental movement. Some of these people occasionally engage in activities which are antithetical to environmentalism (e.g., uprooting of trees).

While the nascent state of Israel was struggling to survive, the Jews in Western industrialized nations were struggling to survive under the pressures of the open society. High rate of intermarriage, secularization, social mobility, and poor Jewish education were among the factors that rendered the spiritual survival of the Jews in the Diaspora a serious challenge in the second half of the twentieth century. Given these challenges, it is not surprising that the survival of the earth appeared a remote issue that need not concern the Jews, who have more pressing problems to solve. The organized Jewish community was more concerned with Jewish education, gender equality, pluralism within Judaism, and the need to support the fledgling state of Israel than with the degradation of the environment. Saving endangered species was not viewed as a Jewish value to which scant resources of the Jewish community should be devoted.

Nonetheless, since the 1970s Jews in the industrialized West could no longer remain oblivious to the environmental movement, because environmentalists, beginning with the famous essay by Lynn White Jr., charged that the Judeo-Christian tradition itself was the primary cause of the current environmental crisis.[16] Presumably, the Bible itself (Gen. 1:28) gave humans the license to "to have dominion" over the earth and its inhabitants, leading to the exploitative practices that resulted in our current ecological crisis. Orthodox thinkers were the first to rise to the defense of Judaism against these charges, attempting to show that non-Jewish readings of the Bible (religious or secular) are either based on misunderstanding of the text or attest ignorance of the postbiblical Jewish tradition without which the Bible could not be understood.[17] In the 1970s and 1980s Jews from all branches of modern Judaism—Reform, Conservative, Reconstructionist, and Humanistic Judaism—reflected on ecological concerns in light of Jewish religious sources, giving rise to a distinctive, albeit still small body of literature. The recent anthology *Judaism and Environmental Ethics* collected some of the seminal essays of this discourse,[18] and another volume, *Judaism and Ecology*, includes essays by Jewish theologians, scholars of rabbinic Judaism, and historians of Jewish philosophy and mysticism in an attempt to think through the interplay between Judaism and environmentalism.[19] Additionally, essays about nature in the Jewish tradition have appeared in various Jewish periodicals, including *Judaism, Tradition, Conservative Judaism, CCAR Journal, Tikkun, Journal of Reform Judaism, The Reconstructionist*, and *Shofar*, signaling the interest in environmentalism among Jews. Finally, overviews of Judaism are now included in major reference works on religion and ecology, such as *A Companion to Environmental Philosophy*,[20] the *Encyclopedia of Religion and Nature*,[21] the revised second edition of the *Encyclopedia of Religion*,[22] and the current volume. Judaism is now a part of the well-established discourse on religion and ecology, and the ecological wisdom of the Jewish tradition is now recognized by many.

In the 1990s Jewish environmentalism also began to make inroads into the life of organized Jewish community. In 1993 the Coalition on the Environment and Jewish Life (COEJL) was founded as an umbrella of twenty-nine Jewish organizations with thirteen regional affiliates.[23] COEJL promotes environmental education, scholarship, advocacy, and action in the American Jewish community. Its mission is to deepen the Jewish community's commitment to the stewardship of creation and to mobilize the resources of Jewish life and learning to protect the earth and all its inhabitants. Beside stewardship, the core principles of COEJL emphasize environmental justice, the prevention of harm, energy independence, equitable distribution of responsibility between individuals, corporations, governments, and nations, pollution prevention, proper treatment of nuclear waste, energy conservation, utility regulation, and promotion of sustainable development. COEJL is part of a larger network of religious interfaith organizations that attempt to shape environmental policies on municipal, state, and federal levels. In Israel too, there are a few, but significant, attempts to anchor environmentalism in the sources of Judaism. The Heschel Center for Environmental Learning and Leadership sponsored many educational programs, and a new forum called Le-Ovdah U-Leshomrah ("To Till and to Tend") is devoted to integrate the values of religious nationalism with environmentalism that is grounded in the Jewish religious tradition. Jewish environmentalism exists today, although it faces challenges.

CONCEPTIONS OF NATURE
IN JEWISH SOURCES

The doctrine of creation is the theological basis for Jewish conceptions of nature.[24] The Bible includes two creation narratives that present different, but not
necessarily contradictory, views of the relationship between humanity and the
natural world.[25] The first creation narrative (Gen. 1:1–2:3) depicts the creation of
the material world as an act of ordering unordered chaos (*tohu va-bohu*). The
order involves the separation of light from darkness, water above from water
below, dry land from the seas, vegetation from animals, aquatic animals from air
animals and land animals, and finally humans from other land animals. Boundary
formation at creation would serve as the rationale for the distinction between the
sacred and the profane, the permitted and the forbidden in the legal parts of the
Bible and in postbiblical Judaism. Thus the prohibitions on mixing different seeds
in the same field, interbreeding diverse species of animals, wearing garments of
mixed wool and linen, and the differentiation between clean and unclean foods
are all traced back to the boundaries set at the moment of creation (Lev. 10:10–11;
19:19; Deut. 22:11).

In the first creation narrative, created nature is not presented as divine and is
not identified with the creator. One animal, namely, the human, is presented as
different from all others, because it was made in the "divine image" (*tzelem
elohim*) (Gen. 1:26). The meaning of this phrase is open to diverse interpretations,
but it is by virtue of the divine image that the human receives the commandment
to have dominion over other animals. The commandment clearly privileges the
human species over others and calls the human to rule over other living creatures,
but does not give license to exploit the earth resources, since the earth does not
belong to the humans but to God. The act of divine creation ends with rest on the
seventh day, the Sabbath, imposing rest on nature.

The second creation narrative (Gen. 2:4–3:24) considers the origin of humanity through the garden of Eden myth and highlights the link between the
human earthling, *adam*, and the earth, *adamah*, from which the human comes
and to which the human will return at death. It is God's breath that transforms
the earthling from the "dust of the earth" into a living being (*nefesh hayah*), thus
establishing the direct link between humanity and God. The second creation
narrative places the human in the garden of Eden "to serve it and to keep it" (*le-
ovdah u-leshomrah*), most likely indicating farming activities such as tilling,
plowing, and sowing. The hierarchical relationship between humans and other
creatures is signified in the act of naming, which culminates in the naming of the
female counterpart, *havah*, who is a "helper fit for him" (*ezer kenegdo*) and "the
mother of all living." The choice of the female to disobey the divine command

"not to eat from the fruit of the tree" results in the expulsion of the human couple from the garden of Eden and the cursing of all involved: the serpent, the human male and female, and even the earth itself. Procuring food will no longer be an effortless task, but will necessitate human toil, sweat, and pain.

The doctrine of creation, which recognizes the gulf between the creator and the created world, facilitates an interest in the natural world that God created. The more one observes the natural world, the more one comes to revere the creator, because the world manifests the presence of order and wise design in which nothing is superfluous. Psalm 19:1 expresses the point poetically: "The heavens are telling the glory of God / and the firmament proclaims his handiwork." Psalm 148 depicts all of creation as engaged in praising God and recognizing God's commanding power over nature. Nature also fears God (Ps. 68:8),[26] and nature observes the relationship between God and Israel and expresses either sorrow or joy at the fortunes of the Israelites (Joel 1:12; Amos 1:2; Jonah 3:7–9; Isa. 14:7–8). In the Psalms, however, awareness of nature's orderliness, regularity, and beauty never leads to reveling in nature for its own sake. Nature is never an end, but always points to the divine creator who governs and sustains nature. The emphasis on orderliness of creation explains why in Judaism we do not find glorification of wilderness (so cherished by the environmental movement) and why the cultivated field is the primary model for the created universe in the Bible.[27]

The Bible abounds with references to the natural world and figurative usage of natural elements to teach about the relationship between God and Israel. Prophetic texts in particular are rich with metaphors and similes from the plant world. In Jeremiah the almond tree represents old age, the vine and the fig tree capture coming desolation and destruction, and the olive tree is a common reference for longevity in biblical parables.[28] In one famous parable, fruit trees and vines willingly serve the human in ritual observance by providing oil, fruit, and wine (Judg. 9:8–13). Conversely, nature does God's bidding when it serves to punish and destroy the people of Israel when they sin; indeed, ungodly behavior leads to ecological punishment. Since God is the sole creator, it is God's prerogative to sustain or to destroy nature (Ps. 29:5–6; Zech. 11:1–3; Hab. 3:5–8). Nature itself becomes a witness to the covenantal relationship between Israel and God and the ongoing drama of righteousness, chastisement, and rebuke. Mostly the Bible emphasizes divine care of all creatures: God provides food to all (Ps. 147:9); God is concerned about humans and beasts (Ps. 104:14; 145:16); God's care is extended to animals that can be used by humans, such as goats and rabbits (Ps. 104:18) as well to lion cubs and ravens that do not serve human interest. Because God takes care of animals, they turn to God in time of need (Ps. 104:21, 27; 147:9; Job 38:41).

That the Bible is replete with descriptions of nature is understandable, since ancient Israel was, as we noted above, an agrarian society in which the farmer was dependent on nature for survival. The relationship to nature became more problematic in the postbiblical period of the Second Temple and after the destruction of the Second Temple, when Judea was part of the Hellenized and rapidly urbanizing Near East. During that time a new movement emerged that generated apocalypses and produced " 'revelatory writings' which disclosed heavenly secrets which could not be deduced by empirical methods."[29] The apocalyptic movement produced an allegorical discourse, the most famous of which is the *Animal Apocalypse*, in which animals symbolize either historical and political entities or eschatological realities.[30]

Apocalyptic literature was largely produced by anonymous Jews, but whether it is part of Judaism is a more complicated question, since the apocalyptic mindset was sectarian. Perhaps in opposition to the apocalyptic speculations, the rabbis considered speculations about the structure of the physical world (*ma'aseh bereshit*) to be esoteric lore to be divulged only to the initiated few. While study of God's creation was not prohibited, speculations about "what is above, what is beneath, what is before, and what is after" (Babylonian Talmud, tractate *Hagigah* 1) were restricted to the intellectual elite. The rabbis debated the details of the biblical creation narrative: were the heavens or the earth created first? what are the dimensions of the firmament? The dominant view was that the earth and the heavens are like "a pot with a cover." The "cover" was identified with the firmament (*raqi'a*), itself composed of water and stars of fire that coexist harmoniously, although it was believed that there was more than one firmament. The sun and the moon were believed to be situated in the second firmament, and beneath the earth there was the abyss. When the rabbis interpreted the creation narratives in Genesis, they mainly insisted that the "Torah reveals the rules that the world follows.... Natural law and revealed truth of the Torah together testify to the same fact: God rules, the Torah and nature together tell us the natural laws by which God rules."[31]

In general, the rabbis were not interested in explaining *how* nature works, but in the sanctification of nature through human action. However, there are a few areas in which the rabbis did express intense scientific interest: astronomy and human physiology. Several rabbis (e.g., Yohanan ben Zakkai, Gamaliel II, and Joshua ben Hanaya) were expert astronomers, using observed data for the calculation and adjustment of the lunar-solar calendar. The rabbinic corpus is also replete with information about the motions of celestial bodies, the four seasons, the planets, the zodiac, and even comets. The picture of the universe in talmudic texts has the earth in the center of creation, with heaven as a hemisphere spread over it. The earth is usually described as a disk encircled by water. Since these cosmological and metaphysical speculations were not to be cultivated in public or be committed to writing, they remained the privileged knowledge of the small rabbinic elite.

Within the created world, the human body was the utmost interest to the rabbis, although their information about human anatomy was shaped by the religious concern for ritual purity. Rich in details about the skeleton, digestive organs, respiratory system, heart, genitals, and other organs, the rabbinic corpus also includes rather fanciful material and is totally lacking in graphic illustration.[32] The discussion is concerned primarily with physical disfigurements that disqualify men from the priesthood, with rules concerning menstruating women, and with other sources of ritual pollution. The rabbinic corpus also includes informative claims about embryology, diagnosis of diseases, and a host of medications and hygienic strategies for prevention of disease. The physician is viewed as an instrument of God, treated with utmost respect, and several talmudic scholars were physicians. To the extent that human body is part of nature, these concerns suggest that the rabbis were quite informed about nature.

More than describing the physical, material universe, the rabbis were interested in the relationship between revealed morality (prescriptive law) and the laws of nature (descriptive laws), but the rabbinic corpus harbors diverse and even conflicting views. One theme highlights the regularity of nature and its indifference to human concerns: "Nature pursues its own course" (*olam ke-minhago noheg*) (Babylonian Talmud, tractate *Avodah Zarah* 54b). Accordingly, nature is independent of the revealed Torah, and the laws of nature are different from the laws of the Torah. A contrary viewpoint holds that the natural world is contingent upon the acceptance of the Torah by the Jewish people; had they rejected the Torah, the world would have reverted to primeval chaos. The link between nature and the moral conduct of humans is expressed in yet a third view that the original natural order was perfect but suffered a radical change as a result of human original sin (Babylonian Talmud, tractate *Qiddushin* 82b). A fourth view posits "the animals of the righteous" as models for human conduct. Presumably these animals do not sin, because they know intuitively what the law is and what is required of them, and they know how to apply the Torah to the world in which they live. Since the "animals of the righteous" live in perfect harmony with their creator, humanity has much to learn from them, not only in terms of the principle of observing God's will but also specific lessons (Babylonian Talmud, tractate *Pesahim* 53b). Finally, there is a rabbinic teaching that animals not only observe the moral laws, but all of nature is perceived as fulfilling the will of God in the performance of its normal functions (Jerusalem Talmud, tractate *Pe'ah* 1.1). Underlying all of these rabbinic reflections is the conviction that the Torah is the wisdom (i.e., Logos) that governs the material universe, an idea that has interesting parallels in Stoicism.[33]

The relationship between nature and Torah were of great concern to medieval Jewish thinkers, who reinterpreted the biblical creation narratives and rabbinic cosmological speculations in light of Greek and Hellenistic science and philosophy. Whether the world was created *ex nihilo* or out of preexisting matter was hotly debated. Moses Maimonides (1138–1204) argued that the origin of the universe is beyond human reason but that the argument in favor of creation is superior to the argument in favor of the eternality of the world.[34] By contrast, Levi ben Gershom (Gersonides) (1288–1344) argued that creation out of preexisting matter is scientifically demonstrable and is in full accord with Aristotelian science. The medieval cosmological picture was a spherical earth lying in the center of a spherical universe. Revolving around the earth in a circular motion are the heavens and the celestial spheres, which are governed by different set of physical laws. Whereas on earth heavy bodies fall and light bodies tend to rise, so that the natural motions are either toward the center or away from it, nothing in the heavens comes into being or ceases to exist; the heavens are immune to change. Celestial motion itself is cyclical, and earth of the moving spheres has an internal moving force, a soul, which is set in motion by corresponding incorporeal substances, the Separate Intelligences. Both terrestrial and celestial realms are structures in successive shells, like an onion. The four elements are arranged so that earth is in the center surrounded by water, air occupies the regions above the water, and fire occupies that above air. Composed of a single element, ether, the heavens—the visible moon, sun, and five planets—are carried around the earth by transparent, invisible spheres. They too are nested one within the next around the center, with each celestial body being embedded in the side of a distinct sphere. They form a compact whole in which there are no gaps or inner vacuum. According to Maimonides' *Guide of the Perplexed* 2.6, the ultimate source of motion is God, the prime mover, who moves the universe insofar as he is the most perfect substance and therefore the object of love of all other substances.[35]

Reflections on the interplay between God, the natural world, and human nature were central to the systematic Jewish theology that flourished in the Middle Ages. Jewish rationalists reflected on the laws of nature in order to fathom how God governs the created world.[36] They regarded the study of God's created world a theoretical activity whose reward was the immortality of the rational soul, or the intellect, in the world-to-come (*olam ha-ba*). The study of nature was a religious activity that enabled the philosopher-scientist to imitate God, an intellect engaged in eternal self-contemplation, and to understand the mind of God. The study of nature, however, was never divorced from the study of the revealed Torah. Even though medieval Jewish philosophers did not use biblical verses as premises of their philosophical reasoning, they all presupposed that in principle there could be no genuine contradiction between the truths of the revealed text and scientific knowledge about the world; both were believed to manifest the wisdom of God. In premodern Judaism the doctrine of creation and the doctrine of revelation functioned as the matrix within which Jews speculated about the natural world.

If orderliness and stability were the main manifestations of the rationality of the natural world according to medieval Jewish rationalists, the Jewish pietists in Germany during the late twelfth and early thirteenth centuries accentuated the extraordinary and the miraculous. Without differentiating between creation and emanation, the pietists of Germany believed that the hidden, invisible God was mysteriously and paradoxically revealed in the natural order. In his goodness God implanted in nature events and signs that violate nature, that is, the miraculous and the fantastic.[37] It is the supernatural, unique, and nonrepeatable phenomena that inform the believer about the hidden God and directs him or her toward fulfillment of God's hidden will.

German pietism influenced the teachings of medieval kabbalah, although the picture of the universe adopted by kabbalah was largely Aristotelian, albeit with a strong dose of Neoplatonic emanationism. In kabbalah the focus was neither on the stability of nature nor on its extraordinary features, but on the linguistic aspect of the creative act. God created the world through speech, and the language that God spoke was the holy tongue, Hebrew. These ideas were articulated in the anonymous *Sefer Yetzirah* ("Book of Creation") and its cognate literature, which posited that the letters of the Hebrew alphabet are the building blocks of the created world.[38] The infinite permutations of the letters account for the diversity of nature.[39] This linguistic theory was developed in medieval kabbalah, yielding the textualization of nature: nature was a text that could be decoded and manipulated by anyone who grasped its grammar. The code itself was known only to the initiate few; the one who knows how to decode nature could manipulate not only physical phenomena but the inner life of God, namely, reunify the feminine and masculine aspects of God that were separated because of human sin. As nature became a myth, esoteric knowledge about the Torah assumed theurgic dimensions, enabling the knower to impact the inner life of God.

The textualization of the nature in kabbalah resulted in two different attitudes toward the natural world. On the one hand, kabbalists had little interest in collecting empirical data about the natural world because for them natural phenomena were ultimately symbolic. Even though kabbalists employed organic symbols (i.e., pomegranate, palm date, rainbow, river, lily of the valley) in their symbolic language about God, the ultimate referent of these symbols was divine reality rather than material reality.[40] To the kabbalists, nature as a physical reality was but the lowest rung of the hierarchical Great Chain of Being, in which each realm of existence occupies its natural place but derives its vitality from the divine efflux (*shefa*) that permeates all levels of existence, depending on their degree of corporeality. The corporeal world is the battleground between divinity and the forces of evil, and to attain religious perfection the kabbalist had to transcend the physical veil of reality. While kabbalistic texts abound with symbols derived from the plant world, corporeal nature was regarded as evil to be transcended or spiritualized.

On the other hand, the knowledge of the linguistic matrix of the natural world meant that the knower could manipulate the forces of nature, namely, engage in magic. Kabbalists claim to draw spiritual energy from the supernal world into the corporeal world, to heal the sick, cause rainfall, and ease childbirth. These forms of practical kabbalah manifest a hands-on approach to nature: it is an activist attitude that aligned kabbalah with magic, alchemy, and astrology. Such wisdom was considered effective only because the kabbalists claimed to possess the knowledge of invisible occult forces of nature created by divine speech. Thus the kabbalists affirmed the human capacity to activate a divine energy that pulsates throughout the universe, and they remained committed to the primacy of humans in the created order (contrary to the views of many environmentalists).

During the early modern period (from the sixteenth to the eighteenth centuries) medieval cosmology remained intact, but Jewish thinkers became increasingly more interested in the flora and fauna of their natural environment, and Jewish philosophical-scientific texts abound with information about minerals, plants, and animals.[41] Nonetheless such information was still framed by the theological assumptions of medieval rationalism: the natural world could be understood in light of the revealed Torah, since it was the blueprint of creation. Jewish thinkers were also rather slow to respond to the scientific revolution of the seventeenth century. Although David Gans (1541–1613) was personally familiar with Johann Kepler and Tycho Brahe and praised Copernicus, most Jewish thinkers rejected Copernicus's heliocentric theory on religious grounds. While a small cadre of Jews earned doctorate degrees at European universities, especially in medicine,[42] interest in natural sciences remained marginal among Jews. Instead, the study of halakah (Jewish law) and kabbalah preoccupied Jewish intellectual interests, and both endeavors were textual, self-referential, and abstract.

Kabbalah gave rise to east European Hasidism in the eighteenth century. Hasidic theology treated all natural phenomena as ensouled: divine sparks enlivened all corporeal entities, and not just human beings. The divine sparks sought release from their material entrapment. Through ritual activity, the Hasidic master attempted to draw closer to the divine energy, the liberation of which will result not only in the sanctification of nature but also in the redemption of reality and its return to its original, noncorporeal state.[43] The worship of God through the spiritualization of corporeal reality (*avodah ba-gashmiyut*) became a major Hasidic value, complementing the general deemphasis on Torah study in Hasidism. Hasidic tales were situated in natural rather than urban settings, encouraging the Hasidic worshiper to find the divine spark in all created beings. Yet Hasidic masters were not concerned with the well-being of the natural environment or with the protection of nature. In fact, to reach their spiritual goals, Hasidic meditative practices attempted to dissolve the corporeality of existing reality (*bittul ha-yesh*) and eliminate the selfhood of the one who meditates on nature (*bittul ha-ani*).

Kabbalah and Hasidism contributed to the bookishness of Jewish culture and the alienation of traditional Jews from the natural world. With the rise of modernity, the very lack of Jewish interest in nature was cited by the Jewish Enlightenment (Haskalah) as the reason for Jewish backwardness. Only the return to nature could modernize the Jews, enabling them to recover their lost vitality and integrate as equals into modern society. The literature of the Haskalah movement in the nineteenth century is full of descriptions of nature, emphasizing its beauty, wisdom, and moral power.[44] Interestingly, while the Haskalah's approach to nature shaped the Zionist movement's desire to return to nature, the current environmentalism in North America is inspired not by nineteenth-century Haskalah literature but by kabbalah and Hasidism.

As Jews were integrated into modern society, many Jews no longer regarded the written and oral Torahs as the source of truth about the physical universe; cosmology now belongs to science rather than to religion. As a result, modern Jewish philosophers no longer reflected about the origin of the universe, but instead focused on explicating the religious and existential meaning of the doctrine of creation in relationship to the doctrines of revelation and redemption.

Thus Franz Rosenzweig (1886–1929) explained the relationship between God, the world, and humanity and saw creation as a dynamic process in which these three elements intersect in continuous relation.[45] Creation describes an infinite, atemporal process, an asymptotic, end-directed process whose duration is endless because its goal is infinitely remote. Rosenzweig emphasized human awareness of creatureliness as dependence on God, echoed by the language of Jewish liturgy. Rosenzweig's colleague and collaborator in the translation of the Bible, Martin Buber (1878–1965), viewed creation as an act of "communication between Creator and created," and humans who imitate God redeem the world through their action. Humanity is thus a participant in the process of creation, completing God's work and initiating redemption.

The moral implication of the fact of creation was also emphasized by Mordecai M. Kaplan (1881–1984), who, under the influence of process philosophy, spoke about creativity as the continuous emergence of aspects of life not prepared for or determined by the past. For Kaplan, creativity constitutes the most divine phase of reality, as each is possible and is still in the process of being created, and humanity has the power to realize these possibilities. These philosophical ideas shaped Jewish thought in the twentieth century, but only recently did Jewish thinkers (mostly Orthodox and Conservative) began to examine the creation story in light of contemporary physics, especially the anthropic principle.[46] This principle shows that the world is not a neutral entity, empty of purpose and meaning, and that the Big Bang theory can lead to intellectual and emotional enthusiasm about the creator. Conversely, contemporary physics could lead to rethinking the doctrine of creation, especially creation in the image of God, and the problem of evil.

Environmental Ethics in Jewish Legal Sources

The core of Jewish environmental ethics is the notion that human beings are responsible toward the natural world.[47] In this regard humans can be said to be stewards or caretakers of nature, even though the term itself does not appear in the Bible.[48] The ethics of responsibility follows from the dual aspect of the doctrine of creation: on the one hand, the human species is part of nature, but, on the other hand, the human is able to transcend nature by virtue of the "divine image" that makes humans like God in some way. The ethics of responsibility is manifested in broad legislation toward various aspects of nature that constitute sound conservation policy.

Land-based Commandments

Various land-based commandments in the Bible express the belief that "God is the rightful owner of the land of Israel and the source of its fertility; the Israelites working the land are but God's tenant-farmers who are obligated to return the first portion of the land's yield to its rightful owner in order to insure the land's continuing fertility and the farmer's sustenance and prosperity."[49] Accordingly, the first sheaf of the barley harvest, the first fruit of produce, and two loaves of bread made from the new grain are to be consecrated to God. In the Mishnah these gifts are to be made only from produce grown by Israelites in the land of Israel, in contrast to all other cereal offerings and animal offerings, which may be brought to the temple from outside the land (Mishnah, tractates *Menahot* 8.1 and tractate *Parah* 2.1). Some of the consecrated produce is to be given to the priests and Levites, whereas others are to be eaten by the farmer himself.

Protection of Vegetation

Scripture and the rabbinic sources pay special attention to trees.[50] Leviticus 19:23 commands that during the first three years of growth, the fruits of newly planted trees or vineyards are not to be eaten (*orlah*), because they are considered to be God's property. When Israel conducts itself according to the laws of the Torah, the land is abundant and fertile, benefiting its inhabitants with the basic necessities of human life—grain, oil, and wine—but when Israel sins, the blessedness of the land declines and it becomes desolate and inhospitable (Deut. 11:6–11).[51]

When alienation from God becomes egregious and injustice overtakes God's people, God removes them from the Holy Land. Thus the well-being of God's land and the moral quality of the people who live on the land are causally linked and both depend on obeying God's will.

Bal Tashchit ("Do Not Destroy")

Protection of fruit-bearing trees in wartime is another important biblical legislation about nature. In war, fruit-bearing trees must not be chopped down while a city is under siege (Deut. 20:19).[52] This commandment is undoubtedly anthropocentric, but it indicates that the Torah recognizes the interdependence between humans and trees, on the one hand, and the capacity of humans to destroy natural things, on the other. To ensure the continued fertility of the land, human destructive tendencies are curbed by scriptural law. In the Talmud and later rabbinic sources, the biblical injunction "do not destroy" was extended to cover all destruction, complete or incomplete, direct or indirect, of all objects that may potentially benefit humans. A sweeping series of environmental regulations is legitimized by appealing to the principle "do not destroy": the prohibition on cutting off of water supplies to trees; overgrazing the countryside; unjustified killing of animals or feeding them harmful foods; hunting animals for sport; species extinction and the destruction of cultivated plant varieties; pollution of air and water; overconsumption of anything; and the waste of mineral and other resources. These environmental regulations indicate that the Jewish legal tradition requires that one carefully weigh the ramifications of all actions and behavior for every interaction with the natural world; it also sets priorities and weighs conflicting interest and permanent modification of the environment.

Speciation and Prohibition on Interbreeding

The Bible recognizes the diversity species (literally, "kinds") in the natural world (Gen. 1:11–25), even though the Bible lacks the philosophical analysis of the concept of species that one finds in Greek philosophy of nature, especially in Aristotle and his school. The concern for protection of diversification is expressed in biblical legislation such as Leviticus 19:19: "You shall not let your cattle breed with a different kind; you shall not sow your field with two kinds of seeds" (repeated in Deut. 22:9–11). The Bible prohibits mixing different species of plants, fruit trees, fish, birds, and land animals, and the prohibition is clarified and further elaborated in tractate *Kil'ayim* in both the Mishnah and the Jerusalem Talmud.[53] While rabbinic rulings about the main grains of the land of Israel—wheat, rye grass, barley, oats, and spelt—and about other species of vegetation do not indicate that the rabbis understood the principles of genetic engineering, it does suggest that they were keen observers of the natural world and that they had respect for diversification of nature.

Limits on Human Consumption: Clean versus Unclean Animals

Placing limits on human consumption of animals and regulating all food sources is a major concern of the Bible and its holiness code. The laws of Leviticus 11 and Deuteronomy 14 are part of "an elaborate system of purity and impurity affecting the sanctuary and the priesthood as well as the lives of individual Israelites."[54] In general, the Torah prohibits eating the meat of certain living creatures that are classified as impure or unclean, the ingestion of blood of any animals, the consumption of animal fat (*helev*), and the eating of meat of the carcass (*nevelah*) of dead animals and fowls. More particularly, the Bible spells out which animals are permitted and which are forbidden for human consumption. The differentiation between clean and unclean animals, which is the core of the Jewish dietary laws (*kashrut*), has generated a lot of discussion about their internal logic. Some scholars explained that the unclean animals were viewed as a threat to life, whereas others suggested that forbidden animals were those regarded as deities in neighboring cultures.[55] Still others considered the means of locomotion as the crucial classification principle.[56] But it is also possible to explain the prohibition on consuming certain animals as ecologically motivated.[57] The Bible permits the husbandry and consumption of ruminants, namely, animals which were able to make the most efficient use of vegetation. Other animals (horse, mule, camel), which were domesticated, could be kept by farmers for transportation and work on the field but not for consumption. The cow was used for work and for milk and meat, and the sheep and goat for milk and meat only. Water animals that could be eaten must have fins and scales (i.e., fish), but frogs, toads, and newts were not to be eaten, perhaps because the authors of the Bible were aware that they are beneficial to the ecosystem and cut down on the mosquito population. Lobsters, oysters, and mussels are also forbidden, most likely because the coast of Palestine is not suited for them. All birds of prey, including owls, were forbidden for human consumption as well as all storks, ibises, herons, and species of bats. Once we realize that many of the forbidden species were actually common in the land of Israel, it is possible to look at these prohibitions as extended protection of birds that are important to "maintaining the ecological equilibrium and serve as the most efficient biocontrol agents of species."[58]

Concern for Future Generations

The Torah (Deut. 22:6–7) attests to concern with the perpetuation of life of non-human animals. If one finds a nest on the ground or in a tree with young ones or eggs in it and "the mother sitting upon the young or upon the eggs, you shall not take the mother with the young; you shall let the mother go, but the young you make take to yourself, that it may go well with you and that you may live long." By saving the mother, the law enables the species to continue to reproduce itself and avoid potential extinction. This law is elaborated in *Deuteronomy Rabbah* 6.5, Babylonian Talmud, tractate *Hullin* 138b–42a, and *Sifre Deuteronomy* 227, specifying that the person who finds the nest is allowed to take the nestlings only if they are not fledged. This concern intimates a notion of sustained use of resources and could provide Jewish support for the concept of sustainability. This reason led the rabbis to prohibit raising sheep and goats that graze, even though the rabbis were aware that these animals generated a very profitable business in the Roman Empire (Babylonian Talmud, tractate *Hullin* 58b). The ban was imposed after the devastation of Judea in the Bar Kokhba revolt (132–35 CE) in order to enable the land to heal from the devastation of the war; thus short-term hardship was traded with long-term gains. This kind of environmental legislation was legitimated by appeal to the holiness of the land, but it also indicates attention to the particular physical conditions.[59]

Tza'ar Ba'alei Hayyim ("Distress of Living Creatures")

Although the Bible and the rabbinic tradition place the responsibility for management of God's creation in human hands, the tradition also recognizes the well-being of nonhuman species: humans should take care of other species and be sensitive to the needs of animals.[60] Cruelty toward animals is prohibited because it leads to other forms of cruelty.[61] The ideal is to create a sensibility of love and kindness toward animals in order to emulate God's attribute of mercy and fulfill the commandment "to be holy as I the LORD am holy" (Lev. 19:2). Thus Deuteronomy 22:6 forbids the killing of a bird with her young because it is exceptionally cruel and because it can affect the perpetuation of the species. This commandment is one of seven commandments given to the sons of Noah and is therefore binding on all human beings, not just upon Jews. In Deuteronomy 22:10 scripture prohibits against yoking an ass and an ox together, because their uneven size could cause unnecessary suffering. The prohibition on "seething a kid in its mother's milk" (Exod. 23:19; 34:26; Deut. 14:21), which is the basis for an elaborate system of ritual separation of milk and meat products in rabbinic Judaism, is explained by the rabbis as an attempt to prevent cruelty in humans (*Deuteronomy Rabbah* 6.1). While scripture does not forbid slaughtering animals for consumption or sacrifice or using eggs for human use, it curtails excess cruelty. Kindness to animals is a virtue of the righteous person, which is associated with the promise of heavenly rewards (Prov. 12:10).

Merciful treatment of animals is but one way through which Israel is separated from the surrounding pagan culture and becomes a holy nation. Most tellingly, scripture forbids cutting off a limb from a living creature (*ever min ha-hai*), even to feed it to the dogs and even in the case of animals that are not to be eaten at all, because they are unclean. The tradition prescribes particular modes of slaughter which are swift because they are performed with a sharp, clean blade. In Hasidism this principle was combined with the belief in the transmigration of souls into nonhuman bodies and the development of very elaborate slaughtering practices designed to protect the human soul that has transmigrated into the body of the animal about to be slaughtered.[62] The concern for unnecessary suffering of animals is applied today to the farming of animals for human consumption and to the use of animals in scientific experimentations.[63]

Social Justice and Ecological Well-Being

The most distinctive feature of Jewish environmental legislation is the causal connection between the moral quality of human life and the vitality of God's creation. The corruption of society is closely linked to the corruption of nature. In both cases, the injustice arises from human greed and the failure of human beings to protect the original order of creation. From the Jewish perspective, the just allocation of nature's resources is indeed a religious issue of the highest order. The treatment of the marginal in society—the poor, the hungry, the widow, the orphan—must follow the principle of scriptural legislation. Thus, parts of the land's produce—the corner of the field (*peah*), the gleanings of stalks (*leqet*), the forgotten sheaf (*shikhekhah*), the separated fruits (*peret*), and the defective clusters (*olelot*)—are to be given to those who do not own land. By observing the particular commandments, the soil itself becomes holy, and the person who obeys these commandments ensures the religio-moral purity necessary for residence in God's land. A failure to treat other members of the society justly, so as to protest the sanctity of their lives, is integrally tied to acts extended toward the land. This aspect of Jewish ecological ethics is the foundation of the concept "ecokosher" coined by Zalman Schachter-Shalomi and popularized by Arthur Waskow. (We will return to it below.)

Imposing Rest on Nature

The connection between land management, ritual, and social justice is most evident in the laws regulating the sabbatical year (*shemittah*). The sabbatical year is an extension of the laws of the Sabbath to the earth.[64] On the Sabbath humans create nothing, destroy nothing, and enjoy the bounty of the earth. Since God rested on the seventh day, the Sabbath is viewed as the completion of the act of creation, a celebration of human tenancy and stewardship. Sabbath teaches that humans stand not only in relation to nature but in relation to the creator of nature. Most instructively, domestic animals are included in the Sabbath rest (Deut. 5:13–14). There are specific cases in which it is permissible to violate the laws of the Sabbath in order to help an animal in distress. Thus one must alleviate the suffering of an animal that has fallen into a cistern or ditch on the Sabbath, bring food or pillows and blankets to help it climb free. The normal restrictions against such labors on the Sabbath are waived. Cattle must be milked and geese fed, lest the buildup of milk in the one case or hunger in the other cause suffering to a living being (Maimonides, *Mishneh Torah* 87.9). The observance of the Sabbath is a constant reminder of the deepest ethical and religious values that enable Jews to stand in a proper relationship with God.

During the sabbatical year it is forbidden to plant, cultivate, or harvest grain, fruit, or vegetables or even to plant in the sixth year in order to harvest during the seventh year. Crops that grow untended are not to be harvested by the landlord but are to be left ownerless (*hefqer*) for all to share, including poor people and animals. The rest imposed during the sabbatical year facilitates the restoration of nutrients and the improvement of the soil, promotes diversity in plant life, and helps maintain vigorous cultivars. On the seventh year, debts contracted by fellow Israelites are to be remitted (Lev. 25; Deut. 15:3), providing temporary relief from these obligations. In the Jubilee year all Hebrew slaves are manumitted, regardless of when they were acquired (Lev. 25:39–41), in order to teach that slavery is not a natural state. The laws of the sabbatical years were practically reversed in the rabbinic period when a written document (*prozbul*) assigned the debt to the court prior to the sabbatical year with the intention of collecting the debt as a later time. In modern Israel some religious kibbutzim selectively revived the practices of the sabbatical year.

Environmental Virtues

The Jewish ethics of responsibility focuses on duties, including duties toward nature, but it is complemented by ethics of virtues, namely, character traits that one must cultivate in order to be able to stand in a proper relationship with God and observe God's commands. The very .virtues that rabbinic Judaism found necessary for standing in a covenantal relationship with God are the virtues that enable Jews to be the stewards of God's creation. The rabbinic tradition highlights the merits of humility (*anavah*), modesty (*tzni'ut*), moderation (*metinut*), and mercifulness (*rahmanut*)—all of which are ecologically beneficial.[65] Rooted in the awareness that humans are created by God and must submit to God's will, humility is the most effective antidote against the human hubris that fuels exploitative practices and destructive behaviors. Similarly, modesty leads one to shun conspicuous consumption and avoid greed, developing habits of consumption that are compatible with the diminishing resources of the world. Modesty is intrinsically linked to moderation, a virtue that encourages humans to avoid excesses and live rightly in accord with nature. And the virtue of mercifulness enables humans to be concerned about the needs of all creatures (human and nonhuman) and avoid vain pride or brute physical force. The Jewish legal tradition, then, has rich ecological wisdom and ethics.

MODERN JEWISH REFLECTIONS ON NATURE

Modern Judaism did not generate ecotheology per se, but nature figures prominently in the thought of several modern Jewish thinkers. In general, modern Judaism is pluralistic, subdivided into Orthodoxy, Conservative Judaism, Reform Judaism, Zionism, and the Jewish Renewal Movement. These strands differ in their attitude toward the normative power of Jewish law and in their attitudes toward modernity and its secularist orientation. Yet, there is no direct and simple correlation between theological reflections on nature and one's openness toward modernity or one's view on the normative power of Jewish religious law. The most creative Jewish thinking about the natural world comes not from avowed secularists, who do not pay attention to Jewish religious sources, but from those who believe that all Jewish views, attitudes, and practices have a religious dimension, namely, from those who wish to perpetuate the dialectical relationship between Torah and nature, between revelation and creation.

The rabbinic ethics of responsibility toward nature is advocated by the founder of modern Orthodoxy, Samson Raphael Hirsch (1808–88). Supportive of the emancipation, Hirsch noted the negative impact of exilic life on the Jewish people. In contrast to exile, "Nature meant us to be men of the fields and flocks," he asserted.[66] In the context of reflecting on Tu Bishevat, the Jewish festival that celebrates the new year for trees,[67] Hirsch goes on to declare that "Jewish law continually invites us to the observation of the laws and ways of nature, and how it is ever leading us from Nature to the life of man and there teaching us to use the products of the soil for bringing to ripeness the still nobler blossoms and fruits of a free human life permeated with the idea of God." For Hirsch, nature has a theological significance because nature is not only a model for the observance of its laws, but also nature places on humans its own demands or commandments.[68] Nonetheless, Hirsch has no qualms speaking about the "conquest of nature" and refers to God as "Creator, Lawgiver, and Controller of Nature."

The ethics of responsibility toward nature also informs the thought of Joseph Dov Soloveitchik (1903–92), the spiritual leader of modern Orthodoxy in the twentieth century. In his famous essay "The Lonely Man of Faith," Soloveitchik interpreted the two creation narratives in the Bible as two paradigmatic human postures toward nature.[69] The first narrative presents the "majestic man" (Adam I) who celebrates the unique position of the human in creation. Adam I is creative, functionally oriented, enamored of technology, whose aim is to achieve a "dignified" existence by gaining mastery over nature. By contrast, the second creation narrative presents the "covenantal man" (Adam II), the human who was commanded "to till and tend" the earth. Adam II eschews power and control; he is a nonfunctional, receptive, submissive human type who yearns for a redeemed existence, which he achieves by bringing all his actions under God's authority. The two postures exist simultaneously and remain permanently at war with each other within every religious Jew. Soloveitchik thus warned against the modern glorification of humanity (Adam I) that brought about the destruction of nature and pointed to religious commitment (Adam II) as the only response to our ecological and existential crisis.

The view that nature can be a resource for the revival of the Jewish people was, as we noted above, the cornerstone of Zionism. Indeed, Zionist thought is a rich source for Jewish environmental thinking, although its inspiration is often non-Jewish sources, be it romanticism, the *élan vital* philosophy of Henri Bergson, or evolutionary biology. Religious Zionists thinkers, by contrast, rooted their theology of nature in the emanationist cosmology of kabbalah and the immanentist theology of Hasidism. While these ideas cannot be reconciled philosophically and religiously, they were all inspired by the same impetus: the Zionist return to the land of Israel.

For some Zionist thinkers, such as Moses Leib Lilienblum (1843–1910), the founder of the "Love of Zion" movement, nations are like races with physical and mental characteristics that are transmitted through inheritance. Nations persist because they have a biological desire to persist. The Jews are a "natural nation" that persists despite anti-Semitism, but they will be able to thrive only if they return to be their natural homeland, the land of Israel. Only there can the Jews produce the culture that will be natural for the Jewish people. These ideas, which manifest the influence of Darwinism prevalent in Russian literature of the late nineteenth century, were further developed by the Zionist thinker known by his penname Ahad Ha-am (Asher Ginzberg) (1856–1927). For Ahad Ha-am Judaism is rooted in the Jewish people, and the collective identity of the people is dictated by biological, natural laws. As a living organism, the Jewish people desire to exist but, in the modern period the Jewish people experience the disease of assimilation, in addition to the suffering of persecutions, social discrimination, and physical threats to Jewish life. The only way to cure the disease is to return to the land of Israel, in which modern Jews will recreate their national culture through the revival of the use of the Hebrew language, the creation of Hebrew literature, and the study of the historical past. Ahad Ha-am envisioned the land of Israel as a cultural center in which a new Jewish national culture will emerge.[70]

A more religious version of cultural Zionism was articulated by Martin Buber, already mentioned above, who reinterpreted traditional Jewish values in order to address the dilemmas of modern Jewish life, especially for acculturated European Jews.[71] If the rabbinic tradition understood the covenant to be law centered, Buber insisted that the covenantal relationship culminating in revelation means a direct, nonpropositional encounter with the divine presence. According to Buber, humans relate to the world either directly and unconditionally ("I-Thou") or indirectly, conditionally, and functionally ("I-It"). The "I-Thou" modality means a direct encounter that encompasses all of one's personality and treats the other as an end rather than as a means. The "I-It" relationship has a purpose outside the encounter itself and involves only a fragment of the other, not the entire person. Buber's ideas became ecologically relevant and very influential, because he extended the "I-Thou" relationship to an encounter with nature. He spoke about his encounter with a horse when he was a boy and extended the possibility of having such a relation with a tree. In treating nature as a "Thou" rather than an "It," Buber personified natural phenomenon and recognized not only the need of humans to communicate with natural objects but also the inherent rights of nature. Nature is a waiting Thou, waiting to be addressed by the wholeness of our own being. Buber's dialogical philosophy has influenced contemporary environmental thinking, among Jews and non-Jews,[72] although Buber's attempt to derive his dialogical philosophy from Hasidic sources has been challenged by historians of Hasidism.[73]

The most ecologically interesting Zionist thinker was Aharon David Gordon (1856–1922), the spiritual leader of labor Zionism.[74] He was keenly aware of the crisis of modernity and the causal connection between technology and human alienation from nature. Settling in Palestine in 1904, Gordon joined the agricultural settlements in order to create a new kind of Jewish life and Jewish person. He viewed humans as creatures of nature but warned that humans are in constant danger of losing contact with nature. For Gordon, the regeneration of humanity and the regeneration of the Jewish people could come only through the return to nature and the development of a new understanding of labor as the source of genuine joy and creativity. Through physical, productive labor, humanity would become a partner of God in the process of creation. Rejecting the traditional Jewish focus on Torah study, Gordon viewed labor as a redemptive act, provided that the means that humans employ are in accord with the divine order of things, that is, with nature.

Gordon distinguished between "consciousness" and "experience." Whereas the former separates humanity from nature, the latter links humanity to nature. Secular culture, its science and technology, and its utilitarian approach to social life are secular products of human consciousness, whereas religion is the result of a primordial experience that unites humanity with the infinite cosmos. Secularism is thus the result of selfish utilitarianism that causes wars and conflicts between peoples and classes, creates a barrier between humanity and its natural environment, and causes the alienation of the individual from society. The ultimate expression of secularism is heresy, because it sees the human in isolation from God and thus as isolated from and within the infinite universe. In the case of Jews this heresy is expressed most acutely in assimilation, the erasure of Jewish identity under the pressure of surrounding civilization. Assimilationism is a typical product of exilic life—the highest manifestation of alienation from the land, from nature, and from one one's authentic identity. The only way for Jews to overcome the pressure of a utilitarian, selfish, and alienated environment is to return to the land of Israel and there not only acquire the external characteristics of a nation but revive authentic Jewish life by reconnecting with nature. The new relationship with nature will not be based on dominion and exploitation but on reunification with the natural world through labor. Labor would culminate in the religious act of *devequt*, the union of the worker with God.

Very similar to A. D. Gordon, but deeply anchored in the Jewish religious tradition was Abraham Isaac Kook (1865–1935), the first Ashkenazi chief rabbi of the Jewish community in Palestine. His thought was rooted in the symbolic worldview of kabbalah, but he was surprisingly open to the secular Zionists and was able to understand their predicament.[75] For Kook, the world of nature is an expression of divinity; nature is the most concrete, material expression of divine reality, but there is more to reality than the natural world. Kook believed that the Zionist insistence on the return to nature will lead only to the revival of religious life, even though the Zionist pioneers were themselves secularists. Secular Zionism helps to heal the collective body of the Jews, their physical existence, but it will serve only as a foundation for spiritual healing. This will mean recognizing that God alone is real and that God is the source from which all particulars come forth. Through the life of nature, the renewed nation could return to its divine source, God.

Zionism, especially the cultural Zionism of Ahad Ha-am, was also influential among American Jews as well, inspiring Mordecai Kaplan, the founder of Reconstructionist Judaism. Born into an Orthodox family, Kaplan came to America as a young boy, and his life and thought reflected the struggle of eastern European Jewry to Americanize and integrate into American culture without losing Jewish identity. In college, Kaplan came under the sway of sociologist of religion Émile

Durkheim and developed his own sociological analysis of Judaism that was particularly averse to the supernaturalist outlook of traditional Judaism. Kaplan posited a naturalist reading of Judaism which was "people focused but not God denying."[76] Kaplan saw morality, as expressed in the teaching of biblical prophets and the rabbis, as a commanding voice that transcends nature. It is the moral stance that enables humans to stand apart from nature and calls on the human to be more than just an animal. Defining Judaism as an "organic unity" that manifests a civilization (i.e., a culture of a people), Kaplan believed that in the land of Israel the Jews could create a majority society that will constitute a new kind of Jewish life. Like Ahad Ha-am, he also held that the Diaspora will not disappear, and he encouraged Jews to recreate their communal structure as holistic Jewish communities in order to facilitate Jewish creativity. Kaplan's main contribution to American Judaism, however, was organizational rather than theological. It was Kaplan who created the Jewish centers as a holistic focus of all Jewish cultural activities that would replace the synagogue.

Creative ecological thinking in America was the contribution of Abraham Joshua Heschel (1907–72), who was Buber's colleague and successor as the leader of adult education in Germany.[77] A scion of a Hasidic family who received modern university training, Heschel was rescued from the Nazis and settled in the United States in 1944. Until his death he inspired scores of alienated American Jews to find their way back to the sources of Judaism in order to heal the atrocities of modernity that culminated in the Holocaust. Heschel's ecologically sensitive Depth Theology spoke of God's glory as pervading nature and leading humans to radical amazement and wonder, viewed humans as members of the cosmic community, and emphasized humility as the desired posture toward the natural world. Recognizing human kinship with the visible world, Heschel celebrates God's presence within the world but also insists that the divine essence is not one with nature. God is simultaneously transcendent and immanent. Heschel did not provide systematic ecotheology, but it is possible to translate some of his ideas into ecotheology. This is precisely what Eilon Schwartz, an American-born environmentalist who settled in Israel, has attempted to accomplish in the educational programs of the Abraham Joshua Heschel Center for Environmental Learning and Leadership in Tel Aviv.[78]

Attempts to link moral consciousness to environmentalism characterize the Jewish environmental movement that emerged in the 1970s and 1980s when some of Heschel's disciples followed his call to return to Jewish sources and gave rise to the Jewish Renewal Movement. The movement involves various strands and intellectual sources, but on the whole it has been very instrumental in putting ecological awareness on the map of Jewish consciousness. Some environmental activists who were born Jews found their way back to the sources of Judaism by recognizing their ecological wisdom. Founded by Ellen Bernstein, the organization Shomrei Adamah ("Keepers of the Earth") popularized the idea of Jewish environmentalism,[79] revived nature-based Jewish rituals, such as the ritual meal for the minor holiday Tu Bishevat, and organized wilderness trips with a strong Jewish component.[80]

The most significant ecological thinker in the Jewish Renewal Movement is Arthur Waskow, who popularized the concept "ecokosher" to highlight the connection between human mistreatment of the natural world and social mistreatment of the marginal and the weak in the society.[81] His concern for ecology is part of a deep passion for justice, and his recommendations include the cultivation of self-control, moderation in material consumption, sustainable economic development, and communitrianism.

While Waskow's environmentalism is linked to Heschel's social activism and indebted to social ecology,[82] another disciple of Heschel, Arthur Green, has attempted to anchor Jewish ecological thinking in kabbalah and Hasidism, the other dimension of Heschel's legacy. Adopting the ontological schema of kabbalah, Green maintains that all existents are in some way an expression of God and are to some extent intrinsically related to each other.[83] Contrary to those who hold that in Judaism nature per se is not sacred, Green wishes to obliterate the ontological gap between the creator and the created. Instead, he adopts the monistic and immanentist ontology of kabbalah and blurs the distinction between creation and revelation. The world and the Torah are both God's self-disclosure, and both are linguistic structures that require decoding, an act that humans can accomplish because they are created in the image of God. From the privileged position of the human, Green derives an ethics of responsibility toward all creatures that acknowledges the differences between diverse creatures while insisting on the need to defend the legitimate place in the world of even the weakest and most threatened of creatures. For Green, a Jewish ecological ethics must be a *torat hayim*, namely, a set of laws and instruction that truly enhances life.

To some extent, all late-twentieth-century Jewish ecological thinking can be viewed as a belated response to the catastrophe of the Holocaust, a determination of the Jewish people to renew themselves so as "not to give Hitler a posthumous victory," to use the famous formulation of Emil Fackenheim. The Nazis' attempt to exterminate the Jewish people because they were supposedly subhuman was the most distorted application of evolutionary theories. Treating the Jews as vermin, the Nazis used Zyklon-B gas to eradicate the Jews, using bureaucratic efficiency and the most advanced science and technology for demonic purposes. The struggle between Nazism and Judaism raises some poignant questions about paganism as a worldview that does not allow for the possibility of transcendence and that takes the world of the senses to be ultimate reality. It is not trivial that the Nazis were ardent environmentalists and that they "abolished moral distinctions between animals and people by viewing people as animals. The result was that animals could be considered 'higher' than some people."[84] The horrendous results of Nazism for Jews need not be rehearsed, but they remind us that Nazism was the most consistent assault on the Jewish notion that nature is not inherently sacred and that only when humans act in accord with divine commands can nature become holy. While nature can be a source of spiritual inspiration, it is important to remember that nature is also violent, competitive, ruthless, and destructive. Nature does not care about the sick, the weak, and the deformed; it disposes of them in the relentless struggle for survival. Nature does not establish moral values that can create a just society in which the needs of the sick and the poor are addressed. These moral values, which constitute the Jewish ethics of responsibility, are revealed by God and implemented by humans who wish to sanctify nature as they strive to become like God. In the ghettos when Jews continued to pray and observe the rhythm of Jewish life despite Nazi oppression, Jews exercised spiritual resistance of the highest order.[85] In terms of physical Jewish survival, the experience of the Holocaust explains why Jews today welcome advanced reproductive technologies and why the state of Israel is much more lax about regulating biotechnology.

PENDING TENSIONS AND CONUNDRUMS

Jewish environmentalism today is a growing phenomenon, especially on a grassroots level. Jewish individuals are raising environmental issues and organizing educational activities to bring the ecological insights of Judaism to the attention of Jews. Tu Bishevat is now celebrated in many communities as a "Jewish Earth Day," and Jewish newspapers regularly report on environmental issues in connection with this festival. The small but growing body of scholarly literature written by rabbis, Judaica scholars, and educators makes it possible to teach college-level courses on Judaism and ecology. Awareness of environmentalist concerns is also shaping the programs of many synagogues, leading to the "greening" of Jewish institutions. Under the leadership of COEJL, environmental education, scholarship, advocacy, and action are clearly on the rise in the American Jewish community. Environmentally concerned Jews can find new ways to express their spirituality through concern for the needs of the natural world.

Yet it is important not to exaggerate the scope and significance of these activities. The number of people involved in these activities is still relatively small (COEJL has a mailing list of about ten thousand people), and the Jewish establishment still has environmental issues low on the list of Jewish priorities. On the academic level, engagement with questions of clean water, nuclear waste, biological diversity, climate change, and sustainable development is still not considered a bona fide academic interest for the simple reason that these issues do not fit into the textual and historical orientation of the academy, and the activism of the Jewish environmental movement appears to conflict with the detached, objective inquiry of Jewish studies. From the perspective of the academic study of Judaism, Jewish environmentalism is still not taken seriously precisely because it does not focus on the study of texts but is about social activism.

Ironically, the main challenges to Jewish environmentalism come from within.[86] In Israel and in America, the religious sources of Judaism do not inform the identity of most Jews, and secular Jews do not appeal to them in their attempt to address environmental concerns. Furthermore, the Jewish ethics of responsibility presupposes a sense of belonging to a community which is larger than the individual self. But the successful integration of Jews into modern society entailed the disintegration of the Jewish community and the erosion of Jewish communal solidarity. In industrialized countries, social mobility meant accumulation of wealth often accompanied with a consumerist lifestyle that undermines sound ecological conduct. And if this were not enough, Jewish environmentalists themselves are not unanimous on the recommended course of action and its justification within Judaism.

Jews who come to environmentalism from a Jewish religious commitment face other challenges. Whereas Jewish ecological thinking is necessarily religious, contemporary environmental philosophy and ethics are predominantly secular. To bridge the gap between these two discourses is not easy, since it requires considerable interpretative skills on the part of Jews and a willingness to understand Jewish legal and textual reasoning on the part of non-Jews. Religiously committed Jews must become familiar with a vast literature, whose worldview and philosophical assumptions not only conflict with the beliefs of Judaism but are in some cases self-consciously neopagan.[87] This is especially evident in nature-based feminist spirituality, which promotes goddess worship in order to overcome the deterioration of nature allegedly caused by the masculinist "Judeo-Christian tradition."[88] Likewise, the biocentrism of deep ecology stands in conflict with the anthropocentric stance of Judaism, which is the basis of its ethics of stewardship and responsibility toward nature.

Can Jews develop a Jewish ecotheology for the twenty-first century? I think so, but it will take a lot more creative thinking than finding certain texts that speak about nature or appropriation of kabbalistic language to express ideas that are quite contrary to kabbalah. What is required is a wholesale reinterpretation of Judaism in light of contemporary science, especially the biological sciences, which are in the midst of a major transformation. One example how to begin to think innovatively in this direction was offered by Hans Jonas (1903–93), a German Jewish philosopher who fled Nazi Germany and found his way to the United States after a brief period in Palestine.[89] Jonas wrote a pathbreaking research on Gnosticism, but after the experience of World War II shifted his interest from the history of philosophy and the history of religions to philosophy of nature and bioethics, extending his existential philosophy and phenomenological analysis to include all forms of life. Unique among twentieth-century Jewish philosophers, Jonas argued for the possibility of a genuinely symbiotic relationship between humanity and nature, which he believed had been suppressed by modern technology. On the basis of Jewish sources, especially the Bible and Lurianic kabbalah, Jonas spoke against the human domination of nature, and he was among the first to articulate the ethical challenges that modern technology poses to humanity. Jonas was critical of genetic engineering and cloning because he believed that life itself has the capacity for moral responsibility and because he considered the very emergence of life as an "ontological revolution in the history of matter." Jonas was one of a handful of Jewish philosophers to systematically incorporate evolutionary biology into his philosophical and ethical reflections. Jonas offers us a model of how Jewish ecological thinking can fuse philosophy and science to meet the challenge of the environmental crisis.

In sum, the Jewish religious tradition includes theological principles that can be very useful in contemporary attempts to think about ecology from a religious perspective and to articulate sound environmental policies on the basis of Jewish religious sources. Judaism can make a very useful contribution to ecoreligious discourse and can serve as insightful inspiration for ecological policies. However, if Jewish environmentalism is to grow, Jews will have to engage the large body of ecological philosophy and ethics in greater depth and explore the areas in which Judaism converges with environmental philosophy and the areas in which Judaism also poses a theoretical challenge to environmentalism. Conversely, if a dialogue between Judaism and environmentalism is to ensue, non-Jewish environmentalists will have to be more informed about Judaism and be willing to rethink some deeply held convictions in light of the Jewish critique. This means that environmentalists will have to recognize that the Bible is not the cause of our ecological crisis but a treasure of environmentally sound ideas, that Judaism is not to be reduced to the Bible, and that postbiblical Judaism was not simply superceded by and fulfilled in Christianity. Judaism need not be viewed as the cause of our current environmental crisis but as a possible path toward addressing the crisis.

NOTES

1. See Moshe Halbertal and Avishai Margalit, *Idolatry* (Harvard University Press, 1992), esp. 256–66 n. 8, where the authors discuss Judaism as a critique of nature religions in the ancient world, i.e., paganism.

2. For discussion of the ecology of the land of Israel and its representation in the Bible, see Yohanan Aharoni, *The Land of the Bible: A Historical Geography* (trans. and ed. Anson F. Rainey; London: Burns & Oates, 1979). For discussion of the various approaches

to the representation of nature in the Bible, see Theodore Hiebert, *The Yahwist's Land-scape: Nature and Religion in Early Israel* (New York: Oxford University Press, 1996), 3–29.

3. See Jacob Milgrom, *Cult and Conscience: The Asham and the Priestly Doctrine of Repentance* (Leiden: Brill, 1976).

4. The social causes of the rebellion against Rome are discussed by Martin Good-man, "The First Jewish Revolt: Social Conflict and the Problem of Debt," *Journal of Jewish Studies* 33 (1982): 417–27.

5. For succinct expositions of the notion of dual Torahs in rabbinic Judaism, consult Jacob Neusner, *The Oral Torah, the Sacred Books of Judaism: An Introduction* (San Francisco: Harper & Row, 1986).

6. A detailed discussion of this mishnaic text is available in Jeremy Benstein, "'One, Walking and Studying...': Nature vs. Torah," in *Judaism and Environmental Ethics* (ed. Martin D. Yaffe; Lanham, MD: Lexington, 2001), 206–29.

7. See Steven S. Schwartzchild, "The Unnatural Jew," in *Judaism and Environmental Ethics* (ed. Martin D. Yaffe; Lanham, MD: Lexington, 2001), 267–82. For Schwartzchild, a follower of Hermann Cohen's Neo-Kantianism, being Jewish is an ethical stance rather than a biological fact of birth.

8. See Arthur Schaffer, "The Agricultural and Ecological Symbolism of the Four Species of Sukkot," in *Judaism and Environmental Ethics* (ed. Martin D. Yaffe; Lanham, MD: Lexington, 2001), 112–24. For the practice of the holiday and its significance to contemporary Jews, consult Arthur I. Waskow, *Seasons of Our Joy: A Guide to the Jewish Holidays* (Boston: Beacon, 1990).

9. See Jeffrey L. Rubinstein, *The History of Sukkot in the Second Temple and Rabbinic Periods* (Atlanta: Scholars Press, 1995).

10. For comparative discussion of Jewish economic life in medieval Islam and Christendom, consult Mark Cohen, *Under Crescent and Cross: The Jews in the Middle Ages* (Princeton: Princeton University Press, 1994).

11. Moneylending positioned the Jews between the secular rulers (who were the major beneficiaries of this activity), the papacy (which recognized the necessity of lending at interest but regarded usury as immoral), the landed aristocracy (which con-sidered the Jews the extension of the monarchs), and the populace (which feared and hated the Jews but needed them for consumption loans). For exposition of this complex social dynamic, see William Jordan, "Jews on Top: Women and the Availability of Consumption Loans in Northern France in the Mid-Thirteenth Century," *Journal of Jewish Studies* 29 (1978): 39–56.

12. For further discussion, consult Hava Tirosh-Samuelson, "Theology of Nature in Sixteenth Century Jewish Philosophy," *Science in Context* 10.4 (1997): 529–70.

13. For an overview of the emancipation of European Jews and the transformation it has brought about in all aspects of Jewish life, consult Jacob Katz, *Jewish Emancipation and Self-Emancipation* (Philadelphia: Jewish Publication Society, 1986); idem, *Out of the Ghetto: The Social Background of Jewish Emancipation, 1770–1870* (New York: Schocken, 1978 [orig. 1973]).

14. For an overview of Zionist ideology and its various voices, consult Shlomo Avineri, *The Making of Modern Zionism: The Intellectual Origins of the Jewish State* (New York: Basic Books, 1981).

15. See Stephen C. Lonergan and David B. Brooks, *Watershed: The Role of Fresh Water in the Israeli-Palestinian Conflict* (Ottawa: International Development Research

Center, 1994); and Susan H. Lees, *The Political Ecology of the Water Crisis in Israel* (Lanham, MD: University Press of America, 1998).

16. Lynn White, "The Historic Roots of Our Ecological Crisis," *Science* 155 (1967): 1250–55.

17. Norman Lamm, "Ecology in Jewish Law and Theology," in his *Faith and Doubt: Studies in Traditional Jewish Thought* (New York: Ktav, 1972), 162–85; Jonathan Helfand, "Ecology and the Jewish Tradition: A Postscript," *Judaism* 20 (1971): 330–35; idem, "Consider the Work of G-d: Jewish Sources for Conservation Ethics," in *Liturgical Foundations of Social Policy in the Catholic and Jewish Traditions* (ed. Daniel F. Polish and Eugene J. Fisher; Notre Dame: University of Notre Dame Press, 1983), 134–48; and idem, "The Earth Is the Lord's: Judaism and Environmental Ethics," in *Religion and Environmental Crisis* (ed. Eugene C. Hargrove; Athens: University of Georgia Press, 1986), 38–52.

18. Martin D. Yaffe, ed., *Judaism and Environmental Ethics: A Reader* (Lanham, MD: Lexington, 2001).

19. Hava Tirosh-Samuelson, ed., *Judaism and Ecology: Created World and Revealed Word* (Cambridge: Center for the Study of World Religions, Harvard Divinity School, 2002).

20. Eric Katz, "Judaism," in *A Companion to Environmental Philosophy* (ed. Dale Jamieson; Oxford: Blackwell, 2001), 81–95.

21. Bron Taylor, ed., *Encyclopedia of Religion and Nature* (London: Continuum, 2005). The overview essay on Judaism was written by me; other essays on various aspects of the Jewish tradition were written by Manfred Gerstenfeld, David Seidenberg, and Alon Tal.

22. Lindsay Jones, ed., *Encyclopedia of Religion* (2nd ed.; New York: Macmillan, 2005). The overview essay entitled "Ecology and Judaism" was written by me.

23. The best information is available on the COEJL website: www.coejl.org. For discussion of the history of the organization and other Jewish environmental activism during the 1980s and 1990s, consult Mark X. Jacobs, "Jewish Environmentalism: Past Accomplishments and Future Challenges," in *Judaism and Ecology: Created World and Revealed Word* (ed. Hava Tirosh-Samuelson; Cambridge: Center for the Study of World Religions, Harvard Divinity School, 2002), 449–80.

24. For an overview of the doctrine, see Ronald Simkins, *Creator and Creation: Nature in the Worldview of Ancient Israel* (Peabody, MA: Hendrickson, 1994).

25. For close reading of the creation narratives written by a non-Jewish scholar mainly on the basis of German biblical scholarship, consult Jan J. Boersema, *The Torah and the Stoics on Humankind and Nature: A Contribution to the Debate on Sustainability and Quality* (Leiden: Brill, 2001), 47–112. There are, of course, many Jewish discussions of the two creative narratives. An interesting reading that shows that the rabbis privileged the second, more ecological, narrative is offered by David Kraemer, "Jewish Death Practices: A Commentary on the Relationship of Humans to the Natural World," in *Judaism and Ecology: Created World and Revealed Word* (ed. Hava Tirosh-Samuelson; Cambridge: Center for the Study of World Religions, Harvard Divinity School, 2002), 81–92. The best close reading of the second narrative is offered in Hiebert, *Yahwist's Landscape*, 30–82.

26. Hebrew Bible references are numbered according to English Bible numbers.

27. For the history of the notion of wilderness, see Max Oelschlaeger, *The Idea of Wilderness: From Prehistory to the Age of Ecology* (New Haven: Yale University Press, 1991).

For discussion of cultivated land versus wilderness in the Bible, consult Jeanne Kay, "Concepts of Nature in the Hebrew Bible," in *Judaism and Environmental Ethics* (ed. Martin D. Yaffe; Lanham, MD: Lexington, 2001), 86–104.

28. Yehuda Felix, *Nature and Man in the Bible* (London: Soncino, 1981), 113–14.

29. David Bryan, *Cosmos, Chaos, and the Kosher Mentality* (Sheffield: Sheffield Academic Press, 1995), 24.

30. The *Animal Apocalypse* is part of the so called Ethiopic Enoch, which was first written in Aramaic and later incorporated into collections of Enochic works. Today it survives in complete form in the Ethiopic Bible. For overview of this literature, see John J. Collins, *The Apocalyptic Imagination: An Introduction to Jewish Apocalyptic Literature* (2nd ed. Grand Rapid: Eerdmans, 1998 [orig. 1984]). For close analysis of this zoomorphic imagery, consult Bryan, *Cosmos, Chaos, and the Kosher Mentality*.

31. Jacob Neusner, *Confronting Creation: How Judaism Reads Genesis: An Anthology of Genesis Rabbah* (Columbia: University of South Carolina Press, 1991), 44.

32. For further information on medical knowledge in the rabbinic corpus, see Immanuel Jakobovits, *Jewish Medical Ethics* (rev. ed.; New York: Bloch, 1975); and David Feldman, *Health and Medicine in the Jewish Tradition* (New York: Crossroads, 1986).

33. The similarities are noted in Boersma, *Torah and the Stoics*, 200–210.

34. These arguments are discussed in Norbert M. Samuelson, *Judaism and the Doctrine of Creation* (Cambridge: Cambridge University Press, 1994), esp. 81–106.

35. For a more detailed discussion of the medieval cosmological picture in the writings of Maimonides, consult Tzvi Langermann, "Maimonides and the Sciences," in *The Cambridge Companion of Medieval Jewish Philosophy* (ed. Oliver Leaman and Daniel H. Frank; Cambridge: Cambridge University Press, 2003), 157–75.

36. See David Novak, *Natural Law in Judaism* (Cambridge: Cambridge University Press, 1998).

37. For further discussion of the doctrines of German pietists, see Ivan Marcus, *Piety and Society: The Jewish Pietists of Medieval Germany* (Leiden: Brill, 1981), esp. 23–52; and Joseph Dan, *The Esoteric Theology of Ashkenazi Hasidism* [Hebrew] (Jerusalem: Bialik Institute, 1968). To my knowledge there is no study of the esoteric theology of German pietism in relation to ecological thinking, because the scholars of German pietism (all of whom are historians of Jewish mysticism) are not interested in ecology.

38. The date and composition of *Sefer Yetzirah* are still disputed among scholars. I join those who maintain that it was probably composed in late antiquity (sometimes between the second and fourth centuries) by Jews immersed in Neo-Pythagorean and Hellenistic magic, but the text was edited in the ninth century in a Muslim setting, as part of the revival of the science in Islam. I discuss these issues and the significance of *Sefer Yetzirah* in my "Kabbalah and Science in the Middle Ages: Preliminary Remarks," in *Science in Medieval Jewish Communities* (ed. Gad Freudenthal; Leiden: Brill, forthcoming).

39. See Elliot R. Wolfson, "Nature as a Mirror of God," in *Judaism and Ecology: Created World and Revealed Word* (ed. Hava Tirosh-Samuelson; Cambridge: Center for the Study of World Religions, Harvard Divinity School, 2002), 305–32; and Hava Tirosh-Samuelson, "The Textualization of Nature in Jewish Mysticism," in *Judaism and Ecology: Created World and Revealed Word* (ed. Hava Tirosh-Samuelson; Cambridge: Center for the Study of World Religions, Harvard Divinity School, 2002), 389–404.

40. For an attempt to decode kabbalistic organic imagery, see Pinchas Giller, "The World Trees in the Zohar," in *Trees, Earth, and Torah: A Tu B'Shvat Anthology* (ed. Ari

Elon, Naomi Mara Hyman, and Arthur Waskow; Philadelphia: Jewish Publication Society, 2000), 128–34.

41. See David B. Ruderman, *Kabbalah, Magic, and Science: The Cultural Universe of a Sixteenth-Century Jewish Physician* (Cambridge: Harvard University Press, 1998).

42. This group and its impact on Jewish culture are discussed by David B. Ruderman, *Jewish Thought and Scientific Discovery in Early Modern Europe* (New Haven: Yale University Press, 1995).

43. The doctrine of "uplifting the divine sparks" (*ha'ala'at nitzotzot*) originated in the kabbalah of Isaac Luria in the sixteenth century, which Hasidism transformed and elaborated. Whether or not this doctrine is ecological is discussed by Jerome (Yehudah) Gillman, "Early Hasidism and the Natural World," in *Judaism and Ecology: Created World and Revealed Word* (ed. Hava Tirosh-Samuelson; Cambridge: Center for the Study of World Religions, Harvard Divinity School, 2002), 369–88, who reached a negative conclusion. Hasidism, he concludes, was not interested in the natural world per se and cannot be used to anchor contemporary Jewish ecological stance. A similar view is shared by other Orthodox scholars such as Norman Lamm and Moshe Sokol.

44. This material, which bristles with the impact of romantic literature, still awaits systematic discussion from an ecological perspective. I will undertake it in my forthcoming *Judaism and Nature* (Rowman & Littlefield).

45. For an overview of Rosenzweig's interpretation of creation, consult Samuelson, *Judaism and the Doctrine of Creation*, 29–67.

46. For example, Larry Troster, "From Big Bang to Omega Point: Jewish Response to Recent Theories in Modern Cosmology," *Conservative Judaism* 49.4 (1997): 17–31; and idem, "Love of God and the Anthropic Principle," *Conservative Judaism* 40.2 (1987/88): 43–51.

47. The term *ethics of responsibility* characterizes the ethical posture of the Jewish tradition in general and not only in regard to nature. See Walter Würtzberger, *Ethics of Responsibility: Pluralistic Approaches to Covenantal Ethics* (Philadelphia: Jewish Publication Society, 1994).

48. The ethics of stewardship is derived from Gen. 2:15. For Jewish reflections on the principle, see David Ehrenfeld and Philip J. Bentley, "Judaism and the Practice of Stewardship," in *Judaism and Environmental Ethics* (ed. Martin D. Yaffe; Lanham, MD: Lexington, 2001), 125–35. For overviews of this environmental concept, consult Douglas John Hall, *The Steward: A Biblical Symbol Come of Age* (Grand Rapid: Eerdmans, 1990).

49. Richard Sarason, "The Significance of the Land of Israel in the Mishnah," in *The Land of Israel: Jewish Perspectives* (ed. Lawrence A. Hoffman; Notre Dame: University of Notre Dame Press), 144. For a modern reworking of this biblical view, see Samuel Belkin, "Man as Temporary Tenant," in *Judaism and Human Rights* (ed. Milton R. Konvitz; New York: Norton, 1972), 251–58.

50. See Yosef Orr and Yossi Spanier, "Traditional Jewish Attitudes towards Plant and Animal Conservation," in *Judaism and Ecology* (ed. Aubrey Rose; London: Cassell, 1992), 54–60. The major protection of trees, especially fruit-bearing trees, is discussed under the principle "do not destroy" (*bal tashchit*) below.

51. This biblical text is part of the Shema, the affirmation of the Jewish faith in every public prayer service. However, Jewish Reform rabbis and theologians found this passage most problematic because it does not conform with modern science, and took it out of

the liturgy. There is now a realization among some Reform rabbis that perhaps this was a mistake and that the deep insight of the biblical text should be retained.

52. For explication of this ruling in its development in Judaism, consult Eilon Schwartz, "Bal Tashchit: A Jewish Environmental Precept," in *Judaism and Environmental Ethics* (ed. Martin D. Yaffe; Lanham, MD: Lexington, 2001), 230–49.

53. See Aloys Hüttermann, *The Ecological Message of the Torah: Knowledge, Concepts, and Laws Which Made Survival in a Land of "Milk and Honey" Possible* (Atlanta: Scholars Press, 1999), 55–68.

54. Baruch A. Levine, *Leviticus* (JPS Torah Commentary; Philadelphia: Jewish Publication Society, 1989), 243.

55. See Hüttermann, *Ecological Message of the Torah*, 82.

56. Mary Douglas, *Purity and Danger: An Analysis of the Concepts of Pollution and Taboo* (London: Routledge & Kegan Paul, 1966), offered an anthropological, structuralist analysis of the dietary laws that has been accepted by many other scholars of Judaism. For example, consult Leon Kass, "Sanctified Eating," in *Judaism and Environmental Ethics* (ed. Martin D. Yaffe; Lanham, MD: Lexington, 2001), 384–409.

57. This is the gist of Hüttermann's analysis; see *Ecological Message of the Torah*, 71ff.

58. Ibid., 76.

59. The intrinsic holiness of the land of Israel was highlighted by the rabbis precisely because of the political defeat of Judea by Rome. Some Jewish environmentalists take this notion and generalize from it that the earth as a whole (and hence nature) is intrinsically holy. See, e.g., Bradley Shavit Artson, "Our Covenant with Stones: A Jewish Ecology of Earth," in *Judaism and Environmental Ethics* (ed. Martin D. Yaffe; Lanham, MD: Lexington, 2001), 168. This view conflicts the way I presented the Jewish position on nature. My view follows the more traditional line of thinking explored by Michael Wyschogrod, "Judaism and the Sanctification of Nature," in *Judaism and Environmental Ethics* (ed. Martin D. Yaffe; Lanham, MD: Lexington, 2001), 289–96, esp. 294.

60. For a comprehensive analysis, consult Noah J. Cohen, *Tza'ar Ba'ale Hayim: The Prevention of Cruelty to Animals, Its Bases, Development, and Legislation in Hebrew Literature* (2nd ed.; New York: Feldheim, 1976).

61. For further discussion of this principle, see Lenn Evan Goodman, "Respect for Nature in the Jewish Tradition," in *Judaism and Ecology: Created World and Revealed Word* (ed. Hava Tirosh-Samuelson; Cambridge: Center for the Study of World Religions, Harvard Divinity School, 2002), 227–60. The paragraph is based on his discussion in 245–52.

62. Hasidic slaughtering was a major contributing factor to the split between Hasidism and their opponents (Mitnagedim). For general overviews, see Samuel Dresner, "Hasidism and Its Opponents," in *Great Schisms in Jewish History* (ed. Raphi Jospe and Stanley Wagner; New York; Ktav, 1981), 118–76.

63. See J. David Bleich, "Judaism and Animal Experimentation," in *Judaism and Environmental Ethics* (ed. Martin D. Yaffe; Lanham, MD: Lexington, 2001), 333–70.

64. On the sabbatical year, see Gerald Blidstein, "Man and Nature in the Sabbatical Year," in *Judaism and Environmental Ethics* (ed. Martin D. Yaffe; Lanham, MD: Lexington, 2001), 136–42. The sabbatical year could not be observed while the Jews were in exile, since it is a land-based commandment, but its observance was renewed in the modern state of Israel; see Benjamin Bak, "The Sabbatical Year in Modern Israel," *Tradition* 1.2 (1959): 193–99. For contemporary reflections on the relevance of this biblical legislation, see Arthur Waskow, "From Compassion to Jubilee," *Tikkun* 5.2 (1990): 78–81;

and Eric Rosenblum, "Is Gaia Jewish? Finding a Framework for Radical Ecology in Traditional Judaism," in *Judaism and Environmental Ethics* (ed. Martin D. Yaffe; Lanham, MD: Lexington, 2001), 183–205.

65. See Moshe Sokol, "What Are the Ethical Implications of Jewish Theological Conceptions of the Natural World," in *Judaism and Ecology: Created World and Revealed Word* (ed. Hava Tirosh-Samuelson; Cambridge: Center for the Study of World Religions, Harvard Divinity School, 2002), 261–82.

66. Samson Raphael Hirsch, *Judaism Eternal: Selected Essays* (ed. and trans. I. Grunfeld; London: Soncino, 1959), 1.36.

67. Tu Bishevat is a postbiblical holiday that originated in the practice of taxing fruit-bearing trees. The rabbis made it into a celebration of the new year for trees, analogous to Rosh ha-Shanah, which is the beginning of the new year. The holiday continued to evolve during the Middle Ages, especially in the communities of north African Jewry, who were steeped in kabbalah. Kabbalists created a new ritual modeled after the Passover Seder to fathom the mystical meanings of nature symbolism, and nature-based rituals, associated with this festival. See Ellen Bernstein, "A History of Tu B'Sh'vat," in *Ecology and the Jewish Spirit: Where Nature and the Sacred Meet* (ed. Ellen Bernstein; Woodstock, VT: Jewish Lights, 2000), 139–41.

68. For further discussion of Hirsch's view, consult Shalom Rosenberg, "Concepts of Torah and Nature in Jewish Thought," in *Judaism and Ecology: Created World and Revealed Word* (ed. Hava Tirosh-Samuelson; Cambridge: Center for the Study of World Religions, Harvard Divinity School, 2002), 189–226, esp. 214–18.

69. The essay appeared first in *Tradition* 7 (1965): 5–67. For discussion of Soloveitchik's paradigms, see Eilon Schwartz, "Response, Mastery and Stewardship, Wonder and Connectedness: A Typology of Relations to Nature in Jewish Text and Tradition," in *Judaism and Ecology: Created World and Revealed Word* (ed. Hava Tirosh-Samuelson; Cambridge: Center for the Study of World Religions, Harvard Divinity School, 2002), 93 108, esp. 96–99.

70. The best discussion of Ahad Ha-am's philosophy is to be found in Eliezer Schweid, *Judaism and Secular Culture: Chapters in Jewish Thought in the Twentieth Century* [Hebrew] (Tel Aviv: Kibbutz ha-Meuchad, 2001); and idem, *New Ways in Jewish Religious and National Thought* [Hebrew] (Jerusalem: Akademon, 1991). His studies of Ahad Ha-Am, A. D. Gordon, Mordecai Kaplan, and Yehezkel Kaufman are most instructive and relevant to our topic.

71. The literature on Buber's philosophy is extensive. A good exposition is Paul Mendes-Flohr, *From Mysticism to Dialogue: Martin Buber's Transformation of German Social Thought* (Detroit: Wayne State University Press, 1989).

72. See Brian J. Walsh, Marianne B. Karsh, and Nik Ansell, "Trees, Forestry, and the Responsiveness of Creation," in *This Sacred Earth: Religion, Nature, Environment* (ed. Roger Gottlieb; New York: Routledge, 1996), 423–35.

73. See n. 63 above.

74. The following discussion is based on Schweid, *Judaism and Secular Culture*, 157–81. Also helpful is idem, "A. D. Gordon: A Homeland That Is a Land of Destiny," in Schweid's *The Land of Israel: National Home or Land of Destiny* (trans. Deborah Greniman; Rutherford: Fairleigh Dickinson University Press, 1985), 157–70.

75. Although Rav Kook was not a systematic thinker, all aspects of his thoughts are internally linked and all are grounded in his own interpretation of kabbalah. For dis-

cussion of various aspects of Kook's religious thought, consult the studies in Lawrence J. Kaplan and David Shatz, eds., *Rabbi Abraham Isaac Kook and Jewish Spirituality* (New York: New York University Press, 1995). For a sustained exposition of his philosophy, consult Yosef ben Shlomo, *Poetry of Being: Lectures on the Philosophy of Rabbi Kook* (Tel Aviv: MOD, 1990). Contrary to most modern Jewish philosophers who followed the Neo-Kantianism of Hermann Cohen, Kook perpetuated the Platonic worldview of kabbalah.

76. Eugene B. Borowitz, *Choices in Modern Jewish Thought: A Partisan Guide* (2nd ed.; West Orange, NJ: Behrman, 1995 [orig. 1983]), 100.

77. For an comprehensive biography of Heschel, consult Edward Kaplan and Samuel Dresner, *Abraham Joshua Heschel: Prophetic Witness* (New Haven: Yale University Press, 1998).

78. For information about this organization and its activities in Israel, see their website: www.heschelcenter.org. The site includes many links to environmental organizations in the world, COEJL being but one of them, making it clear that the Israeli environmentalist movement is part of a larger green politics.

79. Shomrei Adamah was instrumental in publishing the early writings of Jewish environmentalists in the United States; see Mark Swetlitz, *Bibliography, 1970–1986* (Philadelphia, 1987).

80. See Matt Biers-Ariel, Deborah Newbrun, and Michael Fox Smart, *Spirit in Nature: Teaching Judaism and Ecology on the Trail* (n.p.: Behrman, 2000).

81. Arthur Waskow, "What Is Eco-Kosher," in *This Sacred Earth: Religion, Nature, Environment* (ed. Roger S. Gottlieb; 2nd ed.; New York: Routledge, 2004), 297–300.

82. Social ecology is associated with the name of Murray Bookchin, who was raised as a Jew and whose anarchistic views are probably inspired by his Jewish background, but they are not rooted in the religious sources of Judaism.

83. See Arthur Green, "Great Chain of Being: Kabbalah for an Environmental Age," in his *Ehyeh: A Kabbalah for Tomorrow* (Woodstock, VT: Jewish Lights, 2003), 108–19; a version of this essay is published in *Judaism and Ecology: Created World and Revealed Word* (ed. Hava Tirosh-Samuelson; Cambridge: Center for the Study of World Religions, Harvard Divinity School, 2002), 3–16.

84. A. Cockburn, "A Short, Meat Oriented History of the World: From Eden to Matthole," *New Left Review* (1996): 16–42, esp. 30. For a Jewish critique of the Nazi association with environmentalism, see Wyschogrod, "Judaism and the Sanctification of Nature."

85. See Roger Gottlieb, "The Concept of Resistance: Jewish Resistance during the Holocaust," in *Thinking the Unthinkable* (ed. Roger Gottlieb; New York: Paulist Press, 1990), 327–44. Gottlieb further explored this theme in his *A Spirituality of Resistance: Finding a Peaceful Heart and Protecting the Earth* (Lanham, MD: Rowman & Littlefield, 2003).

86. This point is developed by Jacobs, "Jewish Environmentalism," 471–77.

87. See Margot Adler, *Drawing Down the Moon: Witches, Goddesses, Worshippers, and Other Pagans in America Today* (Boston: Beacon, 1979).

88. The best example of this version of ecofeminism is Carol P. Christ, *Rebirth of the Goddess: Finding Meaning in Feminist Spirituality* (New York: Routledge, 1997). I have reservations about nature-based feminist spirituality and express them in my "Religion, Ecology, and Gender: A Jewish Perspective, *Feminist Theology* 13.3 (2005): 373–397."

89. For an overview of Jonas's biography and its relationship to his philosophy of life, see Richard Wolin, *Heidegger's Children: Hannah Arendt, Karl Löwith, Hans Jonas, and Herbert Marcuse* (Princeton: Princeton University Press, 2001), 101–33; and Lawrence Vogel, "Editor's Introduction: Hans Jonas's Exodus: From German Existentialism to Post-Holocaust Theology," in Hans Jonas, *Mortality and Morality: A Search for the Good after Auschwitz* (ed. Lawrence Vogel; Evanston: Northwestern University Press, 1996), 1–40.

CHAPTER 2

CATHOLICISM

JOHN HART

CATHOLICISM and other branches of Christianity rarely reflected on earth *in se* for almost two millennia. Earth was viewed ordinarily as the setting for human existence, the stage on which human protagonists worked toward their salvation in a better world to come. As God's creation, part of an integrated universe, earth was to be appreciated as evidence of divine creativity and divine compassion: the wonderful works of God were to be admired; the role of those works, as intended by God, was to provide for human needs. All was created for "man," made in God's "image," who was to avail "himself" of earth's goods, shaping them to meet "his" needs. Indeed, earth and its forests and valleys were imperfect in their pristine state: they reached perfection only when altered by human hands. A tree was better as a table; a meadow was better as a farm. Catholics and other Christians who thus instrumentalized (and commodified) nature did not see any intrinsic worth in other creatures.

Periodically, some individuals within Christianity (such as Francis of Assisi) became known for their appreciation of pristine nature and their attitude toward nonhuman species. But most (clerical and lay) Christians were preoccupied more with life to come than with life on earth.

Toward the end of the second Christian millennium, the Catholic church, through its hierarchical institutional structures and leadership and through insights of thinkers in the lay community, began to promote care for earth. At first, particularly within the hierarchy, Catholic concern was limited to regard for earth as the provider of life's necessities for humankind. Then it was minimally extended to advocate conserving earth as home and provider for human communities and the broader biotic community, although, with regard to the latter, primarily (if not solely) for their *instrumental value*: their role to satisfy humans' needs and wants. Finally, Catholic church leaders and scholars expressed their appreciation for and advocacy of the *intrinsic value* of earth and all earth's life: other creatures had their own worth inherent in themselves as parts of God's creation, their value was not assigned to them by humans. Such developments in Catholic thought rarely, if ever, originated from the church hierarchy; their primary source was the theologians, ethicists, and other scholars who became aware of scientific studies that indicated a developing global environmental crisis and in response began to promote care for earth and respect for earth's biosphere.

When this understanding emerged, environmental and ecological well-being was promoted more vigorously. (*Environmental* well-being refers here to the good of earth as a whole, as the setting of life. *Ecological* well-being refers to the good of species living interdependent and interrelated lives in *ecosystems*, particular earth places characterized by the integration of specific species and individuals of the biotic community—the community of all life—with each other and with a particular abiotic [nonliving] locale that serves as their *habitat*: the context of their existence and the provider of necessary air, land, water, and minerals to support their quests for survival.) Although a two-pronged approach of caring for creation and caring for community emerged, the former was usually, though not exclusively, subsumed into and subordinated to the latter.

The development of Catholic thought will be considered in three stages: caring for the common good (Catholicism as it traditionally viewed nature as instrumental and the way the church related to it over the centuries); concern for creation in crisis (Catholicism in the twentieth century confronted by environmental and ecological degradation related to such human factors as heightened population growth in poorer countries and rampant consumerism in highly industrialized nations); and creation concern and community commitment (Catholicism strongly advocating conservation and compassion, with some in the church proposing the extension of "community" to otherkind; church consideration of proposals for stabilizing or reducing the rate of human population growth; church advocacy of reduced consumption of earth goods; and church proposals for a better distribution of earth's land base). Issues of water ownership, water use, and water purity will serve as a summary focal point for discussing church teachings on environment, ecology, and economics.

As will be seen, the most significant developments of Catholic environmental thought emerged from the Americas, where the initial foci were, in Latin America, the equitable distribution and ownership of land and the just distribution of the land's goods; and, in North America, respect for earth and earth's creatures and justice in the allocation of earth's goods. Eventually, issues from north and south became intertwined across the hemisphere and within each region, while incorporating throughout a special interest in the plight of the poor.

The Catholic church may be viewed in two distinct, though not exclusive, ways: as an institutional body controlled by a clerical hierarchy and as a community of believers, many of whose lay members are becoming more reflective about church teachings, more analytical about church practices, and more assertive about advancing their own positions and advocating actions that should flow from them. (Institutional church leaders reject such a distinction; they view themselves as the guardians of authentic Catholic doctrine and often expect—an expectation more often satisfied among the clergy and the less educated laity—the "church" to be identified with them as its authoritative teaching body or "magisterium.") When looking at developments in environmental thought within Catholicism, therefore, it is helpful to distinguish among official teachings by the popes and other ranking members of the hierarchy; perspectives presented by clergy and clerical theologians; and, more recently, insights offered by members of the laity and religious orders—theologians, ethicists, community activists, and others reflecting on or engaged with environmental issues.

While a plethora of statements and books have been published by Catholics,[1] including the pope, national and regional conferences of bishops, and individuals (laity, clergy, and members of religious orders), only a few representative perspectives can be considered here.

The sections that follow flow from a presentation of some historical bases for the development of traditional Catholic environmental thought to the evolution of that thought, to the presentation of salient and potentially enduring ideas, and then to a vision of possibilities for future developments of earth consciousness and earth commitments within Catholicism as a whole. In the latter two decades, because of the interaction and complementarity between dynamic "official" church teachings and creative, visionary lay presentations, between the institutional church and the church of the faithful, Catholic environmental thought as a whole has begun to break the constraining confines of past eras, while (perhaps) paradoxically becoming more faithful to core foundational understandings of its biblical origins.

CARING FOR THE COMMON GOOD

Initially, creation was viewed narrowly in Christian thought solely as the provider for the human common good. The provider role was understood in two respects: earth was the *setting* for humans (usually meaning exclusively Christians) striving for salvation, primarily through their relationships with God and each other (relationships that were to be consonant with biblical and church commandments and counsels); and earth was the *supplier* both of life's requirements and of livelihood's resources that were needed for humans' earthly existence. (It should be noted that the concept of the "common good" in Catholicism traditionally has been reserved to describe the human community's well-being.) In the light of interpretations of sacred scripture, it was believed that humans were to subdue the earth, that is, subject it to meet human needs by bringing human order to it. ("Subdue" was understood to be under the ultimate dominion of God: it was God's earth with which people were working and which they were obliged to respect. This biblical phrasing emerged from an agricultural context in which there were no tractors, bulldozers, or other heavy industrial equipment or explosives that could dramatically alter the landscape. To take "subdue" out of context to justify massive alterations to topography, as has been done over the centuries as machines became ever more destructive and ubiquitous, is to do violence to the biblical understanding.)

Augustine and Aquinas

Elements of the teachings of Augustine (354–430) and Thomas Aquinas (1225–74) have been particularly influential in Catholic doctrines regarding creation and the goods of creation. Both men believed in a hierarchy in nature—people on earth had dominion over all other earthly creatures. Both taught respect for other living beings, ordinarily not because they had intrinsic value, but because they had instrumental value for humankind. Still, regarding the inherent worth of otherkind as God's creations, as contributors to the integrity of the universe, Aquinas approvingly cites Augustine's statement that, while some people find fault with aspects of creation that seem to serve no beneficial purpose, they should bear in mind that "though not required for the furnishing of our house, these things are necessary for the perfection of the universe."[2] Aquinas stated additionally that "less honorable parts exist for the more honorable. . . . So, therefore, in the parts of the universe also every creature exists for its own proper act and perfection, and the less noble for the nobler, as those creatures that are less noble than man exist for the sake of man, whilst each and every creature exists for the perfection of the entire universe."[3] Aquinas affirms a pyramidal hierarchy in nature, with humans at the apex, but also declares that all creatures are needed as integral members of an integrated universe.

Augustine advocated community of goods, care for creation, and even intergenerational responsibility when he stated that Christians should regard themselves as pilgrims on a journey who paused for a while at an inn. On their departure they would not take the inn's dishes or bedding with them, but leave them for the next travelers to use for dining or sleeping. Similarly, in this life they were to use earth's goods carefully so that others also might have them to use.

For his part, Aquinas advocated community of possessions and provision of the necessities of life for all people when he discussed the seventh commandment ("you shall not steal" in the Catholic list of ten). He declared that natural law, by which all property was in common so that all people's needs would be met, took precedence over human laws in cases of necessity. It was not theft, he said (although it would be so by civil laws), for a person in need to go quietly in the night to take from the barn of the nobleman what was needed for survival; what they would be taking would actually be their own property, since their need in their time and place made all property come under the natural law of common property. Similarly, a person might quietly take from the lord's barn necessities that their neighbor lacked, if they could not themselves provide for their neighbor.[4]

Augustine and Aquinas, then, did not teach that earth and earth's creatures had an independent value beyond their utility for human well-being and the working of the universe, but they did teach that earth's goods should be distributed in such a way that the needs of all people would be met. Their perspective persists in Catholicism today among most Catholic leaders and laity.

Francis of Assisi

A startlingly distinct perspective entered Catholicism in the time between Augustine and Aquinas through the life and teachings of Francis of Assisi (1182–1226). In terms of attitudes toward creation, the life and teachings of this simple mendicant eventually would have a greater direct influence on Christianity (and beyond) than the writings of academics or the pronouncements of church officials. Saint Francis called all creatures—living and nonliving—"brother" or "sister." He expressed thereby the deep kinship he experienced with all creation.

The best-known work of Saint Francis, "Canticle of Brother Sun" (known also as the "Canticle of Creation" or "Canticle of All Creatures"), clearly suggests a familial creation. Inanimate creatures such as the sun, earth, fire, and water are explicitly evoked in sibling terms. No animate creatures are named in the song, but are present in its melody, and so implicitly (as well as explicitly in stories about Francis and birds, fish, and a wolf, among other living beings) are family members too. Francis borrowed the melody from a period romantic ballad praising various animals and plants. When people heard him sing his canticle, in their minds they added the original words to his, thereby imagining representatives of the totality of creation. The living and nonliving became integrated in the unity of words and music. The canticle, written when the Italian language was developing from French and Latin, is considered one of the classics of Italian literature:

> Most High, all-powerful, and all-good Lord,
>   praise, glory, honor,
>   and all blessing
>   are yours.
> To you alone, Most High, they belong,
>   although no one is worthy
>   to say your name.
> Praised be my Lord, with all your creatures,
>   especially my lord Brother Sun,
>   through whom you give us day and light.
> Beautifully he shines with great splendor:
>   Most High, he bears your likeness.
> Praised be my Lord, by Sister Moon and Stars:
>   in the heavens you made them bright
>   and precious and beautiful.
> Praised be my Lord, by Brother Wind,
>   and air and cloud
>   and calm and all weather
>   through which you sustain
>   your creatures.
> Praised be my Lord, by Sister Water,
>   who is so helpful and humble
>   and precious and pure.
> Praised be my Lord, by Brother Fire,
>   through whom you brighten the night:
>   who is beautiful and playful
>   and sinuous and strong.
> Praised be my Lord, by our Sister Mother Earth,
>   who sustains us and guides us,
>   and provides varied fruits
>   with colorful flowers and herbs.
> Praised and blessed be you, my Lord,
>   and gratitude and service be given to you
>   with great humility. (trans. John Hart)

Despite church leaders' admiration for the life and teachings of Saint Francis, the twofold doctrine that all people's needs should be met and that other creatures exist primarily to meet those needs remained the limiting foundations for Catholic church environmental teachings until the late twentieth century. This doctrine was elaborated and extended more fully as Catholic social thought developed further, particularly through papal social encyclicals that began in the late nineteenth century and through church teachings expressed in the documents of the Second Vatican Council in the last third of the twentieth century.

Leo XIII: First Social Encyclical

In 1891 Pope Leo XIII issued the first papal social encyclical, *Rerum novarum* ("On the Condition of Labor"). The encyclical set forth for the first time in Catholicism a focused analysis of social issues, especially economic injustices.

In *Rerum novarum* Leo XIII writes with regard to the land: "Besides, however it is distributed among individuals, the earth does not cease to serve the needs which are common to all. There is no one who does not feed upon the produce of the fields. [Workers' labor provides] a wage which can have no other source than the manifold fruits of the earth."[5] Leo goes on to state that "the same benefits of nature and gifts of divine grace belong in common to the whole human race, without distinction."[6]

For Leo XIII, society had become divided into two antagonistic economic classes: the rich and powerful, who control industry and trade and influence governments; and the poor majority lacking resources. This division can be overcome if there is a more equitable distribution of the land, which would result in the gap between classes being smaller, and "more abundant supplies of all goods of the earth," since people apply themselves more when they are working for themselves on their own land.[7]

A *just wage* is imperative for workers, as a matter of justice—"to defraud a man of the wage which is his due is to commit a grievously sinful act which cries out to heaven for vengeance"[8]—and to help workers provide the necessities of life for themselves and their families—"the wage ought not to be in any way insufficient for the bodily needs of a temperate and well-behaved worker."[9]

Concerned about the poor, Leo declares that "where the protection of private rights is concerned, special regard must be had for the poor and weak."[10] He observes that "indeed the will of God himself seems to give preference to people who are particularly unfortunate."[11]

When he addresses humankind's place in creation, Leo expresses the traditional view that "man" is over all creatures, since the human soul is impressed with God's image "and it is in it that resides the sovereignty by virtue of which man is commanded to rule over the whole of the lower creation and to use all the earth and the sea for his own needs."[12] As "man" works with the land and gathers its goods, "he makes his own that part of nature's resources which he brings to completion, leaving on it, as it were, in some form, the imprint of himself."[13]

Rerum novarum, then, laments the harsh conditions suffered by workers; calls on employers to act with justice toward their employees, and employees to act responsibly toward their employers; and advocates use of the land to provide for all people's needs.

Father Edward McGlynn

The major catalyst for the encyclical's development was probably the work of a New York City priest, Edward McGlynn (1837–1900), who viewed the impoverished condition of the laboring masses as an affront to God and a negation of biblical social teaching. To better their lot, he urged redistribution of private property in land, and a just distribution of Earth goods, since "All men, inalienably, always, everywhere, have a common right to all the general bounties of nature.... The land as well [as] the sunlight, and the air, and the waters, and the fishes, and the mines in the bowels of the earth, all these things that were made by the Creator through the beautiful processes of nature, belong equally to the human family, to the community, to the people, to all the children of God. The law of labor requires that these natural materials shall be brought into such relations with men that they shall afford to them food, raiment, shelter."[14] He asked, in words reminiscent of Aquinas's teaching on common property, "How are we going to give back to the poor man what belongs to him?"[15] McGlynn supported the New York City 1886 mayoral campaign of Henry George and was excommunicated by Rome the following year at the instigation of New York's Archbishop Corrigan, who had supported the Tammany Hall candidate. McGlynn continued his work among the poor and eventually, after his writings had been examined by Catholic theologians who found nothing in them contradictory to Catholic doctrine, was restored to the Catholic church and to his priesthood in 1892.

The influence of McGlynn's thought and action—by itself and linked to the George campaign—has continued into the twenty-first century. Periodically since 1931, when Pope Pius XI issued an encyclical addressing economic issues and commemorating the fortieth anniversary of *Rerum novarum*, the popes have issued social encyclicals at ten-year intervals, commemorating that first social encyclical and amending it according to changing theological, economic, and political understandings. Encouraged by this papal focus on Christian social ethics, national and regional bodies of Catholic bishops have issued pastoral letters on such issues as peace, economics, and ecology. Recent statements on environmental issues by popes and other church officials are direct descendants of the initial engagement of Leo XIII with social injustices from a Catholic perspective on justice.

Vatican II

The Second Vatican Council (1962–65) ended five years before the first Earth Day and neither anticipated nor discussed the pressing environmental concerns that would dramatically attract international notice beginning in 1970. During its deliberations the Council continued a basic thrust in Catholic social teaching: concern for and commitment to workers and the poor. The Council also retained traditional attitudes toward earth, earth's other creatures, and earth's goods ("resources"): the planet was the setting provided for humans to work out their salvation in a better life to come; other aspects of creation, while fulfilling at times God's unknown purposes, ultimately were provided by God to meet human needs.

At the Council, the *Pastoral Constitution on the Church in the Modern World* (*Gaudium et Spes*), issued in 1965, frequently referred from an anthropocentric perspective to human superiority to and supremacy over the rest of nature and other living creatures. The document declared, for example, that "humanity can and should increasingly consolidate its control over creation";[16] that "all things on earth should be related to man as their center and crown.... For sacred Scripture teaches that man was created 'in the image of God,' is capable of knowing and loving his Creator, and was appointed by Him as master of all earthly creatures";[17] and that "the divine plan is that man should subdue the earth, bring creation to perfection, and develop himself."[18]

At the same time, however, the Council affirmed that this "subduing" of earth was not to be for individualistic human gratification, but to meet human needs in an equitable way. The *Pastoral Constitution* noted that while "modern economy is marked by man's increasing domination over nature," simultaneously "an enormous mass of people still lack the absolute necessities of life; some, even in less advanced countries, live sumptuously or squander wealth."[19] This contradicted the understanding that

> God intended the earth and all that it contains for the use of every human
> being and people ... created goods should abound for them on a reasonable
> basis ... attention must always be paid to the universal purpose for which created
> goods are meant. In using them, therefore, a man should regard his lawful
> possessions not merely as his own but also as common property in the sense that
> they should accrue to the benefit of not only himself but of others. For the rest,
> the right to have a share of earthly goods sufficient for oneself and one's family
> belongs to everyone.... If a person is in extreme necessity, he has the right to take
> from the riches of others what he himself needs.... According to their ability, let
> all individuals and governments undertake a genuine sharing of their goods.[20]

While creation is for "man," then, "he" is to be a collective and compassionate "man." Available earth goods are to be compassionately shared, not acquisitively hoarded. The church uses a "natural law" argument (including Aquinas's concepts of common property and the rights of the poor in his discourse on the seventh commandment) to reinforce this point and even goes so far as to advocate government intervention in property relationships and property distribution to ensure that the "common good"—the well-being of all humans—is provided for: "By its very nature, private property has a social quality deriving from the law of the communal purpose of earthly goods . . . insufficiently cultivated estates should be distributed to those who can make these lands fruitful. . . . Still, whenever the common good requires expropriation, compensation must be reckoned in equity after all the circumstances have been weighed."[21]

The *Pastoral Constitution* also teaches that there should be a sense of inter-generational responsibility. People should "look out for the future and establish a proper balance between the needs of present-day consumption, both individual and collective, and the necessity of distributing goods on behalf of the coming generation."[22]

The Council did note the problems associated with expanding human populations. It advocated international cooperation to deal with the issue and declared that "the question of how many children should be born belongs to the honest judgment of parents," who should keep themselves informed about scientific advances so that "spouses can be helped in arranging the number of their children" and use reliable methods whose "harmony with the moral order should be clear."[23] It is noteworthy that in 1965 the bishops assembled in Council left to prospective parents the responsibility of deciding on the number of progeny they wanted and of keeping abreast of the available scientific data for "arranging" them. Three years later, in his encyclical *Humanae vitae* and contradicting the advice of consultants he had selected, Pope Paul VI opposed contraceptives and instructed Catholics not to use them. Thereafter, in official church teachings, strains on earth's space and earth's goods would be attributed (rightfully so, to a certain extent) to consumption by affluent and acquisitive people, particularly in industrialized nations.

In its teachings prior to and following Vatican II, then, the church advocated some form of human responsibility for earth's well-being and an equitable distribution of earth's goods so that at least the necessities of life were available to all people. In the closing decades of the twentieth century church teachings would develop along new lines, as the beginnings of a regard for the intrinsic value of earth and of all creatures would enter Catholic consciousness.

CONCERN FOR CREATION IN CRISIS

As people around the world—scientists, government leaders, environmentalists, and members of the media—became aware of growing environmental devastation and threats to the biotic community, to air, land, and water, and to the earth as a whole, Catholicism began, in official circles and in its general constituency, to engage environmental issues. In the United States, concerns were expressed in regional statements by bishops and then, after a major statement by Pope John Paul II, by the U.S. bishops as a group and then by regional bishops. From the clergy and the laity further insights were offered.

Appalachia

In 1975 the Catholic bishops in Appalachia issued the first regional pastoral letter on land issues: *This Land Is Home to Me*.[24] The bishops expressed concern for the plight of coal miners who received little compensation for their work, whose labor enriched others, whose quest for work put them in conflict with each other, and whose region thereby suffered extremes of poverty despite the riches extracted from it. Responding to the "cries of powerlessness from the region," the bishops declared that "the poor are special in the eyes of the Lord." They linked Appalachian exploitation with national affluence and called profit maximization "a principle which too often converts itself into an idolatrous power." The bishops linked themselves with clergy of earlier eras who "called for a more just social order, where property would be broadly distributed and people would be truly responsible for one another." They warned about "the presence of powerful multinational corporations" and called for multinational labor efforts to counter it. The letter was a bold step for concerned members of the hierarchy, as it challenged not only the sources of regional poverty, but also the economic structure and corporate greed and power that used it to benefit a few at the expense of the many.

Heartland America

The Appalachian statement inspired farmers from South Dakota to call upon the bishops of the Midwest to issue a regional letter focused on land consolidation and lack of land conservation. The bishops agreed to author such a document, and while it was in process they invited the recently elected Pope John Paul II to travel to the Midwest to support their developing land pastoral.

In response, John Paul II traveled to Iowa in October 1979. In a homily drafted for him at the Diocese of Des Moines,[25] the pope stated at a Mass held at Living History Farms just outside the city that the land was a divine gift and a human responsibility and should be "conserved with care" in order that it might be fruitful and enriched through generations to come; he linked land stewardship to intergenerational responsibility. The pope's homily in Iowa, focused as it was on human responsibility for earth's well-being and for the common human good, was his first foray into discussing land themes and provided a foundation for future statements on diverse related issues.

The Midwestern bishops issued their own document, *Strangers and Guests: Toward Community in the Heartland*, on 1 May 1980. (The document's title came from Lev. 25:23, which states that God declares that "the land belongs to me, and to me you are strangers and guests"; this idea is related to Augustine's understanding that people are to regard themselves as pilgrims on an earth journey and must care for their home and be concerned that it meet the needs of succeeding generations.) The particular crisis to which the bishops initially responded was the loss of owner-operated farms in the Midwest. As a result of public listening sessions in the twelve states of the bishops' regions, additional concerns were added to the issue of consolidation of land into fewer hands, such as harms caused by mining and forestry, disruption of local communities, and Native American treaty rights. Several of these concerns are expressed in the document's ten "Principles of Land Stewardship," equivalent to "ten commandments" for care and use of the earth:

1. The land is God's.
2. People are God's stewards on the land.
3. The land's benefits are for everyone.
4. The land should be distributed equitably.
5. The land should be conserved and restored.
6. Land use planning must consider social and environmental impacts.
7. Land use should be appropriate to land quality.
8. The land should provide a moderate livelihood.
9. The land's workers should be able to become the land's owners.
10. The land's mineral wealth should be shared.[26]

The divine ownership expressed by the first principle provides the basis for the succeeding principles. Since the land is God's and God wants it to provide for the needs of all people, land must be (re)distributed in such a way that its ownership is dispersed among the many rather than the few; those who have civil title to land must use the land well to benefit not only themselves but their local community and region; and the goods of earth must be disseminated fairly to meet people's needs.

After the first two regional bishops' statements, from Appalachia and the Midwest, the U.S. Catholic bishops addressed land issues in their economic pastoral letter *Economic Justice for All*, released in 1986.[27] The sentiment at the time among members of the U.S. hierarchy was that there would be no need to continue with regional documents once issues had been addressed in a definitive national statement.

In regard to environmental issues, the U.S. bishops discussed land and resource use by stating that "food, water and agriculture are essential for life" and that their abundance in the United States should not be "taken for granted." People should remember that God reminded the people of Israel that "the land is mine, for you are strangers and guests with me," and should collaborate with God in ensuring that earth's resources "meet human needs." The bishops observed that "while Catholic social teaching on the care of the environment and the management of natural resources is still in the process of development, a Christian moral perspective clearly gives weight and urgency to their use in meeting human needs."[28]

The catalyst for new environmental statements from regional U.S. bishops was the promulgation of two documents: Pope John Paul II's Message, *The Ecological Crisis* (1990); and the U.S. bishops' national pastoral letter the following year: *Renewing the Earth*.

John Paul II

A significant shift in attitude (and yet also, in part, a retention and affirmation of traditional perspectives on private property and on creation) was signaled by Pope John Paul II in 1990. Previously, while on his January 1979 papal visit to Puebla, Mexico, to address the Latin American bishops' conference, he had declared that all property has a "social mortgage." To this affirmation of the church's position on promoting the common good as exemplified in land redistribution, in his 1990 World Day of Peace Message, *Peace with God the Creator, Peace with All Creation* (also titled *The Ecological Crisis*),[29] he added: "Christians, in particular, realize that their responsibility within creation and their duty toward nature and the Creator arc an *essential part* of their faith."[30] His first statement had reasserted the idea of the "universal destination of goods," supplementing it with the idea of a "mortgage" on private property that was implicitly granted by the human community, whose shared ownership of property is taught in natural law. His second declared that Christians were to see their responsibility for creation not as an adjunct to their Christian life, but as a necessary component of their identity as Christians.

In other parts of *The Ecological Crisis*, John Paul II deplores a lack of respect for, and the plundering of, nature. He declares that the earth is "ultimately a common heritage," and so it is "manifestly unjust that a privileged few should continue to accumulate excess goods, squandering available resources, while masses of people are living in conditions of misery at the very lowest level of subsistence."[31] He linked ecology and economics by noting that ecological problems will not be resolved without addressing the structural forms of poverty.[32]

John Paul II's statements might have been influenced by perspectives emerging from the Catholic church in the United States, much as Leo XII had been so influenced nearly a century before when he wrote *Rerum novarum*.

National Conference of United States Catholic Bishops

The U.S. bishops in *Renewing the Earth* declare that people should care for the earth in the present with a sense of intergenerational responsibility. They state that earth and other creatures must be treated with respect not just because they are useful for humans, but also because they are "God's creatures, possessing an independent value, worthy of our respect and care."[33] The bishops advocate "an environmentally sustainable economy" that includes "a just economic system which equitably shares the bounty of the earth and of human enterprise with all peoples."[34] They add that "Christian love forbids choosing between people and the planet."[35] One teaching expressed in the letter elevated consciousness of divine immanence in creation. The bishops observe that "the Christian vision of a sacramental universe—a world that discloses the Creator's presence by visible and tangible signs—can contribute to making the earth a home for the human family once again."[36] The latter phrasing was significant because it extended the Catholic understanding of the meaning of "sacrament" beyond the seven liturgical rituals usually performed in a church by a member of the clergy; and it acknowledged the possibility of personal encounters with God-immanent through the mediation of creation, an understanding that mirrored the concept of panentheism which was beginning to be developed in Catholic thought.

Shortly after the national bishops' document, regional pastoral letters on environment-related themes emerged again. In 1995 the Appalachian bishops issued a new statement, *At Home in the Web of Life*, to commemorate and update *This Land Is Home to Me* on the occasion of its twentieth anniversary. Written in a poetic framework as its predecessor had been, the document celebrates earth, the relationship among living creatures, and the relationship between them and earth, while recalling the earlier theme of justice for the poor. The bishops advocate sustainable communities, a sense of human kinship with other members of the biotic community, and that private property be rooted in and in the service of the local community, so that the common good might be advanced. The common good is extended beyond humankind's well-being; it includes "the common good of all people, the common good of the entire ecosystem, the common good of the whole web of life."[37] This extension has not been expressed or elaborated in subsequent church statements.

The Columbia River Watershed

In the western United States and western Canada, an international coalition of bishops developed and issued the first bioregional statement promulgated by a church body. *The Columbia River Watershed*, released in early 2001,[38] addressed such issues as salmon extinction, human cultural and community needs, alternative energy development, mining and logging impacts, and economic development. In their proposals for church and community collaboration (and a complement to the Midwestern bishops' statement) the Columbia River Watershed bishops offered a bioregional "ten commandments":

1. Consider the common good.
2. Conserve the watershed as a common good.
3. Conserve and protect species of wildlife.
4. Respect the dignity and traditions of the region's indigenous peoples.
5. Promote justice for the poor, linking economic justice and environmental justice.
6. Promote community resolution of economic and ecological issues.
7. Promote social and ecological responsibility among reductive and reproductive enterprises.
8. Conserve energy and establish environmentally integrated alternative energy sources.
9. Respect ethnic and racial cultures, citizens, and communities.
10. Integrate transportation and recreation needs with sustainable ecosystem requirements.[39]

The bishops earlier in the document explicitly expressed their vision that "people will recognize the inherent value of creation and the dignity of all living beings as creatures of God."

In the watershed document, environment, economics, and ethics are intertwined. The human "common good" theme is reiterated, but the bishops do express concern for other species (particularly salmon, whose ongoing species extinction has harmed the subsistence efforts and cultural activities of native peoples and caused the loss of fishing jobs and cannery employment). Compassion for and support of the poor native peoples and racial and ethnic minorities are evident. Human needs are considered within the context of issues such as environmental degradation (by logging and mining interests, for example), overconsumption of earth's goods, and efficient use of energy.[40]

Voices of the Faithful

The Catholic church, as mentioned earlier, is more than its institutional manifestations. Vibrant ecological thought emerged from laity and clergy working in academia and with grassroots organizations serving the human poor and the impacted and imperiled biotic community. Rosemary Radford Ruether, Leonardo

Boff, and Thomas Berry represent unofficial viewpoints in the church. Their ideas—or parts thereof—might eventually become part of institutional church teachings, much as did the ideas of Edward McGlynn a century earlier.

Rosemary Radford Ruether integrates ecofeminism and ecojustice. She discusses how male domination of women and of nature is interconnected. She states that earth will be healed only when ecojustice is effected, when justice and love characterize relationships among women and men, races, nations, and social classes. She advocates a transformation of the Christian tradition (one of the causes of conflict between the groups mentioned) so that this might be possible, but rejects "replacing a male transcendent deity with an immanent female one."[41]

In the biotic community, all species have intrinsic value and are interrelated.[42] Ecology and justice are linked in the biblical tradition, which does not separate existence into spheres of creation and redemption; there is a single lived reality.[43] Humans should imagine a good society permeated by equity: gender equity, regional and global human equity, biotic community members' equity, and intergenerational equity not just among humans but across the community of all life.[44] Human population growth must be curtailed, and human consumption of earth's goods and degradation of earth's places must be diminished.[45] When considering the impacts of affluent citizens of wealthy nations on environmental and social conditions in poor nations, Ruether chastises those northern hemisphere ecofeminists who "fail to make real connections between their own reality as privileged women and racism, classism, and impoverishment of nature.... [They] must make concrete connections with women at the bottom of the social-economic system."[46]

Leonardo Boff has worked for decades for the liberation of the poor in his native Brazil, where he formerly served as a Franciscan priest. In the 1990s he began to integrate his theology of liberation work with environmental issues. He links land redistribution with land conservation and the economic and political plight of the poor with ecological degradation and devastation, which has had particularly devastating impacts on the poor. For Boff, "social injustice leads to ecological injustice, and vice-versa,"[47] and the current dominating paradigm is "a social sin (the rupture of social relations) and an ecological sin (the rupture of relations between humankind and the environment)."[48] He expresses concern about population growth and questions whether earth's ecosystem would be able to sustain the human population at its current rates of growth.[49]

Boff declares that rights should not be limited to humans or to nations; they belong also to other beings in creation, since "all things in nature are citizens, have rights, and deserve respect and reverence.... Today, the common good is not exclusively human; it is the common good of all nature."[50]

In his discussion of population issues, rights for all members of the biotic community, and extension of the common good to include benefits for all life, Boff offers his new paradigm for liberation while departing from his traditional

Catholic and Latin American perspectives. He does concur, however, with traditional understandings of intergenerational responsibility, stating that "future
generations have the right to inherit a conserved earth and a healthy biosphere."[51]

Creation is sacramental, since all creatures are God's representatives and mes-
sengers and revelatory of divine presence.[52] The Spirit is immanent in creation, an
understanding that is not pantheistic but pan*en*theistic: "Not everything is God, but
God is in everything. . . . God flows through all things; God is present in everything
and makes of all reality a temple. . . . The world is . . . the place where we meet God."[53]

Boff's ecojustice perspective, then, links theology, ethics, ecology, economics,
and liberation of the oppressed.

Thomas Berry is a Passionist priest who has been regarded as the Catholic
guru in ecotheology. Like Boff and other contemporary Catholic thinkers, he
advocates a panentheistic appreciation of and engagement with the universe and
rights for all members of the biotic community. People should be integrated with
cosmic dynamics and recognize that there is "a single community of life" in which
every living being "has inherent rights . . . that come by existence itself."[54] In a
world characterized by environmental degradation and destruction and denial of
nature's and species' rights, the rights of members of the biotic community and of
abiotic creation must be legally established,[55] so that "every component of the
Earth community would have its rights in accord with the proper mode of its
being and its functional role."[56]

Berry declares that while people should reject anthropocentrism, they should
remember that they are extraordinary as "that being in whom the universe celebrates itself and its numinous origins in a special mode of conscious self-
awareness";[57] they "activate one of the deepest dimensions of the universe."[58]
They should recognize, too, that "the spiritual and the physical are two dimensions of the single reality that is the universe itself."[59]

For Berry, indigenous peoples are especially aware of the sacred in the cosmos. They are present to "the numinous powers of this continent expressed
through its natural phenomena."[60] In the natural world, people can engage "the
manifestation of a numinous presence that gave meaning to all existence."[61]

Earth, in Berry's view, is a commons in which "each individual being is
supported by every other being in the earth community. In turn, each being
contributes to the well-being of every other being in the community."[62]

The spiritual dimension of the universe is not limited to experience or awareness of divine presence. People who engage the sacred in the cosmos should not
limit themselves to an experience of divine presence or allow it to detract from the
sacredness of earth and earth's creatures in themselves; people should not just be
drawn through creation to the creator, but experience the spiritual dimension of
all creatures.[63]

Berry integrates anthropology, theology, philosophy, science, and spirituality
in his proposals for promoting respectful relationships among all members of the

biotic community and an awareness of the unique role of humans as the reflective consciousness of the cosmos.

Reflections, attitudes, and proposals for positive change that emerged in the Catholic church in the twentieth century bore some fruit in church statements in the twenty-first. Consideration of institutional church teachings on water issues will illustrate the current stage of official Catholic environmental thought.

CREATION CONCERN AND COMMUNITY COMMITMENT

The most eloquent statements issued by religious bodies, if they lead to no policies or projects that implement them in a concrete way, will be merely intellectual abstractions from the real worlds of life and community in which all individuals and evolving species exist with their particular organismic and social histories. Expressions of environmental concerns, without commitment and concrete actions to resolve them, will not effect positive changes to the situations they describe and lament, nor will they lead to the realization of the vision and hopes they express. The Catholic church in its various forms has offered significant insights into environmental issues from a religious and ethical perspective. It has also sought, through actions by its official bodies and through support for projects undertaken at the grassroots and parish levels, to foster care for earth, understood as God's creation.

The growing global scarcity of potable water, due to privatization and other injustices, and pollution impacts, has presented an opportunity for the church to link deeds to words. The Vatican has promoted concern for water through its various documents on care for creation and also offered testimony, at international meetings, to stimulate national governments and the United Nations to resolve water issues. The Columbia River Watershed bishops of the United States and Canada have promoted concern for water issues and stimulated involvement with them in their region.

Water: Privatization, Pollution, and Conservation

One of the major issues of the twentieth century is and will remain to be a growing crisis in the availability and distribution of potable water. Recently, water consolidation into fewer controlling hands through water privatization schemes and water contamination by a variety of industrial, agricultural, and urban effluents have diminished the availability of water to individuals and communities, particularly in impoverished areas of the industrialized nations and throughout underdeveloped countries. ("*Under*developed" in today's global reality means that a country is politically and economically subordinate to the development and power of the G8 nations, not at an earlier stage of development.) Water issues present an integrated case study for understanding the contrasting perspectives on environmental teaching currently prevalent in Catholicism.

The principal institutional church bodies discussed in this essay have been the Vatican and representative regional bishops' groups. In separate statements but in complementary ways, water issues have been addressed by the Vatican in the statement *Water: An Essential Element for Life*,[64] a Note presented by the delegation of the Holy See (the Vatican's nation designation for the United Nations) at the 2003 "Third World Water Forum" held in Kyoto, Japan; and by the Columbia River Watershed Catholic bishops in their discussion of "living water" in their international bioregional document.

The Vatican: Water Rights

The Holy See's Note declares in its introduction not only that "water is an essential element for life," but also that "the *human being* is the centre of the concern expressed in this paper and the focus of its deliberations" (emphasis original). That humans are the center of concern limits primarily to *human* life, the "life" for which water is an "essential element." As noted earlier, the anthropocentric concept of a hierarchy in nature, with humans above all, over all, and served by all, has continued in Catholic thought, Galileo and Darwin notwithstanding. However, also as noted earlier, the dominant human is viewed not as an individual but as a community, whose common good is to be provided by the rest of the natural world and whose poorest members are not to be denied dignity and the necessities of life. The Note restates that traditional position as well, that *all* human beings need access to water as a matter of justice: "The management of water and sanitation must address the needs of all, and particularly of persons living in poverty"; the current situation "is characterized by countless unacceptable injustices."[65]

In the discussion of ethical considerations, the Note states that "earth and all that it contains" are for all humans' use, in the present and future: water is "a common good of humankind."[66] The just distribution of water should be promoted domestically by the state, "the steward of the people's resources which it must administer with a view to the common good,"[67] and internationally by advanced nations helping developing peoples.[68] Water, particularly for poor people, is "*a right to life issue.*"[69]

The Holy See asserts that water scarcity results not from human *population* growth, but from excessive *consumption* in developed nations.[70] The political power and wasteful lifestyle characteristic of wealthy individuals and nations is threatening the lives and livelihoods of poor people in poor nations.

Water is an essential element for life, a common good for all people, and a right to life issue. It is also a right in itself: "Without water life is threatened, with the result being death. The right to water is thus an inalienable right."[71]

The Columbia River Watershed: "Living Water"

The Catholic bishops of the Columbia River Watershed discussed this theme of "living water" two years before the Holy See Note was issued. In *The Columbia River Watershed*, the bishops drew upon the biblical tradition as the foundation for their ideas, noting that "in the Hebrew Scriptures, living water meant water that is flowing free and pure; it is contrasted with water from wells or cisterns, which tended to be stagnant and undesirable"; and that "in the Christian Scriptures, Jesus appropriated the term 'living water' to refer to himself as the source of genuine spiritual life."[72] They link scriptural understandings by declaring that "the living water offered by Jesus for our spirit and the living water in God's creation for our body are both life-giving waters—one natural, one supernatural. The Columbia River and its tributaries are intended by God to be living water: bountiful and healthy providers for the common good. The water itself is to be a clear sign of the Creator's presence."[73]

Water cannot be a "clear sign" if it is polluted. The implication, of course, is that degraded waters should be cleaned and that clear waters should be conserved: for humans and all living creatures. The bishops hope to see in the future "living waters of God's creation flowing from meadows and mountains to the ocean while providing for the needs of God's creatures along the way."[74] Water's benefits, then, are not solely for humans, but for all "God's creatures."

The pastoral letter uses water concerns as a basis for raising other questions about the biotic community and its bioregion. The bishops wonder: "How will we be images of God and care for that part of creation entrusted to us? How can we ensure that a rich sense of God's presence prevails? How can we assure that spiritual living waters, as well as clear and pure literal living waters, continue to flow in our region?"[75] These questions in the text of a watershed statement are based upon several church understandings. Humans should not seek to control all creation, but only the part entrusted to them: the bioregion in which they live and with which they work to meet their needs. Humans should have a sense of the immanence of God (conveyed by thinkers such as Thomas Berry as panentheism, as seen earlier) and seek to engage the creator's presence in creation. Humans should be committed to assuring life-giving water in their region, which means that polluted waters must be purified and that clean waters must be conserved. When humans so think and act, the revelatory nature of earth becomes possible, so that "each portion of creation can be a sign and revelation for the person of faith, a moment of grace revealing God's presence to us."[76]

The Vatican and Columbia River Watershed documents complement and contrast with each other. They overlap in that both advocate better care for earth and better distribution of earth goods, want enhanced water quality and equitable distribution of water quantity, indicate concern for indigenous peoples and for the poor of all cultures, call for commitments to local communities, and advocate appropriate governmental action to eliminate water inequities and impurities. But the Vatican document is clearly much more anthropocentric, with humans, its primary concern, regarded as recipients of earth as a gift from God which they are to manage wisely, while the Columbia River Watershed document advocates recognition of the inherent value of all individuals and species in the extended biotic community, views people more as stewards of those areas of God's land entrusted to their care (they are not to seek to control all the earth), and has a greater sense of the immanence of creator in creation.

Tradition in Transition: Transformation?

As Catholic environmental teaching evolved, most church leaders still maintained belief in a natural hierarchy, with humans at the top as the designated recipients of divine favor, which was understood to mean a divine dedication of the universe to meet human needs and a divinely ordained subjection of the universe to human authority.

An ongoing transition from tradition in to transformation of Catholic thought is evident in a gradual shift in understandings of humanity's authority over or managerial role in creation. While traditionally the focus had been on human dominion over nature, which became for some domination of nature, gradually dominion became diminished through the introduction of the concept of stewardship, that people were to care for creation on behalf of God, its ultimate and only absolute owner: human ownership was secondary and subordinate to divine ownership, and humans were tenants on God's property, which they were to care for, rather than absolute owners of their own property; therefore they had a greater responsibility to act well on behalf of the real owner. That idea of the relative independence of humankind in creation as caretakers of God's work has begun to be replaced, in some perspectives, with a sense of kinship with earth and all living creatures. Inroads were made in the traditional church perspective when some bishops (as in Appalachia) and individual scholars began to go beyond stewardship understandings to a greater sense of the interdependence of members of a "web of life"; a "relational consciousness" has begun to replace a caretaker or managerial consciousness.[77]

In the closing years of the twentieth century, then, significant changes in perspective occurred in the Catholic church (in the diverse writings of distinct clergy and laity, not by ecclesiastical or theological consensus). They include teachings that care for creation is an "essential part" of Christian faith (John Paul II's *Ecological Crisis*); that humans live in a "sacramental universe" (U.S. Catholic bishops' *Renewing the Earth*); that relationships among members of the biotic community and between that community and abiotic nature should be analyzed in bioregions rather than isolated places or within artificially delineated political boundaries—state, national, or international (Columbia River Watershed bishops' pastoral letter); that the creator-Spirit is Spirit-immanent as well as Spirit-transcendent and might be experienced most often in nature by those who have a panentheistic perspective and consciousness (Leonardo Boff and Thomas Berry); that all creatures have intrinsic value in a common web of life (Appalachian bishops' and Columbia River Watershed bishops' statements); that "natural rights" should not be used to designate solely what are in actuality "human rights," but be extended to all creatures (Boff and Berry); that patriarchal attitudes and practices are harmful to earth and the human and broader biotic communities (Rosemary Radford Ruether); that environmental degradation both flows from (when caused by desperate need) and fosters poverty (caused by deprivation of life's necessities) and ecojustice projects should confront both simultaneously (regional and national bodies of bishops); and that anthropocentrism is unbiblical and antithetical to the well-being of humans, other living creatures, and earth itself and therefore dominion attitudes, where present, should give way to a stewardship mentality and eventually to a relational consciousness, where people understand better their place and their role in creation and evolution (Ruether, Boff, and Berry).

The U.S. Catholic bishops have gone beyond merely issuing statements on environmental issues. They have testified before congressional committees on these issues, and they established an Environmental Justice Program, under the auspices of the Office of Social Development and World Peace, at their national headquarters in Washington, DC. The Environmental Justice Program hosts conferences focused on environmental issues and publishes scholarly papers resulting from them; works on leadership development on environmental issues with Catholic dioceses and state Catholic conferences; develops educational resources for parishes and dioceses; prepares policy statements on environmental concerns; honors parishes committed to environmental responsibility with "St. Francis Model Parish Awards"; and has awarded more than 150 regional and diocesan grants to assist projects that promote ecojustice. Representative projects funded by the program include the Columbia River Watershed Pastoral Letter Project; the Southwest Environmental Equity Project (to assist people to organize in their communities and propose economically sound environmental policies to the Arizona state legislature); the Coalition to Restore Coastal Louisiana (to organize fishers and others to oppose destruction of, and propose restoration of, state wetlands); Mothers of East L.A. (to help people successfully organize to prevent construction of a hazardous-waste incinerator near Los Angeles, California); and the Lower Anthracite Project (to promote economic justice and environmental preservation in hard-coal fields of northeastern Pennsylvania). The grants were intended both to assist local faith-based environmental effort and to create model programs that could be replicated across the United States.

To a greater extent than has been the case for almost two millennia, Catholic environmental thought now re-presents and reinforces the basic tenets of biblical "environmental" thought: earth is permeated by the presence of God-immanent; all creation is very good; and the goods of creation are intended by their creator to provide for all these good creatures. These understandings are complemented by an emerging consciousness: humans are not superior to and over all creatures, nor is the universe created solely to satisfy human needs. Humankind is integrated in, interdependent within, and interrelated with all of the community of life, with its own complementary niche and role. While this view, a transition from tradition, is currently primarily emerging from the church of the faithful, the seeds it plants might yet transform not only institutional Catholicism, but the diverse social contexts in which it is situated.

NOTES

1. A comprehensive presentation and analysis of developing Catholic environmental thought may be found in John Hart, *What Are They Saying about Environmental Theology?* (Mahwah, NJ: Paulist Press, 2004).

2. Thomas Aquinas, *Summa theologica* (trans. Fathers of the English Dominican Province; New York: Benziger, 1981), 1.352, part I, Q. 72: "On the Work of the Sixth Day." (These words might be interpreted as recognizing the worth or goodness of all creatures or, by contrast, that their unknown role in creation should not be lost because it was necessary for the well-being of creation as a whole, and thus for humans' benefit, that all things function well; this might be interpreted also as an unscientific early recognition of ecosystems.)

3. Ibid., 1.326, part I, Q. 65, art. 2.

4. Ibid., 2.1479, part II-II, Q. 66, art. 5.

5. Pope Leo XIII, *Rerum novarum*, in *Proclaiming Justice and Peace: Papal Documents from Rerum novarum through Centesimus annus* (ed. Michael Walsh and Brian Davies; Mystic, CT: Twenty-Third Publications, 1991), 19 §7.

6. Ibid., 26 §24.

7. Ibid., 34 §47.

8. Ibid., 23 §17.1.

9. Ibid., 33 §45.

10. Ibid., 30 §38.

11. Ibid., 25 §23.

12. Ibid., 31 §41. The encyclical then cites Gen. 1:28 to support this statement.

13. Ibid., 19 §8.

14. "The Cross of a New Crusade" speech from which these quotations are selected was delivered by Dr. McGlynn (as he was known internationally), to an overflow audience at the Academy of Music, New York City, March 29, 1887, three months before his excommunication by Rome. Sylvester L. Malone, *Dr. Edward McGlynn* (New York: Arno Press [a New York Times company], 1978), 31. The Arno edition reprints the 1918 edition, published by the Dr. Edward McGlynn Monument Association (New York: McAuliffe & Booth, Inc. Printers).

15. Ibid., 34.

16. Second Vatican Council, *Pastoral Constitution on the Church in the Modern World* (*Gaudium et Spes*), in *The Documents of Vatican II* (ed. Walter M. Abbott; New York: Herder & Herder/Association Press, 1966), 206 §9.

17. Ibid., 210 §12.

18. Ibid., 262 §57.

19. Ibid., 271 §63.

20. Ibid., 278–79 §69.

21. Ibid., 281–82 §71.

22. Ibid., 280 §70.

23. Ibid., 302 §87.

24. Appalachian Catholic Bishops, *This Land Is Home to Me: A Pastoral Letter on Powerlessness in Appalachia by the Catholic Bishops of the Region* (4th ed.; Webster Springs, WV: Catholic Committee of Appalachia, 1990). The letter is written in free-verse style, without paragraph numeration.

25. The present writer was asked by Bishop Maurice Dingman of Des Moines to write the draft of the papal homily, which was then sent to the Vatican to be put in its final form.

26. Midwestern Catholic Bishops, *Strangers and Guests: Toward Community in the Heartland* §50. The writers of the original draft of the bishops' document, which was circulated for comment and was the subject of hundreds of grassroots meetings, were Marty Strange, David Ostendorf, and Stephana Landwehr (who was the sole Catholic in the original trio). The present writer served as the editor and principal writer of several subsequent drafts and the final version of the pastoral letter. The document is available from the National Catholic Rural Life Conference (Des Moines, IA, 1980). The U.S. Catholic Bishops' weekly, *Origins*, prints texts of statements by popes, bishops, and other church members; *Strangers and Guests* is in *Origins* 10.6 (26 June 1980): 81–96.

27. U.S. Catholic Bishops, *Economic Justice for All: Catholic Social Teaching and the U.S. Economy* (Washington, DC: United States Catholic Conference, 1986).

28. Ibid., 106 §216.

29. Pope John Paul II's message was published as *Peace with God the Creator, Peace with All Creation* in *Origins* 19.28 (14 Dec. 1989): 465–68; and as *The Ecological Crisis: A Common Responsibility* (Washington, DC: United States Catholic Conference, 1990). This and other papal and Vatican documents are available at the Holy See website: www.vatican.va/phome_en.htm.

30. Ibid., §15 (emphasis added).

31. Ibid., §8.

32. Ibid., §11.

33. U.S. Catholic Bishops, *Renewing the Earth: An Invitation to Reflection and Action in Light of Catholic Social Teaching* (Washington, DC: United States Catholic Conference, 1991), 7.

34. Ibid., 8.

35. Ibid., 11.

36. Ibid., 6.

37. Appalachian Catholic Bishops, *At Home in the Web of Life* (Webster Springs, WV: Catholic Committee of Appalachia, 1990), 57.

38. U.S. and Canadian Catholic Bishops of the Columbia River Watershed, *The Columbia River Watershed: Caring for Creation and the Common Good* (2001). The present writer served as the project writer and editor of this document. See www.columbiariver .org for versions in English, Spanish, and French. The document is available in print form from the Catholic Archdiocese of Seattle, WA; and was published in *Origins* 30.38 (8 March 2001): 609–19. The document attracted international attention. Stories in the *New York Times*, *Sierra*, and *Outside* magazines and on national PBS, among other media, and recognition from the World Wildlife Fund's Alliance of Religions and Conservation in the form of a "Sacred Gifts for a Living Planet" award, helped to disseminate and promote the progressive Catholic environmental thought of the pastoral letter. The document has four major parts: I. "The Rivers of Our Moment" (the current environmental and social situation); II. "The Rivers through Our Memory" (biblical and church teachings about nature); III. "The Rivers of Our Vision" (imagined changes in the watershed region: spiritually, socially, and ecologically); and IV. "The Rivers as Our Responsibility" (suggestions for making the vision a reality). Citations of this writing refer to those parts; there are no further numerical references (e.g., numbered paragraphs) in the document.

39. Ibid., §IV.

40. Other regional U.S. Catholic bishops' environmental statements to date have included *Reclaiming the Vocation to Care for the Earth* (New Mexico, 1998); *And God Saw That It Was Good* (Boston Province [which included the states of Maine, Massachusetts, New Hampshire, and Vermont], 2000); and *A Catholic Perspective on Subsistence: Our Responsibility toward Alaska's Bounty and Our Human Family* (2002). Other bishops' conferences' environment- and ecojustice-related pastoral letters in the United States and other countries have included *Pastoral Letter on the Relationship of Human Beings to Nature* (Dominican Republic, 1987); *The Cry for Land* (Guatemala, 1988); *What Is Happening to Our Beautiful Land?* (Philippines, 1988); *Ecology: The Bishops of Lombardy Address the Community* (northern Italy, 1988); *Christians and Their Duty towards Nature* (Australia, 1991); and *Global Climate Change: A Plea for Dialogue, Prudence, and the Common Good* (United States, 2001). The majority of these statements are published in *And God Saw That It Was Good: Catholic Theology and the Environment* (ed. Drew

Christiansen and Walter Grazer; Washington, DC: United States Catholic Conference, 1996).

41. Rosemary Radford Ruether, *Gaia and God: An Ecofeminist Theology of Earth Healing* (San Francisco: Harper, 1994), 2–4.

42. Ibid., 101.

43. Ibid., 207–8.

44. Ibid., 258.

45. Ibid., 263–65.

46. Rosemary Radford Ruether, ed., *Women Healing Earth: Third World Women on Ecology, Feminism, and Religion* (Maryknoll, NY: Orbis, 1994), 5.

47. Leonardo Boff, *Ecology and Liberation: A New Paradigm* (Maryknoll, NY: Orbis, 1996), 25.

48. Ibid., 27.

49. Ibid., 15, 18.

50. Leonardo Boff, *Cry of the Earth, Cry of the Poor* (Maryknoll, NY: Orbis, 1997), 133.

51. Boff, *Ecology and Liberation*, 88.

52. Ibid., 46.

53. Ibid., 51.

54. Thomas Berry, *The Great Work: Our Way into the Future* (New York: Bell Tower, 1999), 115.

55. Ibid., 161.

56. Ibid., 80.

57. Ibid., 19.

58. Ibid., 32–33.

59. Ibid., 49–50.

60. Ibid., 39.

61. Ibid., 44.

62. Ibid., 61.

63. Thomas Berry, *The Dream of the Earth* (San Francisco: Sierra Club, 1998), 81, 113.

64. The Holy See website (http://www.vatican.va/phome_en.htm) has the complete document at http://www.vatican.va/roman_curia/pontifical_councils/justpeace/documents/rc_pc_justpeace_doc_20030322_kyoto-water_en.html. The statement was presented by Archbishop Renato Martino.

65. *Water: An Essential Element for Life*, introduction. As with the Columbia River Watershed document, this document contains only section headings, and the paragraphs are unnumbered.

66. Ibid., §II: "The Water Issue: Some Ethical Considerations."

67. Ibid., §IV: "Water: An Economic Good."

68. Ibid., §II.

69. Ibid., §I: "A Far-Reaching Question" (emphasis original).

70. Ibid., §VI: "Other Issues Impacting Water Supply."

71. Ibid., §V: "Water: An Environmental Good."

72. *Columbia River Watershed* §II.

73. Ibid.

74. Ibid., "Conclusion."

75. Ibid., §III.

76. Ibid., §II.

77. The concept of a "relational consciousness" was formulated in John Hart, "Salmon and Social Ethics: Relational Consciousness in the Web of Life," a presentation to the annual meeting of the Society of Christian Ethics in Vancouver, WA, in Jan. 2002. The paper was published under the same title in the *Journal of the Society of Christian Ethics* 22 (2002): 67–93.

CHAPTER 3

THE EARTH AS SACRAMENT

Insights from Orthodox Christian Theology and Spirituality

JOHN CHRYSSAVGIS

If there is one image that presents itself as unique and fundamental in contemporary religious experience, it is that of the earth as sacrament. This is a central feature of the sacramental ethos of the Orthodox church. The conviction is that "the earth with all its fullness" (Ps. 24:1; Deut. 33:16) presents an undeniable theological truth, a truth that "springs from the earth" (Ps. 85:12). Indeed, if there exists today a vision able to transcend and transform all national and denominational divisions, social and political tensions alike, it may well be that of our world understood as sacrament. This is a subject, albeit tentative in nature, which is being studied more seriously in recent times.

Unfortunately we have been conditioned to consider the sacraments in a narrow, reductionist manner: a fixed number of sacraments, so that all else assumes a nonsacramental tone; minimal requirements for the validity of sacraments, so that all else becomes unsacramental in nature; and an overemphasis on the hierarchical structure of the church or the ritualistic nature of liturgy, so that all else falls outside the margins of salvation and sacredness. We need to recall the sacramental principle, which ultimately demands from us the recognition that nothing in life is profane or unsacred. There is a likeness-in-the-very-difference between that which sanctifies (the creator God) and that which is sanctified (the creation), between uncreated and created. The Divine Liturgy of Saint John Chrysostom, celebrated each Sunday in Orthodox churches throughout the world, expresses the conviction about God in relation to creation: "You are the one who offers and is offered, who receives and is distributed, Christ our God, and to you we offer glory."

This chapter explores the understanding of sacrament as including and embracing creation. It endeavors to sketch a theology of creation in light of our dilemma before and response to the current environmental crisis.

A THEOLOGY OF SACRAMENT

One may debate to eternity about sacraments and doctrines as either uniting or dividing believers of diverse faiths. However, is it perhaps not more purposeful to consider the common ground that we all tread as a source of solidarity and as an ultimate sign of the communion that we all share? The fundamental principle according to which the church should exist for the life of the whole world and not simply to satisfy certain religious needs of particular individuals can hardly be overemphasized here. In order, however, to appreciate such a comprehensive view of the church, one must be prepared to acknowledge the sacredness or, rather, the sacramentality of creation. At the same time, one must also confess that the dominant view of the world even, and at times especially, by people within the church is all too often the mechanistic concept, which arrogantly subjects everything to the individualistic desires and conquests of humanity, indeed of man. This contrast between the mechanistic and a more spiritual worldview is eloquently articulated by Wendell Berry in relating the consequences of a very mundane event:

> The figure representative of the earlier era was that of the otherworldly man who thought and said much more about where he would go when he died than about where he was living. Now we have the figure of the tourist-photographer who, one gathers, will never know where he is, but only, in looking at his pictures, where he was. Between his eye and the world is interposed the mechanism of the camera—and also, perhaps, the mechanism of economics: having bought the camera, he has to keep using it to get his money's worth. For him the camera will never work as an instrument of perception or discovery. Looking through it, he is not likely to see anything that will surprise or delight or frighten him, or change his sense of things. As he uses it, the camera is in bondage to the self-oriented

assumptions that thrive within the social enclosure. . . . He poses the members of his household on the brink of a canyon that the wind and water have been carving at for sixty million years as if there were an absolute equality between them, as if there were no precipice for the body and no abyss for the mind. . . . He is blinded by the device by which he has sought to preserve his vision. He has, in effect, been no place and seen nothing; awesome wonders rest against his walls, deprived of mystery and immensity, reduced to his comprehension and his size, affirming his assumptions, as tame and predictable as a shelf of whatnots.[1]

HISTORY AND HEAVEN

The distinction of these worldviews goes back a long way. Already by the first century of the Christian era, Judaism had been profoundly infiltrated and influenced by Hellenism. The encounter of these two influential cultures and powerful worldviews had already been brought about after the rapid spread of Hellenism in the Palestinian and Alexandrian regions, particularly during the time of Alexander the Great and his successors. However, the Greek mind and the biblical spirit were not totally compatible. The former sought truth in the harmony and beauty of the world, even if the material and historical were ultimately just an image of the spiritual and eternal. The latter searched for God in the immediacy of history, even if God is "the one who is and who was and who is [yet] to come" (Rev. 1:4). The early Christian community was in fact deeply marked by an eschatological orientation, by a sense of fervent expectation for the immediate and final revelation of God in history. Thus the early church prayed "maranatha" (i.e., "the Lord is near") whenever it celebrated the Eucharist as well as whenever its believers were faced with the ultimate test of martyrdom.

Subsequently, the creative synthesis of the Greek interest in the cosmos and the Jewish sense of God's immanence emerges in the liturgical and monastic developments of the early Christian tradition. In the area of worship, the Eastern church fathers underlined the Eucharist as the foretaste of the final kingdom and as an act whereby material creation is transfigured by Christ. In the area of ascetic life, the Eastern church fathers again played a crucial role of reconciliation in interpreting the function of the church in the world (i.e., the community of believers as citizens of society; Mark 12:17) and the goal of the church not of this world (i.e., the Christian believers as citizens of heaven; Heb. 13:14).

Another similar distinction becomes evident in the theological discourse of the formative Christian centuries. From the time of Tertullian (died ca. 225), theology in the Western part of Christendom has been characterized by a deep sense of history and a deep concern with the historical. The emphasis is on God working in time, as we understand it, and on an Aristotelian concept of time, marking beginning, middle, and end. This has resulted in a preoccupation by the West with the institutional and the more material or moral aspects of Christianity. By contrast, Eastern theological thought has normally been concerned with the metahistorical, namely, with the more spiritual or mystical dimensions of the Christian life. The worldly or material reality is seen in light of the kingdom of heaven and the eternal nature of time. The spirituality of the East has always searched for some ultimate theological reason behind and justification of historical events and situations. Facts and figures are interpreted in terms of the Holy Spirit; power is understood from the perspective of the sacrament of the Eucharist; the world around is considered in relation to the heavens above. This illuminating teaching about the last things, or the last times, has continually been at the forefront of theological and spiritual reflection.

To the minds of the early Christians, the "end" was far from any immediate concern for human life, as evidenced by the perception of the church fathers: "If the farmer waits all winter, so much more ought you to wait the final outcome of events, remembering who it is that plows the soil of our souls.... And when I speak of the final outcome, I am not referring to the end of this present life, but to the future life and to God's plan for us, which aims at our salvation and glory" (John Chrysostom, *On Providence* 9.1 [Sources chrétiennes 79.145–46]).

Unfortunately, throughout the history of Christian doctrine, many theologians assumed that the last times (known in religious jargon as "eschatology") implied an apocalyptic or even an escapist attitude to the world. It took a long time even for theologians to cease treating the last times as the last, perhaps unnecessary, chapter in a manual of Christian theology. Eschatology is not simply the teaching about the last things that follow everything else; rather, it is the teaching about the relationship of all things to the last things. In essence, it concerns the "lastness" or "lastingness" of all things. This is certainly how Gregory of Nyssa understood the lasting beauty of this world, which nevertheless was never intended as an end in itself or for worship, in the sense of idolatry: "Since the conclusive harmony in the world has not yet been revealed" (*On the Creation of Man* 23 [PG 44.209]).

Gradually, however, the Omega came to be interpreted as giving meaning to the Alpha, the eschatological vision of the present was perceived as the way of liberation from the evils of provincialism and of narrow confessionalism, and the sacrament of the Eucharist became accepted as the only true perspective of reality inasmuch as it was rooted in the eternal present. Indeed, the ultimate purpose of all that exists—the end of all—is the eucharistic offering of all people and all things to the creator. And that is also the beginning, the original principle of the entire creation. In order to appreciate this, we require

> an attitude of mind sustained by a constant awareness of an End intensely present and powerful in the here and now of our historical existence and which imbues

this existence with meaning. When and where we are ready or bold enough to think and live consistently to the end, we reach out every time to that final boundary where our lives are transcended into life eternal, to the Lord of Time, and in so doing we are living eschatologically: our History becomes not merely a series of happenings but the disclosure and consummation of divine and human destiny, that is, apocalypse. . . . Human existence remains a temporal existence; its temporal character, however, contains the seed of the Kingdom: it is destined to end from within through self-transcendence and thus "prepare" the coming of the Kingdom. It must end from within as well as from without. The end from without cannot but be destruction, but the End from within is "construction" or "reconstruction" and transfiguration. "Verily, verily I say unto you, that there are some of them that stand here, who shall not taste death, till they have seen the Kingdom of God come with power."[2]

This sense of time as eternal and eternally present is very significant for grasping the notion of the world as sacrament. for it is in the sacraments that the world not only looks back in historical time to the moment of creation and to the event of the incarnation, but also simultaneously looks forward in sacramental time to and even anticipates the redemption and restoration of all things—of all humanity and of all matter—in Christ on the last day. In the sacraments, everything visible assumes an invisible dimension; everything created adopts an uncreated perspective; everything purely mundane becomes deeply mystical, for in addition to being timely, it is also rendered timeless.

WHOLENESS AND HOLINESS

Our times demand a positive encounter. Negative responses to critical issues always remain a temptation, but they constitute a heresy inasmuch as they isolate one part of the truth; such is the literal implication of the Greek term for "heresy" (*hairesis*). Nonetheless, while there is a narrow path that exists between the heresy of fanaticism or fundamentalism (that cannot embrace any sacramentality of creation) and the heresy of relativism (that can only idolize creation as sacred in itself), this path is not so narrow (Matt. 7:13) that it cannot include the whole world and the abundance of life. The broader, holistic outlook that accepts the land and the world as crucial to and as essential for our relationship with God bespeaks a reverence for the Holy Spirit; it illustrates the connection between wholeness and holiness. If one can visualize the activity of the Spirit in nature, then one can also perceive the consubstantiality between humanity and the created order; then one will no longer envisage humanity as the crown of a creation, which it is able or called to subdue. To regard and relate to the world without reference to the action of the Holy Spirit is also to worship a philosophical, transcendent God incapable of being or becoming involved with human hearts and history. Only an unreserved affirmation of the Holy Spirit permits us to understand how God is able to "move outside" of the divinity and enter into creation without either disrupting divine unity or abandoning divine transcendence.

Affirming the action of the Holy Spirit further safeguards the intrinsically sacred character of creation, its sacramental dimension, for a sacrament remains, in all its transcendence, a historical event, demanding material expression. When God is manifest in time and space and the Eucharist is revealed as God's revelation in bread and wine, the world becomes the historical and material sacrament of the presence of God, transcending the ontological gap between created and uncreated. The world articulates and relates in very tangible terms the cooperation between divine and human in history, denoting the presence of God in our very midst. Were God not present in the density of a city, in the beauty of a forest, or in the sand of a desert, then God would not be present in heaven either.

Nature speaks a truth scarcely heard and, up until recent years, insufficiently formulated among theologians. In our minds, we have eliminated or excluded the role of created nature as central to the salvation of the world. I say "in our minds" advisedly, because if God is revealed in the created world, then God is present "in all things" (Col. 3:11). In other words, there is an invisible dimension to all things visible, a "beyond" to everything material. All creation is a palpable mystery, an immense "incarnation" of cosmic proportions. For C. S. Lewis, it is as if the divine word was written out in large letters across the body of the world—in letters too large for us to read clearly. Augustine's celebrated phrase about himself might accordingly be aptly expanded to state that God is "the most intimate interior and the supreme summit" of the whole world (*Confessions* 3.6.11).

A Theology of Creation

The word *sacrament* (deriving from Latin *sacramentum*) can signify either a result or the means of consecration. The word was originally used in the third-century Latin West as the equivalent to the Greek term *mystery* (*mystērion*), which denotes a reality hidden but at the same time gradually revealed through initiation (*myēsis*). In the sense that everything in the cosmos in some way reflects the divine, the actual number of Christian sacraments or mysteries, as they are called in the Orthodox church, can be limited to seven, still less restricted to two or three. The sacramental principle is the way that we are able and, indeed, called to understand the world around us as being sacred. Such an essential Christian vision consists of three fundamental intuitions concerning creation.[3] When any one of these is either isolated or violated, the result is an unbalanced and destructive vision of the world.

The first is the intuition that the world is innately good and naturally beautiful inasmuch as it was created by a loving creator. From the beginning of time, creation has personified the biblical words "and God saw that it was good" (Gen. 1:25). In the Septuagint translation of the Hebrew Bible, the Greek word for "good" is *kalos*, which also means "beautiful" and etymologically derives from the verb "to call," implying that the world has been called by God to become beautiful. In short, the world's vocation is beauty. Nothing is intrinsically evil (Niketas Stethatos, *Spiritual Paradise* 3 [Sources chrétiennes 8.64–65]), except perhaps the refusal to see God's work as beautiful. The entire world has been created for our enjoyment and admiration. In the words of Saint John Chrysostom: "Creation is beautiful and harmonious, and God has made it all just for your sake. He has made it beautiful, grand, varied, and rich" (*On Providence* 7.2 [Sources chrétiennes 79.109–10]).

In a naïve acceptance of this worldview, one would surrender to the world on the terms of the world and embrace a spirituality that assumes the conditions and criteria of a world that is absolutized. Those who would secularize Christianity need to be reminded that there is an incalculable cost for the process of cosmic transfiguration because of the reality of evil. The glory of Mount Tabor cannot be separated from the suffering of Mount Calvary; the two hills are spiritually complementary, because creation "groans" and "travails" in search of and in expectation of deliverance (Rom. 8:22).

The second intuition is the understanding that the world is subject to evil. This more negative alternative affirms that creation is "fallen," that the world lies entirely within the realm of the prince of evil. Modern thinkers have trouble with the notion of the fallenness of creation, as if it somehow lays blame on inanimate or animal creation. Yet this principle concerns the consequences of "original sin" or of "the fall of Adam" (as it is known in Orthodox theology). The results of the fall are inevitably felt also on the level of the created world. This should not surprise us, writes John Chrysostom: "Why should you be surprised that the human race's wickedness can hinder the fertility of the earth? For our sake the earth was subjected to corruption, and for our sake it will be free of it.... Its being like this or that has its roots in this destiny. We see proof of this in the story of Noah.... What happens to the world, happens to it for the sake of the dignity of the human race" (*On Isaiah* 5.4 [PG 56.61]).

This second intuition might explain, at least in part, the radical dualism of the Manicheans and Gnostics in the early Christian church, according to whom the visible creation has not fallen from perfection, but is simply the work of an inferior deity. Of course, this is not always expressed so crudely: it may be that the natural world is regarded as merely a stage for the more important human drama to be played out. This almost apocalyptic worldview dictates withdrawal or escape from the created world, followed by an inevitable condemnation of the vast majority of humanity who wallow in this "inferior" creation and a belief in the salvation of only "the few." Furthermore, such a pietistic attitude isolates a narrow band of human experience as sacred, while all else is relegated to the realm of the secular. The goal, however, is not to set the sacramental over and against, or even outside, the profane domains of life, but to realize that nothing is intrinsically nonsacred (Mark 7:19; Rom. 14:14), that nothing is excluded from the sacramental principle. By the same token, we must emphasize that because everything is fallen, absolutely everything—including the natural and animal, as well as all inanimate, material creation—requires transformation.

The third intuition is the belief that affirms the world as redeemed. The incarnation, crucifixion, and resurrection of Jesus Christ have effected a re-creation of the world. This redemption, however, can be fully appreciated only in light of the other two intuitions of reality mentioned above: createdness and fallenness. Otherwise, the emphasis on human progress and achievement, together with any optimistic development of civilization, will lead to the post-Christian determinism that has influenced so much of Western technology and culture in recent centuries. The result of this would be a renunciation of the center of Christian faith, that liberation from sin and transformation of evil are wrought by God both in the unique nature of creation, as well as in the total assumption of creation at the divine incarnation. Human salvation and cosmic transfiguration can be achieved only through the cooperation between creator and creation, never by an imposition by one over the other.

The "honest to God" and the "death of God" debates of the 1960s brought about a total devotion to the world and a fervent commitment to its concerns, as well as a total resistance to any self-centered sense of individualism or pietism in religious thought and practice. Theologians during those years alerted us to the danger of disenchanting nature, of stripping it of any mysterious quality. This disenchantment of the natural world provides the precondition for the later development of natural science, rendering nature itself available for our use. Today's lack of regard for and detachment from God's creation could be signs of indifference toward the world and its goodness and a refusal to engage with its fallen aspects. Any definition of the proper Christian attitude toward the world must involve some form of synthesis, a dialectical approach that reflects the nature of reality as a world of opposites. The world is both good and fallen, and these opposites are rooted in both revelation and the experience of the mystics. The world emerges as a sacrament, at least in the mystical and ascetical traditions of the Orthodox church, where the relationship of humanity to the environment continues to be perceived in sacramental terms.

SACRAMENT AND SYMBOL

Ironically, it is the paradoxical or antinomical character of creation understood as sacrament that preserves the balance between the two aforementioned worldviews and dismisses the slightest suspicion of theism or pantheism, for a sacrament can be a symbol of both the transcendence and the immanence of God, transcendence implying not merely divine aloofness but primarily active involvement. Sacraments, therefore, reveal not only the dimension of depth but above all the abyss of mystery in God, recalling not an absent God but one who is present everywhere.

When we refer to a sacrament as symbol, we must rid ourselves of the notion that symbols are merely reminders or signs, which symbolize something else, having an ulterior meaning. The ultimate significance of any symbol is its gratuitousness. A symbol almost resides in its "uselessness": a symbol quite simply means, and does not mean something else. This, of course, is a difficult concept to grasp in a world that expects what-you-see-is-what-you-get productivity and usefulness, in an age where everything is measured in terms of consumer value.

Creation as sacrament is symbolic, like life, art, and poetry. In this regard, a symbol is understood in its original etymological sense of being a "syndrome," namely, an image that brings together (*symballō*) two distinct and even different, although not divided realities. Clearly, certain symbols are more appropriate than others, particularly those which bear no apparent similarity with the archetype or original, since there can be no pretense or disguise of the archetype. The more veiling and distant the symbol, the greater the unveiling (or the "apocalypse") of the invisible God. Consequently, the earth may become the very image of heaven: the ground reveals the abyss above, while the way we treat the world around us reflects our relationship with the God we worship. Our role is to discover or uncover—"at every hour and every moment, both in heaven and on earth,"[4] "in all places of God's dominion" (Ps. 103:22)—the inwardness of the outward, just as visionaries and mystics have done through the ages, for the "end" is already inaugurated among us, while the "there" may be foretasted in the here and now. A sacrament is that which reveals life and the world as a movement incorporating Alpha to Omega, as transition from old to new, as "Pascha" or "passover" from death to life.

That is not to say that sacraments are isolated from the world or from people. Though open to the mysterious and eternal, they retain a material and temporal nature, articulating a divine glory that is present and tangible in all things. It is we who often do not see clearly: "If your eye is healthy, your whole body will be filled with light" (Matt. 6:22). We are called to discern the paradoxical things in unity and not in contradiction, to bring about the same reconciliation of all things, if you will, which is brought about by art, literature, and religion: "Who but Shakespeare could bring the airy nothing of heaven into consonance with the heavy reality of earth, and give it a form that ordinary humans can understand? Who but the Shakespeare in yourself?...When one is truly a citizen of both worlds, heaven and earth are no longer antagonistic to each other.... [It is] only the optical illusion of our capacity—and need—to see things double."[5]

I included religion precisely because the purpose of religion is the task of "rerelating," of "rebinding" (the literal meaning of the term *religion*); its aim is to put back together again, to heal the wounds of separation. Any religious insight into the natural environment invariably connects and bridges by restoring and reconciling the apparent opposites that cause us suffering and torment—namely, to adopt therapeutic jargon, the neurosis that is caused by such dualism.

Thus it is that in Orthodox spirituality, the ultimate symbol or sacrament is the Divine Liturgy, constituting—or, rather, celebrating—as it does the perception and very presence of heaven on earth. This is how the Eastern Orthodox tradition presents a firm spiritual and intellectual framework for a balanced affirmation of the holiness of this world. In the words of Saint Symeon of Thessalonika (died 1429), the Divine Liturgy constitutes "the mystery of mysteries . . . the holy of holies, the initiation of all initiations" (*On the Holy Liturgy* chap. 78 [PG 155.253]). Just before the time of Symeon, Saint Nicholas Cabasilas (died ca. 1391) commented: "This is the final mystery. Beyond this it is not possible to go, nor can anything be added to it" (*The Life of Christ* book 2 [PG 150.548]).

CREATION "OUT OF NOTHING"

According to the Judeo-Christian tradition, the world was created "out of nothing," namely, from no other principle and for no other reason than for God's unconditional love. The teaching that God created something out of nothing is, from an Orthodox Christian perspective, closely related to the element of asceticism and restraint. On every occasion of divine revelation and economy—such as during the moment of creation and at the moment of incarnation—God is seen to exercise some degree of self-limitation and constraint, which in themselves become a model of spiritual behavior for those faithful to the divine covenants. These dimensions are a central part of the discussion that follows. But first, it is helpful to consider the teaching itself about creation out of nothing in order to appreciate its place within and impact upon theological development.

In terms of the history of Christian doctrine, the teaching about creation "out of nothing" (*ex nihilo*) is not part of the Genesis story per se. It is a doctrine that emerged in the intertestamental period, the earliest reference being found in 2 Maccabees 7:28 and therefore not available to the writers of the Hebrew scriptures.

It is entirely unknown to the world of classical philosophy during the pre-Christian and early Christian times, developing rather slowly and even uncertainly within Christian theology as a response to classical Greek cosmology. Tertullian expresses this vagueness in the second century: "I say that, although Scripture did not clearly proclaim that all things were made out of nothing—just as it does not say either that they were made out of matter—there was not so great a need to declare that all things had been made out of nothing as there would have been, if they were made out of matter" (*Against Hermogenes* 22.2 [Ante-Nicene Fathers 3.489–90]).

The doctrine of creation out of nothing assumes a degree of prominence in the subsequent Christian tradition,[6] where it is introduced in order to safeguard creation as an act of freedom and not of necessity, as a product of love and not of nature, and as a result of will and not of essence. Indeed by the early fourth century there is general agreement on this topic, among orthodox and heretical writers alike. For example, at the Council of Nicea (325), the first ecumenical council that marks a watershed in the history of Christian thought, Athanasius and Arius—opponents on many other issues, especially with regard to the incarnation of the divine Word—share a clearly articulated doctrine of creation out of nothing.

Indeed, Athanasius refers to earlier Christian authorities as well as to scripture in order to distinguish between the Trinitarian relationship of the Father to the Son (which he describes in one word as *gennēsis* or "birth") and the relationship of the Son within the Trinity to the world (which, again in a single term, he describes as *genesis* or "creation").[7] One problem that perhaps arises from such a definitive and ontological distinction between God and the world is that there is no room for any intermediate zone between the two. The only exception, as we have already observed, is the sacrament of the Eucharist. Nevertheless, the implications and conclusions from this crucial doctrinal formulation remain dramatic not only for theology (as the understanding of God), but also for cosmology (as the understanding of the world) as well as for mystical theology (as the encounter of the two and the experience of God in the world).

GOD AND THE WORLD

In certain pagan creation myths, both matter and the divine beings assume preeternal and eternal existence. The Christian church, by contrast, claims and proclaims that God alone is the eternal being, "maker of all things visible and invisible" (to quote from the Nicene-Constantinopolitan Creed or Symbol of

Faith). The distinction is that God is omnipotent and independent, while the world is limited, dependent, always understood in reference to and in communion with God, without whom it is incomplete. As already intimated, the inherent danger of the Christian doctrine is the temptation to press the sovereign independence of God to the point of separation from the world. No one is today seriously threatened by any form of deism, namely, by a God entirely unrelated to the world. Today, people prefer to speak in terms of theism, namely, of a God related and relevant to our world. Yet most of us are reluctant to advance this relatedness any further, lest we be criticized of pantheism. Some prefer to speak of panentheism, in order to ensure or safeguard some distinction between God and the world. However the term *panentheism* is neither always clear in meaning nor always consistent in usage, inasmuch as it is adopted by thinkers and theologians of all persuasions.

The question now arises as to the precise content of the "nothing" (*nihil*) from which God creates. Is it an emptiness or void, a vacuum that is deprived of God? Or is it that which is nothing inasmuch as it represents the energies of God?[8] The Western mystic Eriugena noted that *nihil* is another name for God, an alternative description for the ultimate abyss of the divinity. And the notion is not unknown among the Eastern mystics, which is perhaps where Eriugena received and repeated it. In his work *On Divine Names*, Dionysius the Areopagite refers to God as being "at one and the same time in the world (encosmic), and around the world (pericosmic), and above the world (hypercosmic) . . . as being everything and nothing. . . . Nothing contains and comprehends God . . . and nothing exists that does not share in God."[9] The same understanding may also be found in the spiritual teaching of a more contemporary Orthodox saint, John of Kronstadt, a popular parish priest and renowned spiritual director in Russia at the turn of the twentieth century: "The Lord fills all creation . . . preserving it down to the smallest blade of grass and grain of dust in his right hand, and not being limited either by the greatness or smallness of things created; he exists in infinity, entirely filling it, as a vacuum."[10]

Therefore, the "nothing" from which God creates neither denotes nor implies the absence of God. It is in fact another way of understanding and underlining the very presence of God. Such is the depth of the spirituality of the "sponge" that is promoted in the fourteenth century by writers such as Saint Gregory Palamas. Gregory adopts this image in order to describe the way in which God's energies permeate human and material nature. Modern physics, too, supports such a view of matter as nothingness. Seen through the eyes of a physicist, the human body itself is 99.99% void, and even the little that appears to be dense matter is itself empty space. The whole thing is made out of nothing. We may think of creation *ex nihilo* as a form of "quantum theology" or "quantum spirituality" that holds together both the absence and the presence of God in the world.

GOD AS TRANSCENDENT AND IMMANENT

An emphasis, therefore, is definitely given to the transcendence of God in Orthodox theology and spirituality. Yet in the radical otherness of God, the oneness of God in relation to the world continues to be maintained, even though the temptation always remains to divide God from the world. Creation out of nothing, as a central Christian doctrine, demands reconsideration, if not revision, in light of the essential Orthodox doctrine of the distinction between divine essence (that is totally unknowable and inaccessible) and divine energies (that constitute God's revelation and reflection in the world). Together, these two concepts combat the notions either that the material world exists independently of God (which was early condemned by the Christian church as dualism) or that it is identical to God (which is still criticized by many thinkers as pantheism).

Judeo-Christian scripture itself also appears to lay greater stress on the transcendence of God and on the dependence of humanity and the world upon God. There is less emphasis on separation or independence between the two. Creation is never considered as being self-sufficient. Nonetheless, the gap between creator and creation is not irreconcilable, even for scripture, which endeavors to hold the two together in at least a couple of ways. First, there is the twofold interpretation of creation with the use of two distinct Hebraic verbs. Second, there is the concern in the Hebraic tradition to preserve the immanence of God and the extensive relationship between God and the world through the quality of divine "image and likeness" (Gen. 1:26) according to which humanity was created. It is because God has a face that the world too can be said to possess a face, and eyes, and a voice; and it is because humanity has a face that it is able to discern the face of God in the very face of the world (Augustine, *On Psalm* 148, 15 [PL 37.1946]).

At this point, a brief parenthetical remark should be made to correct the unbalanced reference, or reduction, of this "image and likeness" to the spiritual or rational aspect of humanity. Unfortunately the indivisible unity of the human person has often been undergirded in Christian thought, although this was not the case in scriptural literature. Similarly, for the Greek fathers, the human person always retained an element of mystery, due largely to the belief in the human person as an irreducible being consisting of body, soul, and spirit. The entire world, therefore, reflects the beauty and glory and unity of its creator: "The heavens tell the glory of God, / and the firmament proclaims God's handiwork" (Ps. 19:1). The whole of creation comprises "a mystical harmony that constitutes a song of praise." Despite its unity and beauty, creation is not unidimensional. Instead it reveals a loving God who cooperates in a relationship of Trinity in order to create. It is this love of God (1 John 4:16) which undergirds the relationship of humanity with the rest of creation. The Orthodox sacrament of marriage underlines this link between creation, beauty, and union: "O God, you have in your strength created all things, established the universe, and adorned the crown of all things created by you.... Blessed is your name and glorified is your kingdom: Father, Son, and Holy Spirit."

Again we must disabuse ourselves of a commonplace misinterpretation of the Old Testament command: "Be fruitful and multiply; and fill the earth and subdue it; and have dominion over...every living thing that moves upon the earth" (Gen. 1:28). For centuries this text has provided the license to dominate and abuse the world according to our selfish needs and purposes. Yet how can such an interpretation ever be reconciled with Paul's advice in the New Testament that we are to "use the world without abusing it" (1 Cor. 7:31)? The Genesis passage should instead be understood in the context of Adam's naming of the animals: "So out of the ground the LORD God formed every animal of the field and every bird of the air, and brought them to Adam to see what he would call them; and whatever Adam called every living creature, that was its name. Adam gave names to all cattle, and to the birds of the air, and to every animal of the field" (Gen. 2:19–20).

This event itself implies a loving and lasting personal relationship on the part of Adam with the environment, while it is also indicative of the same dialectical (literally, "in dialogue") relationship between Adam and the creator. The radical dedivinization of the world, wrought by our inability to desire or maintain a relationship with God, may be the cause of our lack of caring responsibility toward both heaven and earth. The "dominion" texts denoting the "power" of humanity over the world must in fact always be interpreted in light of human responsibility toward nature, whereby one is called to care for the land (Lev. 25:1–5), for domesticated animals (Deut. 25:4), and even for wildlife (Deut. 22:6).[11] We fall short of our vocation when we fail to care for creation. In fact, by refusing that call, we cause the world to remain a wasteland, unable to come alive by the compassion of a vital human being, who becomes the conscience, eyes, voice, and ears of the earth.

THE MYSTERY OF INCARNATION

A sacramental consciousness—namely, the recognition of and respect for the sacramental principle—further requires an awareness of the centrality of the incarnation in its historical, spiritual, and cosmic dimensions. Eastern Christian writers have usually viewed the incarnation more as a normative spiritual movement than as an isolated moment. For instance, Gregory of Nyssa uses such terms as "sequence," "consequence," or—his favorite expression—"progression." That is to say, God always and in all things wills to work a divine incarnation. The

Word assuming flesh two thousand years ago is one—though arguably the last, and even the most unique and most striking—in a series of incarnations or theophanies, for divine self-emptying (or *kenōsis*; Phil. 2:7) into the world is an essential, and not an exceptional, characteristic of the divinity. In fact, it is an invitation to humanity and the entire creation for transformation: "He [Christ] emptied himself, so that nature might receive as much of him as it could hold" (Gregory of Nyssa, *On the Psalms* 3 [PG 44.441]).

Furthermore, the incarnation is considered as being a part of the original creative plan and not simply as comprising a response to the human fall. In this regard, it is perceived not only as a revelation of God to humanity but primarily as a revelation to us of the true nature of humanity and the world itself. Such is the line of thought from Isaac the Syrian (seventh century) in the East to Duns Scotus (thirteenth century) in the West. This is also the basic theological focus of Maximus Confessor (died 660), who insists on the reality of Christ's presence in all things (Col. 3:11). Christ stands at the center of the world, revealing its original beauty and restoring its ultimate life.

The incarnation is thus properly understood only in relation to creation. The Word made flesh is intrinsic to the very act of creation, which came to be through God's uttering or divine Word. Cosmic incarnation is, in some manner (though the distinction is scholastic), independent of the historical incarnation that occurred two thousand years ago. From the moment of creation, the world is assumed by the Word and constitutes the body of this Word. The historical incarnation is in effect a reaffirmation of this reality and not an alteration of reality. Indeed, according to the first letter of Peter: "Christ . . . was destined before the foundation of the world, but was revealed at the end of the ages for our sakes" (1 Pet. 1:20).

From the moment of its creation, the world comprises the living body of the divine Logos. For "God spoke the word [*logos*] and things were made, / God commanded and they were created" (Ps. 33:9). In this sense, creation may be regarded as a continuous process where the divine Logos is manifest in time and space according to the particular personal energies of the incarnate Word. Thus the incarnation assumes cosmological, and not simply historical, significance. This is already intimated in the fourth century by Athanasius, who is unafraid of broadening these implications of the doctrine of the incarnation of Christ:

> And what is more strange, being Logos, he is not constrained by anything; rather he himself constrains everything; and just as being in the whole of creation, he is on the one hand outside everything according to essence but within everything with his powers, setting all in order, and extending in every way his providence to everything, and vivifying alike each and every thing, embracing all while himself not being embraced, but being wholly and in every way in his Father alone, so also in this manner is he in the human body, himself vivifying it, while he likewise vivifies all and is in everything being outside everything. (*On the Divine Incarnation* chap. 17 [45–46])

LOGOS AND LOGOI

For the Eastern fathers, Christ is the new Adam who realizes the sacrament that was rejected by us. That rejection is the old Adam's original sin; the realization by Christ is the new Adam's original blessing. Therefore, if we reject the world of darkness and accept living in the light of Christ, then each person and each object becomes the embodiment of God in the world. The divine presence is revealed to every order and every particle of this world; the divine Word (or Logos) converses intimately with every word (or *logos*) of creation. For the divine Logos always and everywhere "wills to effect this mystery of divine embodiment" (Maximus Confessor, *Ambigua* 7 [PG 91.1084]). This, of course, implies a contemplation of the principles (*logoi*) of creation, an insight into the meanings (*logoi*) and causes (*logoi*) of the world, a vision of the flesh and blood of Christ in the soil of the earth and the cells of all matter. Such a vision is already expressed in the Alexandrian tradition by Philo and later by Origen, who envisioned the eternal Logos being manifested in history and in the world under various guises: "There exist diverse forms of the Logos, under which he reveals himself to his disciples, conforming himself to the degree of light in each one, according to the degree of their progress in saintliness" (*Against Celsus* 4.16 [Ante-Nicene Fathers 4.503]).

Maximus Confessor refers to these divine *logoi* as being "conceived" only in the eternal Logos of God, constituted by the Son and instituted by the Spirit in historical time and place. We might dare to speak of the "uncreated createdness" of all things, which preexist in God's will (or *logos*) and are brought by the divine will into existence and being: "He is mysteriously concealed in the interior causes [*logoi*] of created beings ... present in each totally and in all plenitude. ... In all diversity is concealed that which is one and eternally identical; in composite things, that which is simple and without parts; in those which had one day to begin that which has no beginning; in the visible that which is invisible; and in the tangible that which is intangible" (Maximus Confessor, *Ambigua* [PG 91.1085; see also 1081 and 1329]).

However, the words (*logoi*) of creation always demand deciphering. So the *Macarian Homilies* speak of interpreting the language spoken by creation (*Homily* 32.1).[12] These "words" require silence on our part in order for us to hear and to dialogue (*dialogos*) with the Word (Logos) of the creator. The incarnate Word of God is, therefore, the basis of harmony and union, while sin and division blur our vision of everything as proclaiming the beauty of God. As preeternal Word, the Son of God is the giver of this gift; and as incarnate Word, Jesus Christ is the receiver of this gift. The Orthodox liturgical rite summarizes this in the words of Basil the Great: "For you, Christ our God, are the one who offers and is offered, the one who receives and is received."

The Word of God is the one who relates and reconciles the world to God: "In relation to the benevolent, creative and sustaining progress of the One toward created things, the One is multiple [*Unum est multa*]; but considered in relation to the return of things and their tendency toward the principle sustaining all things, the Center who has by anticipation in himself the beginnings of all the lines which have come from him, considered in this aspect, the multiple are One [*multa sunt Unum*]." This unity is confirmed in the joy of creation; this unity is joyfully celebrated in the liturgy of the Orthodox church. In this way, the church becomes an image of the world, just as the world bespeaks a cosmic liturgy. In the ecclesial vision of the world of Maximus Confessor, the church represents the world and the world reflects the church in, as he calls it, microcosm (*Mystagogy* 2 [PG 91.669]).[13] The sense of heaven reflected on earth is nowhere more evident than in the splendid architecture and symbolic liturgy of the Byzantine churches, especially that of the Hagia Sophia (Church of the Holy Wisdom) in Constantinople. When Orthodox Christians enter a church, they feel that they are in the comfort of their own home; and when they leave the building of the church, the feeling remains that they are still within the church because the whole world is the church. Indeed, the world is called to do more than merely to reflect or represent the church. It is supposed to respect and reveal the unified structure of the body of Christ, transcending all divisions between created and uncreated, heaven and earth, paradise and world, as well as between male and female.[14]

The Mystery of the Cross
and the Resurrection

In the mind of the Eastern mystics, everything in this world is required to undergo crucifixion in order to achieve resurrection; everything must die in order to rise (John 12:24–25). Or, in the words again of Maximus Confessor, himself a renowned ascetic and profound mystic, "all phenomena must be crucified" (*Theological Chapters* 67 [PG 90.1108B]). Like Christ, everything requires incarnation (the tangible nature of materiality), crucifixion (the testing through death in order to be raised to the vertical level of God), and descent into hell (and to the deepest and darkest recesses of the heart) before it can awaken to the light and arise in the life of Christ. It is what the Greek monastic writers like to call "the little resurrection" (e.g., Evagrius of Pontus [died 399]) or "the resurrection before the resurrection" (e.g., John Climacus [died ca. 649]). We read of the radiance of the popular and charismatic Seraphim of Sarov, who greeted his visitors with the

Easter salutation: "My joy, Christ is risen." We cannot forget, of course, the long, hard years of physical struggle, solitude, and silence that precede any such resurrection. Yet the Eastern ascetic tradition, at least in its more authentic expressions, perceives the cross more as a way of transforming the world than as a means of tolerating it.

The Orthodox liturgical tradition combines this scandalous mystery of the cross with the luminous majesty of the resurrection—the crucifixion prepares the resurrection, while the resurrection presupposes the crucifixion. Together, they visualize and materialize the redemption and sanctification of the whole world. Thus on Good Friday, the Orthodox liturgy sings: "All the trees of the forest rejoice today. For their nature is sanctified by the body of Christ stretched on the wood [of the cross]." And on Easter Sunday the celebration reaches a climax: "Now everything is filled with the light [of the resurrection]: heaven and earth, and all things beneath the earth."

Athanasius of Alexandria understood this universal dimension: "Christ is the first taste of the resurrection of all . . . the firstfruits of the adoption of all creation . . . the firstborn of the whole world in its every aspect" (*On the Divine Incarnation* 20 [48–49] and *Against the Arians* 2.64 [PG 26.281–84]). In the resurrection of Jesus Christ, in the abyss of the empty tomb, and in the depth of joy discovered in the mystical encounters between the risen Lord and his disciples, both women and men, the inner meaning of creation is revealed. In a sense, the Genesis account of creation can be understood only in the light of the resurrection "that enlightens every person coming into the world" (John 1:9). Through the resurrection one perceives the end and intent of God for all (Prov. 8:31); one senses in Christ a new creation, a new earth, and a new joy. Jesus is recognized (Luke 24:31) as the meaning and life of the whole world and not simply as the moral redeemer of individual souls. The entire creation belongs to God; and in God one discerns the destiny of all creation, for "he came to what was his own" (John 1:11). He came "to make everyone see what is the plan of the mystery hidden for ages in God, who created all things, so that . . . the wisdom of God in its rich variety might now be made known . . . in accordance with the eternal purpose that God has carried out in Jesus Christ our Lord" (Eph. 3:9–11).

The two feast days of the early Christian church that signify new life and new light are Easter and Epiphany; both of these feasts were the principal baptismal days for those wishing to be received into the church. The Orthodox church still preserves these powerful images of resurrection and regeneration, and the services for these feasts abound in images expressing the way in which "the entire universe" and "all created matter" contribute to the cosmic liturgy: "Now everything is filled with light, heaven and earth, and all things beneath the earth; so let all creation celebrate the Resurrection of Christ on which it is founded" (Paschal Canon, 3rd ode). "Today creation is illumined; today all things rejoice, everything in heaven and on earth" (Epiphany, *Sticheron* [Jan. 6]). "Today the earth and the sea share in the joy of the world" (Prayer, Great Blessing of the Waters [Jan. 6]). It is not by chance that both of these central feasts of the Orthodox church underline the creation of the world "in the first days" (Gen. 1) and understand the significance of the recreation accomplished "in the latter days" (Heb. 1:2) in light of the restoration of all things "in the last days" (2 Tim. 3:1).

THE ASCETIC DIMENSION

A Christian is supposed to consider seriously the reality and tragic dimension of evil. Christianity describes the world as it should be, while at the same time refusing to accept it as it is. Thus, all expressions of pantheism and paganism have from the earliest years been refuted and refused. Moreover, the sacrament of initiation, baptism, is preceded by a long series of exorcisms: for the individual, for the world around, and for the water of baptism itself. In the East, the "sin of Adam" is always understood as being a "fall" from the level of authentic life to mere survival. The chief consequence of this fall was mortality and corruption, rather than guilt or remorse. Death is seen as a cosmic disease that infects humanity and all of creation; and the reversal of death is effected through baptism, which is a participation in the resurrection of Christ, a real communion in his victory over death. According to Maximus Confessor, the world was created as a dynamic movement (see *Ambigua* [PG 91.1057]), and this movement toward transfiguration in the light of the resurrection is its very *raison d'être.*

There is, then, no greater estrangement from the world than its use in a manner that fails to restore the correct vision of the world in the light of the resurrection. Any other vision is an abuse of the world, whereby "the truth is exchanged... for a lie... and the natural use of things for an unnatural abuse" (Rom. 1:25–26). As such, the discipline of ascesis becomes a crucial corrective for any excess in consumption, for we have learned all too well and all too painfully that the ecological crisis both presupposes and builds upon the economic abuse and injustice in the world.

Nevertheless, when Orthodox Christianity refers to asceticism, it does not imply an individualistic or puritanistic life of self-discipline; nor is it promulgating an idealistic or sentimentalistic form of deprivation. Asceticism is regarded as a social event; it is a communal attitude that leads to the respectful use of material goods. The underlying conviction is that we are never alone, never isolated from one another and from our earth of origin (Gen. 3; Ps. 103:14). The Old Testament prophets warned us against this kind of social indifference and injustice, which is epitomized by consumption: "Ah, you who join house to house, and add field to field, until there is room for no one but you, and you are left alone in the midst of the land" (Isa. 5:8).

Therefore, in spite of all historical and human aberrations (and Christians have often been the primary culprits for the scars on the beauty of both the human body and on the body of the world), what is in fact avoided in authentic ascetic practice is not so much "the world or the things in the world . . . [but] the desire of the flesh and the lust of the eyes and the pride in riches" (1 John 2:15–16). "Flesh" of course signifies the whole of one's life in the state of conflict with oneself, with the world, and with God. "Lust of the eyes" implies the blurred vision of the world as created and as intended by God. And "pride" constitutes the ultimate hubris of humanity that usurps the role of God and seeks to dominate the world.

The diverse expressions of religious or monastic life sometimes deviate into a form of "angelism" that verges on the point of disincarnation and dematerialization, culminating in an aggressive attitude toward the body. Yet the ascetic life is ultimately a way of tenderness and of integration with oneself, with each other, and with creation. The genuine Christian ascetic is the universal person, who is freed from narrowness, limitations, and divisions. The vision of the ascetic is both mystical and personal: the ascetic is consciously aware that the problem of pollution cannot be distinguished from the problem of inner alienation. In the way to transfiguration, the Christian ascetic by no means leaves creation outside but in fact unites the whole cosmos that has been discarded and divided by sin.

In his letters, Saint Paul speaks of a certain division or dualism between flesh and spirit, but never does he intimate that there should be any separation or severance between body and soul or between humanity and the material world. Later Christian writers commonly fail to observe this distinction. For Paul, matter is not an enemy to be conquered, but rather a means whereby one can glorify the creator. "The flesh is the hinge of salvation," wrote Tertullian (*On the Resurrection of the Flesh* 8 [PL 2.852]). And so transfiguration through asceticism entails a revelation of the spiritual potentiality of the world, rather than its suppression. So the transfiguration through ascesis foreshadows the ultimate resurrection, securing in this life the firstfruits of cosmic redemption and resurrection. Through ascetic toil, the monastic recognizes and responds to the brightness of all people and all things. Indeed, through ascesis, nature itself responds more compassionately toward humanity, as Saint John of Kronstadt writes:

> When God looks mercifully upon the earth-born creatures through the eyes of nature, through the eyes of bright, healthful weather, everyone feels bright and joyful. . . . So powerful and irresistible is the influence of nature upon mankind. And it is remarkable that those who are less bound by carnal desires and sweetness, who are less given up to gluttony, who are more moderate in eating and drinking; to them nature is more kindly disposed, and does not oppress them—at least, not nearly so much as those who are slaves of their nature and flesh.[15]

Any philosophy of disembodiment, of dematerialization, and of separation is a philosophy of death. Extreme asceticism and excessive spiritualism alike represent a partial truth; and to assert these solely is to destroy the harmony of the world.

CONCLUSION: THE SEED OF GOD

It is a tragedy that, in spite of the destruction caused and the suffering inflicted, we have not yet apparently learned our lessons from nature. We are still far from a balanced perspective of creation as deriving from, dependent upon, and deified by God. The world remains a human-centered reality; indeed, it is still more narrowly conceived as man-centered. We are preoccupied with ourselves, our problems, our destructiveness, our death. Yet this is precisely what led us in the first place to our fateful predicament. Contemporary deep ecology emphasizes the fact that the correct perspective and relationship between humanity and creation has been distorted, almost destroyed. Yet little significance is attached to the reality that all things are coherent not just in their interrelatedness and interdependence, but also in their relation to and dependence on God. In fact, to be estranged from God is to lose touch also with created reality and with what really matters in this world. It is to be enslaved within this vicious cycle of death and destruction. Symeon the New Theologian expressed this poetically in tenth-century Constantinople: "What is more painful than to be separated from Life? . . . For it also means to miss out on all good things. The one that moves away from God loses all that is good" (*Hymn* 1 [Sources chrétiennes 156.164]). According to an earlier, fourth-century poet and theologian of the same city, Gregory Nazianzus: "All things dwell in God alone; to God all things throng in haste. And God is the end of all things" (*Dogmatic Poems* 29 [PG 37.508]).[16]

The ancient Greeks too had a similar worldview, discerning the presence of God in all things. Thales exclaims: "Everything is full of God" (*Fragment* 22).[17] In their cosmology, the transcendence of God was perceived as a characteristic that rendered the divinity more accessible and more familiar. The slightest detail of creation was seen to bear some mark of the creator. The same conviction inspired the Cappadocian theologians in the fourth century: "Look at a stone, and notice that even a stone carries some mark of the creator. It is the same with an ant, a bee, a mosquito. The wisdom of the creator is revealed in the smallest creatures. It is he who has spread out the heavens and laid out the immensity of the seas. It is he also who has made the tiny hollow shaft of the bee's sting. All the objects in the world are an invitation to faith, not to unbelief" (Basil of Caesarea, *On Psalm* 32.3 [PG 29.329]).

Indeed, the same mystical vision is also articulated outside of established theology and poetry. Well-known twentieth-century Greek writer Nikos Kazantzakis experienced a tumultuous life, and his works equally caused tumult among pious believers who misunderstood him. Yet he had a powerful religious image of the seed of God in the world. This helps to understanding the theological notion of the connection between the divine Logos and the created *logoi*. Seed requires soil, the earth on which it may be sown and on which it may grow (Matt. 13; Mark 4; Luke 8). And a seed also requires an almost cosmic passion for its cultivation. The Little Prince, too, knows that "seeds are invisible. They sleep deep in the heart of the earth's darkness, until some one among them is seized with the desire to awaken."[18] This is the passionate desire that inspires Kazantzakis in his *Ascetic Exercises*, a compelling description of the connection between God and the world:

> Everything is an egg, and within it lies the seed of God, calmlessly and sleeplessly active.... Within the light of my mind and the fire of my heart, I beset God's watch—searching, testing, knocking to open the door in the stronghold of matter, and to create in that stronghold of matter, the door of Gods heroic exodus.... For we are not simply freeing God in struggling with and ordering the visible world around us; we are actually creating God. "Open your eyes," God is crying; "I want to see! Be alert; I want to hear! Move ahead; you are my head!"... For to save something [a rock or a seed] is to liberate the God within it.... Every person has a particular circle of things, of trees, of animals, of people, of ideas—and the aim is to save that circle. No one else can do that. And if one doesn't save, one cannot be saved.... The seeds are calling out from inside the earth; God is calling out from inside the seeds. Set him free. A field awaits liberation from you, and a machine awaits its soul from you. And you can no longer be saved, if you don't save them.... The value of this transient world is immense and immeasurable: it is from this world that God hangs on in order to reach us; it is in this world that God is nurtured and increased.... Matter is the bride of my God: together they struggle, they laugh and mourn, crying through the nuptial chamber of the flesh.[19]

NOTES

1. Wendell Berry, *The Unforeseen Wilderness* (San Francisco: North Point, 1991), 16–17.

2. Cf. E. Lampert, *The Apocalypse of History* (London: Faber & Faber, 1948), 14, 164. See A. Schmemann, *For the Life of the World* (New York: St. Vladimir's Seminary Press, 1973), esp. 23–46. The importance of an eschatological, as well as a cosmological, vision of history is underlined by Metropolitan John (Zizioulas) of Pergamon in "The Book of Revelation and the Natural Environment" [Greek], *Synaxi* 56 (Athens, 1996): 13–21; for an English translation, see *Creation's Joy* 1.2–3 (Cumberland, RI, 1996).

3. For this threefold vision, see A. Schmemann, *Church, World, Mission* (New York: St. Vladimir's Seminary Press, 1979), 77; and idem, *Liturgy and Tradition* (New York: St. Vladimir's Seminary Press, 1990), 98–99.

4. Prayer from the Ninth Hour of the Orthodox daily liturgical cycle.

5. R. A. Johnson, *Owning Your Own Shadow* (Harper, 1991), 108–9.

6. See, e.g., Hermas, *Shepherd* book 2, mandate 1 (PG 2.913); Theophilus of Antioch, *To Autolychus* 2.4 (PG 6.1052); Irenaeus, *Against Heresies* 2.10.4 (PG 7.736); Origen, *On First Principles* 1.4 (Ante-Nicene Fathers 4.256); Basil, *Homily on the Hexaemeron* 7.7 (PG 29.180C); and John Chrysostom, *Homily on Genesis* 2 (PG 57.28).

7. Athanasius, *On the Divine Incarnation* chaps. 2–3 (New York: St. Vladimir's Seminary Press, 1982), 26–28.

8. Cf. P. Sherrard's argument in *Human Image-World Image*, already adumbrated in his earlier *The Greek East and the Latin West* (Cambridge University Press, 1959).

9. Dionysius the Areopagite, *On Divine Names* 1.6 (PG 3.596) and 12.1–2 (PG 3.977) (my translation). See also 13.3 (PG 3.980–81). For the works of Dionysius, see *Pseudo-Dionysius: Complete Works* (Classics of Western Spirituality; New York: Paulist Press, 1987).

10. Cf. John of Kronstadt, *My Life in Christ* (repr. Jordanville, NY, 1977), 140.

11. See, e.g., Gregory Nazianzus, *Homily* 38.11 (PG 36.324A). For the interpretation of these "kingship" passages in the church fathers, see Gregory of Nyssa, *On the Creation of Man* 2 (PG 44.132); Basil of Caesarea, *On Psalm* 44.12 (PG 29.413); and Ambrose of Milan, *On the Gospel of Luke* 4.28 (PL 15.1620).

12. For English translation, see the volume in the Classics of Western Spirituality series published by Paulist Press (New York, 1982).

13. Maximus did not in fact invent but he was the first to systematize the notion of the church as the image of the world. Germanus of Constantinople (died ca. 733) writes in similar fashion in the prologue of his *On the Divine Liturgy*: "The Church is an earthly heaven, in which the heavenly God dwells and moves" (trans. P. Meyendorff; New York: St. Vladimir's Seminary Press, 1984), 56–57. The concept of the temple as microcosm is also found in other religious beliefs; cf. Mircea Eliade, *Images and Symbols* (New York, 1969).

14. See Maximus Confessor, *Ambigua* 41. Also see his *Questions to Thalassius* 40, 63 (PG 90.396, 665).

15. John of Kronstadt, *My Life in Christ*, 141–42.

16. For this radical quality of asceticism, see Metropolitan John Zizioulas, "Ecological Asceticism: a cultural revolution," *Our Planet* 7.6 (Nairobi, 1995): 7–8.

17. See also Xenophanes, *On Nature* 27; Homer, *Hymn* 30; and Aeschylus, *Prometheus Bound* 88.

18. Antoine de Saint-Exupéry, *The Little Prince* (New York: Harcourt Brace, 1971), 16–17.

19. Nikos Kazantzakis, *Ascetic Exercises* (5th ed.; Athens, 1971), 85–89 (my translation).

CHAPTER 4

THE WORLD OF NATURE ACCORDING TO THE PROTESTANT TRADITION

H. PAUL SANTMIRE

JOHN B. COBB JR.

THE sixteenth-century Protestant Reformation in Europe, sparked by Martin Luther and John Calvin, who are sometimes called "the Magisterial Reformers," is of more than historical interest, since the trends of thought and social formation which it set in motion are still of wide-ranging consequence, particularly with regard to the theology of nature. Other theologians and church leaders were also involved in the Reformation, to be sure, from humanistic thinkers like Huldrych Zwingli to popular revolutionary leaders like Thomas Müntzer and advocates of a peaceful imitation of Christ like Menno Simons. Many of these leaders emphasized that one could become a Christian only through a conscious personal decision, so that infant baptism should be replaced by believer's baptism. (Today's Baptists continue that tradition.) A complete account of Reformation theology and piety would take into account the voices of what is often called "the left wing of the Reformation." However, in general, later Protestant attitudes toward nature, including those of most Baptists, have not been greatly affected by the distinctive teachings of this group, whereas large numbers of Baptists and others who trace their heritage to these left-wing sources, as well as those who stand more directly in the traditions of Luther and Calvin, have been influenced in this regard by the legacy of the Magisterial Reformers. Hence it is appropriate to focus here on the teachings of Luther and Calvin.

THE TEACHINGS OF LUTHER AND CALVIN

While the Magisterial Reformers shared many of the same convictions, their theologies of nature differ enough to make a detailed review of each man's thought instructive and help us understand how and why Protestant thought about nature unfolded as it did.[1] Luther's theology was shaped by a deep commitment to what he thought of as the living word of God, a theological fundamental that he discerned first and foremost as an Old Testament scholar, pondering what was for him the written word of God, the holy scriptures. Luther's focus on "the word," in turn, had far-reaching implications for what he and his followers would have to say, or not say, about the world of nature. A theology of *hearing* the word of God would as a matter of course tend to make any theological reflection about nature suspect, since speaking is the characteristic way that humans interact with each other—and with God—whereas humans characteristically interact with nature by contemplation or *seeing*. Persons speak. Trees do not. Persons encounter trees by seeing them. Therein lies the problematic of Luther's thought with regard to nature.

Luther may or may not have actually had the celebrated "tower-experience" (*Turmerlebnis*), when the revolutionary message of justification by faith alone dawned upon him as he pondered the first chapter of Romans. Nevertheless, that story accurately portrays the unfolding experiential logic that lay behind his discovery. The medieval culture into which he was born had been built as a kind of glorious cathedral, according to highest aspirations of human reason and spirituality. But wherever Luther looked he saw corruption and the power of the devil. When he looked at the church, he saw much worldly pomp and little theological substance. The best and the brightest seemed to have sold their souls in order to gain the accouterments of power. There were signs, too, as Luther perceived them, that the untutored and often impoverished masses were lusting for the establishment of some visible kind of kingdom, perhaps by initiating a violent uprising, in the name of Christ the king. In the meantime, many sensitive souls, among them most notably Luther himself, could find no lasting consolation, no peace before God, no certainty of their salvation, as they suffered under the relentless pressure and the frequent abuse of the medieval penitential system. They found themselves adrift and bereft, terrorized by the thought of an avenging deity on high, in a world that in so many respects seemed to them to be damned. The glorious cathedral of the Middle Ages had thus become for Luther, as for many of his contemporaries, a kind of prison, which, Luther believed, could only leave the souls of the faithful forever tormented.

Luther's response to this crisis was in essence this. He shut his eyes—hence the suggestiveness of the image of the isolated tower-experience—and he listened. He refused to countenance the signs of worldly glory and churchly vainglory. He distanced himself radically from what he perceived to be the world of sin, the flesh, and the devil in order to listen anew to the word of God in its radical freshness, freed from the overlay of worldly speculations which had, in his view, muffled its witness for too long. In this sense, *hearing* was the generating experiential matrix, the *Urerlebnis*, of the Reformation. Although the emergence around him of a new, much more verbal culture, whose symbol was the printing press,[2] and his own training as a scripture scholar certainly contributed to Luther's work, this *Urerlebnis* was the fundamental reason for his development of a theology of the word of God. The Reformation began, theologically speaking, with an act of closing eyes and opening ears.

It was while listening to the word of God, disengaged from the claims of his culture and his spiritual heritage, that Luther heard—as if for the first time—the great consolation: "The just shall live by faith" (Rom. 1:17). With this experience of hearing he found liberation and peace. It was an experience of the *mysterium tremendum*. Its impact never left him. Its impact not only delivered him from despair, but also provided him with a sensibility which was then to be the matrix of his every theological utterance. Luther's theology was forever to be a theology of hearing, a theology of the word of God.

Luther virtually said as much himself, on numerous occasions. "A right faith," he once wrote, "goes right on with its eyes closed; it clings to God's Word; it follows that Word; it believes that Word."[3] Again, he stated: "The ears are the only organ of the Christian."[4] And: "Nothing can be more beautiful in the eyes of God than a soul that loves to hear his Word."[5] And then this: "Among all gifts the gift of the Word is the most valuable. For if you take this away, it is like taking the sun away from the earth.... For only the Word keeps a joyful conscience, a gracious God, and all of religion, since out of the Word, as from a spring, flows our entire religion."[6]

This was no less true of Luther's teaching about the sacraments, where one might have expected, if anywhere, to encounter a shift of sensibility, from the moment of hearing to the moment of seeing. The sacraments, after all, have visible elements and actions. But Luther took up the Augustinian motif that the sacraments are "visible words"—and took it up with a passion. "You should know," he stated in one characteristic utterance, "that the Word of God is the chief thing in the sacrament." The word is to the elements, he explained, as the soul is to the body.[7]

On the basis of the *Urerlebnis* of hearing faith, Luther also turned against his theological opponents, both those on the right and those on the left. On the right he saw the theological rationalists, the defenders of the ecclesiastical status quo. If there ever was such a thing as a "medieval synthesis" and if that synthesis could be illustrated architecturally in terms of the Gothic cathedral,[8] then we might venture to suggest that speculative reason was the cement that held all the stones of the cathedral together. Luther challenged the validity of speculative reason, again and again, on the basis of his theology of the word, and to this extent he contributed to the destruction of the grand edifice of medieval thought. "The Gospel truly shines in the dead of night and in darkness," he once stated, "for all human reason is mere error and blindness and the world itself is nothing but a realm of darkness. Now in this darkness God has kindled a light, namely, the Gospel, to enable us to see and walk, as long as we are on earth, until the dawn comes and the day breaks forth."[9]

Luther took on his opponents on the left in much the same manner, on the basis of the word. Some of these so-called enthusiasts, it will be remembered, were espousing claims that by special inspiration they had been called to establish the kingdom of God visibly on earth. "I fear that more of these enthusiasts will come," Luther observed, in a characteristic statement: "Confused, wrapped up in their thoughts, foolish, stubbornly clinging to their own view, and despising the Word of God. This is why I have always with the greatest diligence exhorted men to read Scripture and to hear the spoken Word that we may deal with the God who has revealed Himself and is speaking to us, and may in every way avoid the God who is silent and hidden in majesty."[10]

Luther also approved of the removal of images from the church buildings of his time, although not of the use of violence in doing so. On occasion he could speak approvingly of images in the churches, but he generally did so only if those images had an evident pedagogical purpose. Some of the faithful learn to grasp the word more fully, in other words, with the help of pictures that illustrate biblical events and stories. But for the most part, Luther preferred to think of the church building, in his words, as a "mouth-house."

The earliest example in all Europe of a surviving church building designed and built within the Reformation milieu, the small Schloss Kapelle in Torgau, Germany, was dedicated by Luther in 1534.[11] It would not be too much to say that this building is a superbly designed "mouth-house," with the space cleared to facilitate proclamation and listening, while the few images that were introduced were of obvious pedagogical, not contemplative, purpose. The elevated pulpit that dominates that space surely and dramatically announces the whole building's *raison d'être.*

Luther characteristically juxtaposed the act of hearing over *against* the act of seeing. Seeing, he often seemed to assume, provides access only to the *larvae Dei,* to the masks of God in this world, whereas hearing opens the way for the believer to receive an immediate kind of communication from God, the "living voice of the Gospel" (*viva vox evangelii*). Seeing, as such, was not a way to encounter God. On the contrary, Luther thought it was the way of self-assertion.

Luther apparently took it for granted, with many of his time and before, that hearing is the passive sensibility, while seeing is the active sensibility.[12] Hearing, then, is in this respect appropriate to the experience of grace, according to the fundamental assumptions of Luther's thought, for it undergirds the prevenience of grace. Seeing, on the other hand, in this perspective, is an active outreaching of human sensibility and therefore will constantly prompt the soul to trust in the fruitless speculations of reason or in the futile righteousness of works. Seeing, as Luther thought of it, apparently was a way for the subject to take control, not a way to bow down helplessly, relinquishing all aspirations to control one's own destiny or to establish one's own righteousness. For many reasons, then, he wanted to discourage or even to deny any substantive influence on his thought from the side of the sensibility of seeing.

The apex of Luther's polemic against seeing, and in defense of hearing, was undoubtedly his charged teaching about—and against—the "theology of glory" and in favor of "the theology of the cross." The theology of glory, for Luther, stood for everything that he so vehemently opposed: worldly grandeur in the church instead of biblical substance, visible pretensions of power in the world instead of the weakness of the cross, dependence on works for salvation rather than on faith, the rule of reason and speculation in the church rather than trust and faithfulness, an ethics of hedonistic self-aggrandizement rather than an ethics of neighborly self-giving. What the eyes should rightly see is the suffering and the death of Christ, while the ears must hear the word and the heart must believe.

Given this attack on seeing, it comes as a total surprise that we find in Luther a rich theological apperception of *nature*. He depicts God as being "with all creatures, flowing, and pouring into them, filling all things."[13] He marvels at a grain of wheat: if we really understood it, he says, we would die of wonder.[14] "When I truly grasp the significance of the incarnation of the Son of God in this world," according to Luther, "all creatures will appear a hundred times more beautiful to me than before."[15] He even imagined the world of his present experience to be the time when the new life of the world to come had begun to dawn. Now, with a renewed knowledge of the creation made possible by our faith, he says, by God's mercy "we can begin to recognize his wonderful works and wonders also in the flowers when we ponder his might and his goodness."[16]

The last word of Luther about the theology of nature is therefore an ambiguous word. It would appear that the exclusive dominance of the sensibility of hearing in his thought would not allow him to entertain any encounters with God in the world of nature beyond the walls of the church. But Luther's spirituality and his reflective theology show signs of a rich apperception of God throughout the world of nature. Luther also takes the human body very seriously, along with the tangibility of the sacramental elements. He affirms that the finite is capable of the infinite (*finitum capax infiniti*). That is a fundamental principle of his theology of creation, his Christology, his soteriology, and his eschatology. God is wondrously and gloriously "in, with, and under" all things.

John Calvin's thought offers a striking contrast to Luther's, notwithstanding the profound influence on Calvin of Luther's revolutionary teaching about justification by faith.[17] If Luther's paradigmatic moment could be said to have been his hearing the word of grace, alone, in the tower, Calvin's could be said to be a passionately delivered public oration. In this instance, we would imagine Calvin addressing an audience accustomed to such orations, people who themselves, like Calvin, had been deeply influenced by the humanistic studies of the Renaissance and who were also conversant with the classical theological tradition, particularly with the comprehensive theological insights of Augustine and Thomas Aquinas. While Calvin proclaimed and taught justification by faith with great fervor, he projected that teaching dramatically in the context of a narrated vision of God's unfolding providence in and through the whole creation. He joined together, or held in tension, the primal, existential experience that so captivated Luther and the ancient, sapiential vision of the classical theological tradition.

Calvin developed his thought this way, significantly, not out of speculative interest, but because he perceived the foundations of his cultural world to be crumbling. His theology had its existential genesis not so much in the experience of the quest for a gracious God as in distress about the social, political, and ecclesiastical chaos all around him. For this reason, he accented the theme of providence more than justification by faith. "Ignorance of providence," Calvin said in one characteristic utterance, "is the ultimate misery; the highest blessedness is knowledge of it."[18]

Calvin's universalizing, sapiential interests allowed the sensibility of *seeing* to shape much of what he wrote. This aspect of his thinking is not often highlighted by commentators, probably because Calvin himself so stridently opposed the employment of *images* in the life of the church. Thus the title of chapter 11 of his *Institutes*: "That it is wicked to ascribe a visible form to God, and that all those who set up idols for themselves defect from the true God." Calvin often writes like a classical Christian iconoclast: "Without exception [God] rejects all images, pictures, and other signs by which the superstitious imagine they will have him near them." He sometimes can even sound like Luther, emphasizing hearing over against seeing. Those who seek "visible forms" of God are departing from God, for God is known not by sight but by the hearing of his voice or word (as in Deut. 4:15: you heard a voice; "you did not see a body").[19] The only things he allows to be painted or sculpted are those "which can be seen by the eye."[20]

A careful reading of Calvin in this respect shows, however, that what he opposes so fervently are *humanly constructed images of God*. God himself can be thought of as the image maker par excellence! The invisible God makes himself visible—God images himself—in many ways, in the created world, in the scriptures, and in the face of Christ. For Calvin, *that* is the only way the invisible God may be known. The sensibility of seeing shapes everything else.

Calvin understood the scriptures themselves, the written word of God, under the rubric of seeing. Whereas for Luther the scriptures are God's address to us, for Calvin they are "the spectacles" that show us how to seek God.[21] He thought of God's providential ordering of things throughout the whole creation as a manifestation of God. He even referred to the individual works of God in creation as "represented, as in a painting."[22] One of his favorite images was of the whole creation as "the theater of God's glory."

Calvin regularly used vividly visual terms to describe the believer's encounter with this manifest divine glory. Thus he observes of the Spirit's working throughout the creation: "He quickens and nourishes us by a general power that is *visible* both in the human race and in the rest of the living creatures."[23] Again, with regard to faith's true object: "The pious mind does not dream up for itself any god it pleases, but *contemplates* the one and only true God. And it does not attach itself to whatever it pleases but is content to *behold* him as he *manifests* himself."[24] Indeed, Calvin commented on Second Isaiah's proclamation of God's power in all things that invoking the creation in this way is *necessary for faith*: "For unless the power of God, by which he can do all things, *confronts our eyes*, our ears will barely receive the Word or not esteem its true value."[25] Then, in a summary statement, he says of nature: "Let us not be ashamed to take pious delight in the works of God *open and manifest in this beautiful theater*. For as I have elsewhere said, although it is not the chief evidence for faith, yet it is the first evidence in the order of nature, to be mindful that wherever *we cast our eyes*, all things they meet are works of God, and at the same time to ponder with pious meditation to what end God created them."[26] Calvin's most cherished image of nature was a "*sort of mirror in which we contemplate God, who is otherwise invisible.*"[27]

Calvin's theology of nature might be thought of as contemplative. He was awestruck by the beauties of nature, which he encountered as the very manifestations of God himself and God's glory. So a recent interpreter of Calvin instructively calls the Reformer "the creation-intoxicated theologian."[28] Justly so, as this acclamation by Calvin shows:

> In grasses, trees, and fruits, apart from their various uses, there is the beauty of appearance and pleasantness of odor. For if this were not true, the prophet would not have reckoned them among the benefits of God, "that wine gladdens the heart of man and oil makes his face to shine. . . ." Has the Lord clothed the flower with the great beauty that greets our eyes, the sweetness of smell that is wafted upon our nostrils, and yet will it be unlawful for our eyes to be affected by that beauty or our sense of smell by the sweetness of that odor? What? Did he not

endow gold and silver, ivory and marble, with a loveliness that renders them more precious than other metals or stones? Did he not, in short, render many things attractive to us apart from their necessary use?[29]

Calvin also made every effort to project a vision of what might be called theological balance in nature, thus encountered. On the one hand, he rejected the deists of his time, those who taught that God is totally transcendent and more or less uninvolved in the creation, apart from setting it in motion. On the other hand, Calvin just as fervently rejected the pantheists of his time, those who tended to identify the immanence of God in creation with the creation itself. Rather, he espoused the total dependency—each moment, each drop of rain—of the whole cosmos on God's will. If God were to withdraw his creative word for a moment, the creation would all fall into nothingness. Yet, for Calvin, the creatures themselves are not, as it were, swallowed up by the divine providence; they have their own distinct being and beauty in the greater scheme of things. So he concludes at one point: "It must always be remembered that the world does not properly stand by any other power than the Word of God, that secondary causes derive their power from him, and that they have differing effects as they are directed."[30]

In sum, Calvin's theology of nature can be thought of as profoundly visual, eagerly focused on the manifestations of God in every creature and on the goodness and beauty of every creature. Whereas Luther believed passionately that God is in, with, and under all things, powerfully overflowing and shaping every creature, that belief was thoroughly shaped by his conviction that it is the *hidden* God who is there. Luther also tended to think that any encounter with the hidden God in nature would be an encounter with the wrath of God. For Calvin, in contrast, the beauties of God's providential works are apparent and manifest everywhere, as in a great and glorious theater or in a beautiful painting. Still, whether accenting God hidden (Luther) or God manifest (Calvin), both Reformers espoused in this respect a remarkably vivid theocentric theology of nature. For both, the encounter with nature is first and foremost an encounter with *God's* creation.

A corollary of this theocentrism, for both Reformers, was their teaching about the impact on nature of God's judgment of human sin. Following the second Genesis creation narrative, the story of Adam and Eve in the garden and their eventual expulsion, both Luther and Calvin held that nature itself had been cursed by God, in response to human sin. The earth was less fertile than it had been. Some animals then began to prey on humans. But this postfall predicament by no means excluded what was for both theologians a constant refrain of celebration of God's presence and benevolence in the still good, albeit postfall, creation. Not even the horrors of human sin and the divine judgment on the earth because of that sin, in the Reformers' view, can blot out the goodness and the beauties and wisdom of the creator's handiwork.

But here the plot thickens. For both Luther and Calvin, the whole creation must also be understood in intensely *anthropocentric* terms. Luther remarked in his exposition of Genesis 1 that holy scripture "plainly teaches that God created all these things in order to prepare a house and an inn, as it were, for the future man."[31] Calvin made the same point in his Genesis commentary. The "end for which all things were created," was "that none of the conveniences and necessities of life might be wanting to men."[32] Moreover, although Calvin strongly emphasized that the will of the creator and its providential power rule in and through all things, Calvin's purpose often was to show that the God who elects the believer by grace alone, for eternal salvation, is operative providentially in nature, notwithstanding the apparent evils and experienced tribulations of life in the created order.[33] Luther and Calvin espoused such anthropocentric motifs in the context of their theocentric views of creation, because that is how they read the scripture's own testimonies to the goodness and the graciousness of God. For the Reformers, moreover, the two accents—theocentrism and anthropocentrism—held together harmoniously, however much those accents might appear today to stand in tension.

Calvin heightened the anthropocentric accent even further. He understood the believer's life in the creation much more as the "active life" (*vita activa*) than the "contemplative life" (*vita contemplativa*). For him the believer's moral life, which Calvin thought of in terms of the "imitation of Christ," is dynamic and engaging. The believer is called to transform the world, rather than merely to live in faith and love, as Luther tended to suggest, in the world as it is. For Calvin, the believer is called to work for new social forms and new configurations of the human environment, with a confidence born of an intense sense of chosenness and a divinely ordained historical destiny.

This vocational dynamism separated from its theocentric context later became the captive of forces essentially alien to it—an aggressive spirit of domination driven by a human will-to-power over nature. While it would be going too far, surely, to argue that the theology of Calvin was one of the *causes* of the modern project of domination of nature, undertaken especially by the forces of burgeoning capitalism, it may be recognized that "the spirit of Calvinism" helped *set the stage* for secularized developments in Western mercantile and industrial society.

Such are the ambiguities of the Reformers' theologies of nature. Luther projected a rich theology of divine immanence in nature and of the goodness of material reality in general. But his thought, profoundly shaped by the sensibility of hearing, did not allow for a sustained celebration of God's immanence in nature or of the goodness of nature. Calvin, profoundly shaped by the sensibility of seeing, celebrated the beauty and the goodness and the wisdom of the whole creation, as a manifestation, even a garment of the creator. But Calvin's theology was vulnerable, in a later, more secular context, to destructive ideological constructions.

THE REFORMATION TRADITION
AND THE CULTURE OF MODERNITY

The Reformers developed their theologies in response to wide-ranging cultural trends. The impact of such factors increased dramatically as the Reformation tradition unfolded in the modern era. Accordingly, although the deepest level of the distinctively Protestant ethos came from the Reformers, both popular Protestant thought and sophisticated theological reflection in ensuing centuries were deeply informed by the rise of the natural sciences and the worldview that accompanied those sciences. This kind of cultural influence has probably been greatest in the English-speaking world, where the influence of Calvin predominated over that of Luther.

In the English-speaking sphere, the dominant worldview related to the sciences is often called "Newtonian." However, it actually derives more from René Descartes in the seventeenth century than from Isaac Newton who lived in the same era. Its basic character is well known. Descartes divided the world into two types of substances. There were, on the one side, human minds, characterized by thought. Descartes assumed that these were substances, that is, self-existent entities that endured through time as subjects of human experience. Everything else is to be understood as material substance. Its basic character is extension. *Qua* matter, it has no existence for itself, no principle of action or motion. A bit of matter changes its motion, only as it is moved. What comes to be in the material world can ultimately be understood as changing configurations of material particles resulting from the motions imparted to them.

The Reformers had concentrated attention on the human world and especially on the relationship between God and the believer or the community of believers. When Luther and Calvin did attend to the natural world, they recognized it as having rich qualities in itself. But what came to be known as the scientific worldview stripped the natural world of such qualities. As time went by, those qualities came to be more consistently located in the human mind.

Nevertheless, the scientific worldview formulated in the seventeenth and eighteenth centuries did assign an important religious role to nature. Nature provided the basic grounds for believing in the existence of God. Calvin had already made such claims, although he mainly restricted them to what *believers* could learn about God from nature. Now theologians and philosophers developed what they thought of as a "natural theology," established on the basis of reason alone, not self-consciously dependent on the "spectacles" (Calvin) of scripture.

This is the way the new argumentation was projected. The complex world in which we live with all its patterns of order could not have come into being by chance or accident. It required a creator. This movement of thought from the world to its maker established the reality of God for most people in the early modern period. It was widely understood to confirm the biblical affirmation of creation.

Further, since nature was stripped of all characteristics that would allow for the explanation of its own motions and of the complex forms it has taken, all this had to be attributed to a supreme mind. Hence, it was assumed, the laws of motion are imposed on passive matter by God. Thus God was understood first and foremost as lawgiver, and the existence of laws was thought to be inconceivable apart from such a lawgiver.

Finally, because it was evident that, alongside the material world, there *is* a mental or spiritual world as well, the question arose as to how God was related to this. What was evidently most important in this sphere was morality; and the relation of God to morality was considered to be analogous to the relation of God to nature. Just as God laid down the laws according to which both earthly and heavenly bodies move, so also God laid down the laws for human behavior. The difference was that material entities necessarily obey the laws of God, whereas human beings are free to obey or disobey. However, whereas the laws of earthly rulers are sometimes broken with impunity, the laws of the divine ruler must always be enforced in the end. Hence, it was evident to many that at death all would be judged according to their obedience or disobedience and rewarded or punished. This belief was understood to be both rational in itself and important for the maintenance of good behavior and social order.

This religious view, already contested by Calvin, came to be widely known as deism. Deism was enormously popular in English-speaking Protestant circles. It flourished among English intellectuals from Herbert of Cherbury in the early seventeenth century to William Paley in the late eighteenth. It is a profoundly different form of religious life from that promoted by the Reformers. In some ways it reversed their emphases. They had radically opposed the notion that people are ultimately judged according to their obedience or disobedience to divine law. Only grace could save anyone, and the human side of the reception of grace is faith. The notion of grace, or, indeed, of any action of God directly affecting the present course of events, was absent from pure forms of deism. Deists believed that God's acts occurred at the point of creation and at the death of each individual or at the end of history. The natural world now operates according to the unchanging laws imposed at the beginning. In this life human beings are free to observe or disregard the moral law. God's act will come as judgment in the future.

Many thinkers, such as John Locke in the late seventeenth century and Joseph Butler in the first half of the eighteenth, sought to mediate between deism and biblical thought, chiefly appropriated through the Reformers, especially Calvin. One issue was often formulated in terms of miracles. Since the natural world was understood to operate according to divinely imposed laws, the deists taught that the laws are perfect and God would never interfere. However, much of the biblical narrative assumed that God did act in the natural world in particular ways. In the context of the new worldview, any such act must overrule some natural law, and such overruling of natural law by God was the new definition of a miracle.

Orthodox Protestants typically saw no problem with the idea that God can act in ways that conflict with or suspend natural laws, since, as the author of natural laws, God could break them. Nevertheless, these believers typically held that God interfered with natural laws only for the purpose of communicating with human beings for the sake of their salvation. It was commonly believed that miracles were signs of divine approval of those who revealed the divine will and especially of the divinity of Jesus. Miracles were worked in the early church to establish the authority of the biblical revelation. Once this was accomplished, they ended.

Although more attention was devoted to nature by the deists than by the heirs of the Reformation, this second source of Protestant religious understanding leads no more to ecological sensitivity than does the first. Nature is viewed as a machine made by God. If one has any doubt about God's existence, one can renew one's confidence by observing the law-observant character of nature and its amazing wonders. If its maker causes events to occur that do not follow natural law, this is because the maker has additional purposes to effect. But these purposes have nothing to do with the well-being or intrinsic value of the natural world.

Alongside the continuing influence of the Reformers, on the one side, and the impact of the scientific worldview, on the other, there was the movement of pietism. The pietists accented the inner life of believers. They thought that God's presence and activity within the heart could be felt, that, indeed, such feeling was central to authentic Christian experience. Whereas for the Reformers, the assurance of salvation came from believing the gospel promise, the pietists understood that the assurance itself was worked by God in the human heart. They also emphasized that God's grace purified and developed the spiritual life of the believer, so that the inner life came to be characterized more and more by love of God and neighbor.

Broadly speaking, this emphasis on religious experience and inner transformation came to characterize a great deal of Protestantism, especially in the United States. The two figures who played the largest role in giving it importance are John Wesley in England in the eighteenth century and Friedrich Schleiermacher in Germany in the nineteenth century. Both described and indeed celebrated the transformation of the inner life effected by the experience of God, Wesley as a highly influential practicing evangelist, Schleiermacher as a widely respected systematic theologian.

Although this emphasis on Christian experience did not, in itself, bring about ecological sensibility or teaching or even enlarge the appreciation of the natural world, pietism as a whole did provide part of the context for the romantic movement, which achieved public prominence, especially through the efforts of poets like

William Wordsworth in the nineteenth century. The emphasis on personal experience *could* open persons to their experience of the world of nature, not merely as testimony to the reality of God but also as important and enriching in itself. The romantics generally protested against the rationalism of so much theology, both orthodox and deistic, and the narrowness of the pietists' interest in experience. They enlarged the range of experience to which attention was given and emphasized the esthetic dimension. Schleiermacher's first major book, *On Religion: Speeches to Its Cultured Despisers*, portrayed the Christian faith in highly romantic terms.

Theoretically and esthetically, romantic thinkers opposed the view of the natural world as consisting simply of matter in motion. Nature had its own rich reality in itself. Human beings could commune with nature, not simply marvel at its laws or manipulate it for the sake of human purposes. This appreciation of the natural world as contributing to the meaningfulness of the whole cosmos had resonance in the Protestant community, especially as that community was influenced by pietism. Many members of Protestant churches, to this day, who understand the personal experience of God to be the heart of the Christian life, claim to experience God chiefly in nature. For this reason nature has played a much larger role in popular Protestant piety than in Protestant theology, above all in the United States, where romantic writers like Henry David Thoreau and John Muir influenced popular culture profoundly, during the nineteenth century and beyond.

But the most dominant form of Protestant theology, especially on the European continent, took a different, almost opposite, direction. The direct cause for this was philosophical, although the direction also reflects the influence of the Reformers, especially of Luther. This development led Protestant theology to reduce the role of nature still further.

Although Cartesian philosophy had stripped nature of all attributes except mass, there remained the assumption that the existence of nature was independent of human beings. It was fully real in itself. However, the British empiricists, culminating in David Hume in the mid-eighteenth century, pointed out that our sense experience does not give us any such reality. It provides us only with appearances or phenomena. These provide no grounds for affirming the sort of necessary connection that scientists intended when they spoke of efficient causes.

Hume's formulations seemed to undercut the truth of science as well as of theology. In response, toward the end of the eighteenth century, Immanuel Kant undertook to reestablish science on the basis of the creative activity of the human mind. He acknowledged that the mind can know nothing about what nature is in itself. But he argued that the mind necessarily so orders the phenomena as to provide the grounds for the natural sciences. For Kant, that meant that humans can think only in the categories of the modern scientific worldview. In his philosophy, in effect, the human mind replaced God as the creator of the world. Kant also agreed with Hume that no argument from the order of nature to the reality of God is possible.

This did not mean that Kant dispensed with Christian-oriented religious belief. In fact, his theological views were quite similar to those of the deists. However, whereas they built their strongest case on the law-abiding character of the natural world, Kant held that God can be known only in the sphere of what he called "practical reason," meaning ethics. He analyzed ethical experience so as to show the need to postulate God.

Nineteenth-century German Protestant theology, following the work of Kant, was extraordinarily creative and diverse. One stream followed Kant in associating Christian faith primarily with values and morality. Another followed Schleiermacher in emphasizing religious experience in general and Christian experience in particular. Still another followed Søren Kierkegaard in locating Christianity as the answer to the existential condition of human beings. Still other theologians did their work more as commentaries on the Bible or on the Reformers.

In general all these followed Kant in viewing nature as a construct of the human mind rather than a self-existent and given reality in the context of which human life and thought are located. Nature as something with its own reality and character simply disappears from these theologies. If the idea of creation was considered at all, it was interpreted to refer, in the language of Schleiermacher, to the absolute dependence that people feel in their relation to God.

Next to Kant, the most influential philosopher of the period was G. W. F. Hegel. Hegel also accepted Kant's basic teaching, but instead of viewing the human mind as essentially fixed, he studied it historically. He described how the mind creates the world in ever new ways as a result of a profound dialectic. This process moves toward a climactic end. He and his theological followers could interpret Christianity in these terms and locate its role in world history accordingly. But this did not change the status of the natural world. Indeed, followers of Hegel found it easy to think of nature as a mere stage posited by God to allow the emergence of self-conscious spirit in human history.

The general notion underlying Hegel's system—that the human mind constructs its world in changing ways—has played a large role in the late twentieth and early twenty-first centuries. A number of theologians have argued that historic Christian beliefs have been constructed—invented—by those in power, by affluent North Americans and Europeans, by elites in Third World countries, by males, and by heterosexuals. These theologians have called for new constructions by groups at the margins, who can give a voice for the voiceless, above all the impoverished people of color, women, and gays. For thinkers of this persuasion, the theme "social construction of reality" has thus served as an analytic tool. This theme can imply that nature has no determinate character apart from this social construction. On the other hand, the newly affirmed constructions have often expressed ways in which indigenous people relate to nature that vividly imply its reality and importance.

Even more widespread than the view that the world is socially constructed has been the linguistic turn in philosophy and theology. This has occurred in the fields of influence of both Hume and Kant. In some of its forms this has been highly congenial to the heirs of Luther, for whom faith is awakened by the proclamation of the word—by language. In other forms, such as twentieth-century thinker Ludwig Wittgenstein's concept of "language games," this viewpoint allows for the claim that theology is a language game with its own rules and can be pursued accordingly. Linguistic theology, like linguistic philosophy, takes many forms, but in none of them does it attribute independent reality to nature. In general, language games about nature are left to the natural sciences.

Two maverick theologians deserve mention. The first is German physician, philosopher, and musician Albert Schweitzer. He was influential in theological circles chiefly for his history of the quest of the historical Jesus early in the twentieth century, but later he called for an ethic of "reverence for life" that stood aside from the major currents of the theology of his time. He found all destruction of life to be evil, even though he could not avoid it altogether. This sensitivity to the value of all living things had little influence on theology until much later, but it gained a significant following in the broader public.

The other maverick was German-born American systematic theologian Paul Tillich. During the middle years of the twentieth century, Tillich mounted a thoroughgoing attack on mechanistic thinking about nature, reaching back for theological inspiration to what he considered to be the nature-mysticism of Luther. Tillich produced a monumental *Systematic Theology*, in which both nature and humanity were viewed as being equally rooted in "the Ground of Being" or God. Tillich also preached about the salvation of the whole creation, as in his sermon "Nature, Also, Mourns for a Lost Good."[34] More systematically, Tillich developed a "theology of the inorganic" in the context of a mystical, even romantic fascination with the processes of life on this planet. But Tillich was publicly influential mainly for his "apologetic" encounter with the most prominent secular intellectual trends of mid-twentieth-century Western culture: existentialism, psychoanalysis, and the modernist arts. His rich theology of nature was little noticed or appreciated in his own time. Tillich's influence was also held in check in significant measure by the emergence of a new kind of thought in the Reformation tradition that stood over against the whole effort of post-sixteenth-century Protestant theology to engage and respond to the culture of the modern era, an effort with which Tillich was deeply sympathetic.

During the first half of the twentieth century, in the mainstream of Protestant theology, there was a vigorous reaction against the tendency of nineteenth- and early-twentieth-century theology and philosophy to focus on human experience, thought, creativity, and language and their assumption that God could be spoken of only in that context. This reaction was launched by "the theology of crisis" in Germany, which was spearheaded by Calvinist theologian Karl Barth. Barth adamantly advocated the abandonment of the anthropocentric framework he had inherited and called for a return to the theocentrism of the Reformers. He championed the Reformers' focus on the word of God, independent of the categories offered by the philosophers.

With this turn to the transcendent word of God, fatefully, Barth also passionately rejected the thought that there can be any knowledge of God from nature. Since he did not enter into substantive discussion with the traditions of the philosophers, he did not directly address Kant's denial of reality to the natural world. But in this respect, he was very much an heir of the Kantian tradition. Not only did Barth deny the validity of any natural theology, but in his multivolume *Church Dogmatics*, the great work of his theological maturity, he denied that Christian theology can espouse any substantive theology of nature whatsoever! As a result, the world of nature was, by default if not by intention, treated as if it were merely the scenery for the divine-human drama (a conclusion drawn by Emil Brunner, an early Barth collaborator who became an influential critic). Hence, Barth's shift from anthropocentrism to theocentrism did little to encourage attention to the natural world.

Although Kant was influential in the English-language sphere, his thought was not all controlling there. Other views remained possible. These included various forms of pragmatism, espoused by thinkers such as William James in the early twentieth century, that did not exclude the reality of the natural world. Other theologians and philosophers also continued to think of God quite centrally as the creator of the natural world. At the popular level, much of this theological realism, as we can call it, was a continuation of deistic thinking, based on the materialistic view of nature.

In the late nineteenth century Charles Darwin's theory of evolution was a major challenge to Christian thinking. Responses varied greatly. For Kantians, Darwinian thought was simply another theory about the phenomena and was, therefore, irrelevant to the actual understanding of human existence. But for the heirs of Cartesian dualism and the theological realists, it implied that human beings are fully part of the world machine and totally immersed in the amoral and often violent processes of nature. This was unacceptable to most Christians.

The Kantian separation of science and theology gained in appeal among church leaders trying to avoid this reductionistic view of human beings without rejecting the work of scientists. Among those who resisted this dualism, there were two types of responses. Heirs of the Reformation, who were greatly concerned about the literal inspiration of the Bible (which Luther and Calvin had not stressed), rallied massively, for the first time, against the conclusions of science. They insisted on maintaining the biblical account or, at least, a doctrine of the separate creation of different species, especially the human species. For some it was important to show that the scientific evidence did not preclude separate creation of humans. For others, the literal authority of the Bible simply trumped science. Fundamentalism arose as a reaction against the accommodation of the mainline Protestant churches to evolutionary theory.

Protestant thinkers of the second type, such as William Temple in the first half of the twentieth century, are sometimes called "theistic evolutionists," because they accepted the hypothesis of evolution, but held that it should be interpreted nonreductively. If human beings are part of nature, then nature is not simply matter in motion. Theists of this stripe argued that we can understand that God created the world gradually, bringing one species into being out of another. They did not dispute the facts of evolution, but they understood the process differently. Although this type of theology has played only a minor role in Western theology, its way of relating theology to science has had some popular following, especially in the English-language world, where the controversy over evolution was most intense. It has remained the type of twentieth-century theology most open to new ways of encountering the natural world.

Many of the Protestant thinkers who were drawn into the discussion of the relation of theology and science through the disputes over evolution were also keenly interested in the developments in physics in the early twentieth century. The new physics provided images of nature different from the mechanism that had ruled so long. Nature seemed more active, even alive, than Descartes had allowed. A number of theologians in the English-language world developed philosophical theologies influenced by new developments in the natural sciences.

This renewal of cosmological thinking by physicists and theologians differed from deism in that it worked against the established dualism. It viewed humanity as the climax of a cosmic process rather than a separate creation. Even though the focus remained on human beings, their history, their distinctive needs, and the saving activity of God among them, the idea that God worked creatively in the whole process restored a connection between God and the natural world. The human relation to the rest of nature took on some importance, as well.

The single most influential thinker belonging to this community of thought is Alfred North Whitehead, who wrote in the first half of the twentieth century. He emphasized that reality should be understood in terms of events rather than of material atoms. All unit events have both mental and physical aspects, although the former may often be minimal. Each event comes into being through the inflowing of past events, so that all events are internally related to other events. He believed that this shift of basic metaphysics could lead to better understanding of both relativity and quantum phenomena in physics and of biology as well. He also included what he could learn from religious experience, East and West, in his overall vision. He offered major criticism of traditional theology, especially its ideas of divine omnipotence and impassibility. Much of his influence on theology was mediated by philosopher of religion and metaphysician Charles Hartshorne, who influenced many students at the Divinity School of the University of Chicago and who himself was an ardent naturalist. Whitehead's conceptuality has found support among scattered thinkers in many other fields. It has an obvious affinity with the insights of ecological science and ecological thought more generally, which came into prominence in the second half of the twentieth century.

Alongside these popular and sophisticated theological and philosophical programs was the theology of the ecumenical movement that prospered in the middle of the twentieth century. This movement was an effort to heal the divisions within the church. Some hoped for institutional mergers. More realistic was developing mutual understanding and appreciation that allowed for working together and speaking together to the church as a whole and to the world. Most of the churches interested in this movement were officially committed to the creeds widely accepted in the ancient church, and many gave considerable authority to the thinking of that period. Accordingly, participation led Protestants to give greater attention to the official creeds and confessions of churches and to the reappropriation of traditional formulations that antedated current divisions. It also encouraged careful study of biblical texts accepted as authoritative by all parties.

Since God's creation of the natural world was an important idea in the Bible and in the early church, it played a larger role in the ecumenical movement than it had in the Protestant theology that emerged after the era of Luther and Calvin. Although this did not have much practical effect on the ecumenical movement prior to the 1970s, it may have made the transition to ecological thinking easier there than it has been in most academic theological contexts. But, overall, Protestant theology was ill equipped to respond to what one 1972 church statement called "the human crisis in ecology."

THE REFORMATION TRADITION
AND THE ECOLOGICAL CRISIS

The ecological crisis burst on the consciousness of American Protestantism around 1970. Earth Day that year was an important event for Americans in general, but especially for the churches. It not only raised awareness of the ecological crisis, but many of its promoters leveled *blame* for the mistreatment of nature at the Christian churches. This was due in part to the astonishing influence exercised by the now famous 1966 lecture by historian Lynn White Jr. to the American Association for the Advancement of Science: "The Historical Origins of the Ecologic Crisis." White highlighted anthropocentrism as the culprit, and he held that its deepest sources in the West lay in the way the Western church had interpreted the Bible. This theological anthropocentrism paved the way for the widespread Western view that nature exists for human use only and that technological mastery of nature is in accord with the divine plan. Western science and Western technology, White showed, were nurtured in this theological soil.

The criticism stung. Christians have always tended to assume that Christianity is on the side of the good. Clearly the massive degradation of the natural world was not good. Accordingly, many Christian writers rose to the defense of Christianity. Much of the response, especially from Protestants, was that the Bible is not fundamentally anthropocentric, as it had been interpreted often in the Reformation tradition.

Actually this claim about the Bible was not in direct conflict with White. He spoke mainly of the way the Bible *was interpreted* by the Western church, and the history of Protestant thought, as we have seen, gives strong support to his thesis. Most Protestantism in the modern era has in fact been highly anthropocentric, and even when it has been theocentric, God has been considered overwhelmingly in relation to humanity. Still, the argument that the Bible was not anthropocentric played a valuable role. Almost for the first time, biblical scholars began to search the scriptures looking for the positive role played by the natural world. Theologians began to emphasize the importance of nature, and God's concern for it, throughout the Christian tradition. Defending Christianity against the correct charge that Western Christianity had been anthropocentric led to changes in the self-understanding of Protestants.

This change had been anticipated in a prophetic address by Lutheran theologian Joseph Sittler to the World Council of Churches meeting in New Delhi in 1961. Drawing on scriptural testimony from the letter to the Colossians and from the theology of early Christian thinker Irenaeus, Sittler called for the ecumenical church to develop a new kind of "cosmic Christology." Sittler claimed that the church's understanding of Christ is irrelevant unless it is related to human life as it is immersed in the world of nature, unless it addresses issues like the desecration of the earth and the need for all humans to care of the earth. Sadly, at the time, Sittler's address was mostly greeted with indifference, if not outright hostility.

One reason for the hostile response was the association of a theology of nature with Nazi thinking. During the Hitler era this term had been used by some Protestant theologians who were willing to support the Nazi mythos supposedly based on Nordic traditions by developing a congenial theology of nature. The opposition of Karl Barth and the confessing church to any form of natural theology was formulated against this collaboration with the Nazis. So soon after the Nazi aberrations and the evils of the Holocaust, it is understandable that the representatives of the ecumenical church in New Delhi in 1961, especially Protestants from Germany, should have turned such a deaf ear to Sittler's address in behalf of a new theology of nature.

Even as late as 1972, when the United Nations began to give careful attention to ecological questions, the World Council of Churches, most of whose member churches are Protestant, remained aloof. The reason at this point, however, was not so much that it opposed consideration of the well-being of nature—that concern had begun to subside—as that many Third World members were afraid that this was a way in which the affluent classes in the developed countries could distract attention from the urgency of justice for the world's oppressed. However, by 1975 the theological climate had finally changed. At the World Council General Assembly in Nairobi that year, the theme of *sustainability* took a central place. Australian biologist Charles Birch led the council in adopting this goal.

Previously the World Council had committed itself to the goal of a "just and participatory society." At Nairobi, it committed itself to "a just, participatory, *and* sustainable society." In context, there was no question but that a sustainable society was one that could be supported indefinitely by the available natural resources, including the sinks for its waste and pollution.

Since justice and participation had been discussed for years, attention was devoted in the next seven years to understanding the meaning of sustainability and what was entailed in realizing it. As a matter of fact, the work of the World Council played a significant role in lifting this concept to a central place in the secular world as well. But the goal of a sustainable society is *still anthropocentric*. It reflects a changed view of the proper relation of humanity to the natural environment but not an abandonment of the centrality of the human. Nevertheless, in the global discussions that followed, the issue of anthropocentrism was intensively discussed, and many Christians participated in its criticism. Accordingly there was some preparation for the next step.

This occurred almost accidentally at Vancouver at the next World Council of Churches General Assembly, held in 1982. At that time the Cold War was threatening to turn hot. Many delegates from the First and Second Worlds, sensing the threat of a nuclear holocaust, called urgently for peace. Third World delegates feared that this would again distract the church from the justice for which they longed. Hence the debate in this assembly was about the relation of peace and justice. The old slogan of "a just, participatory, and sustainable society" was largely ignored and the new goal was "peace and justice." Fortunately, someone proposed the addition of the phrase *the integrity of creation*. This motion passed. The goal of "peace, justice, and the integrity of creation" has described the vision of the World Council from that day forward.

Given the fact that this phrase was adopted with little debate, one cannot say that the delegates were consciously moving away from an anthropocentricity of sustainable society. But this was the result: the World Council of Churches since 1982 has been committed to respecting creation as a whole in its own right and not simply in terms of its relation to human beings. What is striking, in view of the dominance of anthropocentric thought prior to 1970, is the rapidity with which this change occurred and the lack of effective opposition.

Undoubtedly this rapid change was supported by the sudden emergence in Protestant theology, especially in the United States, of a new and vigorous kind of "ecological theology" and "ecojustice ethics." Some of the contributors to this movement built on the foundations identified by Sittler in 1961, undertaking to claim the world of nature for Christ by rethinking the meaning of the Bible and the tradition. Others focused on overcoming the philosophical obstacles to a healthy relation to the natural world. Both were supported by the already existing concern within the churches, originating in the 1930s, for stewardship of the soil. Some were influenced by the writings of Aldo Leopold.

In 1964 the Faith-Man-Nature Group was established with loose ties to the National Council of Churches. Four years later the group published the papers of its third national conference as *Christians and the Good Earth*.[35] Many of the themes of later Protestant writers, stimulated by awareness of the global ecological crisis, are anticipated in this book. Some of the participants, such as James Logan, continued the christocentric approach of Sittler. Others, such as Conrad Bonifazi, whose book *A Theology of Things*[36] was published the previous year, spoke of "the inwardness of things." Still others, such as H. Paul Santmire, who wrote a 1966 dissertation on Barth's understanding of nature, emphasized the importance of overcoming anthropocentrism by recognizing the independent value and significance of the natural world. His use of the phrase *the integrity of nature* may have influenced the World Council of Churches' much later affirmation of "the integrity of creation." Santmire published his programmatic study *Brother Earth* in 1970.[37] He has continued his work on ecological theology primarily through his historical study of Christian thinking about nature, especially in *The Travail of Nature*.[38]

Crisis in Eden by Frederick Elder focused attention on the question of whether Christian theology should separate humanity from the natural world or see human beings as included in it.[39] Elder came down strongly on the side of inclusion, and he called for a restraint and a new asceticism to replace the themes of domination and consumption.

Like the Faith-Man-Nature group, Elder had developed his views on nature prior to his encounter with White's challenge to the church and to the alarmist ecological literature that proliferated at the end of the 1960s. The first Protestant theological book responding in detail to White seems to have been *This Little Planet*, published in 1970.[40] Hamilton obtained essays on pollution, scarcity, and conservation from experts in those fields and then had three theologians respond: William Pollard, Roger Shinn, and Conrad Bonifazi.

Ecology appeared the next year, edited by Richard E. Sherrill, a staff member of the National Council of Churches, who had also given leadership to the Faith-Man-Nature group noted above.[41] The participants, including Norman Faramelli of the Boston Industrial Mission, were social activists who realized the importance of incorporating ecological concerns into their frame of reference. As a result of this concern, the book foreshadowed the struggle of the church generally to deal responsibly with the environment while maintaining its concern for justice to the disadvantaged.

Faramelli made use of the term *ecojustice*, which came to be widely accepted in church circles. William E. Gibson devoted much of his life to the cause. In 1973 he began discussions that eventuated in the Eco-Justice Project and Network, which organized conferences and networks from 1974 to 1992. Beginning in 1981 it published a journal, *The Egg*, subtitled "A Journal of Eco-Justice." In 1985 the Eco-Justice Working Group of the National Council of Churches cosponsored this journal, which was one of the main organs of the ecological movement within American Protestantism. In 2004 Gibson published *Eco-Justice*, which includes many of the essays previously published in *The Egg* as well as an account of the ecojustice movement.[42] A close associate of Gibson has been Dieter Hessel, a writer and church activist who played a role with Gibson in leading the Presbyterian Church to an insightful statement on the subject.[43] Hessel also edited *After Nature's Revolt*.[44]

Despite the conjoining of concern for justice with ecological issues, African American theologians remained somewhat aloof, and the topic of ecology has had little play in African American churches. African Americans pointed out that the environmental movement paid little attention to the racist dimensions of current policies, as, for example, the location of noxious waste dumps in minority communities. This struck a nerve among the advocates of ecojustice, and the churches have tried to rectify this failure.

It would hardly be possible to find mainstream Protestants who would abandon or minimize questions of justice for the sake of the natural environment. Nevertheless, there have been theologians who have specialized in the relation to nature. For example, from 1987 to 1990, Richard Cartwright Austin published a four-volume work entitled *Environmental Theology*.[45] The volumes are entitled *Baptized into Wilderness: A Christian Perspective on John Muir*, *Beauty of the Lord: Awakening the Senses*, *Hope of the Land: Nature in the Bible*, and *Reclaiming America: Restoring Nature in Culture*. As these titles suggest, Austin emphasizes the importance of changing Christian sensibility more than concepts or ethical principles. *A Sense of Place* by Geoffrey R. Lilburne supported this esthetic emphasis.[46]

In addition to writers who specialized in the topic of nature, several mainstream Protestant theologians and ethicists have written influential books on this topic. Douglas John Hall, the leading Canadian theologian, has done much to recover the traditional Christian idea of stewardship as an appropriate image for Christian care for the natural world. In 1986 he published *Imaging God*, and four years later, *The Steward*.[47] In 1991 James A. Nash wrote *Loving Nature*.[48] Nash chaired the Churches' Center for Theology and Public Policy and wrote in the broad stream of American Protestant theology. He grounded ecological theology in the doctrines of creation, covenant, divine image, incarnation, spiritual presence, sin, judgment, redemption, and church, concluding with the call to love.

In 1994 James Gustafson, a senior figure in theological ethics, published *A Sense of the Divine*.[49] For Gustafson the basic religious feelings are awe, the sublime, and the sense of limits as well as possibilities. He developed his theology more from these than from distinctively Christian doctrines. He interpreted the Calvinist doctrine of divine sovereignty in a pantheistic direction.

In 1996 Larry Rasmussen wrote *Earth Community, Earth Ethics.*[50] In earlier writings Rasmussen had located his ecological ethics in the context of Luther's theology of the cross. This note is not lost in his book, but it is set into so broad a context as to take on a quite new character. Theological reflection is merged into broadly human and historical reflection on the crisis of our time and the appropriate response in terms of the conceptuality of a global community of life, in which humans live and move and have their being and which has its own integrity and value.

Among the many other voices in the Protestant discussion, some write from their Christian perspective but for the wider culture. Holmes Rolston III usually writes as a philosopher who is particularly well informed about the science of ecology. But he is also a Christian informed by a sophisticated understanding of scripture. Another is Bill McKibben, who is one of the most impassioned voices for saving our degraded earth. Perhaps the most influential such voice is that of Al Gore, whose publication in 1992 of *Earth in the Balance* was a politically important event.

Two topics important to thinking about nature have been largely avoided both in this literature and in official church pronouncements: the rights of individual animals and the population explosion. The most influential theological writer on the first of these topics has been Andrew Linzey, an Englishman, whose long career in behalf of animals culminated in the publication in 1995 of *Animal Theology.*[51] The writer who has dealt most responsibly with the issue of population is Susan Bratten. In 1992 she published her book *Six Billion and More.*[52]

The potential of the churches to reach people at a deeper level than scientific information as such led some outsiders to give support to the efforts of the National Council of Churches to awaken Protestants to the crisis. Carl Sagan and E. O. Wilson gave generously of their time without in any way conceding that there was truth in Protestant beliefs. Others who had been working outside the churches were drawn into them. In 1994 Max Oelschlaeger, who had written a major work on wilderness, published *Caring for Creation.*[53]

One contributor to the 1969 volume *Christians and the Good Earth* was Daniel Day Williams. With Williams's work, we witness the public emergence of a school of theologies of nature fundamentally shaped by process thinking. Williams was interested in overcoming the dominant modern philosophical perspectives that had weakened the ability of the churches to think about nature. He pointed to the American pragmatic tradition, naming Peirce, James, Dewey, and Whitehead as sources for an ecological theology. This tradition rejected the dualism of mind and nature, locating human beings in the natural world. However, it had dealt much more with evolutionary thought than with ecology. On the other hand, writers in this tradition, especially in the wing that came to be known as process theology, had long rejected the dualism of the human and the natural and had made clear that nature's value and significance are not to be viewed anthropocentrically. Further, they understand relationships as intrinsic to the reality of all things, so that nothing exists in and of itself but only in its interconnection with other things, including God.

Substantial contributions of process theologians to the theology of nature came largely in response to the alarmist literature of the late 1960s and the shock of White's charge of Christian responsibility for the crisis. Sharing White's critique of anthropocentrism, this response was affirmative of White's position, agreeing also with Elder's inclusivism. In 1970 John B. Cobb Jr. held a conference in Claremont entitled "A Theology of Survival." Published papers from that conference included Cobb's "The Population Explosion and the Rights of the Subhuman World."[54] His short but incisive book *Is It Too Late? A Theology of Ecology* appeared in 1972 and is still in print.[55] Cobb later teamed up with Australian ecologist Charles Birch to write *The Liberation of Life*[56] and with process economist Herman Daly to write an ecological economics, *For the Common Good.*[57]

Many other process theologians have contributed to the discussion. In comparison with the streams of ecological theology noted above, process theologians have contributed less to the discipline of theology and more to integrating Christian ecological thought into diverse fields in a transdisciplinary way. Of particular note is Ian Barbour, the physicist-theologian who has done more than anyone else to generate discussion between the scientific and religious communities. Already in the 7 October 1970 issue of *Christian Century*, he published an article entitled "An Ecological Ethic." Over the years he has dealt with environmental issues chiefly in his writings on technology, which culminated in volume two of his Gifford Lectures, *Ethics in an Age of Technology.*[58]

David Griffin has promoted ecological thinking both in his own writings and in the SUNY Series in Constructive Postmodern Thought he edited for the State University of New York Press. The first volume, a collection of essays growing out of a conference that Griffin organized in Santa Barbara and appearing in 1988, was *The Reenchantment of Science.*[59] His ecological understanding of nature has informed his work not only on science and religion in general, but also such major, more specialized, studies as his work on the mind/brain problem: *Unsnarling the World Knot.*[60] His awareness of the nature and urgency of the ecological crisis has been a major source of his extensive work on questions of world order.

Frederick Ferre is a process philosopher concerned for Christianity and the church. Like Barbour, his greatest contributions to environmental thinking, such as *Hellfire and Lightning Rods,*[61] have focused on technology. However, he has also published an ecologically informed, comprehensive philosophy in Griffin's SUNY series.

Jay McDaniel has done much work on ecological spirituality. He has also worked to integrate ecological thinking with concern for animals as individuals, building a bridge between ecological thought and that of Linzey. He worked with Charles Birch to introduce the concern for animal welfare into the World Council of Churches' interpretation of the integrity of creation, but that topic was not welcomed by the council leadership.[62] On the other hand, the WCC Press in 1997 did publish *Living with Animals* by Charles Birch and Lukas Vischer.[63]

Most of the women theologians influenced by process thought have contributed to Protestant versions of ecofeminism. Valerie Saiving initiated this discussion well before ecological issues were widely on the table. Later participants include Karen Baker-Fletcher, Donna Bowman, Rita Nakashima Brock, Carol Johnston, Mary Elizabeth Moore, Marjorie Suchocki, and Anna Case Winters. Catherine Keller's important books *Apocalypse Now and Then* and *Face of the Deep*, while richly informed by the methods and insights of a late-twentieth-century movement called poststructuralism, are grounded in a process ecofeminism.[64]

In general, the work of other ecofeminist theologians is highly congenial to process theology. For example, there is extensive overlap in thought with Catholic theologian Rosemary Radford Ruether, who was connecting feminist and ecological concerns as early as 1972 in a chapter on *Liberation Theology*.[65] She has retained her leadership in the field among both Catholics and Protestants with *Gaia and God*.[66] Sallie McFague has done more than any other Protestant writer to interpret Christian theology in view of the insights of ecofeminism, and her work has been widely influential. It may be that, partly because she does not identify herself with process theology, she has done much to make some of its key ideas acceptable in a broader stream of Protestant thought.[67]

European theology has devoted less attention to these questions. However, globally the most influential contribution to a theology of nature has been that of German Calvinist theologian Jürgen Moltmann. He has been the most prominent Protestant voice in the ecumenical church during the latter part of the twentieth century, and he has put his great prestige on the side of developing an affirmative theology of nature.

Moltmann's epoch-making theology of hope was born in the midst of the ashes of World War II, and his mature theological works, above all his magisterial *God in Creation* (published in English and German editions in 1985), signaled the arrival of a postanthropocentric, richly ecological/ecojustice theology within the mainstream of Protestant thought.[68] The theme of hope prompted Moltmann to project a vision of cosmic history, shaped by that which is to come, by God's ultimate future. What had become a kind of tired afterthought in much Protestant thinking, *eschatology* (in Greek *ta eschata* refers to "the last things"), now became the energizing motif of everything else. Moltmann as a matter of course reaffirmed Luther's teaching about justification by faith, as well as Calvin's vision

of the whole creation as the theater of God's glory, but Moltmann also fundamentally and self-consciously rejected the anthropocentric tendencies of the Magisterial Reformers. He envisioned all things as embarked on a kind of cosmic
journey, to be completed in their eternal fulfillment at the very end. Moltmann
also sought to depict the life-giving work of the Spirit of God and the salvation of
Christ in cosmic and, more particularly, ecological terms. Permeating all his
thought, moreover, was a concern for justice, especially for the poor of the earth
but also for all the other voiceless creatures of the earth, and with that a radical
critique of the powers of injustice, both political and economic.

Writing, like Moltmann, in the tradition of Calvin, South African theologian
Ernst Conradie has sought to shape a contextual theology of nature, predicated on
a vision of the church as the liberating center of a new creation amid the brokenness of the present creation.[69] Other African theologians, such as Marthinus L.
Daneel, have explored the rich implications of developments in indigenous African church life, such as the birth of the Association of African Earthkeeping
Churches in 1991.[70]

Given these ecumenical developments in Protestantism, with ecological theology and ecojustice ethics thus claiming prominence in the Reformation tradition in many lands and with similar expressions appearing in the statements of
many Protestant denominations around the world, the judgment that the tradition of Luther and Calvin had, by the early years of the twentieth-first century,
become ecological may seem irresistible. But it may also be premature. Notwithstanding the ecumenical witness and the far-reaching theological and ethical
transformation, a deep-seated habit of focusing attention only on the human
remains strong, especially in North America.

Furthermore, this deep-seated habit comes to vigorous and self-conscious
theological expression from time to time. One of the most sustained and articulate critiques of ecological theology is that of Thomas Sieger Derr, who writes as
a "Christian humanist." In 1996 Abingdon Press published a book entitled *Environmental Ethics and Christian Humanism*, based on an extended essay by Derr
on this topic, combined with a generally supportive response by conservative
Catholic writer Richard John Neuhaus and a sharp critique by James Nash.[71]
There is little doubt that one reason many Protestant theologians and ethicists
have been silent on this topic, and why the church at the grassroots level has been
so slow in responding to these issues, is that there is a great deal of latent support
for Derr's position. Indeed, a strong reaction against both ecological and feminist
thought has been a factor in the movement of a number of prominent theologians
to a neoconservative position in theology and in politics.

In any case, this account of what was once understood to be the mainstream
churches and their theologians gives an incomplete picture of twentieth-century
Protestantism. Early in the century two movements arose within it, which have
burgeoned into major determinants of its twentieth-first-century prospects,

especially in the United States and the Third World: fundamentalism and Pentecostalism. Fundamentalism was mentioned above as the reaction to accommodations made by the mainline churches to evolutionary theory. Pentecostalism built on the heritage of Protestant pietism and its emphasis on inner experience. But the Pentecostal way was much more effusive. It arose as an ecstatic experience of the Holy Spirit. For some time fundamentalists and Pentecostals viewed each other with suspicion, but gradually the boundaries between them eroded. For the most part, the denominations that came into being in these conservative movements have stayed out of the councils of churches formed for ecumenical purposes. Some have formed councils of their own.

Until the 1970s both fundamentalism and Pentecostalism were often viewed with some condescension by the mainstream churches, but in recent decades these twentieth-century expressions of Protestantism have burst on the scene with great power. Many of those who call themselves "conservative evangelicals" work closely with them. Taken together, these theologically conservative groups constitute the great majority of the Christian presence on television. Their work both on college campuses and among the poor has been more effective than that of the formerly mainline churches. They have organized with stunning success in the political arena. By the end of the twentieth century, they had become the most influential voice of American Protestantism, often sidelining what had been the mainline churches.

Within these communities, apocalyptic thinking has a powerful hold. Millions of Americans in these traditions believe that the end of history is coming soon. Many focus on prophesies of the return of Jews to the Holy Land and the coming battle of Armageddon. They share with extremist Zionists the goal of ridding Palestine of Arabs as part of the fulfillment of prophesy, but they expect all Jews who are not converted to Christianity to be destroyed when the end comes. The popularity of this way of thinking is shown by the fact that by the turn of the twentieth-first century, the "Left Behind" series of novels based on that apocalyptic vision had sold more than fifty million books. For those who think in this way, it can seem pointless to celebrate the natural creation or to conserve it. This kind of apocalyptic thinking has played a role in the Republican Party at leadership levels, beginning with the Reagan presidency, and emerged with considerable force within the second Bush administration.

However, the views of conservative Protestants on ecological issues are far from monolithic. Although most affirm some kind of ultimate victory of Christ on earth, numbers of them reject the apocalyptic vision outlined above and find strong reasons in the Bible to care for God's creation. Indeed, some, such as Steven Bouma-Prediger, emphasize the doctrine of God's creation of the natural world more straightforwardly than have many liberal Protestants.[72]

This realistic affirmation of creation is due to the fact that in these conservative communities few have been influenced by the philosophical anthropocentrism that has alienated mainline theologians from the natural world. Hence conservatives take the reality and wonders of the natural creation for granted. A few have even linked themselves with like-minded representatives of Catholic, Jewish, and mainline Protestant traditions in a loose coalition called the National Religious Partnership for the Environment, formed in 1993. Because many participants in the conservative evangelical movement look quite seriously to their religious leaders for practical guidance and because the number of conservative evangelicals and others influenced by fundamentalism and Pentecostalism is so large, the future role of Protestantism in relation to ecological concerns depends greatly on the direction taken by conservative leaders.

The extreme conservatives demonize feminism as goddess worship and theologies of nature as nature worship. For them these are idolatries. Thoughtful leaders recognize, however, that although within the movements there are such tendencies, taking seriously the experience of women and concern for the earth is now important for Christians and does not entail such idolatry. For them it is important to develop a Christian understanding of nature from the Bible. For this purpose they find resources in some of the writers treated above, so that the line between the mainstream discussion and that of conservative evangelicals is by no means sharp.

Loren Wilkinson began publishing on ecological issues in conservative magazines and journals in 1975. Already in 1980 he edited a volume produced by the fellows of Calvin Center for Christian Scholarship at Calvin College, entitled *Earthkeeping*.[73] This was revised and reissued in 1991 as *Earthkeeping in the Nineties*.

InterVarsity Christian Fellowship is the most theologically serious Protestant movement on many college campuses. It is theologically conservative. However, it has published repeatedly on the importance of responding to the environmental crisis. *Bent World* clearly asserted against some in the conservative camp that the crisis is real.[74] It pointed to a distinctively Christian response. Most recently, in 1997, InterVarsity Press published *Self, Earth, and Society* by systematic theologian Thomas N. Finger.[75]

It is significant that other conservative presses have published books on nature. In 1991 Zondervan published John Leax's *Standing Ground*.[76] In 1992 the Southern Baptist Broadman Press published *The Earth Is the Lord's*, edited by Richard D. Land and Louis A. Moore.[77]

Perhaps the leading theological voice from the evangelical side is Wesley Granberg-Michaelson. He has addressed the whole spectrum of Protestants, publishing in the widely respected journal *Sojourners*, beginning in 1981. He also published a series of influential books, beginning with *A Worldly Spirituality*.[78]

One of the contributors to *Earthkeeping* was Calvin de Witt. He has published important books, such as *Earthwise*.[79] However, he has made his greatest contribution through the Au Sable Institute, which relates to a network of evangelical colleges and gives their students hands-on experience of the relation of humanity to the natural world. He is also unusual among theologians of nature for having directly influenced legislation in Congress. When the Endangered Species Act came up for renewal, he persuaded a number of conservative Christian representatives to vote positively on the basis of the meaning of the biblical story of Noah's ark.

Both mainline and conservative Protestant communities, and their theologians, have much to offer the wider public today. It remains to be seen whether these heirs of the sixteenth-century Reformation will marshal their considerable theological resources and public influence to overcome both the opposition to the needed theological, ethical, and spiritual changes and the lethargy within these churches so as to respond to the global ecojustice crisis before it is too late.

NOTES

1. For an overview of Luther's and Calvin's thought about nature, and ensuing developments in the Reformation tradition, see H. Paul Santmire, *The Travail of Nature: The Ambiguous Ecological Promise of Christian Theology* (Minneapolis: Fortress, 1985), chap. 7.

2. For a discussion of the emergence of a verbal culture in Luther's time, see Margaret Miles, *Image as Insight: Visual Understanding in Western Christianity and Secular Culture* (Boston: Beacon, 1985), chap. 5.

3. Martin Luther, *Luther's Works* (ed. Helmut Lehman; Philadelphia: Fortress, 1955), 19.1.1 (quoted by Miles, *Image as Insight*, 152).

4. Ibid., 29.244 (quoted by Miles, *Image as Insight*, 95).

5. Martin Luther, *Werke* (Weimar Ausgabe), 48.48.

6. Ibid., 40.3.75–76.

7. Ibid., 47.219.

8. As Erwin Panofsky once argued in his book *Gothic Architecture and Scholasticism* (Latrobe, PA: Archabbey, 1948).

9. Luther, Weimar Ausgabe 14.29.

10. Luther, Weimar Ausgabe-*Tischreden*, no. 4774.

11. See Miles, *Image as Insight*, 104–5.

12. See ibid., 101.

13. Luther, Weimar Ausgabe 10.143.

14. Ibid., 19.496.

15. Luther, Weimar Ausgabe-*Tischreden* no. 1160.

16. Ibid.

17. See Susan E. Schreiner, *The Theater of His Glory: Nature and the Natural Order in the Thought of John Calvin* (Grand Rapids: Labyrinth, 1991); and Peter Wyatt, *Jesus Christ and Creation in the Theology of John Calvin* (Allison Park, PA: Pickwick, 1996).

18. John Calvin, *Institutes of the Christian Religion* 1.17.11 (cited by Schreiner, *Theater of His Glory*, 35).

19. Calvin, *Institutes* 1.11.1 (cited by T. H. L. Parker, *John Calvin: An Introduction to His Thought* [London: Continuum, 1995], 29).

20. Calvin, *Institutes* 1.11.12 (cited by Parker, *John Calvin*, 30).

21. Calvin, *Institutes* 1.14.1 (cited by Wyatt, *Jesus Christ and Creation*, 104).

22. Calvin, *Institutes* 1.5.12 (cited by Wyatt, *Jesus Christ and Creation*, 96).

23. Calvin, *Institutes* 3.1.2 (cited by Wyatt, *Jesus Christ and Creation*, 64 [emphasis added]).

24. Calvin, *Institutes* 1.2.2 (cited by Wyatt, *Jesus Christ and Creation*, 87 [emphasis added]).

25. Calvin, *Institutes* 3.2.31 (cited by Wyatt, *Jesus Christ and Creation*, 29 [emphasis added]).

26. Calvin, *Institutes* 1.14.20 (cited by Wyatt, *Jesus Christ and Creation*, 19 [emphasis added]).

27. Calvin, *Institutes* 1.5.11 (cited by Wyatt, *Jesus Christ and Creation*, 95–96 [emphasis added]).

28. Wyatt, *Jesus Christ and Creation*, 91.

29. Calvin, *Institutes* 3.10.2 (cited by Schreiner, *Theater of His Glory*, 121).

30. Calvin, commentary on 2 Pet. 3:5 (cited by Wyatt, *Jesus Christ and Creation*, 71).

31. Martin Luther, *Lectures on Genesis (Chapters 1–5)* (Luther's Works 1; ed. Jaroslav Pelikan; St. Louis: Concordia, 1955), 47.

32. John Calvin, *Commentaries on the First Book of Moses Called Genesis* (trans. John King; Edinburgh: Edinburgh Printing Company, 1847), 1.96.

33. Calvin, *Institutes* 1.16.5.

34. Paul Tillich, "Nature, Also, Mourns for A Lost Good," in Tillich's *The Shaking of the Foundations* (New York: Scribner, 1948).

35. Alfred Stefferud, ed., *Christians and the Good Earth* (Faith-Man-Nature Papers 1; Alexandria, VA, 1968).

36. Conrad Bonifazi, *A Theology of Things* (Philadelphia: Lippincott, 1967).

37. H. Paul Santmire, *Brother Earth: Nature, God, and Ecology in a Time of Crisis* (New York: Nelson, 1970).

38. H. Paul Santmire, *The Travail of Nature: The Ambiguous Ecological Promise of Christian Theology* (Philadelphia: Fortress, 1985).

39. Frederick Elder, *Crisis in Eden* (Nashville: Abingdon, 1970).

40. Michael Hamilton, *This Little Planet* (Washington, DC: Washington Cathedral, 1970).

41. Richard E. Sherrill, ed., *Ecology: Crisis and Vision* (Louisville: John Knox, 1971).

42. William E. Gibson, *Eco-Justice: The Unfinished Journey* (New York: State University of New York Press, 2004).

43. Dieter Hessel, ed., *Keeping and Healing the Creation* (Louisville: Committee on Social Witness Policy, Presbyterian Church U.S.A., 1989).

44. Dieter Hessel, ed., *After Nature's Revolt: Eco-Justice and Theology* (Minneapolis: Fortress, 1992).

45. Richard Cartwright Austin, *Environmental Theology* (4 vols.; Atlanta: John Knox, 1987–88/Abingdon, VA: Creekside, 1990).

46. Geoffrey R. Lilburne, *A Sense of Place: A Christian Theology of the Land* (Nashville: Abingdon, 1989).

47. Douglas John Hall, *Imagining God: Dominion as Stewardship* (Grand Rapids: Eerdmans, 1989); and idem, *The Steward: A Biblical Symbol Comes of Age* (rev. ed.; Grand Rapids: Eerdmans, 1990).

48. James Nash, *Loving Nature: Ecological Integrity and Christian Responsibility* (Nashville: Abingdon, 1991).

49. James Gustafson, *A Sense of the Divine: The Natural Environment from a Theocentric Perspective* (Cleveland: Pilgrim, 1994).

50. Larry Rassmussen, *Earth Community, Earth Ethics* (Maryknoll, NY: Orbis, 1996).

51. Andrew Linzey, *Animal Theology* (Urbana: University of Illinois Press, 1995).

52. Susan Bratten, *Six Billion and More: Human Population Regulation and Christian Ethics* (Louisville: Westminster/John Knox, 1992).

53. Max Oelschlaeger, *Caring for Creation: An Ecumenical Approach to the Environmental Crisis* (New Haven: Yale University Press 1994).

54. John B. Cobb Jr., "A Theology of Survival," in *IDOC International: American Edition* (12 Sept. 1970), including Cobb's "The Population Explosion and the Rights of the Subhuman World" (40–62).

55. John B. Cobb Jr., *Is It Too Late? A Theology of Ecology* (Beverly Hills, CA: Bruce, 1972); repr. Denton, TX: Environmental Ethics.

56. John B. Cobb Jr. and Charles Birch, *The Liberation of Life* (Cambridge: Cambridge University Press, 1981).

57. John B. Cobb Jr. and Herman Daly, *For the Common Good* (Boston: Beacon, 1989).

58. Ian Barbour, *Ethics in an Age of Technology* (Harper, 1993).

59. David R. Griffin, ed. *The Reenchantment of Science: Postmodern Proposals* (SUNY Series in Constructive Postmodern Thought; New York: State University of New York Press, 1988).

60. David R. Griffin, *Unsnarling the World Knot* (Berkeley: University of California Press, 1998).

61. Frederick Ferre, *Hellfire and Lightning Rods* (Maryknoll, NY: Orbis, 1993).

62. Charles Birch, William Eakin, and Jay B. McDaniel, eds. *Liberating Life: Contemporary Approaches to Ecological Theology* (Maryknoll, NY: Orbis, 1990).

63. Charles Birch and Lukas Vischer, eds., *Living with Animals: The Community of God's Creatures* (Geneva: World Council of Churches, 1987).

64. Catherine Keller, *Apocalypse Now and Then* (Boston: Beacon, 1996); and idem, *Face of the Deep: A Theology of Becoming* (London: Routledge, 2003).

65. Rosemary Radford Ruether, *Liberation Theology: Human Hope Confronts Christian History and American Power* (New York: Paulist Press, 1972).

66. Rosemary Radford Ruether, *Gaia and God: An Ecofeminist Theology of Earth Healing* (San Francisco: Harper, 1992).

67. For example, Sallie McFague, *The Body of God: An Ecological Theology* (Minneapolis: Fortress, 1993).

68. Jürgen Moltmann, *God in Creation: A New Theology of Creation and the Spirit of God* (San Francisco: Harper & Row, 1985).

69. Ernst Conradie, *Hope for the Earth: Vistas on a New Century* (Bellville: University of the Western Cape, 2000).

70. Marthinus L. Daneel, *African Earthkeepers*, vol. 2: *Environmental Mission and Liberation in Christian Perspective* (Pretoria: Unisa, 1999).

71. Thomas Derr, Richard Neuhaus, and James Nash, *Environmental Ethics and Christian Humanism* (Nashville: Abingdon, 1996).

72. Steven Bouma-Prediger, *For the Beauty of the Earth: A Christian Vision for Creation Care* (Grand Rapids: Baker, 2001).

73. Loren Wilkinson, *Earthkeeping: Christian Stewardship of Natural Resources* (Grand Rapids: Eerdmans, 1980).

74. Ron Elsdon, *Bent World: A Christian Response to the Environmental Crisis* (Downers Grove, IL: InterVarsity, 1980).

75. Thomas N. Finger, *Self, Earth, and Society* (Downers Grove, IL: InterVarsity, 1997).

76. John Leax, *Standing Ground: A Personal Story of Faith and Environmentalism* (Grand Rapids: Zondervan, 1991).

77. Richard D. Land and Louis A. Moore, eds., *The Earth Is the Lord's: Christians and the Environment* (Nashville: Broadman, 1992).

78. Wesley Granberg-Michaelson, *A Worldly Spirituality: The Call to Redeem Life on Earth* (San Francisco: Harper & Row, 1984).

79. Calvin de Witt, *Earthwise: A Biblical Response to Environmental Issues* (Grand Rapids: CRC, 1994).

CHAPTER 5

JAINISM AND ECOLOGY

Transformation of Tradition

CHRISTOPHER KEY CHAPPLE

THIS chapter will investigate the Jaina faith in light of its commitment to environmental values. It will begin with an overview of Jaina history and principles, with attention given to how the observance of the Jaina faith in some ways accords with the worldview of contemporary ecologists. In the latter part of the essay, I will explore the contemporary appropriation of ecological values into a rewriting of the tradition that seeks to proclaim the inherently ecofriendly nature of the Jaina faith, with select adaptations, primarily in the realm of food and consumer habits. This chapter will include my reflections on my own practice of the Jain principles as interpreted through the tradition of classical yoga.

JAINA HISTORY AND PRINCIPLES

Jainism began in the early years of Indian civilization. Greek philosopher Megasthenes documented two forms of religion in the India of 2,300 years ago. The Brahmanical variant observed numerous rituals and followed the lead of the sacrificial priest, relying upon the sacred Vedic texts. The Sramanical variant did not emphasize ritual or the Vedas, but looked to renouncers for spiritual authority and, rather than employing ritual sacrifice to obtain earthly pleasures, followed a strict moral code fundamentally rooted in the practice of nonviolence. Two strands of Sramanical faith have been documented: the Jainas and the Buddhists. The Jaina faith has been traced back as early as 800 BCE, whereas Buddhism arose at least four hundred years later and drew many of its early adherents from the community of Jainas. Both were born and originally flourished in northeast India, though a famine in 300 BCE caused the Jaina community to migrate, with many traveling to south and central India to establish the Digambara sect, and others settling in western India and establishing the Svetambara sect. During long years of separation, the two sects developed different canonical traditions, variant stories of the life of the twenty-fourth Tirthankara, and distinct views on the spiritual potential of women. On fundamentals of Jaina cosmology and ethics, the Svetambaras and Digambaras agree, and both groups assent to the principles of the faith as given in Umasvati's *Tattvartha Sutra*, composed in the mid-fifth century of the common era (Tatia 1994).

The Jainas believe that the universe is suffused with life, has existed eternally, and will continue evermore. The world consists of a countless number of living forces known as *jiva*. These souls, some tiny and some magnificently large, constantly change their shape from birth to birth, due to karmic particles that attach themselves to the soul, depending upon the nature of one's action or karma. In the history of the universe, several souls have ascended the spiritual ladder and have attained a state of transcendence; these persons, known as *siddhas*, have successfully extricated themselves from all karmic fetters. Twenty-four such remarkable beings have actively sought to help other beings rid themselves of harmful karma. These beings, known as the Tirthankaras, are worshiped by the Jainas, along with the *siddhas*, the ascetic leaders, the ascetic teachers, and the ascetics. Jainism emphasizes renunciation and renders homage to those men and women who take up the ascetic life. As we will see, to the extent that ascetic vows may be seen as ecological, Jainism may be considered an ecologically attuned faith.

The first of the Tirthankaras, Rishibha, is credited with the creation of the civilized world. He established marriage, became the first king, established rules of justice, and invented "farming, fire-making, the fashioning of utensils, the cooking of food, and so on" (Babb 1996: 43). Rishibha appeared within a vast "timescape," during the downward cycle of a small space within beginningless infinity. He was the first person after his own mother to achieve liberation or *mukti*. He became the first Tirthankara or teacher of the tradition and has served as the spiritual exemplar par excellence ever since.

Twenty-three other Tirthankars have succeeded Rishibha. The most recent ones, Parsvanath and Mahavira, lived within historical memory of the present age. Lineage charts indicate that Parsvanath lived perhaps as early as 800 or 700 BCE.

Mahavira was a contemporary of the Buddha, placing him at approximately 500 or 400 BCE. These teachers emphasized a moral universe that valued individual life-forms in all their various manifestations.

The cosmology and ethics of Jainism rest upon a worldview that sees the reality suffused with countless living souls or *jivas*. These *jivas* take birth again and again in different forms. According to the tradition, one may take human birth seven or eight times in a row, but must then enter one of the other conditions of existence (*gati*): hell-dwellers (*naraki*), animals and plants (*tiryanc*), and deities (*dev*). In Jain cosmology, even those items generally considered inert and inanimate are said to contain life: earth bodies that dwell in soil and rocks, water bodies that live in water, fire bodies found in fire and electricity, and air bodies that float in the air. In addition, within the plant category, one could be reborn as a microorganism (*nigod*), "little bubble-like clusters that fill the entirety of the space of the cosmos" (Babb 1996: 45). Life-forms are further grouped according to the number of senses they possess. The elemental and plant life-forms possess only the sense of touch. Worms add taste; bugs add a sense of smell. Four-sensed animals such as moths and flies add the sense of sight, while five-sensed beings, which include most mammals, reptiles, and birds, add the sense of hearing.

As one moves through life, harm to any of the life-forms listed above results in a thickening of the veil of karma that shrouds the vibrant nature of the soul. The goal of Jainism is to release the eight inhibiting forms of karma: knowledge-covering, intuition-covering, sensation-producing, deluding, lifespan-determining, body-making, status-determining, and obstructive (Tatia 1994: 33). Due to worldly activities, these karmas shroud and obscure the soul. Their self-perpetuation results in further inflows of karma. The reverse of this binding flow begins through the adoption of a series of vows or *vratas* that restrain one's behavior and allow for the release (*nirjara*) of karmic accretions. These five vows define Jaina practice and also were taken up in the exact same order in Patanjali's *Yoga Sutra*, the Urtext of religious practice in India.

The first of the five vows, nonviolence or *ahimsa*, is said to contain the key to advancement along the spiritual path (*sreni*). This requires abstaining from harm to any being that possesses more than one sense, requiring a strict vegetarian diet. The second vow, truthfulness or *satya*, mandates honesty in speech and action. The third vow, not stealing or *asteya*, urges Jains not only to not take more than is offered, but also not to take more than is needed. The fourth vow, sexual restraint or *brahmacarya*, allows one to avoid harm to other human beings who might be hurt physically or emotionally by a sexual encounter and also spares harm to the microorganisms that congregate in the private regions of the body. The fifth vow, nonpossession or *aparigraha*, requires that one own only the bare necessities of life.

In the contemporary practice of Jainism, as found in the Terapanth mendicant order of the Svetambara branch, additional rules serve to undergird and update these five practices. For instance, a list of rules promulgated by Acarya Tulsi on 18 January 1991 includes the following accommodations to modern life:

- Generally, one should not use a lift for up to three storeys (Fluegel 2003: 19).
- One should not keep more than the prescribed limit of bedding and covering cloths. The prescribed limit is of the following types: five overclothes, two underclothes, three bodices, two uniforms, one shawl, one woolen shawl, two blankets, one wrapping cloth, three small bodices, two handkerchiefs, two mouthmasks, one towel, two glasses, two ball point pens, two pencils, two toothbrushes. One should not keep more than three bowls. One should not eat from bowls made of metal. A bowl must be made of plastic, wood, or clay. (Metal is considered too valuable and requires violence for its production.) (Fluegel 2003: 20).
- One should not watch TV. If some householder shows a community program merely for information, this is another matter. One should not keep clothes, books, etc., in a closed box for a long time. One should not use more than one plastic bag at the time of begging (Fluegel 2003: 22).

This list of rules, excerpted from a total list of ninety, demonstrates the seriousness with which the Jain community regards practices designed to wear away and disperse one's karma. Acarya Tulsi developed a list of special vows to be observed by laypersons in 1945. These included as the fifteenth vow "not to pollute," indicating perhaps that Acarya Tulsi was ahead of his time. During a private interview in 1989, he indicated that his lifestyle and that of his monks and nuns presented the best alternative to resource-intensive consumerism. With just one change of clothes, miniature books that can be viewed with only a magnifying glass, and minimal eating utensils, he indicated that the abstemious practice of his community rendered negligible harm to the environment (Chapple 1993).

Contemporary Jain Environmentalism

From an environmental point of view, adherence to these vows leads to a lifestyle that would rival (and probably excel) that followed by the deepest of deep ecologists. The Jain community itself is quite self-conscious of the ecological implications of their core teachings, and young Jains have been quite enthusiastic about making this explicit connection between their ancient faith and this modern issue. As noted by Anne Vallely, the diaspora community of Jains has developed an identity that in some ways revises the traditional Jain views on renunciation and spiritual practice. She writes that "the teachings of compassion and nonviolence are no longer anchored to a renunciatory worldview. Jain teachings are being redefined according to a different ethical charter altogether—one in which active engagement in the world is encouraged" (quoted in Chapple 1993: 206). In traditional Jainism, the most rigorous vows, such as those listed above, would be followed by monastic men and women. Laypersons also would follow strict dietary rules that increase with age, would observe periodic fasts, and would seek employment in only limited fields that minimize violence. The purpose of these activities would be for the purification of one's own soul, with little if any purposeful direct concern for societal improvement. However, the new generation has adopted a rhetoric that makes the incidental wholesome aspects of Jain practice central to the debate, in what Vallely calls an emerging sociocentric rationale for the practice of the faith. Environmentalism and animal rights stand at the top of the list.

A revision of the traditional Jain practice of dietary nonviolence has been championed by Pravin K. Shah, an influential educator within the American Jain community. Shah, visiting a dairy farm in Vermont, became appalled at the poor treatment of cattle in contemporary agricultural practice. He also personally struggled with an elevated cholesterol count. In seeking to remedy both ills, he posted an influential essay to the internet urging the Jain community to adopt a vegan diet. He notes that in order to maximize milk production cows are kept continually pregnant and that many hormones and drugs are used. Additionally, mother cows are killed after producing milk for four years, and approximately three-fourths of their calves are killed within six months to produce veal and to reduce the overall number of cattle (Shah 2005: 30). Shah also criticizes the harmful effects of dairy farming, including the waste generated by the slaughter process (230,000 pounds per second), the greenhouse gases produced by the world's cattle herds (100 million tons of methane per year), the excessive amount of water required for livestock (2,500 gallons per pound of meat, compared with 60 gallons for a pound of potatoes), and the vast amount of land given over to grazing (Shah 2005: 31). Shah proclaims that "Jainism in Action is an ecofriendly religion which preserves and protects the Earth and Environment, respects the lives of animals, birds, fish and other beings, and promotes the welfare of society through the application of its primary tenets of Ahimsa and Non-possessiveness" (Shah 2005: 31). This modernization of the tradition combines good citizenship with a romantic view of nature, neither of which originate from the tradition itself, but can be seen as presenting a good rationale for becoming vegan.

Another example of a contemporary Jain-influenced environmentalist can be found in Satish Kumar. Kumar became a Jain monk at the tender age of nine and remained within his order until the age of eighteen, when he left the Terapanth Svetambara monks to purse a life dedicated to social change. He worked with the land distribution movement of Vinobha Bhave and then turned to antinuclear activism, walking from Delhi through Pakistan and Afghanistan to the heart of the

Soviet Union, pleading with officials in Moscow to end the production of nuclear weapons. He then walked to Paris and from there spoke with government officials in London and Washington, DC, canvassing the four extant nuclear powers in the 1960s, in an attempt to bring conscience to bear upon the blind building of horrific stockpiles of weapons (see Kumar 1993). Eventually, Kumar settled in Britain in 1991 and helped establish Schumacher College. For more than a dozen years, this Gandhian-inspired center for graduate studies has conducted programs focusing on environmental sustainability.

In an essay titled "Jain Ecology," Satish Kumar interprets the five traditional vows (*vratas*) in light of an ecological application. He writes that *ahimsa* (non-violence) "means avoiding contact with scenes of cruelty and refraining from activities that cause pain and disharmony." He states that "living in truth [*satya*] means that we avoid manipulating people or nature." Not stealing (*asteya*), for Kumar, "means refraining from acquiring goods or services beyond our essential needs.... If you take more from nature than meets your essential need, you are stealing from nature." He states that sexual restraint (*brahmacarya*) "not only recognizes the dignity of the human body but also the body of nature" and that nonpossession (*aparigraha*) allows one to become free "from nonessential acquisitions and from materialism" (in Chapple 2002: 188–89). This interpretation of Jain religious principles in light of environmental concerns reveals what appears to be a natural extension of the tradition. However, the actual application of these principles requires a complex process of decision-making. For instance, Mahatma Gandhi, an advocate of mercy killing of animals in cases of extreme pain, came under severe criticism from the Jain community, who claimed that the violence required to end the life of an animal far outweighed the karma of allowing nature to take its course.

In his work at Schumacher College, Satish Kumar has foregrounded the contemporary rhetoric of environmentalism, though his inspiration draws deeply (if quietly) from his Jain roots. The mission statement expresses a need to bring about social change through a deeper connection with oneself and one's relationship with the earth:

> Schumacher College was founded in 1991 on the conviction that a new vision is needed for human society and its relationship with the earth.... The College offers rigorous inquiry to uncover the roots of the prevailing world view; it explores ecological approaches that value holistic rather than reductionistic perspectives and spiritual rather than consumerists values.... A unified residential education offering physical work, mediation, aesthetic experience and intellectual inquiry creates a sense of the wholeness of life. (Schumacher College, Sept. 2005–July 2006 course program)

The courses offered by the college, generally in a three-week format, include workshops on book-making, art and ecology, sustainable development, food issues, the new science, and other topics. Students come to Schumacher from all over the world and learn indirectly about Jain principles. Additionally, as editor of *Resurgence*, an award-winning "international forum for ecological and spiritual thinking" established in 1966, Kumar has recruited brilliant essayists, from Fritjof Capra to Prince Charles, to enter into the public discussion of issues of ecological concern. The journal, which appears six times per year, includes stunning illustrations and photography. It tackles environmental issues from scientific, religious, literary, philosophical, agricultural, and social perspectives.

L. M. Singhvi, a prominent jurist and onetime member of India's Supreme Court and its Parliament, published a booklet titled *The Jain Declaration on Nature*, which subsequently appeared as an appendix to *Jainism and Ecology*, a volume in the Harvard series on ecology and religion. This declaration, which has often been cited within the contemporary Jain community worldwide, discusses Jain principles in an ecofriendly light, updating and revising the tradition. Rather than being solely concerned with the status of the karma obscuring one's own soul, Singhvi's approach places a positive, almost worldly value on Jain teachings. He emphasizes the centrality of nonviolence and goes on to point out that beings rely upon one another for survival. He writes that "all aspects of nature belong together and are bound in a physical as well as a metaphysical relationship. Life is viewed as a gift of togetherness, accommodation and assistance in a universe teeming with interdependent constituents" (quoted in Chapple 2002: 219). Rather than denigrating the world, Singhvi celebrates its inner connections. He reinterprets the multifaceted philosophies of Jainism (*anekanta, syadvada*) to mean that "Jainism does not look upon the universe from an anthropocentric, ethnocentric, or egocentric viewpoint. It takes into account the viewpoints of other species, other communities, and nations and other human beings" (in Chapple 2002: 220). He goes on to emphasize compassion and charity, interpreting the five vows of Jainism in an ecological light.

CLASSICAL YOGA, JAIN VOWS, AND ECOLOGY

The classical yoga tradition as articulated by Patanjali in the *Yoga Sutra* (ca. 200 CE) includes as its foundational practice the application of the same five vows cited earlier in the earliest extant Jain text, the *Acaranga Sutra*. The *Acaranga Sutra* states that "all beings are fond of life, they like survival; life is dear to all" (trans. Nathmal Tatia in Chapple 2002: 6). To preserve life, this text advocates the application of what Nathmal Tatia translates as "abstinence from violence, falsehood, stealing, carnality, and possessiveness" (in Chapple 2002: 8). Many Jain thinkers have interpreted these vows in light of their application to environmental principles, including Satish Kumar, as noted above.

For several years, from 1972 to 1985, my wife and I trained in classical yoga at Yoga Anand Ashram in Amityville, New York. This process included weekly practice of the five vows (*yama* or *vrata*). In the context of then-current events such as Three Mile Island and Chernobyl, these were often viewed through an environmentalist lens. The founder of Yoga Anand Ashram, Gurani Anjali (1935–2001), was deeply interested in ecological issues and designed homework and special activities for her students that brought the community in touch with the seasons and the elements, as well as with the body and the senses. I will share some reflections on this aspect of my practice of yoga and include excerpts from journal entries from that time.

Each Monday night we were given a new challenge, a tool in the form of an ethical discipline that would shape and guide our week. For instance, if our homework for the week was nonviolence (*ahimsa*) we would avoid harm in various ways. Some students would try to rescue stranded earthworms squirming on the rainy sidewalk. We all would practice strict vegetarianism, seeking out cookies that were not made with eggs. We would keep silent one day and fast one day, minimizing the potential to do harm through speech and through eating. We were encouraged to read sacred texts and reflect on the life and words of nonviolent leaders such as Mahatma Gandhi and Martin Luther King Jr. Because our practice coincided with the oil crisis of the early 1970s, we would try to minimize our driving and carpool together when possible. The connections with environmentalism seemed obvious: by minimizing harm to ourselves through introspection, careful food habits, and careful speech, less damage was done to the earth.

The practice of nonviolence, the first of the vows, was complemented with the practice of nonpossession (*aparigraha*), the fifth vow. A journal excerpt (dated 6 April 1980) follows on this discipline:

> *Aparigraha* means not holding on, not possessing, not hoarding, not appropriating, not collecting or amassing. It is a negative term with an implicit positive meaning. By negating the obvious, the subtle becomes explicit. By eliminating attachment, a freedom arises, a knowledge that life is beyond or more than the accumulation of material goods.
>
> The dictionary definition of *aparigraha* is "no including; non-acceptance, renouncing (of any possession of ascetics); deprivation, destitution, poverty; destitute of possession; destitute of attendants or of a wife." It is the simple negation of the term *parigraha*, which includes the following definitions: "laying hold on all sides, surrounding, enclosing, fencing round; wrapping around, putting on (a dress, etc.); summing up . . . taking, accepting, receiving . . . a gift or present; getting, attaining, acquisition, possession, property." Thus, *aparigraha* is the opposite of all these, a letting go, a release, an "undressing," etc. Whatever

can be held or embraced can also be given up; in the giving up, there is a liberation from the fetters which bound one to the object held.

According to the *Yoga Sutra* of Patanjali, the perfection of *aparigraha* results in an understanding of how things come to be: "*Aparigraha sthairye janma kathamta-sam bodhah*" (2.39). Woods gives the following translation: "As soon as he is established in abstinence-from-acceptance-of-gifts, a thorough illumination [arises] upon the conditions of birth." Thus, by the giving up of objects, the nature of objects is revealed.

Vyasa gives the following definition of *aparigraha*, as quoted by Eliade in *Yoga: Immortality and Freedom*, 50: "Absence of avarice [*aparigraha*] is the non-appropriation of things that do not belong to one and it is a consequence of one's comprehension of the sin that consists in being attached to possessions, and of the harm produced by the accumulation, preservation, or destruction of possessions."

Thus, a twofold process is discerned. First, an individual ascertains that possessions involve a suffering in the recognition that they are perishable. Upon the renunciation of possessions, the second dimension is revealed: the renouncer is shown the nature of existence itself: "He receives the answers to the question: Who was I? How was I? On what [can] this birth be? Or how [can] this [birth] be? Or how shall we become?" (Vyasa translated in Woods 2.39, 187).

Thus the nature of life itself is revealed to the practitioner. In effect, by stripping away the "fluff" of life, the cloud of possessions which generally surrounds the common person, the essence of life is revealed. By eliminating or reducing the physical commodities needed to maintain the body, the body itself takes on a new importance and commands a new respect. In the process of giving up, an understanding arises as to why the previous possession is not needed and an appreciation and understanding arises for what remains.

Our training in yoga entailed a conscious reflection on the ethical and ecological implications of our lifestyle. In a variation on monastic living, we would periodically be asked to make lists of our possessions, to give away something each day, to refrain from eating certain foods, and to occasionally abandon all utilization of electricity, including lightbulbs. In one particularly interesting exercise, we were asked to hand wash our clothes at the end of each day for a week. This made us particularly mindful of the amount of water consumed in what otherwise would be a mechanized process.

Concentration on the Elements and Senses as Vehicle for Ecoawareness

Our Ashram training included concentration exercises on the elements and the senses. This training took place over a period of several months. In addition to practicing one of the vows listed above, we were asked to sit for twenty minutes in the morning and in the evening, gazing upon one element for several weeks, and eventually turning inward to understand the operation of the senses. We discovered that by concentrating on the earth (*prthivi*) one gains a sense of groundedness and a heightened sense of fragrance. By reflecting on water (*jal*), one develops familiarity with fluidity and sensitivity to the vehicle of taste. Through attention to light (*tejas*) and heat (*agni*), one arrives at a deep appreciation for the ability to see. Awareness of the breath (*prana*) and wind (*vayu*) brings a sense of quiet and tactile receptivity. All these specific manifestations occur within the context of space (*akasha*), the womb or container of all that can be perceived or heard. Through these exercises, one progressively works at smelling earth with the nose and excreting solid waste; tasting water with the mouth and eliminating fluids from the body; seeing radiance and color with the eyes and warming our work through the action of our hands; feeling the caress of the air upon our skin and boldly walking in the breeze with our legs; hearing our way within space and filling space with our voice. Recognition of and reflection on the elements and the senses became a gateway into higher ecological awareness. The hills and forests and beaches and bays and open ocean of Long Island took on a Whitmanesque clarity. A connection was made that brought us into an intimacy with the place, with the season, and with our thoughts and senses. Although these practices that focus on the great elements and the senses and the body are generally associated with the Samkhya school of Indian thought, these same categories or *tattvas* can be found in Jain philosophy.

WATER IN JAINISM

Water provides an example of how the elements, so foundational to the Samkhya system, also play a fundamental role in Jainism. The Jaina tradition, noted for its rigorous observance of nonviolence (*ahimsa*), brings unique perspectives to the same five elements revered in all Indic religious systems. The important role of water in Jaina cosmology was first expressed in written form in the *Acaranga Sutra* (ca. 400 BCE) and later in the *Tattvartha Sutra* of Umasvati (ca. 400 CE) and the medieval biological texts such as Santi Suri's *Jiva Vicara Prakaranam* (ca. 1000 CE). According to the earliest texts, water, like the other three great elements of earth, fire, and air, possesses consciousness and the sense of touch. In other words, according to Jainism, when we touch water, reciprocally, water (or any other element) feels the intensity of our contact, the warmth of our hand, the squeeze of our esophagus. We need to treat all elements gingerly to avoid causing pain or harm.

The *Jiva Vicara Prakaranam*, a text that explicates the various forms taken by the innumerable discrete souls or *jivas* that pervade the universe, lists seven varieties of water: subsoil or groundwater, rainwater, dew, ice, hail, water drops on green vegetables, and mist. The commentator goes on to state that groundwater comes from wells, that sky-water is rainwater, and that the oceans consist of "dense water spreading all over the earth" (Suri 1950: 26). A further elaboration is given regarding qualities attributable to these forms of water: pure, cold, hot, alkaline, slightly acid, acid, salt, "water with the taste of wine," milky water, "water having the taste of ghee (melted butter)," sweet water, and nectar (Suri 1950: 27). The text goes on to state techniques of purifying water by boiling to remove parasites, defined as 36,450 possible specific two-sensed living beings who might inhabit the water, with specific time lengths for boiling given according to season. This text specifies that one must not consume snow, ice, or ice cubes, presumably because of the harm that would befall the solidified water as it becomes mashed by our teeth or compressed by our tongue and palate. It also states that a soul (*jiva*) in the form of a water body can live for a maximum period of 7,000 years before taking another form. As a one-sensed being, all water bodies are said to possess the sense of touch, to engage in a process of breathing, to possess life and bodily strength (Suri 1950: 28, 57, 81, 143, 163). As a final flourish, the *Jiva Vicara Prakaranam* states that water arises from 700,000 different points of origin (Suri 1950: 166).

This extensive and detailed treatment of water indicates the care and attention given by Jainism to water. Water is used extensively for ritual purposes in Jainism, particularly the bathing of the Jain image, and it is used sparingly for purposes of personal hygiene, particularly for members of religious orders, who bathe infrequently and with a minimal amount of water. Respect for all forms of life, in the Jaina tradition, includes respect for the elements themselves.

Several years ago, reflecting upon the connection between the human body and the natural order, I wrote the following:

> Without the sense of smell, there could be no crisp scent of autumn leaves, no gentle wafting of the ocean breeze. Without the sense of taste, there could be no soothing drink of water or startling spiciness. Without the eyes and the light that bathes them, there could be no color or form. Without the garment of skin that cloaks our bodies, there could be no caress of human contact. Without the ears, no sound could be heard or sent forth as communication with others. In the most intimate of ways, the world cannot exist without the body in which we dwell.
> (Chapple in Tobias 1991: 146)

By engaging in a regular spiritual practice that includes acknowledgement of the elements, a greater sensitivity may be cultivated toward the natural world. Enhanced consciousness of water is essential in the process of environmental healing. Meditation practices on water, from any tradition, help bring one out of what Thomas Berry calls the technological trance. As our intimacy with water increases, our ability to be informed about and responsive to such issues as the waste and pollution and privatization of water will be enhanced. Reflecting on the intimacy of our body with the elements can help deliver us from the prevailing and increasingly globalized autism that humans experience in regard to nature.

CONCLUSION

Jainism emphasizes the living nature of the world and the importance of purifying oneself through respecting the many life-forms that exist. At the time of its inception and until recent decades, the protection of ecology and environment did not play a significant role in the Jain way of life beyond a care to minimize any harm that one might inflict on others that would harm one's own soul. In contemporary times, Jainism has been reinterpreted as enhancing human-earth relations. The precepts of Jainism, particularly nonviolence and nonpossession, generally associated in their most rigorous forms with monastic men and women, have now been refigured and recast in an ecological mold.

REFERENCES

Babb, Lawrence A. 1996. *Absent Lord: Ascetics and Kings in a Jain Ritual Culture.* Berkeley: University of California Press.

Berry, Thomas. 1988. *The Dream of the Earth.* San Francisco: Sierra Club.

Chapple, Christopher Key. 1993. *Nonviolence to Animals, Earth, and Self in Asian Traditions.* Albany: State University of New York Press.

———. 1998. "Jainism and Nonviolence." In *Subverting Hatred: The Challenge of Nonviolence in Religious Traditions.* Edited by Daniel L. Smith-Christopher. Cambridge: Boston Research Center for the 21st Century.

———. 2001. "The Living Cosmos of Jainism: A Traditional Science Gounded in Environmental Ethics." *Daedalus* 120.4:207–24.

———, ed. 2002. *Jainism and Ecology: Nonviolence in the Web of Life.* Cambridge: Center for the Study of World Religions, Harvard Divinity School.

Cort, John, ed. 1998. *Open Boundaries: Jain Communities and Cultures in Indian History.* Albany: State University of New York Press.

Dundas, Paul. 1992. *The Jains.* London: Routledge.

Fluegel, Peter. 2003. "The Codes of Conduct of the Terapanth Saman Order." *South Asia Research* 23.1.

Jaina Sutras. 1968. Part 1: *The Akaranga Sutra; the Kalpa Sutra.* Translated by Hermann Jacobi. Reprint New York: Dover (orig. 1884).

Jaini, Padmanabh S. 1989. *The Jaina Path of Purification*. Berkeley: University of California Press.

——. 1993. *Gender and Salvation*. Berkeley: University of California Press.

——. 2000. "Fear of Food: Jaina Attitudes on Eating." Pp. 281–96 in *Collected Papers on Jaina Studies*. Delhi: Motilal Banarsidass.

Kumar, Satish. 1993. *No Destination: An Autobiography*. Totnes: Green Books.

——. 2002. *You Are, Therefore I Am: A Declaration of Dependence*. Totnes: Green Books.

Matthews, Clifford N., Mary Evelyn Tucker, and Philip Hefner. 2002. *What Science and Religion Tell Us about the Story of the Universe and Our Place in It*. Chicago: Open Court.

Saletore, Bhaskar Anand. 1938. *Medieval Jainism*. Bombay: Karnatak.

Shah, Pravin K. 2005. "Jainism and Environment." *Jain Digest* (spring).

Settar, S. 1990. *Pursuing Death*. Dharwad: Institute of Indian Art History, Karnatak University.

Suri, Santi. 1950. *Jiva Vicara Prakaranam along with Pathaka Ratnakara's Commentary*. Edited by Muni Ratna-Prabha Vijaya. Translated by Jayant P. Thaker. Madras: Jain Mission Society.

Tatia, Nathmal, trans. 1994. *Tattvartha Sutra: That Which Is: A Classic Jain Manual for Understanding the True Nature of Reality*. San Francisco: HarperCollins.

Tobias, Michael. 1989. *Ahimsa*. JMT Productions. Public Broadcasting Corporation.

——. 1991. *Life Force: The World of Jainism*. Berkeley: Asian Humanities Press.

CHAPTER 6

HINDU RELIGION AND ENVIRONMENTAL WELL-BEING

O. P. DWIVEDI

THE CONTEXT

EARTH is the only planet which supports all known forms of life. It did so for millennia by maintaining a natural balance between all biotic and abiotic organisms, which sustained each other and released energy which formed a cycle of regeneration and reproduction. Various agents, such as producers, consumers, and decomposers, performed their task in unison and complete harmony. The ecosystem used the matter of earth and solar energy to produce living substance and also to continuously cycle matter to support it. However, when human beings appeared on earth, the biosphere started to be influenced by their presence. For three million years or so, they lived in a reasonable balance with the organisms around them while considering the entire creation to be a divine "adventure." According to *Taittariya Upanishad*, creation, then, became an offspring of Supreme Power (*parmatman*) because it was He who created the entire universe and entered into it. *Isavasyopanishad* gives the following overall perspective of this feature: *Isavasyamidam sarvam yatkinchat jagtyaam jagat* ("by the lord enveloped must this all be whatever moving thing there is in the moving world").

For centuries, people in India believed in and acted on this concept (the world around them was divinified); and as such environmental damage was either small or manageable. People utilized plants and other natural products; but their needs were minimal, and they also exercised great care in using natural wealth, thereby maintaining natural balance among species. However, since the advent of industrialization and the triumph of science and technology, we humans have acted as masters; and in the process we have created ecological imbalance, environmental degradation, fast-vanishing flora and fauna, and deteriorating human health. Humanity faces these and many other associated environmental problems today.

Since the late 1980s, not only has there been a growing awareness among the people in India about the ecological challenges facing their society but also the country has instituted a variety of regulatory and administrative mechanisms to deal with the problems of pollution and environmental conservation. However, the mounting pressures of population, expanding urbanization, and growing poverty have led to the ecologically unsustainable exploitation of natural resources, which continues to threaten the fragile balance between ecology and people in India. Environmental challenges, which may appear insurmountable, can be addressed by combining indigenous and technological solutions and impressive religious traditions. The task is no doubt stupendous, for the hitherto unimaginative developmental strategies and largely imitative developmental schemes would have to be replaced by environmentally sound sustainable development. It is here where ancient wisdom depicted through the "Hymn to Mother Earth," practical use of such wisdom through Chipko Apiko, and the Gandhian way of self-reliant life could be of a great use. It can be a kind of environmental *sarvodaya*, respecting the extended family of Mother Earth while being least wasteful in using natural resources. To live in harmony with nature, as Hindu sages used to do, both modern technology and environmental sarvodaya are needed.

During the past three decades, it has been amply demonstrated that if environmental problems are to be solved, then a change in the way that individuals think about and interact with nature must occur. It is for this reason that a society's cultural and spiritual underpinnings of environmental stewardship can be a solid source of strength as well as a benefit to that society. One does not have to go too far to locate such underpinnings, because each society's spiritual heritage can be used to provide new ways of valuing, thinking, and acting that are necessary to nurture the respect for nature and to be prepared to avert future ecological disasters. This essay will examine the Hindu concept of divinity being present in creation and as such exhortations for Hindus to treat nature with respect, the concept of an extended family of Mother Earth, our dharma and karma to the environment, environmental challenges facing Hindus and India, various interpretative explorations and general propositions, and concluding observations.

IN THE CREATION, GOD IS PRESENT EVERYWHERE: ISHWARAH SARVABHUTANAM

Hindus contemplate divinity as the one in many and many in one. The concept of "God is one and is everywhere present" is found in Vedas. For example, *Rig Veda* says: "He is the one God, producing heaven and earth; wields there together with His veins and wings" (10.81.3).

In *Yajur Veda* (32.10), He is described as "our Friend, our Father, our Creator; who knows all positions, and all existing things." Furthermore, in *Gita* 13.13 the Lord Krishna says: *Sarvam avritya tishthati* ("he resides in everywhere"). Thus, it is not surprising that Hindus are enjoined to respect all elements of creation, as stated in *Srimad Bhagavata Mahapurana* (11.2.41): "Ether, air, fire, water, earth, planets, all creatures, directions, trees and plants, rivers, and seas, they all are organs of God's body; remembering this, a devotee respects all species." This is further elaborated in *Gita* 5.18, where it is stated that a learned person is the one who treats a cow, an elephant, a dog, and an outcaste with the same respect that he will show to the Brahman endowed with a great learning. Thus, the foundation of all this is the prescription *vasudevah sarvam iti* ("see the presence of God in all, and treat all species with respect"). *Srimad Bhagavata Mahapurana* (11.2.45) confirms this fundamental principle in the following verse: *Sarvabhuteshu yah pashyed bhagvadbhaavamaatmanh* ("my devotee is the one who sees in all creation my presence"). Furthermore, the lord says to Arjuna (*Gita* 10.32): "I am the beginning, the middle and the end of all creations [*sargaanaamaa-dirantashcha madhyam chaiv aham arjuna*]." It is this veneration, respect, and acceptance of the presence of divinity in nature which is expected from Hindus in order to maintain and protect the natural harmonious relationship between human beings and nature (Dwivedi 1997b). Thus, for Hindus, both God and *prakriti* (nature) are one and the same.

In *Shvetashvatara Upanishad*, it is stated (6.10): *Eko devah sarvabhuteshu gudhah sarvavyaapee sarvabhutaantaraatma; karmadhyakchshah sarvabhutaadhivaasah saakchshi chetaah kevalo nirgunashcha* ("there is only one God, present in all beings, resides in the soul of everyone, apportions Karma to all, observes what everyone does, exists as a pure, and cannot be characterized by any attributes [devised by humans]"). Furthermore, the Hindu belief in *punarjanma* (the cycle of birth and rebirth), wherein a person may come back as an animal or a bird, means that the Hindus give other species not only respect, but also reverence. This provides a solid foundation for the doctrine of *ahimsa*—nonviolence (or non-injury) against other species and human beings alike. It should be noted that the doctrine of *ahimsa* presupposes the doctrines of karma and rebirth (*punarjanma*). The soul continues to take birth in different life-forms, such as birds, fish, animals, and humans. Based on this belief, there is a profound opposition in the Hindu religion (as well as in Buddhist and Jain religions) to the institutionalized breeding and killing of animals, birds, and fish for human consumption. The pain that a human being causes other living beings to suffer will eventually be suffered by that person later, either in this life or even in a later birth. It is through the transmigration of the soul that a link has been provided between the lowliest forms of life and human beings. From the perspective of Hindu religion, the abuse and exploitation of nature for selfish gain is considered unjust and sacrilegious.

The Hindu concept of divine being as the one omnipresent, pervading in all species (and thereby sanctifying all forms of life), is beautifully expressed in *Yajur Veda* 32.8. From these, one sees how this entire universe and every object in it (which has been created by the same supreme God) is meant for the use and benefit of all species. Human beings therefore may enjoy such benefits, but only by forming a close relationship with other species and by not encroaching upon the space of others. Thus, for the Hindus, both God and nature (*prakriti*) are to be one and the same. The lord is the creator of sky, the earth, oceans, and all other species, He is also their protector and eventual destroyer. He is the only lord of creation. Human beings have no special privilege or authority over other creatures; on the other hand, they have more obligations and duties (Dwivedi, Tiwari, and Tripathi 1984).

TREATMENT OF ANIMALS AND BIRDS

Almost all the Hindu scriptures place a strong emphasis on the notion that God's grace cannot be received by killing animals or harming other creatures. That is why not eating meat is considered both appropriate conduct and one's dharma. For example, as mentioned in *Vishnu Purana* 3.8.15, "God, Kesava, is pleased with a person who does not harm or destroy other non-speaking creatures or animals." Even when a human being causes pain and suffering to other living beings, that person will eventually suffer similar pain either in this life or in a later rebirth. This is a karmic justice which is meted out through the transmigration of the soul, a link between the lowliest forms of life with the human beings. The Hindu code *Manusmriti* warns that "a person who kills an animal for meat will die of a violent death as many times as there are hairs of that killed animal" (5.38). Even animal sacrifices were prohibited, as narrated in *Mahabharata*, when Rishis and Gods debated on the merits of offering grain or the sacrificial lamb (goat) in Yajna; the Rishis insisted that according to Vedas, the sacrificial material ought to be the grain only, and thus no animal must be killed for the purpose of Yajna

(*Mahabharata, Shantiparva, Mokshadharma* 337.4–5). Rishis also mentioned that a true dharma never sanctions the killing of any animal.

The most important aspect of Hindu religion pertaining to the treatment of the animal and plant kingdoms is the belief that the supreme being was incarnated in the form of various species. The lord says: "This form is the source and indestructible seed of multifarious incarnations within the universe, and from the particle and portion of this form, different living entities, like demigods, animals, human beings and others, are created" (*Srimad Bhagavata* 1.3.5). For example, among the various incarnations of God (numbering from ten to twenty-four depending upon the source of the text), he first incarnated himself in the form of fish, then a tortoise, a boar, and a dwarf. His fifth incarnation was as a man-lion. As Rama he was closely associated with monkeys, and as Krishna he was always surrounded by the cows. In *Srimad Bhagavata*, this aspect is beautifully stated (1.3.5): "This form is the source and indestructible seed of multifarious incarnations within the universe, and from the particle and portion of this form, different living entities, like demigods, animals, human beings and others, are created."

Srimad Bhagavata further states: "One should look upon deer, camels, monkeys, donkeys, rats, reptiles, birds, and flies as though they were their own children" (7.14.9). In addition to this, many animals and birds are revered by Hindus, as these are consorts of various gods, goddesses, and incarnations. In India, Nepal, and Bhutan, there are many sacred forests and groves where green trees and plants are protected. Also, for Hindus, the planting of a tree is still a religious duty. Thus sanctity has been attached to many trees and plants. Through such exhortations and various writings, the Hindu religion has provided a system of moral guidelines toward environmental preservation and conservation. Furthermore, a main premise of Hindu religion is that every entity and living organism is a part of one large extended family system (*kutumba*), which is presided over by the eternal Mother Earth, called *vasudhaiv kutumbakam*.

The Hindu Concept of
Vasudhaiv Kutumbakam:
The Family of Mother Earth

The development of humanity from creation till now has taken place nowhere else but on the earth. Our relationship with earth, from birth to death, is like children and their mother. The mother—in this case earth—not only bears her children but also has been the main source of fulfillment of their unending desires. It is this earth which provides energy for the sustenance of all species. And as one ought not to insult, exploit, and violate one's mother but to be kind and respectful to her, one should behave similarly toward Mother Earth. This relationship is superbly depicted by Rishi Atharva in the *Prithivi Sükta* ("Hymn to Mother Earth") of the *Atharva Veda*.

The *Atharva Veda*, written about or before 1500 BCE is perhaps the first of its kind of scripture in any spiritual tradition where the concept of respect to earth has been propounded. An entire chapter (kanda 12, hymn 1) consisting of sixty-three verses (mantras) was devoted in praise of Mother Earth. These verses integrate much of the thoughts of Hindu seers concerning our existence on earth. A series of verses follow, addressed to Mother Earth, evoking her benevolence. Mother Earth is seen as an abode of a large and extended family (*kutumbakam*) of all beings (humans and others alike):

> *Giryaste parwatà hima vantoranyam te prithvi syonamastu, Babhrum krishnam*
> *   rohïnïm vishvarüpàm dhruvam bhümim prithvï-mindraguptàm,*
> *Ajïtohato akshatodhyasthham prithvïmaham.*

> O Mother Earth! Sacred are your hills, snowy mountains, and deep forests. Be kind to us and bestow upon us happiness. May you be fertile, arable, and nourisher of all. May you continue to support all races and nations. May you protect us from your anger [in the form of natural disasters]. And may no one exploit and subjugate your children. (mantra 11)

This prayer, which is based on the cosmic vision of our planet earth and which also relates to our consciousness toward the environment is based on the fundamental concept of *vasudhaiv kutumbakam*. Every entity and organism is a part of one large extended family system presided over by the eternal Mother Earth. It is she who supports us from her abundant endowments and riches, it is she who nourishes us, it is who provides us with a sustainable environment, and it is she who when angered by the misdeeds of her children punishes them with disasters. As one ought not to insult, unduly exploit, and violate one's mother but to be kind and respectful to her, one should behave in the same way toward Mother Earth. Through such exhortations and various writings, Hindu religion has provided a system of moral guidelines toward environmental preservation and conservation. From the perspective of Hindu culture (as well as from the Buddhist and Jain perspectives), abuse and exploitation of nature for selfish gain is unjust and sacrilegious.

The hymn also exemplifies the relevance of environmental sustenance, agriculture, and biodiversity to human beings. All the three main segments of our physical environment, that is, water, air, and soil, are highlighted and their usefulness is detailed. The various water resources (mantra 3), such as seas, rivers, and waterfalls, flow on earth. It consists of the three layers of brown, black, and red layers of soil, which is used for agricultural purposes. The hymn maintains that attributes of earth (such as its firmness, purity, and fertility) are for everyone and that no one group or nation has special authority over it. That is why the welfare of all and hatred toward none is the core value which people on this planet ought to strive (mantra 18). For example, there is a prayer for the preservation of the original fragrance of earth (mantras 23 and 25), so that its natural legacy is sustained for the future generations. Further, there is a prayer which says that even when people dig the earth either for agricultural purposes or for extracting minerals, let it be so that her vitals are not hurt and that no serious damage is done to her body and appearance (mantra 35); that is, her natural resources and vegetation cover are conserved. Although human greed and exploitative tendencies have been the main cause of environmental destruction, interreligious and intercultural conflicts and wars have also contributed to the environmental problems. In that respect, the hymn enunciates the unity of all races and among all beliefs; further, the concept of secularism, multiculturalism, and multilingualism has been urged with a prayer which says that Mother Earth bestows upon all the people living in any part of the world the same prosperity which has been pleaded by the Rishi Atharva:

Janam bibhratï bahudhà vivàchasam nana dharmànam prithivï yathaukasam,
Sahastram dhàrà dravinasya me duhàm dhruveva dhenurana pasphurantï.

The Mother Earth where people belonging to different races, following separate faiths and religions, and speaking numerous languages cares for them in many ways. May that Mother Earth, like a Cosmic Cow, give us the thousandfold prosperity without any hesitation without being outraged by our destructive actions. (mantra 45)

On the other hand, those who defend and protect the environment are showered by blessings (mantra 7). That is why, in mantra 59, Mother Earth is implored to bless us with all kinds of nourishment and serenity so that we may live in peace and harmony: "May the Mother Earth who is the provider of milk and many nourishing things, grain and other agricultural produce, fragrance; and bless us with peace, tranquility, and riches."

Finally, the hymn is the foremost ancient spiritual text from Hindu religion which enjoins human beings to protect, preserve, and care for the environment. This is beautifully illustrated in mantra 16, which says that it is up to us, human beings, as the progeny of Mother Earth to live in peace and harmony with all: "O Mother Earth! you are the world for us and we are your children; let us speak in one accord, let us come together so that we live in peace and harmony, and let us be cordial and gracious in our relationship with other human beings." This mantra also denotes a deep bond between Mother Earth and human beings and also exemplifies what kind of relationship ought to exist between "mother and

children," as well as between humans and other forms of life. Such a comprehensive exposition is rarely found in other religious traditions and beliefs as
portrayed by these sixty-three mantras. Moreover, the hymn provides us a moral
guide for behaving in an appropriate manner toward nature and our duty toward
the environment. It enjoins us to take care of God's creation in the form of treating
all organisms as a part of our extended family, the family of Mother Earth (*vasudhaiv kutumbakam*).

THE EXTENDED FAMILY
OF MOTHER EARTH

The concept known as *vasudhaiv kutumbakam* (*vasudhà* means this earth, *kutumba* means extended family consisting of human beings, animals, and all living
beings) means that all human beings as well as other creatures living on earth are
the members of the same extended family of *devi'vasundharà*. Only by considering
the entire universe as a part of our extended family can we (individually and
collectively) develop the necessary maturity and respect for all other living beings.

About *vasudhaiv kutumbakam*, Dr. Karan Singh has said "that the planet we
inhabit and of which we are all citizens—Planet Earth—is a single, living, pul-
sating entity; that the human race, in the final analysis is an interlocking, extended
family—*vasudhaiv kutumbakam* as the Veda has it" (Singh 1991: 123). We also
know that members of the extended family do not willfully endanger the lives and
livelihood of others; instead, they first think in terms of caring for others before
taking an action. That is why, in order to transmit this new global consciousness,
it is essential that the concept of *vasudhaiv kutumbakam* is encouraged. The
welfare and caring of all can be realized through the golden thread of spiritual
understanding and cooperation at the global level. Recently, a new focal point has
emerged around which the concept of the extended family of Mother Earth
appears to have aligned in the form of the Earth Charter, a document which
honors and respects Mother Earth. Because there exists a cosmic connection
between the micro and macro, there exists a harmonizing balance between the
planet and the plants, animals, human beings, and birds. They are all part of one
big family. Consequently, people of all faiths as well as those belonging to other
spiritual traditions should view the large existing in the small; and hence their
every act has not just global but cosmic implications (Shiva 1991: 3). In order to
raise the human spirit and create a worldwide family or community of ecologically sound and sustainable order, there is a need to subscribe to an Earth Charter.

It is an instrument through which a new universal consciousness for the healing of creation and a befitting understanding of divine purpose can be created to transform that human spirit which unites material realities and spiritual imperatives. It is that conscience which could restore and nourish a harmonized world through trusteeship, stewardship, and accountability for present and future generations.

DHARMIC ECOLOGY: OUR DHARMA AND KARMA TO THE ENVIRONMENT

Duty toward God's creation is an integral part of Hindu religion. All other species conduct themselves as per their natural instincts; only a human being has the power to act either in a dharmic (righteous) or adharmic (morally wrong) manner. People are urged to behave in such a manner that their acts do not cause undue harm or injury to any creature or plants. In *Mahabharata*, Rishi Markandeya explains to Pandavas: *Sarvani bhutani narendra pashya tatha yathavad vihitam vidhaatraa; svayonitah karma sadaa charanti neshe balasyeti chareddharmam* ("O King! all creatures act according to the laws of their specific specie-behavior as laid down by the creator. Therefore, none should act in the adharmic way, thinking, 'It is I who is powerful'") (*Mahabharata*, *Vanparva* 25.16).

Dharma is one of the most intractable and unyielding terms in Hindu, Buddhist, and Jain religions and philosophy. But, instead of discussing the root and theological or philosophical definitions of the term, let us move directly to the common usages of the term: a religious code of behavior, a divine system of morality and righteousness conforming to one's duty and nature, and so on. In this essay, we use this term to denote not only righteousness, but also the essential nature of any object, without which its entire existence and being cannot make any sense. For example, it is the true nature of a lion to hunt other animals when hungry; it is the dharma of fire to burn; similarly, it is the true nature of a person to act in a dharmic way although he or she may be capable of both the good and evil tendencies. As *Mahabharata* explains: *Dharanaad dharmamityaahur dharmen vidhritaah prajah; yah syaad dhaaran sanyukta sa dharma iti nishchayah.* "Dharma exists for the general welfare [*abhyudaya*] of all living beings; hence by which the welfare of all living creatures *is* sustained, that for sure is dharma" (*Mahabharata*, *Shanti Parva* 109.10).

In addition, dharma consists of two major attributes: first, it enables us to express ourselves in acts of service and impels us to avoid exploiting other human beings as well as other living organisms. In so doing, we have twofold duties: one to the self, whereby we seek inner strength through spiritual action, and the other to the community-at-large, whereby we work for the social good. Second, it provides for *abhyudaya* (attainment and sustenance of universe). As such, dharma regulates human conduct and casts individuals into the right character mold by inculcating in them spiritual, social, and moral virtues. Therefore, dharma can be considered as an ethos that holds the social and moral fabric together, by maintaining order in society, building individual and group character, and giving rise to harmony and understanding in our relationships with all of God's creation.

Can dharma play a role in environmental protection? It is already known that certain basic characteristics of human nature, such as greed, exploitation, abuse, mistreatment, and defilement of nature, have given rise to environmental destruction and other ills facing our world. Such exploitative tendencies must be disciplined and changed. Since it is devoid of institutional structures, bureaucratic impediments, and rituals associated with organized religions, dharma can be used as a mechanism to create respect for nature; moreover, it may serve as both a model and operative strategy for the transformation of human behavior. It may enable people to center their values upon the notion that there is a cosmic ordinance and a natural or divine law which must be observed; it may also provide the necessary incentive for humanity to seek peace with nature (Dwivedi 1994).

But dharma happens to be only one side of the human behavior; the other side is the concept of karma. The term *karma* comes from the root *kri*, which means "to do" and thus has a general connotation of "action," but in its broadest sense it applies also to the effects of such actions. People often confuse the law of karma with the law of destiny. This misunderstanding needs to be rectified because an appropriate understanding is essential to appreciate the use of this precept in Hindu and Buddhist ways of life. A brief definition of the law of karma is that each act, willfully performed, leaves a consequence in its wake. These consequences, also called *karma-phala* (fruits or effects of action) will always be with us, although their impact may not be felt immediately. The law related to karma tells us that every action performed creates its own chain of reactions and events, some of which are immediately visible, while others take time to surface. Environmental pollution is but one example of the karma of those people who thought that they could continue polluting the environment without realizing the consequences of their actions for future generations. Every action creates its own reaction. What is important to know is that right action, that is, dharmic action, generates beneficial results, while adharmic action results in harmful effects. Of course, it is not easy to foresee the consequences of one's actions, but one should be ready either to overcome obstacles that arise or to suffer the repercussions of one's actions.

A story from Hindu scriptures illustrates this point. Once, due to a curse of Maharshi Durbasha, the devas (demigods) were deprived of their divine vigor and strength. Soon, they were defeated by the asuras (demons) and dislodged from their heavenly abode. So they went to pray to Lord Vishnu and seek His intervention. Lord Vishnu suggested that the only course open to them lay in securing the Nectar of Immortality, which would make them invincible. But in order to obtain this nectar, they had to put all herbs in the cosmic sea of milk and churn it by using Mandarachal Mountain as a pestle and the snake-king Vasuki as the noose. But they needed the assistance of the demons to hold one side of the rope. The demons agreed to assist on the condition that the nectar received from this churning would be shared equally. Through this churning, the most sought after nectar was obtained, along with many precious stones, wealth, and worldly riches. But along with these priceless and rare items, a most venomous poison was produced. If that poison was not immediately disposed of, *sarvanash* (total annihilation) of the entire universe could result; however, no one, god or demon, was willing to touch this most fatal toxic waste created due to churning. Finally, all went to Lord Shiva for help to stop the eventual obliteration of the universe. He agreed to take care of the toxic waste by drinking it, and thus the impending disaster was averted. Later, with the assistance of Lord Vishnu, the demigods were able to trick the demons out of their share of the nectar, so that only the gods became immortal. This story instructs us that the consequences of an activity could be both beneficial and disastrous. Both nectar and poison are the effects of the same cause; and in the same way, prosperity and pollution are the two sides of the same cause—the karma of humanity. Both effects must be accepted, and as we crave for the nectar of riches, we must also be ready to face the side effects (i.e., poison) of technology and industrialization.

Once the karma has started, it continues without a break; and although the person may be dead, yet his or her karma survives in the form of a memory on to the next life of a departed soul. This is stated in *Mahabharata* (*Shanti Parva* 139.22): *Paapam karma kritam kimchid yadi tasmin na drishyate; nripate tasya putreshu pautreshu api cha naptrishu* ("an action which has been committed by a human being in this life follows him again and again [whether he wishes it or not]"). Furthermore, it is stated in *Bramhavaivarta Purana* that whatever an action, good or bad, is knowingly performed by a person, he or she must face its consequences (*Bramhavaivarta Purana, Prakriti* 37/16): *Na bhuktam ksheeyate karma kalp kotistairapi; ashvameyva bhoktavyam kritam karma shubhashubham.*

Sometimes the karmic justice appears indecipherable to individuals when his or her family and later descendants are forced to pay for the past crimes and mistakes of that person. This aspect is further explained by the warrior-prince Bhisma to King Yudhisthira in *Mahabharata* (*Shanti Parva* 129.21): "O King, although a particular person may not be seen suffering the results of his evil actions, yet his children and grandchildren as well as great-grandchildren will have to suffer them." This sound advice related not only to an individual action or karma but also to group action, or action taken in the name of government. Any adharmic act has a way of coming back to haunt us later (either within our own lifetime or even after death in another life).

Once our dharma and karma to the environment are appropriately understood, their precepts recognized, and their relevance in environmental protection and conservation accepted, then a common strategy for ecospirituality and stewardship can be developed. Such a strategy will depend much upon how different people together (a) perceive a common future for society, (b) act both individually and as a group toward that end, and (c) realize that each individual has a moral obligation to support their society's goal since their acts will have repercussions on the future of the society and their own destiny.

Finally, dharma requires that one consider the entire universe as his or her extended family, with all living beings in this universe as the members of the same household. This concept is also known as *vasudhaiv kutumbakam*, discussed earlier. It is argued that by considering the entire universe as a part of one's extended family one can develop the necessary maturity and respect for all other living beings. The welfare and caring of all would be realized through the golden thread of spiritual understanding and cooperation, which is based on dharma and karma of individuals and institutions.

ENVIRONMENTAL CHALLENGES BEFORE HINDU RELIGION AND INDIA

There is no doubt that Hindu religion and culture, in ancient and medieval times, have provided a system of moral guidelines toward environmental preservation and conservation. Environmental ethics, as propounded by ancient Hindu scriptures and seers, was practiced not only by common persons, but even by rulers and kings. They observed these fundamentals, sometimes as religious duties, often as rules of administration or obligations for law and order, but always as principles properly knitted with the Hindu way of life. That way of life did enable Hindus as well as other religious groups residing in India to use natural resources, but not to have any divine or secular power of control over the environment. But with such a religious tradition and philosophy of life, why does a serious environmental crisis exist for India? What has gone wrong?

As we have seen, Hindu ethical beliefs and religious values do influence its people's behavior toward others, including our relationship with all creatures and plant life. If, for some reason, those noble values become displaced by other beliefs, that are either thrust upon the society or transplanted from another culture through invasion, then the faith of the masses in the earlier cultural tradition is shaken. As appropriate answers and leadership were not forthcoming from the religious leaders and priests, the masses became ritualistic, caste ridden, and inward looking. However, besides the influence of alien cultures and values, what really damaged India's environment were the forces of materialism, consumerism, individualism and corporate greed, the blind race to industrialize the nation immediately after achieving independence, and the capriciousness and corruption among forest contractors and ineffective enforcement by forest officials. All of them have acted against the maintenance and resurgence of respect for nature in India. Under such circumstances, religious values that acted as sanctions against environmental destruction were sidelined as insidious forces attempting to greatly inhibit the religion from continuing to transmit ancient values which encouraged respect and due regard for God's creation.

How can such ancient values and wisdom be transmitted into public policy? It is not sufficient to extol the ancient wisdom of Hindu seers, to dwell on the Vedic heritage, and then hope that somehow a self-correcting mechanism will take care of all environmental problems facing India. Instead what is more important is how to put into practice that ecocare vision and to make it relevant to modern times. Toward this, I suggest the following strategy. The Hindu religion, its religious leaders, and its followers should . . .

- Become effective advocates and practitioners of the concept of ecocare and dharmic ecology, rather than staying on the sidelines.
- Take the initiative to help secular institutions by providing timely and appropriate advice to encourage greater integration of ecocare heritage into educational curricula.
- Strengthen the capability of secular institutions to meet their goals of sustainable development and environmental conservation.
- Promote the concept of *sarva bhoot hitey ratah* ("to serve all beings equally").
- Take the lead in promoting the concept of *vasudhaiv kutumbakam* and obligation of humanity to accept a world of material limits.
- Protect and restore places of ecological, cultural, esthetic, and spiritual significance.
- Build partnerships across social, economic, political, and environmental sectors, including dialogue with other religions and spiritual traditions. (Dwivedi 2000b)

The choice before the Hindu religion (as well as before all other religions) is to either care for the environment or be a silent participant in the destruction of planetary resources. Partnership with secular institutions must be forged, and cooperation must be fostered at local, regional, national, and international levels. An environmental and sustainable development strategy, based on the lines suggested above, could offer a way of bridging the gap and making the essential link between secular, scientific, and spiritual forces.

The Hindu heritage for ecospirituality can lead us to control our base characteristics such as greed, exploitation, abuse, mistreatment, and defilement of nature. But, before we can hope to change the exploitative tendencies, it is absolutely essential that we discipline our inner thoughts. It is here where the religious exhortations and injunctions may come into play for environmental stewardship. This ecospirituality-oriented environmental stewardship can be a mechanism which strengthens respect for nature by enabling people to center their values upon the notion that there is a cosmic ordinance and a divine law which must be maintained. Furthermore, environmental stewardship can provide new ways of valuing and caring. It can also influence and promote sustainable development. Ecospirituality, if globally manifested, can also provide the values necessary for an environmentally caring world and will not advance a blind belief in economic growth at all costs, creating in its wake greed, poverty, inequality, and injustice as well as environmental destruction. Environmental stewardship, then, drawing upon the Vedic and Puranic precepts, may become a new universal consciousness around which the concept of global stewardship of the environment may develop; and in this way we may usher in new values in line with an environmentally caring world.

Environmental stewardship drawing upon the Hindu concept of dharma and karma to the environment, as discussed earlier, can provide new ways of valuing and acting. It can promote the preparation of policies of sustainable development and the introduction of environmental protection initiatives. Dharma, if globally manifested, will provide the values necessary for an environmentally caring world and will not advance economic growth at the cost of greed, poverty, inequality, and environmental degradation. There is an immediate need to instill among all people the respect for nature as well as to strengthen the decision-making process for environmental protection. This may be a focal point for a new global consciousness toward an environmentally caring world.

In summary, a new universal consciousness will have to be developed which believes in at least (a) two dictums: "what we sow is what we reap" and "everything is connected to everything else"; and (b) our inherent dharma to the environment. But one should note that these two concepts are intertwined with another dictum: *sarva bhuta hite ratah*. The Hindu religion, like other religions and spiritual traditions, has the capacity to move the individual toward the divine because of its belief in divinity in nature; thus, it is imperative that such an inherent capacity is strengthened to its ultimate end. Toward this, we need a new paradigm of thought, perhaps a dharmic ecology. By developing such a paradigm, drawing upon the concept of dharma and karma, and based on the notions of *vaasudeva sarvam* and *vasudhaiv kutumbakam*, we may be able to sustain not only the present generation but also leave a healthy legacy for future generations. Can the Hindu religion take up this challenge and operationalize it as an integral part of its conscious strategy and vision for a sustainable future? Christopher Chapple says: "The Hindu religion, with its vast storehouse of text, ritual, and spirituality, can help contribute both theoretical and practical responses to this [environmental crisis]" (Chapple 2000: xlvii).

What are those responses? Let me offer some interpretative explorations as well as general propositions.

Growing Population and Flight from the Land

In 1971 the total population of India stood at 548 million. But within next three decades, the country exceeded the one billion mark. The effect of this growth is highly noticeable in the further loss and degradation of common property resources. At the same time, population pressure continues to increase in urban areas due to the flight from rural areas. As the country becomes more urbanized, there will be enormous pressure on municipal government infrastructures such as housing, water supplies, sewage treatment and waste management, public transportation, schools, and hospitals. Indian cities and towns are already bursting through their seams; the local governments will be unable to provide necessary health-care and social programs needed by these extra people. These people will suffer from heavy pollution (both airborne and waterborne diseases) and other environmental hazards. The worst sufferers will be the urban poor, who will face malnutrition as well as other environmental hazards. Unless municipal and local governments are given sufficient financial autonomy (revenue and tax reform) and market mechanisms to take care of the ever-expanding demand for their services, and the necessary legal base to regulate urban development, the public will continue suffering. This is a serious challenge before India.

Impact of Accelerated Industrialization

By liberalizing its economy, India has awakened the great spirit of entrepreneurship, with its promises of high economic growth and the stabilization of fiscal imbalances which the country suffered until July 1991, when the reform program started. One of the major blocks to accelerated industrial development are public enterprises, out of which 40% are chronic loss makers; for example, they generated a fifty-three-billion-rupee loss in 1993 alone, thereby putting a substantial burden on the national budget which could otherwise have been used in national development (World Bank 1995: 20). In turn, their losses prohibit the investment of public funds in social and environmental sectors. In time, these enterprises will be either sold off to private firms or simply disbanded. However, the process of accelerated industrialization will also bring in its wake environmental problems such as urban congestion and squatter settlements, toxic industrial effluents and waste, air and water pollution, and depletion of natural resources. In the race to industrialize and urbanize quickly, India will suffer from a high exposure to environmental diseases in the coming decades, unless a simultaneous effort is made to control pollution at the point source.

Dharma and Karma to the Environment

Such a suffering may continue to visit humanity unless we realize that the destruction we are inflicting on our natural surroundings will result in dire consequences for the present and succeeding generations. If we recognize and act on this philosophy of life, then and only then will people start paying due respect to nature as well as taking a stake in the care for the environment. This appreciation is the key to the point being made here, and it draws on the concept that people living in any part of the world are our brothers and sisters. All of our actions are interrelated with and interconnected to what eventually happens in this world. Although we may not face the consequences individually, nevertheless someone else is going to be burdened with or benefit from our actions. It is in this context that the concept of dharma and karma becomes meaningful.

Ecospirituality

As the new millennium begins, it is becoming increasingly clear that many of our values are totally inadequate for long-term survival and sustainable development; that is why it is not surprising that we are witnessing an emergence of a wide spectrum of challenges to the traditional materialistic view. For millennia, guided by Western culture, people have had blind faith in the prowess of science and technology to bring material progress. It is only recently that we have come to understand that so-called material prosperity should not be an end in itself. Slowly, a realization is emerging that spirituality and the control of one's desires can bring a more lasting happiness than acquisitive materialism. However, such a realization has yet to enter the domain of governmental policy or the corporate world, where spiritual perspectives are generally ignored. The economic criteria which place no value on the commons (air, water, oceans, outer space) and which use concepts such as cost-benefit analysis, law of supply and demand, rate of return, land as commodity, and so on, have been based on the delusion that has operated independently of the cultural and spiritual domain. As Daniel Gomez-Ibanez

(1993: 25), organizer of the Parliament of World's Religions in Chicago, observed: "A great danger in this materialistic and mechanistic view of the universe is that even when we see the problems it has wrought, we often assume that the solutions are to be found only in the same material realm, perhaps because we forget to consider any other possibility."

Those solutions, if based (as has been the case thus far) on technical fixes, would not help us unless we change our ways with regard to how we use nature. There is a desire by many to ascribe to a radically new set of values. Without a change in our current value system, there is little hope of correcting the present environmental problems we face today. In order for this to occur, a new consciousness will have to be developed, which draws on the concept of ecospirituality and stewardship, as discussed earlier in this essay.

The Importance of Spirituality in Nations' Well-being

There is a well-known association between religion and happiness, although it is not proven which particular aspect of religiosity correlates with life satisfaction and overall well-being. Koenig, McCullough, and Larson (2001), after reviewing over one hundred studies, found a positive association between some measures of religiosity and well-being, happiness, and fulfillment. Cohen (2002: 288) states that spirituality may be related to satisfaction with life in that religious people may feel close to God, see beauty in the world, and feel that their lives have purpose. One can extrapolate from this that spirituality also has a positive impact on how we treat nature and the environment. Although not all religious beliefs and practices may have the same impact on environmental issues, spirituality does play an important role in shaping our views toward nature. Does this mean that Hindus in general are more environmentally caring people (because of their religious traditions) compared to, for example, their neighbors such as Chinese (with much less religiosity)? The evidence, particularly when compared with the ecosystem well-being index (see table below), does not indicate it. More than fifty-five years have passed since India received independence from the Great Britain, but the rate of environmental stress does not appear to have been abated, although the level of environmental consciousness has increased (Dwivedi 1997a).

Comparing Human and Environmental Well-being

Different cultures aspire to different things pertaining to good life. For example, the Chinese seek the five blessings as a part of good life: long life, riches, health, love of virtue, and a natural death in an old age; the French seek liberty, equality, and fraternity; the British seek health, wealth, and wisdom; Americans seek life, liberty, and the pursuit of happiness; and Indians seek power, pleasure, morality, and final emancipation of the soul (*nirvana*) (Prescott-Allen 2001: 1). Despite what different people seek, when it comes to environmental quality, it is clear that those who have a high standard of living impose excessive demands on the environment, but those who are poor also place heavy demands on the environment because they lack means, infrastructure, and resources to control environmental pollution. In a survey undertaken for preparing an index of quality of life and the environment (see table below), it has been shown that the five countries listed above score differently. Actually, the three wealthy countries show a better score (when human and ecosystem well-being are combined) compared to the two poor nations of China and India:

Country	Human Well-being Index	Ecosystem Well-being Index	Average
United States	73	31	52.0
United Kingdom	73	30	51.5
France	75	29	52.0
China	36	28	32.0
India	31	27	29.0

Total countries surveyed: 158. The highest/lowest scores for each of these three indices are 82/3, 68/14, and 64.0/25.0 respectively. *Source:* Prescott-Allen 2001: 151–53 (table 1), 181–83 (table 9), 267–68 (table 27).

The Earth Charter: An Instrument
for Protecting Mother Earth

The Earth Charter is set apart from many other international agreements in that it "recognizes [that] the successful achievement of the goal of sustainable development requires not only international commitment and legal regulation, but also basic changes in attitudes, values, and behavior of people" (Taylor 1999: 193). In addition to suggesting changes in attitudes, values, and behavior of people, there are policy objectives that could be associated with such a charter. Mackey (2004: 89) identified three such policy objectives: increased protection of ecological systems, through the creation of protected areas; an abatement of anthropogenic activities resulting in climate change; and the maintenance of natural resources, especially those in developing countries. Additionally, according to Norton (2000: 1030) the Earth Charter provides the political framework needed to develop environmental protection strategies on various scales: local, regional, national, and international.

Finally, Earth Charter provides optimism for the development of a political framework in line with its objectives of preserving nature for future generations. However, the Earth Charter should not be seen as a policymaking document, but can be used as a guide for a more sustainable way of life. Finally, the Earth Charter is a reminder of our moral duty to leave a healthy legacy for future generations by not only protecting the environment from the harmful ways of our activities, but also by attempting to restore the status quo of two generations ago. The charter creates a list of duties for us all to see that future generations receive appropriate reparations for the damage which we visited upon them by our careless and exploitative tendencies. It also entreats us to consider flexibility, adaptability, and the ability to reconcile the needs of different communities and local interests by carefully nurturing such differences. The reality of our world is not independence but interdependence. The complexity and scale of environmental problems no longer permit a watertight division of environmental issues within nation-states and among sovereign states. The challenge before us is the reform and renewal of the existing world system and securing a firm place for the rights of future generations. It is to this end—the care of our planet—that much of our efforts must now turn if we are to avoid further damage to the world ecosystem. The Earth Charter may lead us toward a safer, cleaner, and better world.

Dharmic Ecology and Current Ground Realities in India

I have stressed through my writings "an environmental stewardship that draws upon the Hindu concept of *dharma* and *karma* to the environment can provide new ways of valuing and action. It can promote policies for sustainable development and introduce environmental protection initiatives" (Dwivedi 2000b: 19–20). Of course, leaders use ancient wisdom in their speeches and quote various verses and statements as the situation demands, but when it comes to transforming such wisdom and concepts in public policies, nothing appears to have taken place even when the Bharatiya Janata Party was in power at the federal level. But those who have observed India from different perspectives may agree with me that the country presents some paradoxical images: it has profound ecological exhortations (unlike in any other culture), and yet it also has severe environmental destruction; abject poverty against pockets of affluence; prevailing unsanitary conditions in many places and yet competently cleaned and healthy areas; and shantytowns existing close to five-star hotels. But these contradictory images are not going to remain too long with people of India, as the country is poised to become an industrial giant in Asia and the world. A conscious hope is emerging in the nation to strive toward the ideal of a mutually sustaining and ecologically balanced society which supports diversity (of want, culture, religion, and doing things differently), sharing of knowledge and technology rather than erecting fences in the name of intellectual property rights, controlling any further ecological damage, and encouraging a system of moral governance. This is to be achieved by creating an ecologically balanced society against the mounting pressures of population and rapidly accelerating consumerism.

Transformation of Classical Ecology into Public Policy

Transformation of traditions continually take place in a vibrant society, as these open up new opportunities to use older symbols and traditions in the modern context. In the case of Hindu ecology, for example, when the Chipko movement (Chipko means "to embrace trees") started in 1973–74, the movement derived its strength in the old tradition of not cutting a green tree as enunciated in Hindu scriptures, sanctified by Buddha, and exhorted by Emperor Ashok. Later in 1983, Apiko, a similar movement, was started in south India (in the northern Karnataka state). In 1978–79, Sundarlal Bahuguna led a campaign against logging in the Himalayan area; he went on a hunger strike and was jailed. During that movement, spiritual music and readings from *Gita* and *Ramcharitmaanas* were used. *Padyatra* ("pilgrimage on foot") has been used in India by various leaders to protest against environmental destruction. Even in the Narmada dam controversy, people against dams have used ancient exhortations to strengthen their positions. Even Marxist and Hindu-religion-based political groups such as Vishwa Hindu Parishad have used ancient religious symbols and traditions to their respective causes (Gosling 2001). Nevertheless, with such admirable intent, general exhortations do not turn themselves into specific policy actions for governments or organizations to perform. Specifically in India, unless the relationship between secularists (who see any reference to scriptures and religious texts as fundamentalism gone crazy) and religion-based groups is improved (which is now mired with acrimony and distrust), chances are uncertain for some concepts of Hindu ecology being transformed into policy process. Even for many academics of India, these exhortations appear to be either uninteresting or irrelevant to the existing environmental issues faced by their nation. What is needed is an integration of traditional values with modern environmental management systems to create an ecologically healthy and sustainable world. Ancient exhortations and wisdom can provide the environmental movement with coherent ethical guidelines and a strong legitimacy for policy directions at the governmental level. Can there be a strategy which is practical and yet draws on the Hindu heritage for nature conservation and the appropriate respect for the extended family of our Mother Earth? So far, examples like the Chipko and similar movements in India have demonstrated that when appeals to secular norms fail, one can draw on cultural and religious sources for environmental protection. Perhaps, such examples illustrate the role of dharmic ecology in action and a practical impact of Hinduism on environmental conservation.

THE WELL-BEING OF ALL: *SARVE BHAVANTU SUKHINAH*

While in the West, divinity was driven out of nature (when Greco-Roman culture was converted to Christianity) and was concentrated in one unique transcendent god, Hindu religion (and others in the East) kept it because the godhood is considered to be diffused throughout the universe and nature. Divinity, according to Hindu religion, has been and is at present all over, including mountains, in rivers and forests, in plants and grains, in animals and beasts, in stars and planets, in earthquakes, and in lightning and thunder. Thus the Hindu divinity has never been a singular or a unique almighty super being. Human beings have always been made conscious of their obligations and duties toward nature, its creatures, and the universe, although such exhortations were seldom obeyed by many in India.

Hinduism (as well as other world religions in their own way) offers a unique set of moral values and rules to guide human beings in their relationship with the environment. Religions also provide sanctions and offer stiffer penalties, such as the fear of damnation, for those who do not treat God's creation with respect. Although, in the recent past, religions have not been in the forefront of protecting the environment from human greed and exploitation, many are now willing to take up the challenge to help protect and conserve the environment. But this offer of help will remain purely rhetorical unless secular institutions, national governments, and international organizations are willing to acknowledge the role of religion in environmental study and education. And we believe that environmental education will remain incomplete until it includes cultural values and religious imperatives. For this education, we require an ecumenical approach. While there are metaphysical, ethical, anthropological, and social disagreements among world religions, a synthesis of the key concepts and precepts from each of them pertaining to conservation must become a foundation of a global environmental duty.

But that duty is not restricted to Hindus alone. People of all faiths and of various spiritual traditions should view the large existing in the small; and hence their every act has not just global but cosmic implications (Shiva 1991: 3). They have a global duty (*vishwa dharma*) toward the environment; that duty requires us not only to integrate environmental conservation and protection as a part of our universal consciousness, vision, and acceptance, but also to raise the human spirit and create a worldwide community of ecologically sound and sustainable order by subscribing to the Earth Charter (Earth Council 1998). There exists a divine purpose which can transform the human spirit by uniting material realities and spiritual imperatives. It is that conscience which could restore and nourish a harmonized world through trusteeship, stewardship, and accountability for present and future generations. A fitting conclusion to this discussion is Albert Einstein's 1933 speech delivered in Albert Hall, London, when he stated that we human beings have a most important duty: "The care for what is eternal and highest amongst our possessions, that which gives to life its import and which we wish to hand on to our children purer and richer than we received it from our forebears" (Einstein 1950: 151).

REFERENCES

Quotations in Sanskrit for *Isavasya Upanishad, Aitareya Upanishad*, and *Shvetashvatara Upanishad* are from *Upanishad Anka of Kalyan* (Gorakhpur: Gita Press, 1949 [23.1]). Quotations from *Yajyavalkya Smirit, Narsingh Purana*, and *Vishnu Purana* are from the publications of the Gita Press in Gorakhpur, India. All translations and transliterations are my own; because of printing constraints, diacritics are not always used.

Atharva Veda. 1982. Translated by Devi Chand. New Delhi: Munsiram Manoharlal.
Bhagavadgita. 1996. Gorakhpur, India: Gita.
Chapple, Christopher Key. 1993. "Hindu Environmentalism: Traditional and Contemporary Resources." Pp. 113–23 in *Worldviews and Ecology*. Edited by Mary Evelyn Tucker and John A. Grim. Lewisburg, PA: Bucknell University Press.
Chapple, Christopher Key, and Mary Evelyn Tucker, eds. 2000. *Hinduism and Ecology: The Intersection of Earth, Sky, and Water*. Cambridge: Center for the Study of World Religions, Harvard Divinity School.
Cohen, Adam B. 2002. "The Importance of Spirituality in Well-being for Jews and Christians." *Journal of Happiness Studies* 3:287–310.
Deussen, Paul. 1980. *Sixty Upanisads of the Veda*. Translated by V. M. Bedekar and G. B. Palsule. Delhi: Motilal Banarsidass.
Dwivedi, O. P. 1994. "Our Karma and Dharma to the Environment." Pp. 59–74 in *Environmental Stewardship: History, Theory, and Practice*. Edited by Mary Ann Beavis. Winnipeg: Institute of Urban Studies, University of Winnipeg.
———. 1997a. *India's Environmental Policies, Programmes, and Stewardship*. London: Macmillan.
———. 1997b. "Vedic Heritage for Environmental Stewardship." *Worldviews: Environment, Culture, and Religion* 1:25–36.

———. 1998. *Vasudhaiv Kutumbakam: A Commentary on Atharvediya Prithivi Sukta.* 2nd ed. Jaipur: Institute for Research and Advanced Studies (orig. 1995).

———. 2000a. "Classical India." Pp. 37–50 in *A Companion to Environmental Philosophy.* Edited by Dale Jamieson. New York: Blackwell.

———. 2000b. "Dharmic Ecology." Pp. 4–22 in *Hinduism and Ecology: The Intersection of Earth, Sky, and Water.* Edited by Christopher K. Chapple and Mary Evelyn Tucker. Cambridge: Center for the Study of World Religions, Harvard Divinity School.

Dwivedi, O. P., and B. N. Tiwari. 1987. *Environmental Crisis and Hindu Religion.* New Delhi: Gitanjali.

Dwivedi, O. P., B. N. Tiwari, and R. N. Tripathi. 1984. "Hindu Concept of Ecology and the Environmental Crisis." *Indian Journal of Public Administration* 30:33–67.

Earth Council. 1998. *The Earth Charter: Values and Principles for a Sustainable Future.* San Jose, Costa Rica: Earth Charter International Secretariat.

Einstein, Albert. 1950. *Out of My Later Years.* New York: Philosophical Books.

Gomez-Ibanez, Daniel. 1993. "Spiritual Dimensions of the Environmental Crisis." In *A Source Book for the Community of Religions.* Edited by Joel D. Beversluis. Chicago: Council for a Parliament of the World's Religions.

Gosling, David L. 2001. *Religion and Ecology in India and Southeast Asia.* London: Routledge.

Griffith, Ralph T. H. 1889. *The Hymns of the Rig Veda.* Edited by J. L. Shastri. Delhi: Motilal Banarsidass (orig. 1973).

Hume, Robert Ernest. 1977. *The Thirteen Principal Upanishads.* New York: Oxford University Press.

Koenig, H. G., M. E. McCullough, and D. B. Larson. 2001. *Handbook of Religion and Health.* Oxford: Oxford University Press.

Lal, S. K. 1995. "Pancamahabhutas: Origin and Myths in Vedic Literature." Pp. 5–21 in *Prakrti: An Integral Vision.* Edited by Sampat Narayanan. New Delhi: Indira Gandhi National Centre for the Arts.

Mackey, Brendan. 2004. "The Earth Charter and Ecological Integrity—Some Policy Implications." *World Views* 8:76-92.

Mahabharata. 1988. Translated by M. N. Dutta. Delhi: Parimal.

Mahabharata. 1994. Gorakhpur, India: Gita Press.

Manusmriti [The Laws of Manu]. 1975. Translated by G. Buhler. Delhi: Motilal Banarsidass.

Norton, Bryan. 2000. "Biodiversity and Environmental Values: In Search of a Universal Earth Ethic." *Biodiversity and Conservation* 9:1029–44.

Prescott-Allen, Robert. 2001. *The Wellbeing of Nations.* Washington, DC: Island Press/Ottawa: International Development Research Centre.

Rig Veda. 1974. Commentary by Maharishi Dayanand Saraswati. 12 vols. New Delhi: Sarvadeshik Arya Pratinidhi Sabha.

Satvalekar, Pandit Shripad Damodar. 1958. *Prithivi-Sukta: Atharvaveda Book* 12. Surat: Swadhyay Mandal.

Shiva, Vandana. 1991. "The Greening of the Global Reach." *Illustrated Weekly of India* (Oct. 12–18).

Singh, Dr. Karan. 1991. *Brief Sojourn.* Delhi: B. R. Publications.

Srimad Bhagavata Mahapurana. 1982. Translated by C. L. Goswami and M. A. Shastri. Gorakhpur: Gita Press.

Taylor, Prue. 1999. "The Earth Charter." *New Zealand Journal of Environmental Law* 3: 199–203.

World Bank. 1995. *Economic Developments in India.* Washington, DC: World Bank.

World Commission on Environment and Development. 1987. *Our Common Future.* New York: Oxford University Press.

CHAPTER 7

THE GREENING
OF BUDDHISM

Promise and Perils

STEPHANIE KAZA

As a major world religion, Buddhism has a long and rich history of responding to human needs. From the moist tropical lowlands of Sri Lanka to the towering mountains of Tibet, Buddhist teachings have been transmitted through diverse terrain to many different cultures. Across this history Buddhist understanding about nature and human-nature relations has been based on a wide range of teachings, texts, and social views. The last half century, as Buddhism has taken root in the West, has been a time of great environmental concern. Global warming, habitat loss, and resource extraction have all taken a significant toll as human populations multiply beyond precedent.

With the rise of the religion and ecology movement, Buddhist scholars, teachers, and practitioners have investigated the various traditions to see what teachings are relevant and helpful for cultivating environmental awareness. The development of green Buddhism is a relatively new phenomenon, reflecting the scale of the environmental crisis around the world. Thus far the gleanings have followed the lead of specific writers and teachers opening up new interpretations of Buddhist teachings. Western Buddhists, still new to the philosophies and practices of the East, have often sparked the conversations, seeking ways to complement secular approaches to environmental thought.

HISTORY AND DEVELOPMENT
OF GREEN BUDDHISM

One of the earliest voices for Buddhist environmentalism in North America was Zen student and poet Gary Snyder, who illuminated the connections between Buddhist practice and ecological thinking (see, e.g., Snyder 1999 and 1969). Snyder studied Zen in Japan and cultivated an "in the moment" haiku-like form to his poetry, much of which was set in the mountains of the western United States. One of his more lighthearted pieces, "Smokey the Bear Sutra," was handed out by activists urging better protection for U.S. forests. Snyder was associated with the early Beat generation of the 1950s and 1960s, which had a strong influence on the 1960s counterculture. Hippies, communards, and back-to-the-landers took up Snyder's approach, made popular in Jack Kerouac's travelogue *Dharma Bums*. Many early Buddhist students felt that spiritual leadership was crucial in the race toward planetary ecological destruction.

In the 1970s the environmental movement swelled, and Buddhist centers became well established in the West. While Congress passed such landmark legislation as the Clean Water Act, some of the new retreat centers confronted ecological issues head on. Zen Mountain Monastery in New York challenged the Department of Environmental Conservation over a beaver dam and forest protection. Green Gulch Zen Center in northern California had to work out water-use agreements with the neighboring farmers and national park. Some Buddhist centers opted for vegetarian fare at a time when vegetarianism was not that well known. For a few, this reflected an awareness of the environmental problems associated with raising meat. A number of Buddhist centers made some effort to grow their own organic food.

By the 1980s Buddhist leaders were explicitly addressing the ecocrisis and incorporating ecological awareness in their teaching. In his 1989 Nobel Peace Prize speech, His Holiness the Dalai Lama proposed making Tibet an international ecological reserve. Vietnamese Zen monk and peace activist Thich Nhat Hanh invited his followers to join the Order of Interbeing, teaching Buddhist principles using ecological examples. Zen teachers Robert Aiken in Hawaii and Daido Loori in New York examined the Buddhist precepts from an environmental perspective. Buddhist activist Joanna Macy creatively synthesized elements of Buddhism and deep ecology, challenging people to take their insights into direct action. The Buddhist Peace Fellowship, founded in 1978, added environmental concerns to its early activist agenda.

In Thailand, teak forests were being clear-cut at an accelerating rate for foreign trade. This resulted in massive flooding and mudslides, generating a national wave of environmental protest. Buddhist priests in rural villages made headlines with their ritual ordination of elder trees as a symbolic gesture of solidarity with threatened forests (Darlington 1998). As Buddhist environmental activism spread, the "forest monks," as they came to be known, formed an ethical front in the protest against overexploitation. Other monks got involved with activist efforts to question economic development and its environmental impacts. Plastic bags, toxic lakes, and nuclear reactors were targeted by Buddhist leaders as detrimental influences on people's physical and spiritual health. In Burma, Buddhists concerned about the environment drew attention to the impacts of a major oil pipeline and the decimation of tropical forests. In Tibet, the environmental impacts of Chinese colonization were documented and publicized by support groups in the West (see the essays on environmental issues in Tibet, Burma, and Thailand in Kaza and Kraft 2000: 161–235).

Interest in Buddhist views on the environment gained momentum in the 1990s through books, journals, and conferences. For the twentieth anniversary of Earth Day, the Buddhist Peace Fellowship produced a teaching packet and poster for widespread distribution. That same year, 1990, the first popular anthology of Buddhism and ecology writings, *Dharma Gaia* (Hunt-Badiner 1990), was published by Parallax following the more scholarly collection *Nature in Asian Traditions of Thought* (Callicott and Ames 1989). World Wide Fund for Nature brought out a series of books on five world religions, including *Buddhism and Ecology* (Batchelor and Brown 1992). Well-established Buddhist magazines such as *Tricycle, Shambhala Sun, Inquiring Mind, Turning Wheel,* and *Mountain Record* devoted whole issues to the question of environmental practice.

In 1990 two groundbreaking national conferences were held in Seattle, Washington, and Middlebury, Vermont—both focused on ecoreligious approaches to the environment. At the Vermont conference the Dalai Lama was the keynote speaker, urging people to take care of the environment. A few years later at the 1993 Parliament of the World's Religions in Chicago, Buddhists gathered with Hindus, Muslims, pagans, Jews, and Christians from all over the world; one of the top agenda items was the role of religion in responding to the environmental crisis. Parallel interest in the academic community culminated in ten major conferences at Harvard Center for the Study of World Religions, purposely aimed at defining a new field of study in religion and ecology. The first of these conferences, convened by Mary Evelyn Tucker and John Grim in 1996, focused on Buddhism and ecology and resulted in the first major academic volume on the subject (Tucker and Williams 1997).

For the most part, the academic community addressed the philosophy but not the practice of Buddhist environmentalism. This was explored by socially engaged Buddhist teachers such as Thich Nhat Hanh and Bernie Glassman. In Thailand and around the world, Sulak Sivaraksa worked tirelessly for global change, and in the United Kingdom Vipassana teacher Christopher Titmuss ran for Parliament as a Green Party candidate. John Daido Loori committed a substantial portion of his retreat center land in the Catskills of New York to be "forever wild," while Rochester Zen Center founder Philip Kapleau actively promoted vegetarianism. In California nuclear activist Joanna Macy promoted a model of experiential teaching designed to cultivate motivation, presence, and authenticity. Her workshops popularized Buddhist meditation techniques and a Buddhist view of systems thinking. Together with Buddhist rainforest activist John Seed of Australia, she developed a Council of All Beings to engage people's attention and imagination on behalf of all beings (Seed et al. 1988). Thousands of these councils have now taken place in Australia, New Zealand, the United States, Germany, Russia, and other parts of the Western world.

Since 2000 the religion and ecology movement has gathered steam and become a forceful presence at the American Academy of Religion as well as the World Council of Churches. The acceleration of global environmental problems has added to the urgency of the agenda, taken up now by Buddhists as well as Christians, Jews, and all the major religious traditions. Buddhist initiatives have been strongest in Buddhist countries such as Thailand, Tibet, and Burma. Though fewer in numbers, Western Buddhists have contributed texts and academic study to provide a foundation for the new movement. There are now doctoral programs in the United States where a student can earn a graduate degree with a focus on Buddhism and ecology.[1]

CURRENT STREAMS IN BUDDHIST ENVIRONMENTAL THOUGHT

As interest has developed in Buddhism and ecology, the fields of thought have expanded through various writers as well as popular and academic discourses. When a field of thought first coalesces from wide-ranging points of engagement, a common first step is the publication of collected writings on the topic. This then opens the field to newcomers by providing an overview and introduction to the major themes within the field. For Buddhism and ecology, this step was taken with *Dharma Gaia* (Hunt-Badiner 1990), followed by the academic collection of papers entitled *Buddhism and Ecology* (Tucker and Williams 1997), which led to the most complete collection to date: *Dharma Rain* (Kaza and Kraft 2000). This last anthology drew together classic reference texts from a range of Buddhist traditions, along with modern commentaries, exploratory essays, and academic critiques.

With such texts available to academic audiences, professors in religious studies and environmental studies could now offer courses on Buddhism and ecology at the undergraduate level. For students in the West, Buddhism held its own magnetic attraction as the exotic "other" next to Christianity. Young people concerned about the environment and eager for a more congruent spiritual fit with their experience in nature found Buddhist environmental thought very appealing. At a professional level, Buddhist perspectives have been a regular part of the programs organized by the Religion and Ecology Group at the American Academy of Religion.[2] This group has seen a rapid rise in interest, with attendance increasing every year.

Environmental concerns have also been a significant part of interreligious dialogue in the West. At the 2005 international conference of the Society for Buddhist-Christian Studies held in Los Angeles, the theme was "Hearing the Cries of the World," with one session focused on the "cries" of the environment. At Gethsemani II in 2002, a Catholic-sponsored dialogue in Thomas Merton's tradition, speakers addressed structural poverty and violence resulting from global exploitation of environmental resources. Impacts of consumerism were taken up at the 2003 annual meeting of the Society for Buddhist-Christian Studies.

Not all topics of environmental concern have attracted attention from green Buddhists; some key issues such as global climate change have hardly been mentioned in either academic or popular discourse. Arenas requiring technical knowledge such as air and water pollution or pesticide regulation also do not seem to draw much Buddhist commentary. Issues in regional or local ecosystem protection are apparently better handled by a coalition of local groups, more often nonreligious than religious. Buddhism, however, does offer rich resources for immediate application in food ethics, animal rights, and consumerism—areas which are now developing some solid academic and popular literature. The most basic Buddhist tenet of nonharming provides a strong platform for evaluating animal welfare and animal rights issues, since many of these revolve around degrees of harm to human-impacted animals, whether on factory farms or in zoos. Paul Waldau has written extensively on both Buddhist and Christian attitudes toward animals in his book *The Specter of Speciesism* (Waldau 2002).

Food ethics are evolving rapidly in the West as consumers realize the tremendous costs of globally shipped goods and agriculture based on chemical inputs. In the last decade, farmers' markets and community-supported agriculture have gained great popularity and expanded quickly. Fast-food menus were deeply challenged by Eric Schlosser's research in *Fast Food Nation*, as well as the fast-food experiment in the movie *Supersize Me.* In Italy the Slow Food Movement has taken off as a celebration of cultural values for local, homemade food, especially breads, wines, and cheeses. The demand for organic produce in the West has increased steadily, and in some states such as Vermont and Oregon, organic farmers are a significant portion of the farming community. Students at Buddhist meditation retreats in the United States have come to expect high-quality, thoughtfully prepared meals. At Green Gulch Zen Center in California, students have pressed for locally grown produce of all types as well as fair-trade coffee and tea.

Interest in Buddhist food practices was perhaps ignited by one of Thich Nhat Hanh's famous exercises for mass gatherings: the tangerine meditation (sometimes replaced by the apple meditation or the raisin meditation). In this long guided meditation, students practice mindfulness of touch, smell, taste, first bite, swallowing—in short, every moment in the act of eating. This meditation evoked interest in mindful food practice in general: eating slowly, eating as family practice, eating to support a healthy environment. It raised again the issue of vegetarianism, always a concern in a Buddhist setting. For Westerners exposed to both Buddhist and environmental reasons for not eating meat, ethical food practices can vary substantially (Kaza 2005b).

Consumerism, the social emphasis on "stuff" and status, also lends itself well to Buddhist analysis. Since the Four Noble Truths identify desire as the cause of suffering, Buddhist practice offers useful antidotes to the runaway desire that characterizes a consumer society. In the anthology *Hooked!* (Kaza 2005a), a number of Buddhist teachers, scholars, and practitioners take up Buddhist values, methods, and principles to address the all-penetrating tangle of consumerism that dominates social consciousness today. Buddhist meditative practices are quite useful in taming the impulses of desire that lead to shopping sprees and consumer therapy. Zen teachings that focus on taking apart the ego-self make a good foil to skillful marketers who specialize in identity needs. Initiatives in Thailand and Japan indicate what is possible when Buddhist grassroots organizers or temples take on the institutional structures of consumerism (see Williams 2005 and Hutanawatr and Rasbash 2005).

Buddhist environmental thought has found its way into creative writing as well, in both prose and poetry. A number of Buddhist and Buddhist-leaning poets followed in Gary Snyder's footsteps, taking up subjects of nature or human-nature relations in their poetry. A collection of their works, *Beneath a Single Moon*, pulled together work reflecting Buddhist environmental themes (Johnson and Paulenich 1991). Among nature writers, several authors have alluded to their Buddhist practice as part of what informs their intimate relations with the landscape. Peter Matthiesson wrote eloquently of Zen insights in his book *The Snow Leopard*, set high in the Himalayas on a search for this rare endangered cat. Gary Snyder published two collections of essays, *The Practice of the Wild* (1990) and *A Place in Space* (1995), which developed his Buddhist environmental thought in fresh and pragmatic ways. Gretel Ehrlich, in *Islands, the Universe, Home* (1991), wrote of meditation in the open spaces of the western United States and in *A Match to the Heart* (1994) drew on bardo imagery to describe being hit by lightning.

ENVIRONMENTAL THEMES IN BUDDHIST TRADITIONS

Buddhists taking up environmental concerns are motivated by many fields of environmental suffering—from loss of species and habitats to the consequences of industrial agriculture. Informed by different streams of Buddhist thought and practice, they draw on a range of themes in Buddhist texts and traditions. Many of the central Buddhist teachings seem consistent with concern for the environment, and a number of modern Buddhist teachers advocate clearly for environmental stewardship. As Buddhists develop their contribution to environmental caregiving, they tend to reflect the themes and values of the teachings that are most supportive and useful to their work.

The primary themes or values usually cited as foundational to Buddhist environmental thought originate with the major historical developments in Buddhism—the Theravada traditions of southeast Asia; the Mahayana schools of northern China, Japan, and Korea; and the Vajrayana lineages of Tibet and Mongolia. While Buddhists engaged in environmental work in Asian countries may draw primarily on the teachings of their region, Western Buddhists tend to take hold of whatever seems applicable to the work at hand. This list of themes is not a comprehensive review but rather an introduction to the dominant ideas in Buddhist environmental discourse today.

Theravada Themes

In the earliest Buddhist sutras there are many references to *nature as refuge,* especially trees and caves. The famous story of the Buddha's life begins with his mother giving birth under the shelter of a kindly tree. After young Gotama wandered for years in the forests of India, he took refuge at the foot of a bodhi tree, where he achieved enlightenment. For the remainder of his life, the Buddha taught large gatherings of monks and laypeople in protected groves of trees that served as rainy-season retreat centers for his followers. The Buddha urged his followers to choose natural places for meditation, free from the influence of everyday human activity. Early Buddhists developed a reverential attitude toward large trees, carrying on the Indian tradition regarding *vanaspati* or "lords of the forest." Protecting trees and preserving open lands were considered meritorious deeds. Today in India and southeast Asia many large old trees are often wrapped with monastic cloth to indicate this age-old appreciation for nature as refuge.

One of the first Buddhist teachings on the Four Noble Truths explains the nature of human suffering as generated by desire and attachment. Fully embracing the nature of impermanence, the medicine for such suffering is the practice of compassion (*karuna*) and loving-kindness (*metta*). The early Indian Jataka Tales recount the many former lives of the Buddha as an animal or tree when he showed compassion to others who were suffering. In each of the tales the Buddha-to-be sets a strong moral example of compassion for plants and animals (one collection is Beswick 1956). The first guidelines for monks in the *Vinaya* contained a number of admonitions related to caring for the environment. For example, travel was prohibited during the rainy season for fear of killing the worms and insects that came to the surface in wet weather. Monks were not to dig in the ground or drink unstrained water. Even wild animals were to be treated with kindness (see, e.g., de Silva 2000). Plants too were not to be injured carelessly but respected for all that they give to people.

Early Buddhism was strongly influenced by the Hindu and Jain principle of *ahimsa* or nonharming—a core foundation for environmental concern. In its broadest sense nonharming means "the absence of the desire to kill or harm" (Chapple 1993: 10). Acts of injury or violence are to be avoided because they are thought to result in future injury to oneself. The fourth Noble Truth describes the path to ending the suffering of attachment and desire—the Eightfold Path of practice. One of the eight practice spokes is Right Conduct, which is based on the principle of nonharming. The first of the five basic precepts for virtuous behavior is often stated in its prohibitory form as "not taking life" or "not killing or harming." Buddhaghosa explains: " 'Taking life' means to murder anything that lives. It refers to the striking and killing of living beings. 'Anything that lives'— ordinary people speak here of a 'living being,' but more philosophically we speak of 'anything that has the life-force.' 'Taking life' is then the will to kill anything that one perceives as having life, to act so as to terminate the life-force in it" (as cited in Gaffney 2004: 227).

The first precept applies to environmental conflicts around food production, land use, pesticides, and pollution. The second precept, "not stealing," engages global trade ethics and corporate exploitation of resources. "Not lying," the third precept, brings up issues in advertising that promote consumerism. "Not engaging in abusive relations," interpreted through an environmental lens, can cover many examples of cruelty and disrespect for nonhuman beings. Nonharming extends to *all* beings—not merely to those who are useful or irritating to humans. This central teaching of nonharming is congruent with many schools of ecophilosophy which respect the intrinsic value and capacity for experience of each being.

The Eightfold Path also includes the practice of Right View, or understanding the laws of causality (*karma*) and interdependence. The Buddhist worldview in early India understood there to be six rebirth realms: devas, asuras (both god realms), humans, ghosts, animals, and hell beings. To be reborn as an animal would mean one had declined in moral virtue. By not causing harm to others, one would enhance one's future rebirths into higher realms. In this sense, the law of karma was used as a motivating force for good behavior, including paying respect to all life. Monks were instructed not to eat meat, since by practicing vegetarianism they would avoid the hell realms and would be more likely to achieve a higher rebirth. In one sutra it is said, "If one eats the flesh of animals that one has not oneself killed, the result is to experience a single life (lasting one *kalpa*) in hell. If one eats the meat of beasts that one has killed or one has caused another to kill, one must spend a hundred thousand kalpas in hell" (Shabkar 2004: 68).

A third element of the Eightfold Path, Right Livelihood, concerns how one makes a living or supports oneself. The early canonical teachings indicate that the Buddha prohibited five livelihoods: trading in slaves, trading in weapons, selling alcohol, selling poisons, and slaughtering animals. The Buddha promised a terrible fate to those who hunted deer or slaughtered sheep; the intentional afflicting of harm was particularly egregious, for it revealed a deluded mind unable to see the relationship between slaughterer and slaughtered. Proponents of ethical vegetarianism point out that large-scale slaughtering of animals for food production breaks the Buddha's prohibition. Some Buddhist environmentalists speak of their work as Right Livelihood, a path of practice that serves others and cultivates compassionate action.

Though Buddhism generally places little weight on creation stories (since there is no creator god in the Buddhist view), the *Agganna Sutta* contains one parable of creation in which human moral choices affect the health of the environment. In this story the original beings are described as self-luminous, subsisting on bliss and freely traveling through space. At that time it was said that the earth was covered with a flavorful substance much like butter, which caused the arising of greed. The more butter the beings ate, the more solid their bodies became. Over time the beings differentiated in form, and the more beautiful ones developed conceit and looked down on the others. Self-growing rice arose on the earth to replace the butter, and before long people began hoarding and then stealing food. According to the story, as people erred in their ways, the richness of the earth declined. The point of the sutta is to show that environmental health is bound up with human morality (described in detail in Ryan 1998). Other early suttas spelled out the environmental impacts of greed, hate, and ignorance, showing how these three poisons produce both internal and external pollution. In contrast, the moral virtues of generosity, compassion, and wisdom were said to be able to reverse environmental decline and produce health and purity.

Mahayana Themes

As Buddhist teachings were carried north to China, a number of northern schools of thought evolved, emphasizing different texts, principles, and practices, some of which have now been applied to environmental concerns. The Hua-Yen school of Buddhism of seventh-century China placed particular emphasis on the law of interdependence or mutual causality. Because ecological thinking fits well with the Buddhist description of interdependence, this theme has become prominent in modern Buddhist environmental thought (see, e.g., Cook 1989; Devall 1990; Ingram 1990; and Macy 1991a). Hua-Yen Chinese philosophy perceives nature as relational, each phenomenon dependent on a multitude of causes and conditions that include not only physical and biological factors but also historical and cultural factors.

The *Avatamsaka Sutra* of the Hua-Yen school uses the teaching metaphor of the jewel net of Indra to represent the infinite complexity of the universe. This imaginary cosmic net holds a multifaceted jewel at each of its nodes, with each jewel reflecting all the others. If any jewels become cloudy (toxic or polluted), they reflect the others less clearly. To extend the metaphor, tugs on any of the net lines, for example, through loss of species or habitat fragmentation, affect all the other lines. Likewise, if clouded jewels are cleared up (rivers cleaned, wetlands restored), life across the net is enhanced. Because the net of interdependence includes not only the actions of all beings but also their thoughts, the intention of the actor becomes a critical factor in determining what happens.

The law of interdependence suggests a powerful corollary, sometimes translated as "*emptiness* of separate self." Since all phenomena are dependent on interacting causes and conditions, then nothing exists as autonomous and self-supporting. This Buddhist understanding and experience of self contradicts the traditional Western sense of self as a discrete individual. Interpreting the Hua-Yen metaphor, Gary Snyder suggests that the empty nature of self offers access to "wild mind," the energetic forces that determine the nature of life (Snyder 1995). These forces act outside human influence, setting the historical, ecological, and even cosmological context for all life.

T'ien-t'ai monks in eighth-century China believed in a universal Buddha-nature that dwelled in all forms of life. Sentient (animal) and nonsentient (plant) beings and even the earth itself were seen as capable of achieving enlightenment. This concept of Buddha-nature is closely related to Chinese views of Qi or moving energy, ever changing, taking new form. This view of nature reflects a dynamic sense of flow and interconnection between all beings, with Buddha-nature arising and changing constantly. Buddhist scholar Ian Harris suggests that a Mahayana vegetarian ethic was first formulated around the idea of Buddha-nature. In the *Mahaparinirvana Sutra* Buddha-nature is understood to be an embryo of the Tathagata or the fully enlightened being (Harris 2002). Addressing the ethics of meat-eating, Western Zen teacher Philip Kapleau wrote, "It is in Buddha-nature that all existences, animate and inanimate, are unified and harmonized. All organisms seek to maintain this unity in terms of their own karma. To willfully take life, therefore, means to disrupt and destroy this inherent wholeness and to blunt feelings of reverence and compassion arising from our Buddha-mind" (Kapleau 1982: 19). Taking an animal's life, therefore, is destructive to the Buddha-nature within the animal to be eaten. Kapleau taught that to honor the Tathagata and the potential for awakening, one should refrain from eating meat.

Environmental advocates sometimes call themselves "ecosattvas," those who take up a path of service to all beings. They are following the Mahayana model of the enlightened being or *bodhisattva* who returns lifetime after lifetime to help all who are suffering. Where the early Theravada schools emphasized achieving enlightenment and leaving the world of suffering, the northern schools, influenced by Confucian social codes, placed great value on becoming enlightened to serve others. The *bodhisattva* vow to "save all sentient beings" calls for cultivating compassion for the endless suffering that arises from the fact of existence. Such *bodhisattva* acts of environmental service are marked by a strong sense of intention that reflects a Buddhist virtue ethic (Sponberg 1997). Environmentalists apply this ethic to plant and animal relations as well as to people and societies, promoting environmental stewardship as a path to enlightenment.

Monastic temples in the Ch'an traditions of China were often built in mountainous or forested places. Chinese poets from the fifth century on accumulated an extensive body of literature reflecting a spiritual sense of belonging to wild nature on a cosmological level (Hinton 2002). Japanese schools of Zen influenced haiku and other classic verse forms that cultivated a sense of oneness with nature in the moment. Dogen, founder of the Soto sect of Zen, spoke of mountains and waters as sutras themselves, the very evidence of the dharma arising (see Dogen 1985: 97–107). He taught a method of direct knowing, experiencing this dharma of nature with no separation. For Dogen, the goal of meditation was nondualistic understanding or complete transmission between two beings. Dogen taught that much human suffering generates from egoistic views based in dualistic understanding of self and other. To be awakened is to break through these limited views (of plants and animals) to experience the self and myriad beings as one energetic event.

Vajrayana Themes

Tibetan schools drew on all of the historic teachings transmitted to the far north from China, India, and southeast Asia. Kindness for others was emphasized strongly with the encouragement to treat all sentient beings as having been their mother in a former life. In Santideva's classic eighth-century text on the *bodhisattva* path, the practice of compassion for all beings becomes world transforming. He vows: "Just like space and the great elements such as earth, may I always support the life of all the boundless creatures. And until they pass away from pain, may I also be the source of life for all the realms of varied beings that reach unto the ends of space" (Santideva 1971: 25).

For indigenous Tibetans, the landscape was seen as a sacred mandala, a symbolic representation of Vajrayana teachings. Monks and others for many centuries have gone on pilgrimage to specific mountains to demonstrate their spiritual devotion, sometimes taking years to complete their journeys. Heaps of inscribed prayer stones are placed along stone mountain paths, and prayer flags stream in the winds, offering encouragement to pilgrims traversing the sacred lands. *Stupas*, or relictual shrines, are placed at significant points on the land to draw energy and commemorate important religious leaders. Pilgrims make offerings at these sites, linking the points of energy across the landscape with their own footsteps (Govinda 1976).

Contemporary Themes

Today's Buddhists have drawn on a number of the principles above as supportive teachings for environmental work. Several additional themes have also been popularized by modern Buddhist teachers and practitioners. Vietnamese Zen teacher Thich Nhat Hanh promotes mindfulness as a central stabilizing practice for calming the mind and being present. He works with the teachings of the *Satipatthana Sutta*, providing instructions in mindfulness of body, feelings, mind, and objects of mind, sometimes linking these directly with the most basic actions of eating and walking. Thich Nhat Hanh is one of a handful of Buddhist teachers today who has offered retreats for environmentalists. His word *interbeing* has become popular among Western Buddhists as a way to express the dynamic sense of relationship with the earth. He frequently teaches about interbeing through the example of a piece of paper which holds the sun, the earth, the clouds, and all the beings of the forest (Thich Nhat Hanh 1993). Mindfulness practice in Buddhist retreat centers supports thoughtful food practices, from organic gardening to silent cooking.

Environmentally engaged Buddhists are concerned about the ecological consequences of harmful human activities. Buddhist scholar Kenneth Kraft has proposed the term *eco-karma* to cover the multiple impacts of human choices as they affect the health and sustainability of the earth (Kraft 1997: 277–80). An ecological view of karma extends the traditional view of karma to a general-systems view of environmental processes. Eco-karma might be expressed, for example, as one's ecological footprint—the amount of land, air, and water required for food, water, energy, shelter, and waste disposal. Tracing such karmic streams across the land is one way to understand the human responsibility for environmental stewardship.

Among today's Buddhists, environmental work is regarded as a form of social activism, a practice with a component of advocacy for social change. Activism such as this is called socially engaged Buddhism, a practice path mostly outside the gates of the monastery. Taking up environmental work in this way, there is no sense of separation between the activist work and one's practice. Caring for the environment becomes a practice that engages one fully in the core Buddhist practices. Teaching others about the ecological problems and solutions in this context can be seen as a kind of dharma teaching, offered in the spirit of liberating humans from the suffering they are creating for the earth and themselves. Socially engaged Buddhists have taken up the concerns of nuclear waste, animal factory farming, and consumerism, among others (Kaza and Kraft 2000: 237–351). By working with other Buddhist activists, Buddhist environmentalists gain support in keeping Buddhist practice and philosophy at the heart of their work.

Role for Buddhism in the Environmental Crisis

Will green Buddhist activists play a significant role in addressing the multitude of environmental problems in need of creative solutions? Will scholars of Buddhist environmental thought contribute useful insights to understand human motivation and behavior? Will Buddhist priests and teachers take up environmental concerns as part of their work with students and local communities? How will Buddhism stack up compared with other world religious traditions in affecting the outcome of unsustainable environmental trends? This section reviews the strengths and limitations that are apparent at this early stage of Buddhist environmental engagement, looking at three arenas of activity. Because the field of Buddhism and ecology is evolving at a rapid rate, much more may yet be drawn from the Buddhist teachings and be of help in sorting through the difficult environmental choices that lie ahead.

Strengths

How effective is Buddhist environmental action? And what might make Buddhist environmentalism distinctive from other environmental activism or from other ecoreligious activism? Let us first consider the role for activists and what strengths from Buddhism they might bring to bear on their work. First and perhaps most obvious to others, Buddhist activists would ground their work in regular engagement with Buddhist practice forms. Thich Nhat Hanh, for example, has encouraged activists to recite the precepts together to reinforce guidelines for Right Conduct in the midst of challenging situations. Walking meditation is taught regularly as part of activist retreats at Vallecitos Mountain Refuge in New Mexico and Whole Thinking Retreats in Fayston, Vermont.[3] Practicing with the breath can help sustain activists under pressure in the heat of a conflict. At Green Gulch Zen Center, Earth Day celebrations have been woven into the public event for Buddha's birthday. Environmental activists associated with the Buddhist Peace Fellowship include meditation as part of their regular meeting activities.

Buddhist texts recognize a strong relationship between intention, behavior, and the long-range effects of action. Clarifying one's intention in advocacy work helps prevent a sense of being overwhelmed or burnout. Environmental issues are rarely small and self-contained; one problem leads to another, and many parties are often involved in negotiating a lasting solution. Campaigns or public hearings can be toxic with frustration, anger, and power displays. The Buddhist activist may be able to carry some emotional stability in the face of this heated energy by maintaining clear intention, as holding to the *bodhisattva* vow to reduce suffering and help all beings. This can help other activists clarify their motivation and set the stage for more effective collaboration and division of tasks.

Central to Buddhist teaching and emphasized strongly in the Zen tradition is the focus on breaking through the delusion of the false self, the ego that sees itself as the center of the universe. One antidote for this universal human tendency is the practice of detachment. A green Buddhist approach to activism would include some healthy ego-checking work to see if the activist is motivated by a need to build his or her ego identity as an environmentalist or earth saver. Keeping intention strong but letting go of the need for specific results is a practice in detachment. One recognizes that the outcome of any situation will depend on many factors, not just the contributions of one person. Being receptive to the creative dynamics at play and less identified with a particular end result can produce surprising collaborations. Sulak Sivaraksa calls this "small 'b' Buddhism"— downplaying the ego of being a good Buddhist in favor of being an effective friend to others working toward a common goal (Sivaraksa 1992: 62–72).

Key to a Buddhist approach to problem solving is taking a nondualistic view of reality. This follows from an understanding of self as not separate from all others but rather dynamically cocreated. Most environmental battles play out as confrontations between seeming enemies: treehuggers versus loggers, housewives versus toxic polluters, organic farmers versus corporate seed producers. From a Buddhist perspective, this kind of demonizing destroys spiritual equanimity; thus it is far preferable to act from an inclusive standpoint, listening to all parties involved rather than taking sides. This approach has traditionally been quite rare in environmental problem solving but is becoming more common now as people grow weary from the dehumanizing nature of enemy making. In a volatile situation, a Buddhist commitment to nondualism can help stabilize negotiations and work toward long-term functional relationships.

Buddhist practice is grounded in the fundamental vow of taking refuge in the Three Jewels: the Buddha or teacher, the Dharma or teachings, and the Sangha or practice community. Asian activists usually base their work in relations with local sanghas as an effective grassroots base for accomplishing change. Western Buddhists, handicapped by the Western emphasis on individualism, tend to value sangha practice as the least of the three jewels. They tend to be drawn first to the calming influence of meditation and the moral guidelines of the precepts. Practicing with community can be difficult for students living some distance from Buddhist centers and surrounded by a predominantly Judeo-Christian culture. Building community is crucial for Buddhist environmentalists even though they are geographically isolated from each other and sometimes marginalized by their own peers in Buddhist centers. This has been mitigated substantially by internet organizing, and now, for example, the Green Sangha based in the San Francisco area has an international presence through its existence on the web.[4]

Second, let us consider the role for scholars of Buddhist environmental thought and what aspects of Buddhism might inform their work. This new academic field has engaged both traditional scholars who study but do not practice Buddhism as well as those who both study and practice, the scholar-practitioners. Each has strengths to contribute in growing the field of knowledge. Traditional scholars can bring an objective view, placing environmental perspectives in the broader field of Buddhist studies, helping to legitimize these discussions. Academics such as Ian Harris and Alan Sponberg have made such contributions, raising questions about popular green Buddhism and providing accurate historical background (Harris 2002). Scholar-practitioners such as Rita Gross and Kenneth Kraft bring an experiential understanding of the teachings of their lineages to complement their academic training (see Gross 1998 and Kraft 1997). Scholar-practitioners are generally more comfortable and clear about their intention in doing environmental academic work that it is motivated by their *bodhisattvas* vows, for example. However, their work is sometimes challenged by academics who imply that "arm's length" engagement in one's scholarly pursuits is not possible for practitioners.

Scholars of either persuasion can bring their well-trained minds and analytic skills to critiquing green Buddhism and challenging ungrounded idealistic interpretations. As Buddhism grows in popularity in the West, it is vulnerable to mistaken views, blurred with New Age ideas of individually designed spirituality. Scholars grounded in the original texts can check emerging ideas for distortions of Buddhist thought. Tibetan Buddhist texts and training are particularly strong in methods of analysis. Judith Simmer-Brown uses these to understand the "empty" nature of globalization and the possibility for other forms of sustainability to arise (Simmer-Brown 2005). Ian Harris has examined the popular interpretation of Buddhism as the most environmentally friendly of the world religions, arguing that the historical record shows much more ambivalence (Harris 1997). Scholars of Buddhist environmental thought are also in a good position to critique the tenets of monotheistic traditions that act as a deterrent to seeking a sustainable future. Rita Gross, for example, has questioned the strong pronatalist positions of the Christian church as problematic in dealing with exponential population growth and its impacts on the planet (Gross 1997).

The strength of academic work in Buddhist environmental thought lies in legitimizing this new field in the eyes of traditional schools of religion and philosophy. Thus far the list of academic volumes addressing environmental concerns from a Buddhist perspective is still fairly small. Though Buddhist insights are usually included in panreligious commentaries on the environment, entire volumes by single or multiple Buddhist scholars are quite rare. It is likely that new work will build on the first round of anthologies and take up specific aspects of environmental concern, as *Hooked!* (2005) has done with consumerism. Topics that already lend themselves to academic analysis such as food ethics and animal rights may be the next work to emerge from the greening ivory tower.

What about Buddhist priests, monks, and teachers? What strengths do they bring to environmental discourse and action? East or West, ordained Buddhists often are in leadership positions within their local temples. As leaders they can adapt the practice forms to new settings, including concern for the planet as part of their community responsibility. One Zen teacher served only bread and water for a day during a weeklong retreat, surprising the students with their own disappointed expectations and using this as a springboard to raise issues of poverty and inequity around the world. Another teacher regularly holds ceremonies for victims of major disasters such as Hurricane Katrina or the earthquake in Pakistan in 2005. Several lay teachers have developed practice forms that take place in the garden to incorporate the presence of plants in memorial ceremonies. One Japanese Pure Land priest has galvanized his entire sangha to place solar panels on the roof of the temple to help reduce global climate change (Williams 2005).

Buddhist centers that interface with the public can serve as models of environmentally sustainable practices (see Kaza 1997). Through architectural design choices and monastic example, visitors can see the possibilities for energy and water conservation. Through exposure to mindful kitchen practice, retreatants can learn about the food they eat and its origins. The leadership role of the head priest or teacher is often necessary for environmental concerns to be emphasized in everyday practice. Where a Buddhist teacher has shown environmental commitment, the centers tend to reflect that commitment. Vermont Zen Center recently added a new dining hall and housing wing, and head teacher Sunyana Graef led the effort to follow green building principles. With her support, the grounds have been transformed through extensive volunteer efforts from community members, returning trees to a suburban lawn and cultivating spaces for thoughtful reflection (such as the lovely Jiso rock garden). In New York, John Daido Loori and his students at Zen Mountain Monastery lead summer canoeing and wilderness programs in the Adirondacks to deliberately place students in contact with the forces of nature. This has become a hallmark of the center, and now Daido has cemented his concern for the pristine areas of the northern mountains by purchasing a piece of lakeshore where his students can monitor water quality.

More and more, experienced Buddhist teachers are being asked to provide meditation instruction for environmental advocates. When the ecosattva chapter of the Buddhist Peace Fellowship prepared for protests against old-growth redwood logging in northern California, they trained at Green Gulch Zen Center. When high-ranking executives in the Trust for Public Land sought to revitalize their commitment to land conservation values, they developed staff meditation retreats at Vallecitos Mountain Refuge. Thus far Buddhist practice techniques have been applied much more extensively to hospice and health-care settings, prison work, and AIDS assistance. But, bit by bit, the work with environmental issues is adding up. In the West, Buddhist meditation instruction is perceived to be neutral training available for people of any faith or secular persuasion. It is generally not seen as proselytizing. Environmentalists who tend to reject organized religion and find spiritual fulfillment in the outdoors are open to Buddhist support for their environmental aims. The possibility of this work first emerged in a retreat for environmentalists organized by Thich Nhat Hanh's followers at Ojai, California, in 1993. Thich Nhat Hanh oriented the retreat around taking care of the environmentalists, sensing that burnout was rampant for those driven by concern for the plight of so many suffering beings and places.

Limitations

So far this would appear to be a rosy picture, filled with useful options for Buddhists interested in supporting environmental action. But critics have already pointed out significant barriers to any extended Buddhist influence in environmental work, at least in the West. One philosophical problem is that there is no single view of nature or environment that crosses all the Buddhist traditions. David Eckel has described in some detail the difference in Indian Buddhist views of nature compared with Japanese views, for example. These views represent very different time periods and cultures; Eckel finds it problematic that Westerners looking for the "green" in Buddhism blur over these major distinctions (Eckel 1997). Some have called this process "mining" the tradition for what you want from it, a common human tendency among all the religious traditions, not so different than what is done to support fundamentalist Christian interpretations. Green Buddhism could suffer from the same sort of myopic views unless it encourages further understanding of Buddhism itself.

Further, Buddhism is not a nature religion per se, as are pagan or native American traditions which base their spiritual understandings on relations with the land and its living beings. The central principles of Buddhism deal with human suffering and liberation from that suffering; the process of insight awareness is not dependent on the land or any physical forms. It is much more of a mental process, cultivating capacities in the human mind. Thus, at its roots, Buddhism does not immediately lend itself to environmental concern. In fact, since the Buddhist approach can work within any situation, environmental sustainability is not necessarily a prerequisite or a goal for liberation practice. The practice of detachment to hobble the power of desire could actually work against such environmental values as "sense of place" and "ecological identity."

Alan Sponberg critiques the green Buddhist emphasis on interdependence, suggesting that green Buddhists may be stepping too far away from the core spiritual development challenges in Buddhist training (Sponberg 1997). Though the law of interdependence interfaces very well with similar laws of ecology, this alone is not enough, in his opinion, to lead a practitioner to enlightenment. Ian Harris critiques Joanna Macy for taking the metaphor of Indra's jewel net too far and missing the original teaching emphasis, which was on karma, not ecology (Harris 2002). He is wary of Buddhist activists who interpret key Buddhist principles too narrowly, from *only* an environmental point of view. For Harris, the project of "saving the world" is not a central concern, and dragging Buddhist concepts into the process may not be necessary or even helpful. He joins Eckel and Lambert Schmidthausen in exposing the lack of concern for animals and nature in many of the Pali Canon texts.

To this point, green Buddhism has taken up specific environmental problems primarily in countries which already have a significant Buddhist population. Thai forest monks and Sulak Sivaraksa's Grassroots Leadership Training Program have gained some ground in protesting lake pollution, fish die-offs, and clear-cutting of forests. Tibetans in exile in India have been able to undertake environmental education programs with local Tibetans, but they have had virtually no impact on the rampant exploitation of Tibet's natural resources by the Chinese. In the West, green Buddhists such as the Green Sangha have taken on energy conservation and recycling as everyday actions, but their impact has been fairly local. Green Buddhists have not yet been significant players in some of the Western interfaith environmental initiatives which *are* making a difference: the global Jubilee Debt forgiveness campaign, religious advocacy for corporate social responsibility through stockholder actions, and the interfaith power and light movement for alternate energy purchasing. This is partly because green Buddhists are still so few in number, but it may also be because Buddhism as a tradition does not carry the same charge for social justice as the monotheistic traditions. Righting environmental wrongs is often a situation of injustice for those who are harmed, whether plants, animals, ecosystems, or people. Buddhist virtue ethics does address these wrongs, but not with the same fire as Judaism and Christianity.

A further critique of green Buddhism in the West is that it has had so little influence solving real environmental problems in Buddhist countries. For most people it is local environmental problems that catch their attention; possibilities for local action seem more accessible than those on a global scale. As a consequence, few Westerners are actively working to stop or reverse environmental devastation in the countries that spawned their beloved religion. Some members of Buddhist Peace Fellowship have joined in solidarity with the International Network of Engaged Buddhists on their environmental campaigns. But for the most part it is difficult for Westerners to engage Asian problems from afar and from a different cultural perspective. For some Asians, Westerners are seen as part of the problem, due to their disproportionate consumption of planetary resources.

Nevertheless, despite these limitations, interest in Buddhist environmental thought and action is very strong in both the West and East. Misinterpretations, mistaken views, and idealized projections are perhaps inevitable for any young movement as it takes shape. At this point the environmental movement itself is so well established in large and small nonprofit advocacy groups, in state and federal legislation, and in campus sustainability actions that it hardly needs a Buddhist contribution. But in small supportive ways it may be that Buddhism will yet take its place in shaping the direction of environmental problem solving around the globe.

CONCLUSION

Buddhist environmental thought is both ancient and brand new. While many Buddhist principles handed down from centuries ago seem broadly applicable to environmental concerns, articulating those applications is still a very new project of the last few decades. Scholars of Buddhist environmental thought have many topics yet to address. Green Buddhist activists have barely begun to make a unified impact. This is a movement of both thought and action to track over the next few decades.

Has a Buddhist environmental movement coalesced around the globe? Not at all. Only a tiny handful of organizations have been formed to promote Buddhist environmental views and approaches. No clearly defined environmental agenda or set of principles has been agreed upon by any group of self-identified green Buddhists. All this is perhaps too much to expect of a fledgling movement. It may yet be that in ten years many more books will be published offering Buddhist views regarding environmental concerns. It may yet be that green Buddhist centers will be established for the express purpose of fostering environmental sustainability, a sort of green Catholic Worker house model. As more and more serious students in the West become teachers and temple leaders, some may take up leadership roles cultivating mindfulness around environmental issues.

What is completely unknown is what larger forces and events will shape all environmental concern and activity. In 2005 alone the record number of hurricane-strength storms generated more environmental disasters than could be handled effectively. Global climate change, the shrinking supply of oil, and the lack of available drinking water may be much more powerful forces shaping human behavior than any religious tradition. All this is yet to unfold. But certainly Buddhists of all traditions and cultures would be welcome to join the much needed efforts to turn the tide from further planetary destruction.

NOTES

1. See, e.g., the program in religion and nature at the University of Florida website: http://www.religion.ufl.edu/gradprog/field-nature.html.
2. For a history of the group's academic sessions, see www.religionandnature.com/aar.
3. For Vallecitos Mountain Refuge, see www.vallecitos.org; for Whole Thinking Retreats, see www.centerforwholecommunities.org.
4. See the projects listed on www.greensangha.org.

REFERENCES

Batchelor, Martine, and Kerry Brown, eds. 1992. *Buddhism and Ecology*. London: Cassel.
Beswick, Ethel. 1956. *Jataka Tales: Birth Stories of the Buddha*. London: Murray.
Boston Research Center for the 21st Century. 1997. *Buddhist Perspectives on the Earth Charter*. Cambridge: Boston Research Center.
Callicott, J. Baird, and Roger T. Ames, eds. 1989. *Nature in Asian Tradition of Thought: Essays in Environmental Philosophy*. Albany: State University of New York Press.

Chapple, Christopher Key. 1993. *Nonviolence to Animals, Earth, and Self in Asian Traditions*. Albany: State University of New York Press.

Cook, Francis H. 1989. "The Jewel Net of Indra." Pp. 213–30 in *Nature in Asian Traditions of Thought*. Edited by J. Baird Callicott and Roger T. Ames. Albany: State University of New York Press.

Darlington, Sue. 1998. "The Ordination of a Tree: The Buddhist Ecology Movement in Thailand." *Ethnology* 37.1:1–15.

de Silva, Lily. 2000. "Early Buddhist Attitudes toward Nature." Pp. 91–103 in *Dharma Rain: Sources for a Buddhist Environmentalism*. Edited by Stephanie Kaza and Kenneth Kraft. Boston: Shambhala.

Devall, Bill. 1990. "Ecocentric Sangha." Pp. 155–64 in *Dharma Gaia: A Harvest of Essays in Buddhism and Ecology*. Edited by Alan Hunt-Badiner. Berkeley: Parallax.

Dogen. 1985. "Mountains and Waters Sutra." Pp. 97–107 in *Moon in a Dewdrop: Writings of Zen Master Dogen*. Edited by Kazuaki Tanahashi. San Francisco: North Point.

Eckel, Malcolm David. 1997. "Is there a Buddhist Philosophy of Nature?" Pp. 327–50 in *Buddhism and Ecology: The Interconnection of Dharma and Deeds*. Edited by Mary Evelyn Tucker and Duncan Williams. Cambridge: Center for the Study of World Religions, Harvard Divinity School.

Ehrlich, Gretel. 1991. *Islands, the Universe, Home*. New York: Viking.

———. 1994. A Match to the Heart. New York: Penguin.

Gaffney, James. 2004. "Eastern Religions and the Eating of Meat." In *Food for Thought: The Debate over Eating Meat*. Edited by Steven Sapontzis. Amherst, NY: Prometheus.

Govinda, Anagarika. 1976. *Psycho-cosmic Symbolism of the Buddhist Stupa*. Emeryville, CA: Dharma.

Gross, Rita. 1997. "Buddhist Resources for Issues of Population, Consumption, and the Environment." Pp. 291–311 in *Buddhism and Ecology: The Interconnection of Dharma and Deeds*. Edited by Mary Evelyn Tucker and Duncan Williams. Cambridge: Center for the Study of World Religions, Harvard Divinity School.

———. 1998. *Soaring and Settling*. New York: Continuum.

Habito, Ruben L. F. 1993. *Healing Breath: Zen Spirituality for a Wounded Earth*. Maryknoll, NY: Orbis.

Harris, Ian. 1997. "Buddhism and the Discourse of Environmental Concern: Some Methodological Problems Considered." Pp. 377–402 in *Buddhism and Ecology: The Interconnection of Dharma and Deeds*. Edited by Mary Evelyn Tucker and Duncan Williams. Cambridge: Center for the Study of World Religions, Harvard Divinity School.

———. 2002. "Buddhism and Ecology." Pp. 113–36 in *Contemporary Buddhist Ethics*. Edited by Damien Keown. Richmond, Surrey: Curzon.

Hinton, David, trans. 2002. *Mountain Home: The Wilderness Poetry of Ancient China*. New York: New Directions.

Hunt-Badiner, Alan, ed. 1990. *Dharma Gaia: A Harvest of Essays in Buddhism and Ecology*. Berkeley: Parallax.

Hutanawatr, Pracha, and Jane Rasbash. 2005. "No River Bigger than *Tanha*." Pp. 104–21 in *Hooked! Buddhist Writings on Greed, Desire, and the Urge to Consume*. Edited by Stephanie Kaza. Boston: Shambhala.

Ingram, Paul O. 1990. "Nature's Jeweled Net: Kukai's Ecological Buddhism." *The Pacific World* 6.50–64.

James, Simon P. 2004. *Zen Buddhism and Environmental Ethics*. Hampshire, U.K.: Ashgate.
Johnson, Kent, and Craig Paulenich, eds. 1991. *Beneath a Single Moon: Buddhism in Contemporary American Poetry*. Boston: Shambhala.
Kapleau, Philip. 1982. *To Cherish All Life: A Buddhist Case for Becoming Vegetarian*. San Francisco: Harper & Row.
Kaza, Stephanie. 1997. "American Buddhist Response to the Land: Ecological Practice at Two West Coast Retreat Centers." Pp. 219–48 in *Buddhism and Ecology: The Interconnection of Dharma and Deeds*. Edited by Mary Evelyn Tucker and Duncan Williams. Cambridge: Center for the Study of World Religions, Harvard Divinity School.
———. 2005a. *Hooked! Buddhist Writings on Greed, Desire, and the Urge to Consume*. Boston: Shambhala.
———. 2005b. "Western Buddhist Motivations for Vegetarianism." *Worldviews: Environment, Culture, Religion* 9.3:385–411.
Kaza, Stephanie, and Kenneth Kraft, eds. 2000. *Dharma Rain: Sources for a Buddhist Environmentalism*. Boston: Shambhala.
Kerouac, Jack. 1958. *Dharma Bums*. New York: Buccaneer Books.
Kraft, Kenneth. 1997. "Nuclear Ecology and Engaged Buddhism." Pp. 269–90 in *Buddhism and Ecology: The Interconnection of Dharma and Deeds*. Edited by Mary Evelyn Tucker and Duncan Williams. Cambridge: Center for the Study of World Religions, Harvard Divinity School.
Macy, Joanna. 1991a. *Mutual Causality in Buddhism and General Systems Theory*. Albany: State University of New York Press.
———. 1991b. *World as Lover, World as Self*. Berkeley: Parallax.
Martin, Julia. 1997. *Ecological Responsibility: A Dialogue with Buddhism*. Delhi: Tibet House/Sri Satguru.
Matthiesson, Peter. 1978. *The Snow Leopard*. New York: Viking.
Ryan, P. D. 1998. *Buddhism and the Natural World: Toward a Meaningful Myth*. Birmingham: Windhorse.
Santideva. 1971. *A Guide to the Bodhisattva's Way of Life*. Translated by Stephen Batchelor. Dharamsala, India: Library of Tibetan Works and Archives.
Schlosser, Eric. 2001. *Fast Food Nation*. New York: Houghton Mifflin.
Schmithausen, Lambert. 1991. *Buddhism and Nature*. Studio philologica buddhica Monograph 7. Tokyo: International Institute for Buddhist Studies.
Seed, John, Joanna Macy, Pat Fleming, and Arne Naess. 1988. *Thinking like a Mountain: Toward a Council of All Beings*. Philadelphia: New Society.
Shabkar. 2004. *Food of Bodhisattvas: Buddhist Teachings on Abstaining from Meat*. Translated by Padmakara Translation Group. Boston: Shambhala.
Simmer-Brown, Judith. 2005. "Cultivating the Wisdom Gaze." Pp. 89–103 in *Hooked! Buddhist Writings on Greed, Desire, and the Urge to Consume*. Edited by Stephanie Kaza. Boston: Shambhala.
Sivaraksa, Sulak. 1992. *Seeds of Peace*. Berkeley: Parallax.
———, ed. 2001. *Santi Pracha Dhamma: Essays in Honor of the Late Puey Ungphakorn*. Bangkok: Santi Pracha Dhamma Institute.
Snyder, Gary. 1969. *The Real Work: Interview and Talks, 1964–1979*. New York: New Directions.
———. 1990. *The Practice of the Wild*. San Francisco: North Point.
———. 1995. *A Place in Space*. Washington, DC: Counterpoint.

————. 1999. *The Gary Snyder Reader*. Washington, DC: Counterpoint.

Sponberg, Alan. 1997. "Green Buddhism and the Hierarchy of Compassion." Pp. 351–76 in *Buddhism and Ecology: The Interconnection of Dharma and Deeds*. Edited by Mary Evelyn Tucker and Duncan Williams. Cambridge: Center for the Study of World Religions, Harvard Divinity School.

Thich Nhat Hanh. 1993. "The Sun My Heart." In *Love in Action*. Berkeley: Parallax.

Tucker, Mary Evelyn, and Duncan Williams, eds. 1997. *Buddhism and Ecology: The Interconnection of Dharma and Deeds*. Cambridge: Center for the Study of World Religions, Harvard Divinity School.

Waldau, Paul. 2002. *The Specter of Speciesism*. New York: Oxford University Press.

Williams, Duncan. 2005. "Green Power in Contemporary Japan." Pp. 225–36 in *Hooked! Buddhist Writings on Greed, Desire, and the Urge to Consume*. Edited by Stephanie Kaza. Boston: Shambhala.

CHAPTER 8

....

ISLAM

....

RICHARD C. FOLTZ

The world is frozen; its name is inanimate, which means "frozen,"
  O master.
Wait till the rising sun of Resurrection that thou mayest see the
  movement of the world's body.
Since God hath made Man from dust, it behooves thee to rec-
  ognize the real nature of every particle of the universe,
That while from this aspect they are dead, from that aspect they
  are living; silent here, but speaking yonder...
They all cry, "We are hearing and seeing and responsive, though
  to you, the uninitiated, we are mute."

                                              Jalal al-Din Rumi (1207–73)

ABOUT 1.2 billion people around the world today identify at least on some level with the 1,400-year-old Islamic tradition. If religions are understood to be the major source of value systems by which individuals and societies rank possible outcomes and make decisions about their own behaviors, it would seem that the adoption of a caring and nonexploitative ethic toward the environment by Muslims would presuppose the existence in Islamic tradition of principles which accord value to the natural world. This essay attempts to show that such principles are indeed to be found in Islam and discusses some of the ways that contemporary Muslims throughout the world are seeking to apply these principles in response to the global environmental crisis. It will also point out some of the cultural and political obstacles facing those who would implement Islamic guidelines for preserving the environment.

SOURCES OF ISLAMIC
ENVIRONMENTALISM

In recent years a number of Muslim writers, living mainly in the West, have published essays to the effect that, based on the scriptural sources of the tradition, Islam is an ecologically oriented religion. Whereas historically Muslim legal scholars and philosophers, when they addressed issues of the natural world, were concerned primarily with constructing theoretical arguments about social justice among humans, Islamic environmental ethics as articulated by contemporary writers tends more to be rooted in more practical terms, often by way of response to Lynn White's well-known critique of Western Christianity. Iqtidar Zaidi, for example, is clearly paraphrasing White when he states that the ecological crisis is "a crisis rooted in moral deprivation" (Zaidi 1981: 35). Iranian American philosopher Seyyed Hossein Nasr has made the connection between the West's spiritual and environmental crises since the 1950s and anticipated White's 1967 argument in his own lectures given at the University of Chicago earlier the same year (Nasr 2000).

It may be useful to restrict use of the term *Islamic* to that which can be derived from the canonical sources of Islam, as opposed to the activities or attitudes of Muslims, which may or may not be directly motivated by those sources. In other words, there is a distinction to be made between *Islamic* environmentalism—that is, an environmentalism that can be demonstrably enjoined by the textual sources of Islam—and *Muslim* environmentalism, which may draw its inspiration from a variety of sources, possibly including but not limited to religion (Hamed 1993: 146). To date it must be acknowledged that much if not most of the environmentalism that one sees throughout the Muslim world follows Western secular models rather than Islamic ones.

Muslims have always been culturally diverse, and never more so than today when they number a billion or more and inhabit every corner of the globe. Historically the one indisputable source of authority which all Muslims have agreed upon is the will of Allah as expressed in the revealed scripture of the Qur'an. In addition, the Sunni majority (perhaps 80 percent of all Muslims) accepts six collections of reports about the deeds and words of the Prophet Muhammad, called *hadīths*, as supplementary sources of authority. (Shi'ites agree with some, but not all of these reports and have compiled collections of their own.)

Contemporary Muslim environmentalists seeking to articulate an environmental ethic in Islamic terms have therefore attempted to base their arguments on the Qur'an and *hadīths*, giving comparatively little attention to possible cultural contributions from the various societies in which Muslims live. These arguments assume that local or regional attitudes cannot form a basis for any kind of universal Islamic ethic, since they are almost invariably criticized by Islamists as "accretions" and therefore un-Islamic.

The politics of environmental activism among Muslims, where present, have tended to be region specific. From an Islamist perspective, the mere involvement of Muslims does not make an activity or ideology Islamic; this requires a basis in the Qur'an and the *sunna* (the example of the Prophet Muhammad, as attested in *hadīth* reports). This is not to suggest, however, that broader cultural contributions by Muslims living in diverse societies around the world will not be significant in addressing the environmental crisis.

For an idea to achieve anything approaching universal acceptance by Muslims as Islamic, it must be convincingly demonstrated that it derives from, or is at least compatible with, the principles laid down in the Qur'an and the *sunna*. Applying the principles of analogical reasoning (*qīyas*) and consensus among scholars (*ijma'*) to these textual sources, Islamic legal scholars from the eighth–tenth centuries devised an all-encompassing social model known as the *sharī'a*, or Islamic law. The *sharī'a* contains numerous guidelines, restrictions, and injunctions which have environmental implications, in particular regarding the maintenance of preserves, distribution of water, and the development of virgin lands, but historically the Islamic legal code was applied only selectively and is practically not applied at all in most of the Muslim world today.

Muslim environmentalists seeking to articulate a contemporary Islamic environmentalism have tended to begin with the Qur'anic concept of stewardship, expressed by the Arabic term *khilāfa* (lit., "successor"). The following verses are cited: "I am setting on the earth a vice-regent [*khalīfa*]" (2.30); and, "It is he who has made you his vice-regent on earth" (6.165). Similarly, a *hadīth* states that "verily, this world is sweet and appealing, and Allah placed you as vice-regents therein; he will see what you do." One contemporary scholar has gone so far as to suggest that "viceregency forms a test which includes how human beings relate to the environment," but such interpretations have not yet found resonance on a broad scale (Abu-Sway 1998).

Some Muslim writers, like thirteenth-century Sufi poet Jalal al-Din Rumi, play with established Islamic notions of cosmic hierarchy for pedagogical purposes. For example, it would seem that to Rumi that some humans are less than animals: "Wolf and bear and lion know what love is: he that is blind to love is inferior to a dog!" (Rumi 1926: 5.2008). This is a mere literary device, however, since Rumi, like most Muslim philosophers of his time, takes for granted the "great chain of being" model adopted from Aristotle. Thus he can say,

> I died to the organic state and became endowed with growth
> Then I died to vegetable growth and attained to the animal
> I died from animality and became a Man; why then should I fear?
> When have I become the less for dying? (Rumi 1926: 3.3901–2)[1]

The Qur'anic concept of *tawhīd* ("unity") has historically been interpreted by Muslim writers mainly in terms of the oneness of God (in contradistinction to polytheism), but some contemporary Islamic environmentalists prefer to see *tawhīd* as meaning "all-inclusive," a notion that finds its greatest resonance in the Sufi mystical tradition. A few even suggest that the idea of *wahdat al-wujūd* ("unity of being") associated with medieval Sufi philosopher Ibn 'Arabi can be understood in environmentalist terms. Ibn 'Arabi, however, has always been a highly controversial figure for Muslims, since many have accused him of holding pantheist or monist views incompatible with Islam's radical monotheism. Likewise, the notions of certain Sufi mystics are felt by many Muslims to blur dangerously the distinction between the creator and creation. This discomfort appears most often in accusations of *shirk* ("associating partners with God"), which is considered by many Muslims to be "the one sin that Allah cannot forgive." Thus, as in certain Christian contexts, Muslim environmentalists today are prone to accusations that they make of nature a "false idol."

In support of the more inclusive interpretation of *tawhīd*, a Qur'anic verse (17.44) is often cited which states that all creation praises God, even if this praise is not expressed in human language. Based on this verse and others like it, an argument has been made that nature—all of creation, in fact—is "*muslim*," that is, "submits to the (natural) laws of God," with the sole exception of human beings who, having free will, may or may not submit to these laws. Humans thus face a unique danger (i.e., eternal damnation), but also a unique opportunity since mainstream Islam considers that our species alone can be rewarded with eternal life. As Rumi puts in the *Masnavi*, "Man disbelieves in the glorification uttered by inanimate things, but those inanimate things are masters in performing worship" (1926: 3.1497). On the other hand, "All created beings are glorifiers of God, but that compulsory glorification is not wage-earning" (1926: 3.3289).

In the Islamic cosmology, human distinctiveness would not seem to go beyond their unique capacity to gain eternal life through choosing to serve God. The Qur'an states that "there is not an animal in the earth, nor a flying creature on two wings, but they are peoples like unto you" (6.38). There might be a basis here for tempering the hierarchical notion of stewardship implied in the concept of *khilāfa*. Animals, who praise God and whose prayers are listened to by him, are argued to have an existence and a purpose which is independent of their usefulness to humans, although at the same time the Qur'an does mention certain animals—such as cattle and beasts of burden—as having been created to serve humans.

The Qur'an describes Islam as the religion of *fitra* ("the very nature of things"). Humans are considered to be *muslim* by nature, though they can be distracted from this path of submission to the divine will. In Sufism especially, the human aim is seen as a quest to reestablish this primordial union with God. Extending the *fitra* concept, some contemporary thinkers reason that a genuinely

Islamic lifestyle will "naturally" be environmentally sensitive, if only the "distractions" of selfish worldly existence can be removed (Chishti 2003).

Traditional accounts of the deeds and sayings of the Prophet Muhammad, which together with the Qur'an have formed the basis for Islamic law, emphasize compassion toward nonhuman animals. The prophet is reported as having said, "If you kill, kill well, and if you slaughter, slaughter well. Let each of you sharpen his blade and let him spare suffering to the animal he slaughters"; and also, "For [charity shown to] each creature which has a wet heart (i.e., is alive), there is a reward." Muslims are urged to respect plant life as well, as in the prophetic saying, "Some trees are as blessed as the Muslim himself, especially the palm."

The Qur'an contains judgment against those who despoil the earth: "And when he turns away [from you] his effort in the land is to make mischief therein and to destroy the crops and the cattle; and Allah loves not mischief" (2.205); and, "Do no mischief on the earth after it has been set in order" (7.85). Wastefulness and excess consumption are likewise condemned: "O Children of Adam! Look to your adornment at every place of worship, and eat and drink, but be not wasteful. Lo! He [Allah] loves not the wasteful" (7.31). The Qur'an repeatedly calls for maintaining balance in all things (13.8; 15.21; 25.2; and elsewhere). Certain sayings of the prophet seem particularly relevant to contemporary issues of sustainability: "Live in this world as if you will live in it forever, and live for the next world as if you will die tomorrow"; and, "When doomsday comes if someone has a palm shoot in his hand, then he should plant it" (Ibn Hanbal, *Musnad* 5.440 and 3.184).[2]

It is often argued by Muslim environmentalists today that the Islamic legal tradition (*sharī'a*), in both its Sunni and Shi'i variants, if applied to the letter, contains adequate restrictions to ensure a use of natural resources that is both sustainable and just. Resources such as water, air, and wildlife are deemed to be common property, and the utilization of resources is supposed to be determined by the following principles, among others: (1) nonvital needs are subordinated to vital needs; (2) the needs of the poor take precedence over the needs of the rich; and (3) one may not cause harm in order to obtain a benefit. As in any human society, however, these principles remain ideals and have often not been followed in practice.

Moreover, the conditions facing Muslims today are in many important respects very different from those which faced Muslims ten to twelve centuries ago when the legal codes were devised. Direct application of *sharī'a* guidelines to contemporary environmental problems is a matter for interpretation (*ijtihād*) by legal scholars (*'ulema'*).

To date such an effort is in its very early stages. One contemporary legal scholar, Mustafa Abu-Sway, argues that *hadīth* reports which enjoin Muslims from relieving themselves on public pathways or into water sources can be understood "to prevent pollution in the language of today." Since we now know that discharging toxic chemicals and waste into the water supply is harmful to human health, Abu-Sway reasons that "by analogy, from the perspective of the *sharī'a*, this is prohibited" (Abu-Sway 1998). To date, however, few Islamic scholars seem to have taken up the challenge of debating the merits of such interpretations.

STILL A MARGINAL APPROACH

Indeed, one may ask whether the ecological interpretations of the scriptural sources by a small number of Islamic environmentalists, the most prominent of whom live in the West and write for Western audiences, are in any way representative of the attitudes of Muslims on any kind of significant worldwide scale. Islam has not figured prominently in emerging discussions on religion and the environment; rather, the same articles and faces keep appearing in anthologies and at meetings, little more than tokens of Islamic representation.

For the most part contemporary Muslim writers on the environment have characterized environmental degradation as merely a symptom of social injustice. The problem is not, it is argued, that humans as a species are destroying the balance of nature, but rather that *some* humans are taking more than their share. Muslim environmentalists such as Fazlun Khalid, Yasin Dutton, Omar Vadillo, and Hashim Dockrat have argued that the entire basis for the contemporary global economy is un-Islamic and that the singular goal of maximizing profits for investors and speculators leads to massive, environmentally destructive "development" projects which disproportionately benefit wealthy elites, rather than smaller scale projects that would be of more value to the truly needy. If, on the other hand, in accordance with the Qur'anic prohibition of interest taking (*ribā*), the interest-based global banking system is eliminated, then there will be no more environmentally destructive development projects, and there will be plenty of resources for all.

Similarly, Muslim environmentalists often dismiss overpopulation as a non-issue. The problem is stated to be the restriction of movement; if visa restrictions are eliminated, then people will simply migrate from overpopulated areas to "underpopulated" ones. In assigning blame for environmental degradation Muslim writers focus instead on overconsumption by Westerners and by elites in developing countries, pointing out that the average U.S. citizen consumes as much as sixty-five Bangladeshis. Furthermore, in light of the prioritizing of vital needs over luxury needs in Islamic law, the overexploitation of marginal lands by desperate, impoverished masses is seen as more forgivable than the endless mining of resources for the benefit of insatiable Western appetites.

In recent times global initiatives on birth control and women's reproductive rights have been most strongly opposed in Muslim countries. Such efforts are frequently met with accusations that "the West is trying to limit the number of Muslims." Warnings of starvation and deprivation from overpopulation generally elicit the response that "God will provide," which draws its support from the Qur'anic verse (11.6) that reads, "There is no beast upon the earth for which Allah does not provide."

Yet in Islamic law there are no inherent barriers to practicing contraception. Classical theologian Abu Hamid Muhammad Ghazali (1058–1111), who has been called "the second greatest Muslim after Muhammad" and whose writings remain highly influential throughout the Muslim world today, argues in his book *The Proper Conduct of Marriage* (*Kitāb adāb al-nikā*) that birth control in the form of *coitus interruptus* (*'azl*) is permitted in Islam. He suggests, furthermore, that "the fear of great hardship as a result of having too many children... is also not forbidden, since freedom from hardship is an aid to religious devotion." In response to the Qur'anic verse cited above, Ghazali comments that "to examine consequences... while perhaps at odds with the attitude of trust in Providence, cannot be called forbidden" (Ghazali 1998: 79).

In fact the founder of the Family Planning Society of Kenya, Yusuf Ali Eraj, is a Muslim. "Using birth control does not mean following the West," he says. "Muslims can practice family planning, following Islamic principles. It is a *myth* that Islamic doctrine opposes it! All methods are approved, even sterilization" (Hope and Young 2000: 173). Despite these arguments, many Muslims still see arguments against having more children than one can afford as being symptomatic of unbelief (*kufr*). Today, Iran is the only Muslim country where an official policy of birth control and reduction of birth rate is backed up with Islamic rhetoric, though such efforts are beginning to appear on a small scale in some other Muslims societies as well.

The traditional Muslim response to the kinds of doomsday scenarios propagated by Western environmentalists is that of *tawakkul*, or trust in God (Qur'an 5.23; 14.11–12; 65.3; 25.58; 26.217–18), a tendency which is often perceived by Westerners as fatalism. Among Muslim ethicists today there is far greater interest in human-centered issues of justice than in the biosphere as an integral whole. Islam holds that the world is a passing phenomenon, created to serve God's purpose, which will cease to be once that purpose has been fulfilled. Islam likewise emphasizes the relationship between humans and God above all else and has by comparison little to say about the importance of our myriad fellow creatures.

Whether the true essence of Islam is proenvironment or not, in practice throughout most of its history Muslim theologians, philosophers, and laypersons have focused almost exclusively on the relationship between Allah and humanity. Iqtidar Zaidi implicitly confirms this when he states that "we are seeking a religious matrix which maintains man's position as an ecologically dominant being" (1981: 36). Indeed, one Muslim writer recently concluded that "Islamic anthropocentrism negates the claims of Islamic ecology" (Erdur 1997: 160). Given the importance of the petroleum industry and the widespread pursuit of materialistic, consumption-oriented lifestyles in numerous Muslim-majority countries, it would appear that Muslims must now share some of the blame for the present and rapidly deteriorating state of the environmental crisis.

THE TWO-EDGED SWORD
OF DEVELOPMENT

Some of the most severe environmental problems in the world today are found in countries where the majority of inhabitants are Muslim. Even accepting a degree of outside responsibility, these problems would clearly be less pronounced if large numbers of Muslims were shaping their lifestyles according to an interpretation of Islam which strongly emphasized *khilāfa* as applied to the natural environment. The reality is that most are not, and this includes governments for whom development and economic growth are the top priority.

Indeed, Islamic principles such as *maslaha*, which states that laws may be altered to meet the common good, have most often been used to *override* Islamic guidelines for environmental preservation, on the basis that the latter are superseded by pressing needs for development. Thus, even the areas surrounding the sacred cities of Mecca and Medina, which the Prophet Muhammad declared sanctuaries to be protected unto perpetuity, have been turned over to urban development, and the traditional preserves known as *himās* have all but disappeared. Despite the fact that the prophet forbade the wasting of water even when it was plentiful, arid Saudi Arabia now consumes twice as much water per capita as the United States.

The majority of the world's Muslims live in developing countries, where environmental problems are particularly severe and rapidly worsening. And while the Arab Gulf nations are generally ranked among the world's wealthiest, they are often considered to be developing countries as well. In any case the deteriorating state of the environment in the Gulf region is both apparent and severe, so that development appears to be a double-edged sword. The damage to Kuwait's ecosystems as a result of the Iraqi occupation and subsequent war in 1991 is a vivid example of just how much damage to the environment humans can do.

A major issue affecting development policies throughout the world is lack of civil society and democratic institutions. More often than not, autocratic regimes decide on policies in the absence of public consultation. As a result, the interests of the majority of the population are often not addressed, and decisions are based on the short-term interests of elites. What this means is that even on rare occasions when such governments do adopt proenvironment policies (including family planning), the fact that the regimes themselves are seen as fundamentally illegitimate tends to create opposition to any policies they promote. Thus, in cases where governments are seen as un-Islamic, Islamically inclined opposition movements may perceive any environmental initiatives as yet another foreign ideology forcibly imposed by despotic regimes subservient to the West.

Nevertheless, there are signs that recognition of environmental problems in Muslim societies is growing. The 1990s saw a proliferation of environmental nongovernmental organizations throughout the Muslim world, as well as the establishment of environment ministries by governments of various Muslim-majority nations. The level of effectiveness of both governmental and nongovernmental environmental organizations has varied from one country to another, however, and the overall assessment seems to be that throughout the region sustainability, or even a slowing of damaging trends, is a very long way from being achieved. It should also be noted that in the majority of cases these organizations are in fact led by Western-trained environmentalists, who may not always be sympathetic to Islamic sensibilities.

Moreover, in the case of government ministries, it is sometimes the case that those entrusted with environmental policymaking are not themselves environmentalists. For example, when the government of Egypt established an Environmental Affairs Agency in the mid-1980s, it appointed directors who lacked any background in environmental affairs. In 2001, when a new Minister of Environment was first appointed, his opening statement in a highly publicized interview was to call for the planting of trees to line Cairo's boulevards. While tree-planting is in itself a useful and laudable act (and one for which Muhammad's example provides a precedent), it may be questioned how much impact this activity is likely to have when, for example, so many Cairenes lack access to safe drinking water. Egypt's environment ministry has also devoted substantial resources to things like building a national park in the Sinai aimed at attracting foreign tourists.

Indeed it is worthwhile to ask just how various constituents within Muslim societies define things like "the environment," "nature," "pollution," and so forth, since it may emerge that these definitions differ in some respects from those current among Western environmentalists or international agencies. This fact was highlighted in a recent study of environmental attitudes among inhabitants of several Cairo neighborhoods, which noted that

> in the common use of the word *bī'a(t)* [normally translated "environment"] in Egypt, the physical and the social environment are often confounded. In fact, pretty clearly some people understand primarily the social environment by the word. For instance, some cited the moral pollution of youth among

environmental issues.... The word *talāwath*, or "pollution," elicited a more definite response.

The same study found that most Egyptians surveyed defined "pollution" mainly as garbage (Hopkins, Mehanna, and El-Haggar 2001: 72). "Nature," meanwhile, was defined as air or weather or greenery or "everything that God created and had not been changed by human beings" (Hopkins, Mehanna, and El-Haggar 2001: 73). On the other hand, Palestinian water expert Sherif Elmusa "once remarked that the first thing that comes to mind when he thinks of 'nature' is a plowed field, not a tract of wilderness" (Albert, Bernhardsson, and Kenna 1998: 8).

It is worth noting that even in the West these terms have acquired whole new shades of meaning and emphasis in very recent times, with their current meanings dating only from the 1960s or so. The use of the words *bī'a(t)* in Arabic, *mohīt-e zīst* in Persian, and *çevre* in Turkish as translations for the English term *environment* likewise have very short histories.

DISSEMINATING ISLAMIC ENVIRONMENTAL ETHICS

It would seem that in the context of the world's many and diverse Muslim societies, a successful indigenous environmentalism would require a demonstrated compatibility with Islamic norms. An early attempt to do just this was made by a group of legal scholars at the University of Jeddah in Saudi Arabia in 1983. Under the auspices of the International Union for the Conservation of Nature and Natural Resources, four Islamic jurists formulated a thirty-five-page tract which was meant to serve as a basis for environmental policymaking in Saudi Arabia and other Muslim countries (Ba Kader et al. 1983). Among the precedents cited were the historical establishment of *himās* ("protected areas") and *harīms* ("areas in which development was forbidden"). Although several of the authors went on to act as advisors to the Meteorology and Environmental Protection Administration of Saudi Arabia in the 1980s, they have since complained both publicly and privately of the difficulties of implementing—or in the cases of *himās* and *harīms* even preserving—traditional Islamic environment-related institutions.

On the other hand, the governments of both Saudi Arabia and Iran have made formal attempts in the past few years to provide a forum for the articulation of an Islamic environmental ethic. Conferences on Islam and the environment were held in Tehran, Iran, in 1999 and in Jeddah, Saudi Arabia, in 2000. Jeddah was also the site of the First Islamic Conference of Environment Ministers in June 2002. More such initiatives will surely be seen in the future.

Environmental nongovernmental organizations are proliferating throughout the Muslim world as well, both locally generated and as branches of international organizations such as the World Wildlife Fund and the International Union for the Conservation of Nature and Natural Resources. For the most part they are secular in ideology and approach, often attempting to apply models of conservation and sustainable development imported from the West. An exception to this has been the case of Iran, where due to ideological restraints imposed by the government, all initiatives must be articulated in Islamic terms.

Perhaps the most significant effort to date to promote an Islamic environmentalism worldwide has been the work of the Islamic Foundation for Ecology and Environmental Sciences (IFEES), an international nongovernmental organization founded by Fazlun Khalid in Birmingham, England, in 1991. IFEES has sponsored environmental education programs in rural and village communities throughout the Muslim world, including projects in Tanzania, Madagascar, Indonesia, and Saudi Arabia, and has had impressive success in promoting a Qur'an-based understanding of environmental protection in areas where they have operated. Funding and staffing limitations, however, have so far prevented IFEES's initiatives from reaching their full potential.

A similar organization based in Delhi, India, the Islamic Foundation for Science and Environment, has worked locally to promote an Islamic form of environmental education, for example, by integrating environmental discussions into the curriculum of Delhi's religious schools (*madrasas*). Some rural religious schools in Indonesia (*pesantrens*) have also begun to incorporate environmental education into their programs. In Wales, students and faculty associated with Lampeter University have recently founded a Center for the Study of Islam and the Environment.

Muslims everywhere, throughout the developing world as in the West, must deal with a wide array of issues ranging from simple survival to finding political and economic voices within their own countries and within the modern global system. In the face of so many problems and struggles, environmental protection often seems peripheral. Yet it might be argued that environmental protection is perhaps the most central issue of all, whether or not humans recognize this reality. Merely seeking to turn back the clock is not an option; Muslims, like those of other cultural backgrounds, must find ways of living with new technologies and within a global economy that cannot be escaped, but can perhaps be accommodated in ways that are compatible with the value system their cultural heritage provides them:

> Air, earth, water, and fire are God's servants
> To us they seem lifeless, but to God living. (Rumi)

NOTES

1. It should be noted that Rumi is being metaphorical here, as Muslims do not believe in reincarnation.

2. A nearly identical saying exists in the Jewish Mishnah: "Rabbi Yochanan ben Zakkai used to say, 'If you happen to be standing with a sapling in your hand and someone says to you, "Behold, the Messiah has come!" first plant the tree and then go out to greet the Messiah'" (*Avot de Rabbi Natan*, ed. Nossel et al. 12).

REFERENCES

Abdel Haleem, Harfiyah, ed. 1998. *Islam and the Environment.* London: Ta-Ha.

Abu-Sway, Mustafa. 1998. "Towards an Islamic Jurisprudence of the Environment [*Fiqh al-bi'a fi'l-islām*]." Lecture given at Belfast mosque, Feb. 1998; online at http://homepages.iol.ie/~afifi/Articles/environment.htm.

Afrasiabi, Kaveh L. 2003. "The Environmental Movement in Iran: Perspectives from Above and Below." *Middle East Journal* 57.3:432–48.

Ahmad, Akhtaruddin. 1997. *Islam and the Environmental Crisis.* London: Ta-Ha.

Albert, Jeff, Magnus Bernhardsson, and Roger Kenna, eds. 1998. *Transformations of Middle Eastern Natural Environments: Legacies and Lessons.* Yale School of Forestry and Environmental Studies Bulletin Series no. 103. New Haven: Yale University Press.

Ammar, Nawal H. 2000. "An Islamic Response to the Manifest Ecological Crisis: Issues of Justice." Pp. 131–46 in *Visions of a New Earth: Religious Perspectives on Population, Consumption, and Ecology.* Edited by Harold Coward and Daniel C. Maguire. Albany: State University of New York Press.

Ba Kader, Abou Bakr Ahmed, Abdul Latif Tawfik El Shirazy Al Sabagh, Mohamed Al Sayyed Al Glenid, and Mawil Y. Izzi Deen. 1983. *Islamic Principles for the Conservation of the Natural Environment.* Gland, Switzerland: International Union for Conservation of Nature and Natural Resources. 2nd edition online http://www.islamset.com/env/contenv.html.

Chishti, Saadia Khawar Khan. 2003. "*Fitra*: An Islamic Model for Humans and the Environment." Pp. 67–84 in *Islam and Ecology: A Bestowed Trust.* Edited by Richard C. Foltz, Frederick M. Denny, and Azizan Baharuddin. Cambridge: Center for the Study of World Religions, Harvard Divinity School.

Erdur, Oguz. 1997. "Reappropriating the 'Green': Islamist Environmentalism." *New Perspectives on Turkey* 17.

Foltz, Richard C. 2005a. *Animals in Islamic Tradition and Muslim Cultures.* Oxford: OneWorld.

———, ed. 2005b. *Environmentalism in the Muslim World.* Hauppage, NY: Nova Science.

Foltz, Richard C., Frederick M. Denny, and Azizan Baharuddin, eds. 2003. *Islam and Ecology: A Bestowed Trust.* Cambridge: Center for the Study of World Religions, Harvard Divinity School.

Ghazali, Abu Hamid Muhammad. 1998. *The Proper Conduct of Marriage in Islam* [*Adab an-Nikah*]. Translated by Muhtar Holland. Hollywood, FL: Al-Baz.

Hamed, Safei El-Deen. 1993. "Seeing the Environment through Islamic Eyes: Application of Shariah to Natural Resources Planning and Management." *Journal of Agricultural and Environmental Ethics* 6.2.

Hope, Marjorie, and James Young. 2000. *Voices of Hope in the Struggle to Save the Planet.* New York: Apex.

Hopkins, Nicholas S., Sohair R. Mehanna, and Salah El-Haggar. 2001. *People and Pollution: Cultural Constructions and Social Action in Egypt.* Cairo: American University in Cairo Press,.

Ikhwan al-Safa. 1978. *The Case of the Animals versus Man before the King of the Jinn.* Translated by Lenn Evan Goodman. Boston: Twayne.

Izzi Dien, Mawil Y. 2000. *The Environmental Dimensions of Islam.* Cambridge: Lutterworth.

Khalid, Fazlun, and Joanne O'Brien, eds. 1992. *Islam and Ecology.* New York: Cassell.

Masri, Al-Hafiz B. A. 1987. *Islamic Concern for Animals.* Petersfield, Hants, England: Athene Trust.

Nasr, Seyyed Hossein. 1996. *Religion and the Order of Nature.* New York: Oxford University Press.

———. 2000. *Man and Nature: The Spiritual Crisis of Modern Man.* Reprinted Chicago: Kazi (orig. 1967).

Rumi, Jalal al-Din. 1926. *Masnavi-yi ma'navi: The Mathnawi of Jalalu'ddin Rumi.* Translated by R. A. Nicholson. London: Gibb Memorial Trust.

Sardar, Ziauddin, ed. 1984. *Touch of Midas: Scientific Values and the Environment in Islam and the West.* Manchester: Manchester University Press.

Zaidi, Iqtidar H. 1981. "On the Ethics of Man's Interaction with the Environment: An Islamic Approach." *Environmental Ethics* 3:35–47.

CHAPTER 9

DAOISM AND NATURE

JAMES MILLER

THE sciences of evolution, ecology, and environment are ushering in a new understanding of the time, place, and responsibilities of human beings within nature. The story of the fifteen-billion-year evolution of the universe gives an awesome sense of connection through the other animals, plants, microorganisms, complex molecules, carbon, silicon, helium, and hydrogen right back to the zero-event of the big bang. Ecologists have heightened our consciousness as to the place of human life within the contexts of multiple interdependent ecosystems. Environmental science is demonstrating the deep and perhaps irreversible impact of human economic culture on the planetary biosphere, which nourishes the amazing diversity of species with whom we make our home. These three interrelated sciences are together compelling a fundamental rethinking of our worldview, that is, our location in the time and space of the cosmos and our existential orientation and moral obligations within it. Put simply, the anthropocentric humanism of the European Enlightenment mentality is beginning to clash profoundly with the findings of contemporary holistic sciences. This clash is far greater than those between fundamentalists and secularists or between socialists and capitalists, for fundamentalists, secularists, socialists, and capitalists all share the same view of the primacy of humans within the scheme of things. Evolution, however, tells us that humans share the same genetic roots as other animals; ecology tells us that human life depends on plants, trees, bacteria in a whole host of interlocking ecosystems; environmental science makes it abundantly clear why we owe ethical obligations to the nonhuman world. All of this flies in the face of religious views that regard a single god as the focal point of human life and equally in the face of the European, secular, and patriarchal humanism that expelled such a god from the center of the universe, only to replace him with "man." In traditional Christian and modern Enlightenment views, human life is radically incommensurable with animal and plant life, not deeply related to it; human creativity depends on the strength of our mind or soul to transcend nature, not the weakness of our bodies that are inscribed within it; and only putative gods and living humans are owed ethical obligations or endowed with inalienable rights. This anthropocentric worldview, whether conceived in religious or secular terms, we now know to be untrue; that is, it does not correspond to the reality of the physical universe as understood by science. Rather, it has enabled the formulation of an unsustainable industrial, economic culture that threatens the very basis for life itself.

This chapter analyzes the ways in which the religious and philosophical thinking of Daoism intersects more fruitfully than monotheistic religion or liberal secular humanism with the sciences of evolution, ecology, and environment. It demonstrates the possibility for a radically alternative worldview that can help human beings symbolize their time, place, and obligations in a way that accords more closely with science and can help nurture a sustainable future.

NATURE AS EVOLUTION

Daoism, it is often said, is the least understood major world religion. Indeed it is only recently that the English-speaking world has had any conception of Daoism beyond the live-and-let-be world of the *Tao of Pooh* or the force-filled magic and cryptic sagacity of Yoda in the *Star Wars* series. The reasons for this general misunderstanding are historically complex and have only recently begun to be decoded (see Kirkland 2004). However, the largest part of this misunderstanding has revolved around construing Daoism as a individualistic philosophy or a way of life that centers on harmony with nature. Although it is perfectly possible to read this view from certain Daoist texts, notably the classical philosophy of Laozi or Zhuangzi, this notion of harmony with nature has by and large not been the focal point of real Daoists throughout Chinese history. Rather, Daoism has historically taken the form of an evolving series of religious movements that have had a wide variety of religious aims and an equally wide variety of methods to achieve those aims. Such aims have included the liberation of the souls of ancestors from underworld prisons so as to attain a postmortem life of luxury in a richly textured paradise. They have included the refashioning of the subtle energies of the body so as to give birth to an embryo of pure spirit that can transcend the finitude of the human condition. Their methods have included soot-covered rituals of repentance for the forgiveness of sins; yogic meditations involving the breath, saliva, and semen; and elaborate visions of star-dwelling deities who descend to inhabit the spleen and liver. Perhaps such variety is common-place across many traditions, but unlike many religionists, Daoists have rarely espoused a single normative reference point such as a holy book or holy person to serve as the universal reference point for the faithful. Like Hinduism, Daoism is a rich, subtle, diverse, and complex web of movements; unlike the missionary re-ligions of Islam, Buddhism, or Christianity, Daoism has not developed ways of containing that subtlety within a single universal framework such as the five pillars of Islam, the Four Noble Truths of Buddhism, or the Apostles' Creed of Christianity.

Given the historical reality of the Daoism's growth and evolution over the centuries, it is difficult to speak of a classical, authentic, or original period of Daoist history. Nor was there a singular event such as the Buddha's attainment of *nirvana* or the revelation of the Qur'an from which the complex reality of the tradition flowed. Rather, there have been several of these foundational moments, each of which has generated its own movement, its own dynamics and religious contours. The reason this is important for the study of religion and ecology is that the historical process of Daoism's growth is more closely related to the concept of evolution than to the notion of creation. The biblical concept of divine creation suggests that the world is the product of a divine agency. This creation is origi-nated with a singular creative act or "big bang" and continues through the mysterious workings of divine providence so that the history of humankind is bound up with the history of the divine activity. The stories of the Bible reveal human history to be also the history of God's actions toward his chosen people. The interrelated biblical concepts of divine creation, providence, and salvation history thus produce a worldview in which the existential decisions and moral actions of human beings toward each other and within the horizons of eternity are significant precisely because they are actions, that is, the creative products of human will and rationality. Human actions are to be judged in terms of their goodness, that is to say, the degree to which they accord with the will of the creator as expressed in the divine act of creation and in the numerous acts of self-revelation throughout human history.

In contrast to the biblical notion of human and divine agency founded in an original and singular act of creation, Daoist historical self-understanding is construed in terms of the concept of the Way. The Way is not the product of a divine creative act, nor is it a divine agent. Rather, the Way is the emergent process of creativity by which the cosmos becomes what it becomes. The cos-mological premise of one of the earliest Daoist texts, *The Way and Its Power* (*Daode jing*), is that the natural world is not a collection of interacting objects set in motion by a divine being but rather a dynamic system of vital processes whose basic character is that of self-transformation. The universe was not created by any external creative act, but rather subsists as a complex of "ways" that are wholly spontaneous or self-generating. *The Way and Its Power* (chap. 25) summarizes this view rather cryptically in the phrase *Dao fa zi ran*, which may be translated as "ways take as their model their own capacity for self-generation." The principle that this Daoist maxim enshrines, therefore, is the capacity of life to shape itself independent of any external impetus or teleology. Things simply come into being of their own accord; they are not enacted by divine fiat according to some mythic metanarrative or with any external purpose. Nor are they the consequence of some all-encompassing karmic logic of cause and effect. What is ultimately significant for Daoists, therefore, is not the agent (does God exist? who am I?) or the value of the product (is the universe good? did I do a good deed?), but the mode of agency (how do I do?). The Daoist canon, compiled in 1,445 of some 1,487 texts, may be legitimately viewed as a storehouse or repository of Daoist ways to answer to this question. From the cryptic wisdom of *The Way and Its Power* ("do nothing and nothing is not done") to the complex methods, recipes, formulas, and incantations recorded in Daoist scriptures, Daoists are preoccupied with the ways that we do and do not. I want to suggest that the views of agency and creativity espoused by Daoists can accord well with contemporary scientific notions of evolution and can be appropriated to construct a worldview that will be beneficial for life on earth in the twentieth-first century.

In the modern scientific account of cosmic and biological evolution, the explosion of the big bang gives rise to enormous hydrogen fusion (stars) that produce more complex atoms (helium) and increasingly complex atoms, such as lithium, carbon, and silicon that eventually accumulate in the form of planets. Carbon, hydrogen, and oxygen atoms combine to form carbon-based molecular life, and these molecules combine into ever more complex molecular forms such as lipids and proteins and eventually the highly complex DNA-based animal life. This emergence of organic complexity also has a Daoist parallel in the sparse cosmogonies found in *The Way and Its Power* (chap. 42):

> Dao gives birth to one
> One gives birth to two
> Two gives birth to three
> Three gives birth to ten thousand things.

What is instructive about this Daoist cosmogony is the use of the Chinese term *sheng*, which means "generate" or "give birth to." This cosmogony is different from the Neoplatonic account of creation, which sees differentiation as the fundamental cosmogonic process. In that account, the multiplicity and diversity of the universe arise out of the splitting up or differentiation of some primordial unity. The Daoist account, however, is quite different. One does not divide into two, nor two into four. Rather the one becomes, as it were, pregnant with itself and gives birth to two; two becomes pregnant with itself and gives birth to three. In this way we may understand the process of Dao—the ways of our universe—as a sort of recursive, fractal-like complexity in which life takes up itself into itself and emerges into a yet more complex form. The result of this ongoing creative process is the ten thousand, or myriad, things, the complex imbrication of life processes that together constitute the self-organizing collectivity of life.

The linear aspect of the Daoist worldview, however, does not lend itself to modern notions of progress or religious notions of eschatological crisis. Daoists have not historically viewed themselves as inscribed in an overarching linear temporality within which they feel compelled to make progress toward some future horizon. Rather they have generally favored a more complex view that regards progress and return as complements of each other. This view derives chiefly from the theoretical work of Chinese alchemists who sought an elixir of immortality from the decoction or refinement of cinnabar, a bright red ore comprised of mercury sulfide, back to a state of original purity. In this view, progress was made by returning the ingredients to a primordial state. By heating the cinnabar repeatedly in a crucible, ever purer forms of mercury were extracted. The mercury was ingested and was thought to preserve the body in a transformed state beyond death, even though it also killed the person who took it. The body was thus preserved and transformed in a process that can be interpreted as both progress and return. When this alchemical language was internalized into the language of Daoist biospiritual meditation, progress along the spiritual path was also seen as the recovery of a primordial state of energy. This state was symbolized as an embryo of immortality, something both primordial and transcendental, that was generated within the body and eventually born into the world through the crown of the head.

This alchemical view of nature, therefore, whether understood in terms of the physical world or the inner spiritual life of the adept, seems irreducibly complex. When analyzed from one perspective it seems as though Daoist spiritual practice is concerned above all with the return to simplicity, yet Daoism developed more and more complex forms of practice designed to achieve this. The ideal state is achieved through embodying the complex transformative power of nature rather than denying it.

NATURE AS SPONTANEITY

The autopoietic, self-generating character of nature provides the key component of the locative self-understanding of the Daoist. Daoists do not view themselves as inscribed within a single mythic framework of sin and salvation or creation and redemption but rather recognize the multiplicity of transformations within the universe and their place within them. The consequence of this view is an explicitly pluralist theology that recognizes a multiplicity of gods each associated with specific religious functions and geographic spaces. Gods are associated with nature: with mountains, villages, streams, seas, organs of the body, and stars in the sky. The Daoist religious movement known as the Way of Highest Clarity, for instance, comprises numerous methods for correlating the gods of the stars with the organs of the body though meditative visualizations undertaken at cosmologically auspicious moments of the calendar. It is the coordination and correlation of these numerous dimensions of existence that facilitate the transformation of the adept. Transformation comes not from being located within the mythology of creation and salvation associated with a particular deity, but from the alignment of the energies of the body with the energies of the cosmos. When body and cosmos are coordinated, transformation spontaneously and naturally occurs.

Although gods have their own mythologies associated with them, there was never a successful attempt to weave together the stories of all the gods into a singular transcendental metanarrative of creation and redemption. This is true even within the mythology of the god Laozi, the alleged author of *The Way and Its Power* and regarded as the personification of the Dao. According to the *Scripture of the Transformations of Laozi* (*Laozi bianhua jing*), which is perhaps the closest Daoist text to the mythic metanarratives of west or south Asian religions, the overall creativity of the cosmos is understood as the function of the creative powers of the god Laozi. Yet even within this mythological framework, what is central to the character of the deity—and thus to the cosmos that he inspires—is the fact that he transforms himself and appears in various guises as a teacher throughout the history of the world. The text does not understand the god's principal mode of relationship with the world as a singular act of creation or genesis, but one of continuous creative transformation (Kohn 1993: 71). This suggests that what is primordial in the Daoist conception of the universe is diversity. Diversity is explained not by a god who creates and names specifically different things but by assuming as fundamental the process of transformation within the universe, namely, the Dao or Way. All the things that exist within nature exist not because each was separately created (as in the Genesis story) but because nature transformed differently. This process of change is symbolized mythologically in the person of Laozi, who transforms into a variety of guises so as to give instruction to confused human beings in various ways corresponding to various locations and historical periods.

The repeated manifestations of Laozi throughout history suggest important similarities and differences between aspects of the Daoist understanding of evolution and contemporary evolutionary science. If we understand evolution as the nonteleological emergence of a self-organizing complex of life processes, there is a startling overlap with the Daoist conception of the natural self-organizing spontaneity (*ziran*) of the universe. In Daoist terms, nature is natural because it is self-generating, not because it is a creation that follows divinely ordained natural laws.

Once we realize that cosmos is not fulfilling natural laws that aim toward some final goal set in place by some transcendent creator, this has important ramifications for our view of ethics and, in particular, human intention and agency. In the Daoist view, value does not consist in the achievement of some teleological goal or conformity to some transcendental ideal. Rather, the value of a thing consists in the process of transformation that is inherent in its own process of being. Accordingly, the core value espoused by Daoists, namely, spontaneity or the capacity for autonomous self-transformation (*ziran*), derives directly from the Daoist understanding of nature. Nature is both a description of the universe and a value to be espoused by humankind. There is no fact/value distinction, one of the principal hallmarks of the modern worldview (see Neville 1998). In the terms of *The Way and Its Power* this spontaneity is actualized through adherence to an ethic not of action but of nonaction. By practicing an ethic of nonaction, Daoists value the capacity for self-transformation that is inherent within things and generally do not seek to exert influence through external creative acts. In the terms of the religious movements that have flourished for thousands of years of Chinese history, this ethic of nonaction was realized through ritual movements that sought to bring harmony from within rather than impose it from without. Such ritual movements, whether the individual actions of monks and nuns or the communitarian celebrations presided over by Daoist priests, function as transactions within an economy of cosmic power according to which the various dimensions of existence are creatively harmonized through their mutual coordination and correlation.

Nature as Balance

The Daoist view of nature and spontaneity outlined above did not develop out of a scientific evolutionary cosmology but rather in concert with medical theories about the body. In fact, the body remains one of the principal lenses through which Daoists interpret the world. Although the approaches of traditional Chinese medicine and contemporary medical science are in many ways radically opposed to each other, both share an important point of agreement in the concept of homeostasis, the ability of the body to maintain an overall physiological equilibrium. The difference lies in understanding how this homeostasis is maintained. The Chinese view rests on an understanding of the flow of Qi energy through the various meridians associated with the organs of the body. Chinese medicine views pathologies as disruptions in the flow of Qi, and doctors of Chinese medicine intervene in the body only insofar as to restore the normal flow of energy. The various techniques employed, from diet to acupuncture, all seek to restore a healthy balance to the energy flow of the body. This view is consistent with the cosmological principle explained above, that the various processes of the cosmos inherently possess the capacity for harmonious self-transformation and that the correct mode of action is the type of action that seeks to coordinate and enable the spontaneous creative capacities of these various processes, rather than force them to conform to some external norm.

The health of the body is assured when the state of homeostasis obtains among the Qi flows associated with the various organs. It is worth considering in some detail the Chinese view of energy and organs, because it is the basis for Daoist views of nature and harmony. Within the body, there are two kinds of organs or energy systems: yin systems and yang systems. According to the foundational medical text *Simple Questions on the Yellow Emperor's Internal Classic* (*Huangdi neijing suwen*), the function of the yin systems is to store or collect the "essential Qi," defined by a "structive [structuring] potential" (Porkert 1974: 179–80). It is the function of the complementary yang systems to "transmit or transform things" (Unschuld 1985: 286). Thus the body contains two basic physiological dynamics. The yin systems store the potential energy to maintain the dynamic structuring of the body, and the yang systems transmit this energy.

The circulation of energy throughout the body is thus brought about through a continuous interplay of positive and negative forces, comparable to an electrical circuit operating on a positive and negative charge. In the system of traditional Chinese medicine the basis for this circulation of energy is the pattern of yin and yang. Since the *Book of Changes* (*Yijing*), this pattern of yin and yang has been held to be a cosmological pattern, the pattern of potential and actualization in each phase of a cosmic matrix (*dao*). The treatise on yin and yang in the Suwen stresses the cosmic significance of these categories: "The Yellow Emperor spoke: [The two categories] yin and yang are the underlying principle of heaven and earth; they are the web that holds all ten thousand things secure; they are father and mother to all transformations and alterations; they are the source and beginning of all creating and killing; they are the palace of spirit brilliance."

In order to treat illnesses one must penetrate to their source: "Heaven arose out of the accumulation of yang [influences]; the earth arose out of the accumulation of yin [influences]. Yin is tranquility, yang is agitation; yang creates, yin stimulates development; yang kills, yin stores. Yang transforms influences, yin completes form" (Unschuld 1985: 283). It is important to remember that yin and yang are not forces or substances but modes or aspects of the processes of Qi. The nature of yang-Qi (expiration) is to transform, whereas the nature of yin-Qi (inspiration) is to receive and store form.

Nature, then, is unitary in that it is comprised ultimately of one energy matter, Qi. It has a binary characteristic, however, in its positive and negative

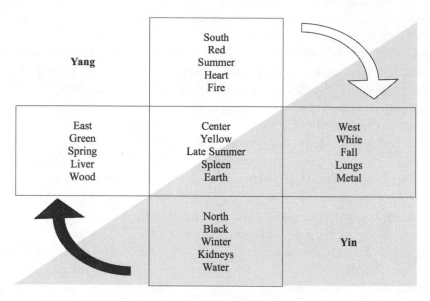

Figure 9.1 Yin, Yang, and the Five Phases

charges. This duality leads energy to flow in a circuit. When it flows in a circuit it achieves a state of balance or homeostasis. This simple pattern, readily understood by analogy to energy, Daoists take to be the fundamental pattern of nature. Everything can be explained in terms of positive and negative forces which produce a circular flow and tend toward a systemic balance. Viewing the body, and by extension nature, as a dynamic system that tends toward an overall equilibrium is a further point of agreement between classical Daoist philosophy and contemporary systems thinking.

This holistic, systemic view of the body and the cosmos leads Daoists to view nature in terms of the correlations between its various dimensions and not in terms of the sequential cause and effect that is a characteristic of Indian theories of karma and classical scientific views of logic. This emphasis on correlation and synchronicity makes Daoism seem premodern or unscientific, because the emergence of modern science depended precisely on analyzing nature in terms of the sequential logic of cause and effect. A scheme for mapping the correspondences of yin, yang, the five directions, and other aspects of nature is detailed in figure 9.1.

Figure 9.1 shows how the Chinese cultural imagination came to depict the achievement of balance in terms of time, space, and the body. The achievement of balance through the harmony of yin and yang occurs through the constant circulation of energy. This principle is the same at the micro and macro levels, and the various dimensions within which this circulation occurs function interdependently with each other.

Such schemes for conceptualizing nature should be regarded not as pre-modern or unscientific but as early forms of systems thinking, which scientists have begun to espouse as fruitful for considering certain aspects of biology. Stephen Jay Gould writes: "The principles of physics and chemistry are not sufficient to explain complex biological objects because new properties emerge as a result of organization and interaction. These properties can only be understood by the direct study of whole, living systems in their normal state" (quoted in Barlow 1991: 103). Though Chinese schemes for conceptualizing nature in terms of synchronous correspondences would not pass the test of science, scientists now see the value of looking at nature in terms of organic systems, as well as breaking nature down into analytical sequences.

What distinguishes Daoists from most modern scientists, however, is their desire to derive values from the facts of nature. Just as Daoists view nature as spontaneous and therefore value spontaneity, so also Daoists view balance and systemic equilibrium as a virtue. In this Daoist worldview, balance or harmony in fact can be regarded as a supreme virtue, not in the sense of some abstract ideal of goodness or justice, but rather as the natural process of balancing out of the diversity of forces so as to attain an equilibrium that proves beneficial for the whole. This value encourages a type of systems thinking that recognizes irreducible diversity and dissuades thinking in the dualistic terms of good versus evil. This is not to suggest that Daoists are amoral. Rather it is to suggest that Daoists view morality in medical terms: goodness consists of the optimal health of a system comprised of various interdependent subsystems. This medical concept of virtue can certainly be useful in constructing an ecological ethics, one that recognizes that humans cannot act for their own good without considering the overall health of the ecosystems in which they are embedded.

Nature as Fluid

The reason why Daoists and practitioners of traditional Chinese medicine espouse a holistic view of the relationship between the body and the cosmos is that their view of the body is more porous than we might expect. This concept can be understood in terms of Qi which, in the early medical texts, was viewed not simply as the system of energy within the body but more generally as a type of "influence" or "wind," that is, a kind of fluid that transmits power (Unschuld 1985: 67–73). Human health, moreover, is maintained not simply through the homeostasis within the body, but by maintaining an equilibrium between the body and its environment. Chinese medical texts thus pay great attention to the nine orifices of the body, for it is through these orifices that the balance between the energies within the body and those without must be maintained.

The most basic procedure for maintaining this harmony is breathing. Thus many Daoist meditative techniques pay close attention to regulating breath, that is, the passage of Qi as air or breath from outside to inside our bodies and back again. Each time we exchange a breath we are exchanging aspects of our physical environment and our internal physiology. Since breath is the basis of life, we can define life in the basic Daoist sense as the continuous exchange of energy between environment and body. When this exchange ceases, then life ceases. Livia Kohn recounts the story of Daoist Master Yinshi, born in the 1870s, who famously cured himself of tuberculosis and went on to expound the virtues of proper breathing. In his autobiography he writes:

> Breathing is one of the most essential necessities of human life, even more so than food and drink. Ordinary people are quite familiar with the idea that food and drink are important to maintain life, that they will starve if left without it for a while. But they hardly ever turn around to think about the importance of breathing and that air is more important to life than anything else. (quoted in Kohn 1993: 136)

This focus on breath as the fluid medium between body and environment lends itself very well to an ecological sense of the body, one that sees our understanding of life and self inextricably linked with the process of life in our surrounding environment.

Daoist texts, moreover, dwell not only on breath but on saliva, semen, menstrual blood, and food as elements in the process of biospiritual transformation. All these may be understood as ingredients in the constant fluid interchange between the body and its environment. Even gymnastic exercises are to be conducted with a sense of awareness as to one's surroundings. Daoist gymnastic guides typically instruct the adept to face in certain directions or perform exercises at certain times of the day (see Kohn 1993: 145). These suggest that it is not simply the exercise in and of itself that is beneficial, but rather the way in which the exercise connects practitioners to their surroundings.

The Daoist view of the porous body can thus be regarded as an ecological view because it tends to mitigate against the perception of the self as an autonomous, isolated mind within a physical container. Rather the effect of Daoist practices is to nurture a sense of self as thoroughly translucent to one's environment. This translucency or porosity of self is even symbolized physically in Daoist hagiographies. The biography of the Daoist perfected person Zhou Ziyang, for instance, relates how his practice of Daoist arts led eventually to a luminous and radiant physical condition in which one could see through his skin to his internal organs (see Porkert 1979 for a complete translation). The Daoist practice that he followed seemed to result in a less opaque, more transparent body. What this concept indicates is not simply the transformation of his body in an amazing way, but the thinning of the membrane between interior and exterior.

The idealization of a increasingly diaphanous and porous body can be interpreted in contemporary terms as an ecological sense of self. My argument here is that this ecological sense is grounded in the Chinese view of nature as a fluid process rather than a material substance. In fact, matter or "stuff" is not the basis of the Daoist view of nature. Rather, its basis lies in water.

The symbolic affinity between water and early Chinese philosophy has been well demonstrated by Sarah Allan (1997). Allan argues that water, "which provides life, gurgles up unbidden from the earth and moves of its own accord, becomes perfectly level and clears itself of sediment when still, takes the shape of any container, penetrates the tiniest opening, yields to pressure but wears down the hardest stone, becomes hard as ice and disperses as steam, was the model for philosophical ideas about the nature of the cosmos" (1997: 4). Allan's work is based on the concept of a "root metaphor" developed by Lakoff and Johnson in *Metaphors We Live By* (1980). Allan argues persuasively that early Chinese philosophy's root metaphors or prelogical conceptual schemes were developed principally from images of nature (1997: 13). In contrast to the Western philosophical emphasis on nature as matter or elements, Chinese philosophy has favored a more fluid view.

In her book, Allan concentrates on understanding the cultural meanings of the terms she is dealing with, treating them, in Lakoff and Johnson's terms, as metaphors. *Shui*, the Chinese term for water, for instance, has a wider semantic range than the English term. It denotes fluid, flowing, and river as well as water. The Chinese character for water depicts a flowing stream. This idea of "streaming" is captured in a range of early Chinese philosophical texts documented in detail by Allan. Perhaps the most famous example of the use of water as a metaphor can be found in chapter 22 of *The Way and Its Power*: "Highest good is like water. Because water excels in benefiting the myriad creatures without contending with them and settles where none would like to be, it comes close to the way" (Addis and Lombardo 1993).

Here water is also employed as a metaphor for goodness. Good is defined as being like water in the sense that water benefits creatures without contending with them and in that it chooses the lowest place. To be good thus means to be of benefit to others and to be humble. All this is straightforward. But it is helpful to consider this symbolic use of water more closely. In this sentence water has two relevant properties. The first is that it benefits the myriad creatures. The second is that is always sinks to the bottom. Of these two properties, the first is uniquely the property of water whereas the second is the property of any liquid. The unique property of water is that it benefits the myriad creatures (*wanwu*) without contending with them—without causing them any harm. The Chinese term *wanwu* is rather unusual in that it includes humans, animals, and plants in a single biological category. The most important ecological connection be-tween these three life-forms is that they are all dependent upon the earth's water cycle. To be alive, to be one of the myriad creatures, is achievable only through water. Thus in Daoist and ecological terms, life is a water-based proposition. Whereas any form of liquid could be used to express the concept of "settling where none would like to be," only water "excels in benefiting the myriad creatures." Thus the symbolic use of water in this text here is based in its natural, biological value for human beings. In the passage above it is water's ability to sustain water-based life that makes it ultimately significant. It expresses a bio-logical dependence that transcends culture and embeds religious meaning within biological life.

Daoism, as a religious tradition, locates this natural fluidity in the mystical aspect of the Dao or Way. *The Way and Its Power* (chap. 4) speaks of the Dao not just as the watercourse or the irrigation channels of cosmic vitality, but as their mysterious ever-full source:

> The Dao is empty [empties], yet using it it does not need to be refilled.
> A deep spring [*yuan*]—it seems like the ancestor of the myriad living things.
>   (quoted in Allan 1997: 76)

The irrepressible flood of life that constitutes the world is a source of mystery. Life's liquid vitality must originate in some watery abyss, some deep well that "does not need to be refilled." For Daoists this unfathomable mystery is funda-mentally fluid, in the mystical, symbolic sense of being ungraspable by human reasoning and in the biological sense of being the wellspring of water-based life.

Daoism and Nature in Contemporary China

From the initial historical research that has already taken place we know that what we would today call environmental ethics was a small but certain concern of the Daoist religious movement known as the Way of the Celestial Masters. A recent analysis of the ethical code entitled the 180 *Precepts of Lord Lao*, an ethical code adopted and transmitted by the Way of the Celestial Masters, has revealed that environmental protection was an intrinsic part of the ethical framework by which leaders of this tradition sought to abide. The code contains specific injunctions against burning vegetation, felling trees, digging holes in the ground, drying up wetlands, hunting, polluting wells, bathing in rivers, disturbing wildlife, and creating artificial lakes (Schipper 2001: 81–82).

To understand the historical context in which this code was adopted we must place ourselves in Sichuan Province in west China about two thousand years ago. This area of China was a key economic engine with large merchant towns, extensive exploitation of the land, a high population density, and commerce with nomadic central Asian tribes. According to Kristofer Schipper, "neither classical nor medieval Daoism developed in primitive surroundings, but in places of highly developed culture" (2001: 83). However, although the Way of the Celestial Masters thrived in this richly developed area, the leaders created their twenty-four parish centers (*zhi*) almost exclusively in mountain areas or natural reservations. One area in the plain that did receive religious sanction, however, was the Dujiangyan irrigation project.

Dujiangyan is located just outside present-day Chengdu, the capital of Sichuan Province. To the east begins the mountain area where many Daoist sites are located and which leads toward the western edge of the Tibetan plateau. To the west lie the rich plains of the central Sichuan. The Dujiangyan irrigation project was begun in 267 BCE and completed in 256. It was designed as a water conservancy project to regulate the flow of the Minjiang River so as to prevent flooding downstream in times of heavy rainfall and to provide a constant flow of water for irrigation. The Dujiangyan irrigation project separates the Minjiang River into three main channels: one provides irrigation to thirty thousand separate irrigation channels downstream; the second receives surplus water in times of flooding; the third provides water for the city of Dujiangyan. The project is significant from a water engineering perspective because it is the world's oldest irrigation project in use today that is not built around a central dam. It celebrated its 2,260th birthday on 4 April 2004. The water irrigation facility and nearby Daoist mountain, Qingcheng shan, were placed on the UNESCO World Heritage List in 2000.

The irrigation project provides a context for understanding the environmental precepts adopted by the Daoist Celestial Masters. Their concern for water, wetlands, and the earth was not simply a theoretical or metaphysical concern but was an economic and ecological concern. This concern was, I think, quite complex. On the one hand the wealth of Sichuan depended on the ability of engineering projects like Dujiangyan to regulate the flow of water in the environment. This is concordant with the classical Chinese ideal of the role of humans as the harmonizers of heaven and earth. Dujiangyan perhaps can be considered a concrete implementation of this ideal. Through environmental engineering humans can never tame or domesticate nature, but they can find a way of harnessing its power so as to provide health and well-being to human society. The religious sanction of environmental engineering can be seen in the fact that a Daoist temple was established at Dujiangyan in honor of Li Bing, its chief architect. On the other hand the fact that the Celestial Masters deliberately established their centers in mountain areas and not in the fertile plains suggests that this type of social-environmental engineering must be balanced by a deference and respectful appreciation for nature that can be experienced in mountains because they are the places where this type of engineering cannot take place. They are in effect nature preserves, sanctuaries from engineering complexes such as Dujiangyan. Although such complexes may be considered necessary for human development, they are not, it seems, sufficient. By interpreting Qingcheng shan and Dujiangyan (natural mountain and environmental engineering) as part of a single complex religious-environmental complex, we can say that the Celestial Masters were attuned to the complexity of the symbolic and environmental value of water.

Such complexity is revealed in the well-documented environmental crisis of present-day China. The aspect of this crisis that most directly impinges on Daoists in contemporary China is the phenomenon of ecotourism. When Mount Qingcheng received its UNESCO world heritage designation, signs in English and Chinese proudly proclaimed to thousands of tourists Daoism's affinity with the way of nature. In December 2003, the Chinese government committed ten billion yuan over five years to make Dujiangyan the number one ecotourism attraction in China. By 2004 some twenty-four million yuan had been spent on environmental protection that has seen some two thousand egrets return to the top of Mount Qingcheng, attracted by ninety-nine Machilus trees some reportedly one thousand years old. The rapid development of this area as a tourist destination brings awareness of environmental issues but also potential harm. A recent plan to construct a new dam 23 meters tall and 1,200 meters wide across the river just upstream of the Dujiangyan project has attracted widespread criticism because of its implications for the cultural and environmental heritage of the area. The dam would be part of the Zipingpu hydroelectric project, which aims to help reduce China's dependence on coal-fired electricity and to supply clean energy to fuel local economic development, one part of which is ecotourism. Based on discussions with local government officials in 2004, it now seems that this project is stalled.

The complex problems of religious and ecological tourism reflect the broader pattern of environmental problems facing China in the twentieth-first century. Government officials recognize, however, that traditional Chinese views of nature may have a positive role to play in China's transition to an ecologically sustainable economy. Pan Yue, the deputy director of China's State Environmental Protection Agency, gave a recent speech in which he called for a "recycling" economy—one not predicated on exhausting the finite resources of the environment. Such an economy, he claimed, would be consonant with the "law of nature," which he connected in his statement to Daoist notions of flow and circulation (Pan 2003). In so doing he recognized and articulated the validity and relevance of the Daoist worldview, one which regards nature as spontaneous, balanced, and fluid. According to this worldview, humans benefit from aligning themselves with the flow of the Dao, the irrigation flow of liquid vitality, and harm themselves and their world by blocking it. This view can be of benefit in the construction of an ecological ethic that respects the view of life that has emerged in the sciences of evolution, ecology, and environment.

REFERENCES

Allan, Sarah. 1997. *The Way of Water and the Sprouts of Virtue.* Albany: State University of New York Press.

Addis, Stephen, and Stanley Lombardo, trans. 1993. *Tao Te Ching.* Bloomington: Hackett.

Barlow, Connie, ed. 1991. *From Gaia to Selfish Genes: Selected Writings in the Life Sciences.* Cambridge: MIT Press.

Campany, Robert Ford. 2001. "Ingesting the Marvelous." In *Daoism and Ecology.* Edited by N. J. Girardot, James Miller, and Liu Xiaogan. Cambridge: Center for the Study of World Religions, Harvard Divinity School.

Gould, Stephen Jay. 1994. *Hen's Teeth and Horse's Toes: Further Reflections in Natural History.* New York: Norton.

Hahn, Thomas H. 2000. "Daoist Sacred Sites. In *Daoism Handbook.* Edited by Livia Kohn. Leiden: Brill.

Kirkland, Russell T. 2004. *Taoism: The Enduring Tradition.* London: Routledge.

Komjathy, Louis. 2003. *Title Index to Daoist Collections.* Cambridge, MA: Three Pines.

Lakoff, George, and Mark Johnson. 1980. *Metaphors We Live By.* Chicago: University of Chicago Press.

Miller, James. 2003. *Daoism: A Short Introduction.* Oxford: OneWorld.

Neville, Robert Cummings. 1998. "Orientation, Self, and Ecological Posture." Pp. 265–71 in *Confucianism and Ecology.* Edited by Mary Evelyn Tucker and John Berthrong. Cambridge: Center for the Study of World Religions, Harvard Divinity School.

Pan Yue. 2003. "Environmental Culture and National Renaissance." Speech given at the Green China Forum on Oct. 25. http://www.cfej.net/htm/Apfej/lunwenji/english/Environmental%20Culture%20&%20National%20Renaissance.doc.

Porkert, Manfred. 1974. *The Theoretical Foundations of Traditional Chinese Medicine.* Cambridge: MIT Press.

———. 1979. *Biographie d'un taoïste légendaire: Tcheou Tseu-yang.* Mémoires de l'Institut des Hautes Études Chinoises 10. Paris: Collège de France.

Schipper, Kristofer M. 2001. "Daoist Ecology: The Inner Transformation; a Study of the Precepts of the Early Daoist Ecclesia." In *Daoism and Ecology.* Edited by N. J. Girardot, James Miller, and Liu Xiaogan. Cambridge: Center for the Study of World Religions, Harvard Divinity School.

Unschuld, Paul U. 1985. *Medicine in China: A History of Ideas.* Berkeley: University of California Press.

CHAPTER 10

MOTIFS FOR A NEW CONFUCIAN ECOLOGICAL VISION

JOHN BERTHRONG

It has been a hard two centuries for Confucians. They have been under attack ever since Western imperial colonial powers began their concerted incursions into China and the rest of east Asia beginning with the first Opium War in 1839. This epochal defeat was followed by a century of unequal international treaties, creating a semicolonial status for China until the middle of the twentieth century. While it now seems a world lost, as W. Theodore de Bary reminds us, we must remember that at the beginning of the twentieth century the Confucian tradition still dominated the Chinese educational system, the imperial civil service, and family rituals.[1] This overt Confucian domination of Chinese elite culture has all been swept away by the beginning of the twentieth-first century. Parallel to all of this, it has been a hard two centuries for the ecological balance of the earth. Put simply, the vast expansion of the human population and the rise of massive technologically driven industrialization has made some scholars wonder if human beings will remain a viable species. In this sense, the size of the population growth and its attendant industrialization does matter: human beings are genuinely facing a crisis generated by their own reproductive and technological and industrial success. These problems parallel the fall of Confucianism from grace within the Chinese political and intellectual world.

By 1949 when Mao Zedong (Tse-tung) proclaimed that the Chinese people had stood up to the challenge of the West, Confucianism appeared to be one of the things that the Chinese people were going to leave behind in their rush toward Maoist modernization. But as friends have noted in conversation about the rise of modern China, the Chinese people not only stood up to the Western colonial powers, but concomitantly they have also gone shopping with a vengeance—vast shopping malls dot the Chinese landscape of any major or even minor Chinese city. Armed with People's Liberation Army MasterCards, the modernized consumerization of contemporary China is nothing short of astounding. Great new superhighways link the country from one end to the other; new airports are built or rebuilt every decade; the greatest dams and buildings in the world are routinely dedicated on the rivers and cities of China; and now a wonderful Swiss cable car will carry you almost to the summit of Mount Tai, the most sacred of China's holy mountains. And notwithstanding all the attacks on Confucianism in the twentieth century, the Confucian tradition shows new signs of life throughout east Asia; but in its renewal it likewise faces new challenges, and one of the most striking of the new challenges is the ecological crisis of the late twentieth century. Standing on Mount Tai these days in Shandong Province, one sees a vast yellowish cloud of pollution that evokes all the ecological problems of modern China.

The renewing Confucian tradition, now called New Confucianism in order to distinguish the modern movement from the classical Confucianism of Confucius (Kongzi), Mencius (Mengzi), and Xunzi and the great Song-Yuan-Ming-Qing Neo-Confucian scholars, faces a whole series of new challenges originating from its confrontation with Western modernity.[2] As is the case with so many of the historic ethico-religious movements of the world, Confucians must come to terms with new roles and intellectual challenges in modern east Asia caused by what sociologist of knowledge Peter Berger calls the second great east Asian industrial and scientific revolution. A great part of the renewal is directed toward figuring out just what role Confucians will play in the modernization of east Asian societies and then deciding what the tradition will want to share globally with other cultures and philosophical traditions.[3] To do this, the New Confucianism will have to review, as some thinkers such as Xiong Shili, Feng Youlan, Qian Mu, Tang Junyi, Liang Shouming, Hsu Fuguan, Mou Zongsan, and many more have already done, what must be revived internally or needs to be added to the tradition from outside resources—and what must be rejected or discarded from the historical inventory of the tradition as inappropriate for the modern world. For instance, the New Confucians have uniformly rejected traditional restrictive views of the role of women in the formation of a modern culture. The first and second generations of New Confucians (from the 1920s to the 1990s) have mostly carried out this revivification in terms of reordering and supplementing a philosophical revival of the best of the Song-Yuan-Ming-Qing Neo-Confucian heritage.

But the world has not stood still for a century while Confucians were first watching their traditional world being destroyed and then revived. New issues have emerged, and one of the most pressing concerns is the question of the ecological crisis. This was not something that was even on the agenda of the late imperial Confucian tradition save for the persistent Confucian concern for economic policies that recognized the need for the conservation of natural resources.[4] Here traditional Confucians could approvingly cite Mengzi's and Xunzi's reflections on the obligation to pay attention to the human cultivation of the natural world via a concern for the proper conservation of natural resources. However, if Confucianism is to play an active role in Chinese and global intellectual life, it must learn how to address new as well as traditional issues that go beyond traditional concerns for conservation, though this is a good place to start. While it is hard to conceive of any form of New Confucianism that does not take personal self-cultivation and social ethics seriously, it is also now impossible to envision a New Confucianism that does not deal with questions of international human rights and responsibilities, the changing roles of women, and the ecological crisis. The crux of the matter is that, along with many other of the global axial-age religions, the primary concern of the tradition was focused on human ethical and social life, whereas New Confucians must preserve the best of Confucian self-cultivation and social ethics while expanding this heritage into the new arena of ecological reflection.[5]

The following essay is a modest attempt to begin to think through what is an appropriate Confucian philosophical response to the ecological crisis of the modern world. New Confucians need to ask, What are the resources for ecological thinking within the tradition and how can they be related to the ecological crisis? How can Confucian insights into the relational sociality of human beings be expanded to include a positive view of the preservation of nature? Furthermore, New Confucians are aware that they need to think these issues through in light of global philosophy. Just as the great Song Confucian revival was stimulated by the immense impact of Buddhist thought on Chinese culture, so too will the New Confucian movement be influenced by its dialogue with modern Western, Islamic, and Indian philosophy—just to mention the most important Eurasian cultural systems.[6] Linked to this global dialogue of civilizations is the recognition of the vast ecological crisis that has arisen over the last two or three centuries around the globe. What is intriguing about this new encounter is the joint perception that all human beings need to work together in order to address the profound question for human survival raised by the ecological crisis of the day. What this essay will try to do is to look at those major themes of the Confucian tradition that will play a role in its response to the ecological situation.

This essay will deal, in sequence, with the question of the nature of the Confucian cumulative tradition as well as inquiring how the New Confucians might address the ecological crisis. However these two historical and philosophical questions are answered, any future Confucian ecology must be based on acceptable elements or motifs of the tradition, ideal types that are amenable to search for a Confucian method for constructing a responsible ecology. Such an ecological vision must be securely situated within the main body of the tradition and must not merely be some kind of ad hoc manipulation of items taken at random from the history of Confucian thought.

COMPARISON

Philosophy and religious studies are becoming more global in scope, and therefore projects constructing comparative philosophy and theology are emerging as thinkers from the historic cultural regions of the North Atlantic world and west, south, and east Asia become aware of each others' histories and concerns.[7] Of course, most great philosophers and theologians have always assumed that their philosophies were world encompassing in terms of truth, coherence, and adequacy. However, the statement that philosophy and theology are becoming world encompassing or truly ecumenical emphasizes the historical fact that at the end of the second millennium Confucians are studying Plato, Aristotle, Aquinas, Kant, Hegel, Tillich; and Western process, pragmatic, continental, and analytic philosophers and deconstructionists are studying Hinduism, Confucianism, Daoism, and Buddhism. Some pluralists like Walter Watson and David Dilworth have even sketched a hermeneutics based on Aristotle's four causes that takes into account as many world philosophies as they can fit into a standard-size academic monograph.[8]

This embrace of the truly global reach of world philosophy and theology makes comparison even more difficult and promising than it was before. Gone are the days when one could state with clarity the essence of any theoretical position simply because of the purportedly easy recognition of commonalties and differences that comes from working within one large philosophical tradition, such as what was called Western philosophy or the Confucian tradition in east Asia. At least these vast sets of interlocking Eurasian traditions did separately control their own internal canons and forms of discourse, allowing for a comparison to be made between thinkers based on common philosophical vocabularies. Essences or tight systems as stable forms of life have a way of falling out of focus when the scope of philosophy and theology is expanded to include Western, Chinese, Indian, and Islamic forms of thought. While we are no longer able serenely to present the timeless essence of a tradition, we can still articulate synopses of prototypical positions that present the outlines of philosophies as collective intellectual narratives sharing some common features between generations. There are enough family resemblances that we can still tell a typical passage from Zhu Xi apart from Aquinas if we have a modicum of global philosophical competence.

While I hold that all Confucians orient their thought in terms of their understanding of the common classical canon (basically the Thirteen Classics) as a necessary condition for membership in the Confucian fellowship, this is not always a sufficient reason to establish Confucian identity.[9] Modern scholars of the tradition such as Zhang Hao, Liu Shuxian (Hsu-hsien), Julia Ching (Jing), Chung-ying Cheng (Zhonging Zheng), Du Weiming (Tu Wei-ming), John Makeham, Umberto Bresciani, and W. Theodore de Bary (among others) have noted that there is yet another dimension to the Confucian tradition that must be acknowledged as giving it its characteristic appeal as a philosophy, form of cultural discourse, and even a way of life. This is the element of the Confucian Way's intrinsic value as something unique in the intellectual, moral, social, technological, artistic, and religious life of humanity. In short, this is what Confucians have recognized that makes the whole tradition as an embodied way of life something more than merely instrumental value. It is the referent of worth or excellence that can be commended to anyone seeking a moral, intellectual, or spiritual orientation to life. It is something that, if properly understood, literally transforms the person by embracing or appropriating its message. The "what" and "how" of this axiological process is the ebb and flow of Confucian discourse as a way of self-cultivation.[10] Furthermore, this ultimate source and process of supreme value has been called the religious dimension of the tradition or a transcendental faith in the ultimate connections of heaven, earth, and humanity. Chang (Zhang) Hao writes that "Confucianism is not the secular humanism that some modern scholars take it to be, for its inner-worldly character was anchored in a transcendental belief that centered on the idea of Heaven [*tian*] or 'the way of Heaven' [*tiandao*]."[11]

Chang Hao's interpretation of the issue is especially pertinent because he has never been identified as a partisan of the New Confucian philosophical revival that seeks to distinguish a unique Confucian sense of religiosity. Nonetheless, Chang clearly states what to him is the obvious transcendent reference of the tradition: "The overriding concern of Confucianism may be how to realize humanity in this world, but this concern was transcendentally ordained."[12] At its most basic reality the value of the tradition lies in the ultimate as the human appropriation of *tiandao*. But as Mengzi recognized so long ago, this appropriation of the ultimate does not happen without human effort defined as the Confucian task of self-cultivation. Nor, as Mengzi also recognized, was this human effort to realize full humanity set within nature as a whole. For both Mengzi and Xunzi this recognition of human embeddedness in nature gave rise to what can aptly be called the Confucian conservationist view of human interaction with nature. Hence, to ask, why bother reading the canon, is to receive the answer that it preserves the message of the sages; yet this message remains mere words unless a person is transformed by its depth meaning. Philosophies and religions often both describe reality and commend a certain approach to it based on their description of what really is.

Du Weiming, in a number of recent lectures, has pointed out that one cannot confine the tradition to the texts, even if the texts are crucial for a historical understanding of the development of the tradition.[13] As Du notes, there is a school within the Confucian tradition, rather like the Chan Buddhists, that would argue for a transmission of the Dao outside of the texts, a teaching that goes from mind-heart to mind-heart without the necessity of a scholarly rendition of all the texts per se. This is what Du calls the extralinguistic referent of the tradition. It is this extralinguistic referent that others call the transcendent dimension of the tradition that must be realized in the mind-heart and not merely recognized by the rationality of the mind-heart, even if there is nothing wrong with the union of reason and realization. In fact, it is precisely this fusion of reason and realization of the Dao that Zhu Xi argues is the essence of the learning of the way or *daoxue*.

Mou Zongsan (Tsung-san) (1909–95), the most systematic New Confucian philosopher writing within the circle of the moral philosophy tradition of Song-Yuan-Ming-Qing thought, argued that any form of thought touched by the Confucian hand would have at least one common root metaphor as part of its interpretive matrix. Mou chose to call this root metaphor concern-consciousness. Because of this sensibility of concern-consciousness, Western scholars have always viewed Confucianism as a form of social ethic, a way of helping human beings live in civility with other human beings. In terms of modern Western speculative philosophy, Mou's ideal type of concern-consciousness is a form of fundamental axiology. At the root of all being or even the not-yet-come-to-be is a concern for values as expressions of thought, action, and passion. Anything, that is, is a value as a pattern of action, thought, and emotion; this is what is called the object-event, the subject of Confucian discourse about the creation of human relational values. As we shall see, because of this axiological view of reality, the New Confucians can expand their root metaphor of concern-consciousness to include a profound concern for nature as well as humanity.

Mou Zongsan himself proposed a list of four traits as a synopsis of the kinds of perennial themes he deemed essential to any fiduciary reading of the Confucian Way as a cumulative tradition and as an expression of ultimate concern-consciousness.[14] It is perhaps better to think of these traits as ideal types, flexible constructions of sets of meaning that extend through the history of the development of the Confucian tradition. For instance, the trait of human nature has been differently understood from century to century though it has remained a perennial topic of analysis and debate within the tradition. The Confucian tradition was highly adept at adding new wine to old wineskins; in fact the wineskins were so modified and repaired over time that it is sometimes difficult to see how they remain the same. Yet they are still wineskins and are hence useful for containing the new wine of thought.

Because Mou identified his own constructive thought more from within the schools of Kongzi and Mengzi in the classical period, and then with Cheng Hao, Hu Wufeng, Wang Yangming, and Liu Zongzhou in the Song, Yuan, and Ming dynasties, I have chosen to supplement his list with an extended list of additional traits and themes drawn from the common store of the classical and Song-Yuan-Ming-Qing and Korean and Japanese Confucian tradition for a grand total of twenty items. Such a list is at least inclusive of the major trends in post-Song philosophy, including the ecumenically minded modern New Confucian intellectuals.[15] Moreover, the expanded list takes into consideration the developments of Confucian discourse in Korea and Japan as well. As we shall see, some of these key Confucian motifs relate directly to New Confucian discourse on ecology. The twenty themes, traits, and motifs of Neo-Confucian discourse may be summarized as follows:

1. *Ren* 仁 is the paramount virtue and marker for all the other virtues, such as *yi* 義 (justice), *li* 禮 (ritual action), *zhi* 知 or 智 (knowledge, wisdom, discernment), and *xin* 信 (faithfulness). These five constant virtues provide the axiological sensibility to the whole Neo-Confucian enterprise. These are linked to *xiao* 孝 (filial piety) as an expression of primordial familial relationships.

2. *Li* 禮 (ritual action) is the social glue that holds society together and in fact helps to constitute the humane person; it is closely linked to *yi* 義 as the appropriate form of equity in moral interchange with the intent to create a form of humane flourishing (*wen* 文).

3. *Tian* 天 (heaven), *tianming* 天命 (the mandate of heaven), and *di* 地 (earth): whether we should use a capital *H* for *tian* is an important question for the Neo-Confucian philosophy of religion. *Tian* (heaven), *di* (earth), and *ren* 人 (human beings) form an important cosmological triad for the Neo-Confucians.

4. *Li* 理 (principle, pattern, order to the whole of the cosmos) is a key Song philosophic term and a little-used early Confucian concept.

5. *Xin* 心 (mind-heart)—the living center of the human person—needs to be cultivated by proper ritual in order to realize true virtue.

6. *Xing* 性 (human tendencies, dispositions, nature) is the *li* 理 (principle) given to each emerging person by *tian* 天 as the mandate for what the person ought to be.

7. *Qi* 氣 (vital force, material force) functions as the dynamic force or matrix out of which all object or events emerge and into which they all return when their career is completed.

8. *Qing* 情 (emotion) and *yu* 欲 (desire) are passion, intimately related to *qi* 氣 (vital force) as the dynamic side of the cosmos.

9. *Daowenxue* 道問學 (serious study and reflection) and *zundexing* 尊德性 (honoring the moral tendencies or dispositions) are designations of two different ways of cultivating *xin* 心 (mind-heart) and as contrasting modes of moral epistemology.

10. *Gewu* 格物 (the investigation of things) was a key (and highly contested) epistemological methodology for the examination of the concrete objects and events of the world, in order to *zhujing qiongli* 主敬窮理 (reside in reverence), in order to exhaust (comprehend) principle.

11. *Cheng* 誠 (sincerity, genuineness, self-actualization of the moral virtues) allows one to achieve a morally harmonious life via various forms of *xiushen* 修身 (self-cultivation) by means of praxis such as *jing* 敬 (mindfulness or attentiveness). This praxis is the "how" of the moral self-cultivation of the five constant virtues.

12. *Nei/wai* 內外 (the inner and outer dimensions of any process) are often used for the "king without, sage within" and are often discussed in terms of the opposition of *si* 私 (selfishness) or *pian* 偏 (partiality, one-sidedness) and *gong* 公 (public spirit).

13. *Ti/yong* and *wen* 體用 (substance and function) and 文 ("literary" manifestation of the substance and function) via the balance of *ganying* 感應 (stimulus and response) describe the reactive movement, generations, productions, and emergence of the objects and events of the cosmos.

14. *Liyi fenshu* 理一分殊 is the teaching that principle is one or unified while its manifestations are many or diverse. It is often seen as the characteristic, holistic, organic sensibility and yet realistic pluralism of Neo-Confucian thought.

15. *Daotong* 道統 (transmission, succession, or genealogy of the way) is Zhu Xi's masterful account of the revival of the Confucian Way by a set of Northern Song philosophical masters.

16. *Siwen* 斯文 (this culture of ours) is the expression of refined self-cultivation and the manifestation of principle from the family to the cosmos.

17. *He* 和 (harmony) and *zhong* 中 (centrality) are designations of the goals or outcomes of the successful cultivation of all the virtues necessary for humane flourishing; they are the goal of *de* 德 (virtue). One seeks harmony but not compete uniformity (*he* 合).

18. *Zhishan* 至善 (the highest good) is the realization of harmony and centrality; the ideal would be to become a *sheng* 聖 (sage) (theoretically possible but in practice very difficult) or a *junzi* 君子 (worthy or noble person).

19. *Taiji* 太極 (the supreme polarity or supreme ultimate) is the highest formal trait of the principle of the whole cosmos and for each particular thing. It is often discussed in terms of *benti* 本體 (the origin-substance or substance and source of all objects and events).

20. *Dao* 道 (the perfect good of all that is, will, or can be) is the totality of the cosmos as the *shengsheng buxi* 生生不息 (generation without cessation). It also usually implies a moral "more" to the myriad things of the cosmos.

These twenty traits will actually form the backbone of a new project, namely, the generation of what I am now calling a Boston *Daoxue* 道學 *Tu* 圖 (Learning of the Way). This is an attempt to restate the Song-Yuan-Ming-Qing and east Asian vision of the Learning of the Way in a contemporary philosophical idiom. However, for the purposes of the discussion of ecology and Confucianism, we will focus on only a selection of the themes and motifs that most closely parallel or sustain conversation with the modern ecumenical and now global ecological discourse. The first case, of course, is a discussion of Qi 氣 or vital force.

It is quite difficult to specify which pre-Han Confucian canonical text best thematizes Qi in its role as matter energy, vital force, or matter energy—all typical attempts to translate this most difficult to name of Chinese philosophical terms into English.[16] Of course, the many layers of the *Yijing* have become the philosophical foundation for Confucian discourse on Qi, yin-yang, and the five phases, all representing the essential vocabulary for the Confucian description of the physical world. The *Yijing* is often the patron text not only of Confucian thought but also of all Chinese thought about any cosmological thinking. Zhang Liwen and his colleagues in their 1991 monograph devoted to Qi, begin the pre-Qin period with selections from the *Zuozhuan* and the *Guoyu*, both major historical texts for the Confucian tradition. They end the chapter with a longer section on the *Guanzi* and its extensive use of Qi.[17] In the *Zuozhuan* we find a list of six Qi, all of which indicate natural elements that represent the vital, life-giving nature of matter energy; in the *Guoyu* we read about the Qi of heaven and earth, further expanding the role of Qi as a crucial philosophical concept. It is perhaps appropriate that the trait of matter energy is first found concretely in the historical sections of the canon, as well as in an eclectic text, because the cumulative Confucian tradition was always at pains to point out how important history becomes because it is only *in* history that the pragmatics of Confucian theory finds its moral ends.

There are at least two general reasons why it is difficult to find one patron text, beyond the *Yijing* itself, for Qi. The first is that, of all Chinese philosophical concepts, Qi is perhaps the most comprehensive term in Chinese cosmology in the sense that Whitehead defined metaphysical concepts as those general ideas that are never absent from our consciousness. Or as Whitehead put it rhetorically, these are concepts so foundational to any cosmic epoch that they are never absent from any conceivable world. Because of the generality of true metaphysical concepts, Whitehead was loath to claim that he could identify just what these ideas really were from the limited perspective of human thought. He often said that dogmatism as to first principles was a sure sign of intellectual failure. The best that we can probably do is to asymptotically approach these general ideas and always retain a healthy dose of skepticism about their finality for even this cosmic epoch.[18] Philosophy and science, Whitehead noted, are littered with failed attempt to capture the absolute.

Whitehead's point is that we normally learn about the world through the method of difference, save for metaphysical concepts that cannot be known through our normal way of learning about the world. Whitehead's own example is that of an elephant; sometimes we see an elephant and sometimes we do not, but given its size, we notice both its presence and its absence—we then make a comparison between the present elephant and its absence and generate the notion of elephant and nonelephant. Contrary to elephants, metaphysical concepts, according to Whitehead, never take a holiday from the backgrounds or foregrounds of our consciousness like elephants are wont to do. True metaphysical concepts are always there and hence we can never contrast presence and absence for metaphysical concepts for all possible worlds. Our normal ways of knowing and imagining simply do not work very well for these kinds of all-encompassing ideals. That is why Whitehead believed that we should be very nervous about proclaiming that we have actually discovered metaphysical ideas. I have thought, for instance, that Wang Yangming's strictures about *liangzhi* (innate moral knowledge) were also aimed at trying to provoke us to think about truly fundamental structures of thought, rather like what Whitehead calls metaphysical principles.

Qi plays this kind of role in Confucian thought; it is always there as the dynamic trait of the cosmos even when it is not the focus of conversation or concern. This was especially the case for the Neo-Confucian cosmologies of the Song-Yuan-Ming-Qing dynasties. Of course, Confucians can and do talk about Qi when they frame their cosmologies, but even then they take for granted a great deal of generality for Qi that does not need to be discussed because it is simply assumed as obvious or clear and distinct or turbid or dense as the particular allotment of Qi under discussion may be.

The second reason why Qi is sometimes not mentioned in various short lists of Confucian defining traits is because of the historical development of classical Confucian discourse. Rather like Greek thought, classical Confucianism, from its beginnings in the middle of the Zhou to its conclusion in the scholasticism of the Han, combined cosmological and ethical reflection in diverse patterns. The germane point here is that the early Confucians, and especially Kongzi as he is captured in the *Analects*, were not overly interested in what we would define as the cosmology as it came to be understood as reflection on Qi, yin-yang, and the five phases. What did concern Kongzi were questions of ethical conduct, ritual, good government, education, and self-cultivation. Such human conduct, of course, could also include a strong conservationist view of how humans should act in the natural world. However, as the tradition developed, later Confucians such as Mengzi and Xunzi would utilize the work of the early cosmologists in order to defend the Confucian Way against other schools. As the scene shifts from the Zhou to the Han, there was a more sympathetic interest in the cosmological speculations of the pre-Han thinkers. The great Han scholar Dong Zhongshu is credited with finally fusing Zhou moral thought with the general background of pan-Chinese cosmological thinking into what has been called Han Confucianism. In many respects this Han Confucianism provides all the later schools of Chinese thought with their philosophical vocabulary as well as a pan-Chinese cosmology that is assumed to be the way the world works.[19]

We need to stress that there is nothing contradictory about this fusion of intellectual horizons between the moral and the cosmological. Modern scholarship has shown that there was a generally accepted series of concepts that were the common property of all the Zhou schools, so much so that many of the most heated battles were fought over just whose concepts were the best in terms of pristine Confucian origins. For instance, the notion of Dao itself is a perfect example of a word that all the major schools would have used in one way or another. Of course, the specification of what the Laoist redactors[20] of the *Daode jing* thought the Dao meant and what Mengzi taught make the study of the classical Chinese thought of the Zhou Dynasty the foundation for all later developments of Chinese philosophy.[21] Ideas such as Dao, yin-yang, and Qi appear to have become the common property of all the various philosophical schools at the beginning of the Han Dynasty. The genius of Han Confucianism was to explain how the commonplace cosmology of the late classical world related in detail to the axiological speculations of the early Confucian masters. Dong Zhongshu believed that all he was doing was explaining how ideas such as humaneness and ritual were linked to yin-yang and the five phases; these cosmological ideas either complemented the Confucian ethical norms or simply served as background to the world in which the specific Confucian values were embodied.

Of course, in the Northern Song Dynasty (960–1126) there was a reflowering of fascination with the cosmological dimensions of the tradition in the work of Shao Yong, Zhou Dunyi, and Zhang Zai.[22] These thinkers are remembered for reintroducing the importance of yin-yang, the five phases, Qi, and even the equivocal concept of the Supreme Ultimate/Polarity as the symbol supreme of the cosmological side of their reflections. Even in the more sober times of Qing evidential scholarship, Wang Fuzhi, hardly an enthusiast for the more speculative reaches of Song cosmology, based his systematic recounting of the original intent of the Confucian Way on a complicated explication of Qi-theory. It is also important to remember that Wang argued that he returned to speculation on Qi because it was only out of matter energy that all object-events emerge, especially those complicated human iconic exchanges known as rituals. Wang went so far as to develop a whole theory of *shen* (spirit) as that which brings harmony to the constant generative action of yin and yang. Wang believed that if we did not understand the nature of the world then we could not become masters of our fate and competent ritual specialists.[23]

Of course, we must always keep in mind that Zhu Xi, as the most famous of the Song *daoxue* philosophers, would never have discussed Qi without a concomitant analysis of the linked role of Li as principle, pattern, or order.[24] Zhu had a highly complex theory of nature, and any modern Confucian hermeneutics of retrieval will have to come to terms with the vast movement synthesized by Zhu known as Neo-Confucianism. While Zhu himself worked hard to reinterpret the classical Confucian texts from the Warring States Period in light of the new social and intellectual exigencies of Southern Song, it is still clear that he did help create a completely new form of Confucian discourse and praxis. The history of Confucian thought after Zhu is a chronicle of those who approved of his vision and those who reacted negatively against it. Like a vast mountain range, it cannot be avoided in any contemporary Confucian ecological reflection. And key to Zhu's thought was his insistence on the critical role of the norm of principle in the emergence of any of the myriad objects or events of the world. Because the ecological crisis is certainly one of values, it is incumbent on any New Confucian to pay attention to the ethical foundations of Song-Yuan-Ming-Qing and Korean and Japanese thought after Master Zhu. This is the style of Confucian discourse that is at the heart of modern Confucian attempts to revive the tradition.[25]

It is also the case that it is within the general confines of Qi-theory that New Confucians will frame their philosophy of ecology, supported by other key traits of the cumulative tradition.[26] For instance, the notions of yin-yang have always been related to Qi, and this is one avenue for arguing about the need for balance between humanity and nature in terms of large ecological systems. One should also recognize that this whole complex of cosmological theory becomes an area of research wherein the New Confucians make use of Daoist as well as Confucian sources. While the New Confucians are committed to Confucianism as they interpret it, they tend to be more ecumenically open to Daoism and Buddhism than were some of their more strict ancestors. Here the typical argument runs that, for instance, the Daoists (and Buddhists as well) may well have been generally more sensitive to human relations to nature than most Confucians and therefore that modern New Confucians need to open themselves to the richness of the thought, art, and symbolism of Daoism in order to restore an ecological dimension to the Confucian Way.[27]

Of course, the most common traditional way to analyze Qi is to present Qi in terms of the equally ubiquitous yin-yang polarities. There is merit in this if we remember that in classical Chinese medical theory, always closely related to the Confucian tradition, yin-Qi has the characteristic, according to Manfred Porkert, of structive action. Porkert means by this that matter energy always provides a structure for reality and that this is seen as a yin function.[28] If principle and form are yang then those elements that relate more to matter energy are structive or yin in Porkert's sense of that term in that they provide the living structure for the form or principles. While it is easier to see how this analogy applies to matter energy and ritual action, it is harder to fit critical reason as reflection and study neatly into the yin-yang categories. However, if we understand the role of *dao wenxue* (self-cultivation via study) as the cognitive preparing the *xin* (mind-heart) for ethical self-cultivation and that the *xin* is the most refined and cognitive part of the human allotment of Qi, then the recognition and realization of principle within the field of Qi is also structive in this yin aspect of reason's role in the Confucian tradition.

THE ECOLOGICAL CONNECTIONS

Stimulated by a growing realization that all is not well with the health of the planet, intellectuals committed to the great religious and philosophical communities of humankind are now engaging ecological self-criticism. How can the earth itself be saved from becoming a toxic wasteland? Du Weiming has even asked if humanity is still a viable species if human beings continue the degradation of nature. The reasons for this concern are both practical and religious in nature. First, breathing the air or trying to look at the sky in almost any major urban area is generally enough to convince intellectuals (religious or secular) that it is high time that they begin to think seriously about ecology. Second, because religions all make the claim that they are "ways of life" sufficient unto every need, it is obvious that the religions had also better attend to the question of ecology because there will be no way of life if the habitat for life is destroyed. This has meant a lively theological debate, begun by Lynn White's classic 1967 article "The Historical Roots of Our Ecologic Crisis," wherein White argued that classical Christian doctrine helped to foster the condition that lead to the modern ecological crisis by promoting a view of the natural world as the realm of the nonsacred or as merely dead matter to be dominated by humankind for the greater glory of God.[29] White's article has spawned a minor cottage industry of scholars seeking to either sustain White's point or discover that Christian theology is really more nature friendly than it seems on the surface. Similar attempts are now under way in other traditions. While each tradition finds its own way to answer to charges of ecological insensitivity, no one wants to be left in a position of defending the continued degradation of the environment. So, we might ask, if Confucianism and Daoism have a better view of nature, why does China also suffer from the same kinds of ecological disasters that plague North America?

The negative and positive charges laid against the axial-age religions are generally of two kinds.[30] The first is whether the basic assumptions of the tradition make it ecofriendly or not. For instance, it has been alleged that ancient, medieval, and early modern Christianity characteristically held a view of matter and the world that was essentially negative because the world was a source of corruption and would be replaced by some other totally reformed state at the end of days. An analogous case has been made that Buddhism is unconcerned about the world because the physical world is mere illusion and that liberation from *samsara* ends with the realization of *nirvana*, a state beyond any worldly concerns. Second, even if the charge of basic hostility toward nature can be avoided, as it often is by those interested in Daoism, for instance, then there is still an urgent need for a search for those principles and practices that can be reaffirmed and expanded to deal with the contemporary ecological crisis. The theory here is that modern technology has transformed the whole scale of debate because for the first time in human history human beings are now capable of literally destroying the environment as a place fit for human habitation.

The Buddhism and Ecology conference held at the Center for the Study of World Religions at Harvard Divinity School on 2–5 May 1996, demonstrated a version of the second strategy. It was first argued that Buddhism is essentially a nature-friendly tradition and that some of its basic teachings, such as the universality of Buddha nature, could become the basis for a carefully crafted Buddhist philosophy of nature. Along with the recognition that there is nothing ultimately antinature in the Buddhist tradition was the concomitant realization that the modern situation was unique because modern technology was qualitatively different from anything that had been the case in the Buddhist world before and that the modern ecological crisis demanded a fresh review of the tradition(s) in light of the contemporary reality. However, some examples were given from the rich history of Japanese Buddhism illustrating that conservationist strains were already part of early modern Japanese Buddhist sensibility. A related Confucianism and Ecology conference held in June 1996 followed this pattern of hermeneutical and historical recovery of the ecological roots and prospective ecological options of the Confucian tradition throughout east Asian, and the results were published in the Forum on Religion and Ecology series.[31]

Following the second strategy of seeking to renew the tradition from within, New Confucians, representing a reformed Song-Yuan-Ming-Qing systematic philosophy, argue that there is nothing intrinsically antinature to be found within the traits and motifs of the composite synopsis of contemporary Confucianism. The only caveat might come from those within and without the tradition when it comes to the fact that Confucians fairly persistently focused their philosophical approach on the human realm without an intense concern for nature. However, Confucians can point to figures such as Zhou Dunyi, Zhang Zai, Cheng Hao, Hu Wufeng, Zhu Xi, Wang Yangming, and Wang Fuzhi from the Song to the Qing dynasties as examples of Confucians who took nature seriously indeed.[32] And modern New Confucians also do not believe that they need to reject as severely as did their forerunners Daoist inclinations for a deep reverence and love of nature. Many of the sectarian quarrels of the early modern past, New Confucians argue, need not constrain the present reformation of the tradition, and dealing with the ecological crisis is a perfect test case for the need for a more ecumenical worldview to be articulated and commended as part of the Confucian Dao. This renewed vision can capaciously include an appreciation of Daoist and Buddhist insights as well as material drawn from dialogue with Western philosophies and religions.

Of course, it should be noted that the trait of *tianming* as the boundless creativity of the cosmos is never merely applied to human life but rather encompasses all that has been, presently is, and can be in the future. One of the criticisms of the religious dimensions of Confucianism is that it has a weak sense of ultimate transcendence, but this is now turned into strength in the sense that Confucians have always had a reverence for and connection to the world, or at least a love of its beauty in their more Daoist moments. For instance, in Japan the Confucian tradition, building on Mengzi, developed a conservationist theory of nature as an integral aspect of solid Confucian learning and administrative lore as witnessed by the forestry policies of the Tokugawa *bakufu*.[33] In this cosmological and pragmatic sense there is no ontological distinction that can be made between humanity and nature.[34] In fact, a case could be made that this foundational account of the cosmos is predicated on a view that makes humanity only one aspect of a much larger reality. The typical Confucian argument would very quickly move to articulate another primordial Confucian claim that heaven, earth, and humanity form a unity of lived experience wherein no one element of the triad can be elided in favor of the others. This is really an organic union of heaven, earth, and humanity in which each must play a role in the proper constitution of the others. As Zheng Zhongying (Cheng Chung-ying) has argued at length in various publications, this is a balanced dialectic of mutual relations and not the sublation or elimination of one element in favor of some new and grander synthesis.[35]

The perennial Confucian affirmation of the world is yet another reason that modern Western scholars have noticed what they take to be an affinity of Confucian thought with American pragmatic and process philosophies. The Confucians are, for the most part, pluralists and realists, who are also committed to a processive and relational view of reality. This means a closure to the world, but only in the sense that the world is the only locus of cosmological interaction between creatures and hence the only place that the creatures can sustain or destroy their own habitat. As the *Zongyong* so long ago argued, and as Wang Fuzhi reaffirmed in the eighteenth century, it is self-conscious humanity that completes the triad of heaven and earth in a harmonious dialectic. From the mainline Confucian viewpoint there is no appeal for assistance from outside the natural triad, though this does not mean that the world is deterministic or without recourse to the comparison of what is with what ought to be. Confucians have made a living being critics of social wrongs;[36] what needs to be added to this persistent Confucian moral vision is an expansion of its range from the typically human to the engaged cosmic via ecological insight.

A quick review of the other traits and motifs also shows that none of them can be directly construed as antinature, although only Qi-theory again deals concretely with human-nature relations. In fact, another case could be made that one of the prime traits of all proper human activity, from embodiment of *tianming* as humaneness to the functioning of the active moral mind-heart to the cultivation and articulation of human nature by means of a broad and comprehensive intellectual quest within a world of civility as expressed in ritual interaction, is an expression of concern-consciousness for every aspect of the world as embodied experience within the matrix of Qi. Although Qi in its ecological role has not been a dominant focus of the tradition, it certainly informs a great deal of nature poetry, landscape gardening, and travel narratives as appreciations of the beauties of the natural world.[37]

Hence a Confucian philosopher must now envision the task of cultivating humanity as set within a larger ecological system wherein human beings must cultivate a sense of reciprocal concern for everything that is the context for human flourishing itself. The specific task of embodying concern-consciousness as the epitome of human self-cultivation must be expanded to a proper sense of what this concern for humanity must mean in order to create a sustainable and balanced ecosystem. Heaven, earth, and humanity must truly learn to cooperate in ways not previously contemplated in the tradition.

The best example, of course, of this kind of ecological sensitivity within the world of early modern Confucian discourse was eloquently expressed in the Song Dynasty by Zhang Zai's famous *Ximing* or "Western Inscription." This essay, for very good reasons, has become recognized as the founding manifesto for any future Confucian ecological vision. Its opening stanzas in laconic yet resonant rhetoric lay out a distinctive and all-encompassing Confucian view of the ecological connections of the cosmos: "Heaven is my father and Earth is my mother, and even such a small creature as I find an intimate place in their midst. Therefore that which extends throughout the universe I regard as my body and that directs the universe I consider as my nature. All people are my brothers and sisters, and all things are my companions."[38] Later in the Ming Dynasty, Wang Yangming even went as far as to say that we should feel sorrow not only for livings things but also for something like a broken roof tile. In doing so the greatest of Ming Confucians was echoing and reinforcing the ecology insight of Zhang Zai.

In a very short compass, the Western Inscription sketches out the basic Confucian ecological program. Even while focusing on the role of the ethical cultivation of the human person in community, Zhang shows how this sense of community must be expanded outward to include all the things of the cosmos. It is not until such a concern-consciousness can be expanded to include the whole cosmos can the Confucian rest worthily of her or his ecological labors. As Zhang wrote at the end of his essay: "In life I follow and serve [heaven and earth]. In death I will be at peace."[39] This is such a typical Confucian vision: it both stresses the hard work needed to achieve the vision of the Dao but also emphasizes the pure joy of a life spent in the cultivation of this cosmic Dao. As modern New Confucian thinkers Du Weiming and A. S. Cua have stressed, the vision begins with the life of the person within the family but does not stop till this anthropocosmic task is expanded to include the proper cultivation of the whole cosmos, from the human ethical, social, and cultural sphere to reverence for even the smallest elements of inorganic nature.[40]

The Confucian tradition has always been interested in the experience of the person who seeks to embody the Way. The praxis of self-cultivation was continuously aimed at helping the person become someone of integrity with the ability to face the trials of the world with a measure of equanimity. Wherein would an ecological sensitivity fit into the range of traditional and modern Confucian experience? Here again the cultivation of the esthetic dimension of Qi is probably the place to begin to look for answers. For instance, Confucians often wrote wonderful travel narratives that focused on the scenic beauty and purity of famous landscapes all over China. What does a modern Confucian make of visits to terribly polluted cities? What about the view of vast pollution spreading in every direction as seen from the summit of Mount Tai (China's most sacred mountain) in Shandong Province? Surely such sights, like Mengzi's observation about how the ruler's mind-heart could not remain unmoved to pity when observing the suffering of sacrificial animals, ought to cause the same feeling or seed of ecological morality to emerge into consciousness. Therefore linking reflection on matter energy and how it interacts with the mind-heart as the seat of human moral sentiment is one promising place to look for the experiential seat for ecological theory and praxis. Mengzi's point still holds good for today, even though it must be expanded to include a scale of the destruction of the environment that would have been beyond even Mengzi's fertile imagination, even though we need again to remember that it was Mengzi who looked at the ruin of Ox Mountain caused by deforestation and was moved to comment on this mindless desecration of north China.

Much more can and should be said about exploring a contemporary elaboration of a Confucian philosophy of nature. Although such a philosophy of nature has never been a major interest of the tradition in the past (and the same could be said for all the Eurasian axial-age religions and philosophies), there is nothing within the tradition that halts addressing the modern ecological crisis. In fact, if Confucians are truly concerned about human flourishing, then they will have to find a way to contribute to the panhuman and pannature conversation that is beginning within and among all the religions and philosophies of the world. New features will be noticed; older points of interest will be modified or even returned to the storehouse of tradition; alliances with other religions and philosophies will be explored; perhaps here will even be recognition of the worth of all those Daoist ancestors lurking around the beautiful grottoes and waterfalls beyond the typical Confucian purview. Even Kongzi once said that he wanted to swim in a stream during the spring with his friend Zengxi (11.26).[41] If the stream is polluted then this Confucian aspiration will be aborted. As the great Qing critics of Song-Yuan-Ming moral philosophy never tire of pointing out, Confucianism should and must be a tradition concerned with the practical nature of human life, and nothing more could be more practical these days than our collective responsibility to address the ecological crisis. All our streams need to be cleansed for their own sake and for the sake of humanity.

APPENDIX: THEMATIC OUTLINE FOR THE COSMOLOGY OF BOSTON *DAOXUE*

1. Root metaphor of concern-consciousness = a fundamental axiology.
2. Form = relational and dynamic notion of *tianming* and *ren*; because of the nature of *ren* as relational, this is a correlational model that leads to the third statement concerning the plurality of object-events.
3. Ontological as well as cosmological pluralism of autotelic object-events.
4. Object-events are self-creating and sustaining = autotelic worldview; there is only this world and no other.
5. Qi provides the field of dynamic energy for the autotelic and relational generativity of the object-events to occur. There is where ecological issues find their most persistent place. Object-events, because they are connected by and in Qi, must respect each other if they are not to cause harm one to the other. This concern-consciousness is made aware of the need for ecological balance and responsibility on the part of humanity.
6. Relationships are not static or dialectical qua a higher sublation but rather a dynamic balance of changing balance and imbalance.
7. The final aim of life is harmony as the highest good, both for humanity and the cosmos.

NOTES

....

1. See W. Theodore de Bary, *The Trouble with Confucianism* (Cambridge: Harvard University Press, 1991). De Bary does a brilliant job of setting the stage for the revival of Confucianism as well as showing just where the problems will be in any such attempt.

2. For accounts of the rise of New Confucianism, see Liu Shuxian (Shu-hsien), *Essentials of Contemporary Neo-Confucian Philosophy* (Westport, CT: Praeger, 2003); Umberto Bresciani, *Reinventing Confucianism: The New Confucian Movement* (Taipei: Taipei Ricci Institute for Chinese Studies, 2001); Chung-ying Cheng and Nicholas Bunnin, *Contemporary Chinese Philosophy* (Malden, MA: Blackwell, 2002); and John Makeham, ed., *New Confucianism: A Critical Examination* (New York: Palgrave Macmillan, 2003). There is, of course, a huge body of excellent material on the rise of New Confucianism in Chinese.

3. For an excellent discussion of these kinds of issues, see Daniel A. Bell and Hahm Chaibong, eds., *Confucianism for the Modern World* (Cambridge: Cambridge University Press, 2004).

4. For reviews in English of the relationship of Confucianism and ecology, see Mary Evelyn Tucker and John Berthrong, *Confucianism and Ecology: The Interrelation of Heaven, Earth, and Humans* (Cambridge: Center for the Study of World Religions, Harvard Divinity School, 1998); Helaine Selin, ed., *Nature across Cultures: Views of Nature and the Environment in Non-Western Cultures* (Dordrecht: Kluwer, 2003), 373–92; the special issue of the *Journal of Chinese Philosophy* 32.1 (March 2005) entitled "Environmental Ethics and Chinese Philosophy"; and, most important, a major new study of the theory and practice of Chinese environmental discourse: Mark Elvin, *The Retreat of the Elephants: An Environmental History of China* (New Haven: Yale University Press, 2004).

5. While it is a completely unscientific methodology (I am a philosopher and not a social scientist), any visit to a university or large urban bookstore in China will demonstrate this growing ecological sensibility: there are simply many shelves devoted to trying to think of a way of constructively coping with the ecological crisis from historical, philosophical, religious, economic, technological, social, and scientific perspectives. The Chinese are no happier about the pollution that covers the vast north China plain as seen from Mount Tai than are Americans living in Los Angles.

6. See W. Theodore de Bary, *Nobility and Civility: Asian Ideals of Leadership and the Common Good* (Cambridge: Harvard University Press, 2004).

7. One can make a good case for Eurasia forming one cultural world with varying degrees of intellectual contact over the last four thousand years. There have been periods, such as during the Roman Empire and the Han Dynasty, when there was a great deal of trade and even some knowledge of other cultures. The golden age of Buddhism in the middle of the first millennium of the common era is another example of a period when people, products, and ideas moved across the Eurasian landscape. Compared to this kind of constant interchange, the role of sub-Saharan Africa, the Pacific beyond Japan, and the Americas is not as sustained or intellectually rich until after the expansion of the European powers in the sixteenth century.

8. See Walter Watson, *The Architectonics of Meaning: Foundations of the New Pluralism* (Chicago: University of Chicago Press, 1993); and David A. Dilworth, *Philosophy in*

World Perspective: A Comparative Hermeneutic of the Major Theories (New Haven: Yale University Press, 1989).

9. Sarah Queen recently argues that it was Dong Zhongshu who first provided the Confucian tradition with a clear and consistent idea of a canon as a way of defining Confucian identity; see *From Chronicle to Canon: The Hermeneutics of the Spring and Autumn, according to Tung Chung-shu* (Cambridge: Cambridge University Press, 1996).

10. In a fine response to the work of Zhang Dainian, Kwong-lai Shun, "On the Idea of Axiology in Pre-Modern Chinese Philosophy," in *Chinese Philosophy in an Era of Globalization* (ed. Robin R. Wang; Albany: State University of New York Press, 2004), takes exception to the use of axiology to describe classical Chinese thought. With due regard to Shun's always insightful arguments, I would still make the claim, especially after the Song revival of Confucianism, that this early modern form of Confucian discourse is profoundly axiological in content and intent.

11. See Chang Hao, "The Intellectual Heritage of the Confucian Ideal of *ching-shih*," in *Confucian Traditions in East Asian Modernity: Moral Education and Economic Culture in Japan and the Four Mini-Dragons* (ed. Tu Wei-ming; Cambridge: Harvard University Press, 1996), 73.

12. Ibid.

13. For his classic statement about the religious dimensions of the Confucian Way, see Tu Wei-ming, *Centrality and Commonality: An Essay on Confucian Religiousness* (rev. ed.; Albany: State University of New York Press, 1989); and idem, *Confucian Thought: Selfhood as Creative Transformation* (Albany: State University of New York Press, 1985).

14. Mou Zongsan offers a short outline of this vision of the Confucian Way and its informing metaphors most succinctly in *Zhongguo zhexue de tezhi* [Special Characteristics of Chinese Philosophy] (Taipei: Student Book Company, 1994). These include the themes of *tianming, ren, xin,* and *xing.* The controlling traits of these themes would be concern-consciousness and creativity itself.

15. Liu Shuxian (Shu-hsien) also offers another synopsis of Confucian themes in "Confucian Ideals and the Real World: A Critical Review of Contemporary Neo-Confucian Thought," in *Confucian Traditions in East Asian Modernity* (ed. Tu Wei-ming; Cambridge: Harvard University Press, 1996), 104–5. Liu's list is more extensive in that he deals with a more extended range of issues including government, politics, psychology, education, and economics. Yet Liu has been enough influenced by Mou Zongsan's reading of the tradition that his longer list supports the basic elements given in my synopsis, even as it goes into more detail in terms of the specifics of the cumulative tradition.

16. Zhang Dainian has an excellent discussion of key Chinese philosophical terms, and it is interesting to note that neither he nor his translator offer an English term for Qi. See Zhang Dainian, *Key Concepts in Chinese Philosophy* (trans. and ed. Edmund Ryden; New Haven: Yale University Press, 2002), 45–63.

17. The *Guanzi* has never been part of the Confucian canon, but modern scholars have recognized that its eclectic collection of materials contains elements from all the major intellectual movements after Confucius. For instance, the medical treatises are just as important to the Confucians as they are to the Daoists. It is perhaps within the protomedical texts in the *Guanzi* that we see most clearly how Qi becomes the trait of matter energy that will play such an important role in the later Confucian tradition. See Zhang Liwen et al., *Qi* (Beijing: Zhongguo renmin daxue chupan she, 1991), 18–43. There

is a complete modern English translation of the text by W. Allyn Rickett. It is always important to keep in mind that our labels for Confucianism and Daoist were invented only in the Han by scholars of that period trying to make sense of and to systematize the wealth of material that they had inherited from the classical age of pre-Han thought. Some texts, such as the *Guanzi*, were called eclectic because they did not conform to the way that Han historians believed the various classical schools evolved.

18. For instance, Whitehead once suggested that perhaps the ideas of identity, overlap, contact, and separation are about as general as we are going to get—though they are hardly rousing candidates for the grander metaphysical inclinations of philosophers. He confessed that he personally could not imagine any cosmic epoch that would not be governed by these kinds of very abstract considerations. See Alfred North Whitehead, *Process and Reality: An Essay in Cosmology* (corrected ed.; ed. David Ray Griffin and Donald W. Sherburne; New York: Free Press, 1978), 66–70, 288.

19. For an excellent description of this development, see A. C. Graham, *Disputers of the Tao: Philosophical Argument in Ancient China* (La Salle, IL: Open Court, 1989); John B. Henderson, *The Development and Decline of Chinese Cosmology* (New York: Columbia University Press, 1984); and John H. Berthrong, *Transformations of the Confucian Way* (Boulder, CO: Westview, 1998).

20. The term *Laoist* has been suggested by Michael LaFargue, *Tao and Method: A Reasoned Approach to the Tao Te Ching* (Albany: State University of New York Press, 1994), to designate the various editors who helped to assemble the *Tao Te Ching* out of earlier teachings of what we have come to call the school of philosophical Daoism.

21. The major exception to this development of Chinese thought was the introduction of Buddhism in medieval China. The arrival of the West also seems to witness a similar addition of novel terms to the Chinese intellectual vocabulary. Over time Buddhism was itself incorporated into, as it in turn enriched, the classical Chinese tradition.

22. For an extended discussion of the rise of the Northern Song Confucian revival, see Peter K. Bol, *"This Culture of Ours": Intellectual Transition in T'ang and Sung China* (Stanford: Stanford University Press, 1992).

23. For an excellent discussion of Wang, see Alison Harley Black, *Man and Nature in the Philosophical Though of Wang Fu-chih* (Seattle: University of Washington Press, 1989).

24. For the best discussion of Zhu Xi's natural philosophy, see Yung Sik Kim, *The Natural Philosophy of Chu Hsi*, 1130–1200 (Philadelphia: American Philosophical Society, 2000).

25. I want to thank an excellent class of undergraduate and graduate students in the departments of religious studies and philosophy at Shandong University, Jinan, China, in June 2005 for arguing that I must mention the role of *li* (principle, order, or pattern) even if only in passing when talking about the ecological interpretation of a Song scholar such as Zhu Xi. The class, of course, was absolutely correct.

26. For instance, one Zhu's favorite expressions—*liyi fenshu* (principle is one, its manifestations are many)—is taken to be an affirmation of the realistic pluralism of the empirical cosmos. Moreover, because this differentiation of the objects of the cosmos is based on the ethical norm of differentiation as embodied in principle itself, this is taken as a way to form a positive evaluation of the rich diversity of the world as witnessed in modern ecological theory.

27. For an excellent discussion of the relationship of Daoism and ecology, see N. J. Girardot, James Miller, and Liu Xiaogan, eds., *Daoism and Ecology: Ways within a Cosmic*

Landscape (Cambridge: Center for the Study of World Religions, Harvard Divinity School, 2002). One should note that many scholars are skeptical about the pristine purity of Daoist ecological visions of harmony. The point is that no premodern religiophilosophical tradition ever had to face the kind of ecological crisis that is now the common fate of modern humanity.

28. See Manfred Porkert, *The Theoretical Foundations of Chinese Medicine: Systems of Correspondence* (Cambridge: MIT Press, 1974), 9–54. Porkert's point is that yin is something that accomplishes something, something that is responsive to other things. Yin is a response to a stimulus. Hence to say that yin is passive is not completely correct. It is better to say that yin is something sustaining.

29. I have found Roger S. Gottlieb, ed., *This Sacred Earth: Religion, Nature, Environment* (New York: Routledge, 1996), to be an excellent collection of primary statements on the issue. For instance, the anthology reprints the short text of White's 1967 article.

30. Judaism, Christianity, Islam, Buddhism, Hinduism, Jainism, Sikhism, Confucianism, and Daoism are the major axial-age traditions that still flourish around the world.

31. For the official website for the Forum on Religion and Ecology, see http://environment.harvard.edu/religion. This is a very useful place to find current information about religion and ecology, including a whole range of current publications on these issues.

32. Moreover, William T. Rowe, *Saving the World: Chen Hongmou and Elite Consciousness in Eighteen-Century China* (Stanford: Stanford University Press, 2001), demonstrates that practical Confucian bureaucrats such as Chen were indeed concerned with the physical welfare of the provincial populations they served. While Chen was a true exemplar of a committed provincial governor, he was certainly not alone in his concern for the well-being of the people he served.

33. See Conrad Totman, *Early Modern Japan* (Berkeley: University of California Press, 1993), for a discussion of conservation polices.

34. One could argue that this reading of the Confucian tradition links it to what Justus Buckler has called "ontological parity." By this Buckler meant that we can no longer privilege any one element of our world over against any other. Hence dreams and ideas must be seen as ontologically "real" as concrete things. This would mean that nature must be seen as real as humanity. See Justus Buckler, *Metaphysics of Natural Complexes* (2nd ed.; ed. Kathleen Wallace, Armen Marsoobian, and Robert S. Corrington; Albany: State University of New York Press, 1990).

35. See Chung-ying Cheng, *New Dimensions of Confucian and Neo-Confucian Philosophy* (Albany: State University of New York Press, 1991), 185–218.

36. One, of course, must be careful about such broad generalizations. For instance, the Confucian traditions, especially in various post-Song varieties, do not comport well with modern feminist theory, and in fact many New Confucians have strenuously argued for a complete reform of gender relations within the Confucian Way. Likewise, many modern Confucians also argue for a much more positive view of the role of law, democratic institutions, human rights, and the place of merchants within the social order. See Dorothy Ko, *Teachers of the Inner Chambers: Women and Culture in Seventeenth-Century China* (Stanford: Stanford University Press, 1994); Victoria Cass, *Dangerous Women: Warriors, Grannies, and Gesihas of the Ming* (Lanham, MD: Rowman & Littlefield, 1999); and Daniel A. Bell and Hahm Chaibong, eds., *Confucianism for the Modern World* (Cambridge: Cambridge University Press, 2003).

37. For a collection of travel narratives, see Richard E. Strassberg, *Inscribed Landscapes: Travel Writing from Imperial China* (Berkeley: University of California Press, 1994).

38. See W. Theodore de Bary et al., *Sources of Chinese Tradition* (2 vols.; New York: Columbia University Press, 1999–2000), 1.682–84, for a translation and short commentary on Zhang's memorable essay.

39. Ibid., 684.

40. For the most comprehensive English statements of Du's and Cua's ecological vision, see Du Weiming, *Confucian Ethics Today: The Singapore Challenge* (Singapore: Curriculum Development Institute of Singapore, Federal Publications, 1984), 1–129; and A. S. Cua, *Moral Vision and Tradition: Essays in Chinese Ethics* (Washington, DC: Catholic University of America Press, 1998).

41. See Confucius, *Analects* (trans. Edward Slingerland; Indianapolis: Hackett, 2003), 122–24.

CHAPTER 11

RELIGION AND ECOLOGY IN AFRICAN CULTURE AND SOCIETY

JACOB OLUPONA

THE complexity of the relationship between environment and religion in indigenous and contemporary African, cultures, and societies requires that we adopt a more multidisciplinary approach that draws from a variety of sources, approaches, and epistemological positions: phenomenology, ecology, geography of religion, indigenous hermeneutics, and traditional anthropological theories under which religion and spirituality is normally studied. However, in my analysis, I shall privilege what I call indigenous hermeneutics, that is, a mode of interpretation that recognizes African imagination, sensitivity, and concerns about the relationship between religion and nature. Based on this premise, I am guided by a variety of conceptual and theoretical frameworks, though not in any systematic fashion at this stage.

First, I am concerned with the environmental referentiality of lived religion, especially rituals. Ritual practice—whether ancestral veneration, processional performance, pilgrimage, rites of passage, and so on—often has no referent, and therefore no hermeneutical foundations or basis of meaning, outside the environment. For example, the symbols, costumes, metaphors, and structure in rituals are intricately linked to nature and environment.

Second, I am concerned with forms of religion and rituals and their contingency upon environmental variations. Do the variations in form and content of rituals and religious beliefs follow the lines of variations corresponding to environmental factors?

Third, and closely linked to the environmental variations, are the environmental imperatives in religion and rituals. Environmental imperatives are those imperatives ranging from comprehension to control and appeasement of nature that are the *raison d'être* of religious behaviors and ritual practices and the essence of almost all aspects of African religions, such as the centrality of the cow in Nuer religion and the importance of sacred hills, mountains, and rivers (Oke-Ibadan, Osun, Olosunta, Orosun) in west African religion.

Based on these three related concepts, I propose that the core of religious worldview and the origins of ritual and the cycles of nature—the regularities and repetitions as phenomenon of nature—may account for the origins of ritual. Indeed, it may not be out of place to speak of the ritualization of the environment as a way to describe the intricate relationship between ritual and environment in African cosmology and religion.

RELIGION AND ECOLOGY: AN OVERVIEW

The study of African traditional religion and the environment can be termed the ecology of religion. The *Encyclopedia of Religion* defines the ecology of religion as "the investigation of the relationship between religion and nature conducted through the disciplines of religious studies, history of religion, and anthropology of religion" (Hultkrantz 1987).

The ecology of religion is a broad topic and in many ways outside the scope of this article. I will focus on ritual; perhaps we can call it the ecology of religious ritual, because it is through ritual practices that people are brought most intimately into contact with the awe-inspiring natural world. Ritual often includes focused thinking and concentration, which tends to inspire mindfulness about nature, mindfulness that is often absent during everyday life and activities.

Contact with the natural world occurs in African cultures when young male and female initiates are separated from their families and brought "into the bush" for elaborate rites-of-passage ceremonies. Novices and devotees visit shrines dedicated to deities that inhabit the mountains, forests, rivers, hills, and lakes surrounding any village or town. Accordingly, ritual objects derived from natural sources are used for divination, and clothes and masks are worn for dances and ceremonies. Devotees are reminded of natural phenomena as stories and myths are told about their origins, explaining how the surrounding world came to be as it is today.

Religious ritual helps Africans to center themselves in the world around them in unique ways that Western religions may neglect. Traditional ritual grounds people in the natural world, where their ancestors have lived and died for many generations before the present. Even Africans who have left their world in Africa far behind can return to their religious and natural roots to find clarity and meaning in their lives, as well as a sense of place. Two fine examples of academics who retain connections with their homeland after having left are Diedre Badejo and Malidoma Patrice Somé. Badejo writes about the Yoruba goddess, Osun Sèègèsí, while Somé writes about Dagara religion and his own experiences of initiation into Dagara culture—after being forcefully educated by French Jesuit missionaries. The stories and research of these two scholars are significant and characteristic examples illustrating the connection between religion and the natural world to practitioners of African traditional religions.

ECOLOGY, NATURE, AND
THE ENVIRONMENT

What is the environment? What does it connote to individuals or communities within the context of African life and traditional religion?

The environment and nature are infused in every aspect of African traditional religions and culture. This is largely because cosmology and beliefs are intricately intertwined with the natural phenomena and environment. All aspects of weather, thunder, lightning, rain, night, day, moon, sun, stars, and so on may become amenable to control through the cosmology of African people. Natural phenomena are responsible for providing people with their daily needs (Mbiti 1991: 46).

With new global environmental crises facing people, animals, and plants every day, it is essential to consider the role of religions in creating, and solving, some of these crises. World religions, and African traditional religions in particular, are very important in creating human views of nature and the ways that people relate to nature and natural phenomenon (Tucker and Grim 1993: 11). Compared with other world religions, African traditional religions are still much more closely tied to the environment where they are practiced, by incorporating natural and ecological ideas more readily into their belief structure. Many practitioners of African traditional religions make their living directly off the land in professions such as farming, ranching, and hunting. In addition, much environmental destruction in Africa today is caused by lucrative economic exploitation exacted in logging, farming, and mining operations. Much is also caused by poaching and popular consumption of "bush meat." These operations are run without any consideration for religious ideologies and concerns of ethnic groups practicing traditional religions. In fact, members of smaller ethnic groups are often preyed upon by larger companies, often quasigovernmental and/or foreign owned, in exploiting environmental resources.

Many phenomena in nature are sacred to followers of African traditional religions; their worldview teaches that almost every living and inanimate object surrounding a person is sacred on some level. There are natural places and objects, however, that are more sacred than others are. Waterfalls, trees, forests, mountains, large or unusual rock formations, and lakes are usually considered the most sacred because they are places where spirits reside, and the spirits can be contacted by humans. Domestic and wild animals also are considered very powerful and sacred. Domestic animals are often used for sacrificial purposes, and certain elements—such as feathers, nails, entrails, horns, beaks, and blood—are used as sacred offerings and divining (Mbiti 1991). Many wild animals are considered sacred because of their wisdom, powers, or the spirits that inhabit them and because in some cases they are sent to earth by God to communicate with humans. For example, the Zulus tell a story in which a chameleon, followed by a lizard, is sent to earth by God to tell humans that God arranged death to be a part of the cycle of human life (M'Timkulu 1977). In the Yoruba creation mythology, a five-toed chicken performs the act of creating the world, while a chameleon serves as the appraiser who ensures that the ground formed from the watery space is solid enough for humans to walk on.

Intrinsically indistinguishable from one another, both animate and inanimate objects exhibit power for believers in African traditional religion. That is, these objects share the same essence in the believer's imagination. Natural forces such as wind, floods, tornado, drought, forests, earthquakes, locusts—and often larger mammals such as buffalo, hippopotamus, lions, and elephants—are deemed more powerful than humans are. These powers are often detrimental to human livelihoods and the cause for much destruction of property and crops. The opposition between humans and nature can be seen in many ritual contexts when humans are symbolically associated with wild animals and other uncontrollable forces (Morris 1998: 120): spirits, deities, and the invisible. Some of the most powerful natural forces are represented by wild animals, whose natural habitat is found around African villages and towns. In Malawi, and presumably in many other parts of Africa, animals and humans are viewed essentially as equals but in opposition. Each has powers superior to the other. This opposition is not, however, an attitude of domination over wild animals, in part because their woodland and grassland habitat is also a source of food, building materials, medicinal substances, and life-generating powers (Morris 1998: 120). Wild animals are seen as representing the bush and are more often antagonistic to human activities.

Significantly, African rituals are concerned with the sacralization of the environment and nature that may take many forms. The gods and spirits are connected to natural objects that serve as hierophanies for the manifestations of these deities. Nature provides the perfect space for worshiping God and practicing ritual. Nature provides the most inspirational, awe-inspiring, and spiritual places and objects. Thus, if it can be used, nature is an obvious place to perform religious ritual.

Within the history and cosmology of African religious life, the supreme being and lesser deities (both gods and goddesses) in different African traditional religions are defined and conceptualized as divinities of nature. Olorun in Yoruba religion connotes the owner or inhabiter of the sky or heaven. These divine agents represent (inhabit, control, live in) natural objects such as bodies of water, hills, mountains, plants, and animals.

What Is Ritual?

Ritual and ceremony distinguish the religious practices of any group of people. Ritual and ceremony are the most important entry points to the religious life of African communities. To the observer of religious practices, rituals are more visible than mythology, but they may relate to myths by conveying and reinforcing the deep meanings and values that groups or communities hold sacred (Olupona 1999). Before discussing the particulars of ritual in African traditional religions and its relationship to nature, there needs to be an understanding of what ritual is, what role it plays in religious life, and what it symbolizes. Ritual can have an extremely broad meaning that covers many aspects of human life even outside the religious realm. For the purposes of this essay, we will understand ritual as applying to "those conscious and voluntary, repetitious and stylized symbolic bodily actions that are centered on cosmic structures and/or sacred presences." Verbal practices such as prayer, singing, and chanting are included in the definition of bodily actions (Zuesse 1987).

All African traditional religions incorporate ritual into their practice, although the forms vary greatly from region to region, from ethnic group to ethnic group, and even from individual to individual within the same religious tradition. Not all rituals are performed by every member of society; instead, a particular ritual may be prescribed for certain members of a community to perform (Mbiti 1990). An example is the Lovedu of South Africa.

Ritual in African traditional religions, as suggested by Parrinder (1976), can be separated into those performed communally and individually. Communal ritual is done, as the term implies, as collective response and quest to the sacred and transcendence, for the good of an entire community or ethnic group. But, in spite of these differences, African religions and especially their rituals and ceremonies share certain commonalities. They "exhibit an astonishing uniformity of emphasis" quite distinct from the variations we find, for example, in Christianity, Islam, and Hinduism (Zuesse 1991: 171). They involve larger groups of people or entire communities. For example, agricultural rituals are performed communally for the benefits of the group and as responses to the cosmic well-being and existential situation of the group. Since many more Africans are still involved in agricultural economies, ritual surrounding widespread agricultural endeavors has much more meaning to devotees compared to devotees found in the minority world (i.e., developed or First World) cultures.

Rituals related to rain, for example, fall under the communal category since rain affects the lives of so many. While many devotees may be involved in performing rituals to induce rain, such as community members helping with dancing, singing, and chanting, sacred ritual objects are manipulated in hierarchical societies by one or a few people who have the necessary knowledge and powers. Other aspects of social ritual consist of libations of water or alcohol, sacrifices of animals, or the presentation of small amounts of food. Sacrifices in which food is shared display a communal bond between ritual participants and God or lesser deities. Shrines, temples, and altars are places in which social ritual can occur (Parrinder 1976).

THE RELATIONSHIP OF RITUAL
AND NATURE IN AFRICAN RELIGION

In order to discuss the interaction and relationship between ritual and the environment, the following question must be answered: What is the impact, meaning, and relationship between the ritual and nature for African people and their spiritual lives? In espousing this relationship, I am guided by a very simple definition and explanation that my former teacher, Peter Berger, once provided: religion can be grasped in terms of its origin, functions, and intrinsic quality or essence. My explication of the ritual-nature relationship will also draw from this notion.

It is almost universally recognized that African traditional religions are inseparable from the social, psychological, and moral dimensions of African life (Ray 1976: 16). From the definitions of rituals and environment we provided, it is clear that many, if not almost all, aspects of African traditional religious rituals exhibit dimensions linked to the environment. First, I will argue, as proposed in the introduction to this essay, that ritual practice has its hermeneutical foundations or basis of meaning outside the environment. The very structure of the African religious universe is mirrored in the structure of the natural world. The three-tiered hierarchy of the universe—heaven, earth and the underworld (land)—is the abode of the sky god, humans/gods, and the ancestral spirits respectively. It provides the framework for African beliefs and practices. It is because of this understanding of direct human relationship with nature that an indirect relationship with God exists in nature. Nature, then, is a "natural" place to be when performing ritual intended to communicate with, to ask, or to appease God. Indeed, the root of traditional African attitude to nature lies in the idea of interconnectedness of the spirit world, nature, and humans (Olupona 1999).

Several African traditional myths illustrate this point. For example, in the Zaramo culture of Tanzania, Nyalutanga is the central mythological character in the myth explaining the origin of life. Nyalutanga appears from the earth, as the origin of all life. Nyalutanga is the first human being and came from the womb of the earth. In this way, the fertility of the ground and of human beings is connected. This myth is manifested in the role/division of agricultural production, which to most Western minds is not a religious act, but to many Africans it is a religious act. Since women are responsible for fertility, physically they are the ones to put seeds in the ground (Swantz 1970: 258–60).

Ritual symbolism is often based upon some of the most basic and intense sensory experiences known to humans, such as eating, sexuality, and pain (Zuesse 1987: 406). These experiences predate any human development of ritual, and ritual probably developed along with other human aspects of social and cultural life such as toolmaking and language. Humans usually do not take action unless that action has cultural value—if it provides food for a family, increases material wealth, or offers spiritual satisfaction. Ritual underscores everyday actions, turns them into symbolic actions, which results in placing them in a realm separate from everyday experience, into the realm of the sacred. Repetition and stylization are usually a large part of ritual acts, and repetition is a way to teach and to memorize. The daily and annual cycles of nature may have led to the origin of rituals. The cycles, while always slightly different, repeat themselves in human perception and timeline. The repetition witnessed in nature must have initially inspired and influenced the repetitive actions of ritual. Almost every African ritual event recreates and renews past human experience in the present (Ray 1999), so that all may better understand what has happened in the past in the context of their present-day lives. Mbiti (1991) observes that African religion "is a living religion which is written in the lives of the people." Since there are no sacred texts in African religion, one could say that ritual is the "text" by which people practice their religion. And, since ritual is timeless, it is a physical connection to the beginnings of a people's beginnings.

For most of human history, no written or codified form of communication existed to which current or future generations could refer to learn how their ancestors performed ritual. Even written documentation is often inadequate for learning information since the specifics of how ritual is performed are rarely recorded. Therefore, repetition still plays a role in the didactic passing on of ritual.

From the beginning of human religious thought, nature has provided inspiration. What separates African traditional religion today from strands of other modern religions is that connections to the natural world have remained as central to religious belief and practice today as they have been in the past. An interesting area for further research in African traditional religions is to look at how urban and rural believers differ in their relationship with nature and how nature is incorporated into religious beliefs. Even in the adopted religions of Islam and Christianity, which on the surface appear to have converted many millions of Africans from their traditional religions, many aspects of traditional religions are still manifest. One of the most common aspects of African traditional religions is the proliferation of shrines throughout the countryside and urban areas. Since shrines provide the basic structure for African religious practice, I will briefly examine the shrine and interaction with nature.

SHRINES, TEMPLES, AND RITUALS

Shrines are the most common kind of religious structure produced by humans. Shrines are found throughout Africa, and they are intended to be used exclusively by family members or for public and community use. They function primarily as the place of contact between humans and the transcendent: for pouring libations, performing rituals, saying prayers, making offerings, and any number of other religious activities. Shrines are usually the center of a family's religious life and the connection between the visible and invisible world. Altars are small structures where offerings can be placed and sacrifices performed. They are found in shrines or temples, or they may stand on their own. Shrines are often built on top of graves, or the grave itself may serve as a shrine. Departed family members can be remembered at their grave, and messages from the living to the departed can be relayed so that they can then carry the message to God or so that the departed can know about the family's sentiments. Graves play a more important religious role for farming communities than for pastoralists, who are constantly moving from one place to another (Mbiti 1991). The location of graves varies from group to group. In most west African communities, burials take place in homes on pieces of land within a compound because they are regarded as secured places where the dead will have peaceful rest. Graves may also be located in a designated sacred forest where the spirits of the ancestors concentrate. But paradoxically, a bad death (suicide or murder) may cause the victim to be buried in the "waste bush" to discourage the spirit from reincarnating or disturbing the peace of the living at home.

Shrines and altars are most often found in natural spaces protected for that purpose or in locations that are powerful places for connecting with the invisible. Frequently, a taboo may restrict the kinds of materials used for building shrines and altars. Often, only local materials found in the environment can be used to build these structures. Nature-based religious places are virtually infinite in number and place, and every African culture has many. They can be forests or parts of forests, rivers, lakes, trees, mountains, waterfalls, and rocks. They are the meeting places between heaven and earth and between the visible and invisible worlds. Natural spaces are usually set aside from their everyday use and circumstances, whether grazing cattle, washing clothes, or growing crops. They are used only for ceremonies, rituals, prayers, and sacrifices and for communication with sprits of the dead and with God and the heavenly world (Mbiti 1991). Shrines are extremely important places for rituals to be performed.

Set apart from villages, shrines in natural places abound in the literature. One outstanding example is in a 1980 study by Ugandan scholar Okot p'Bitek, who wrote that each chiefdom of the Central Luo has a

> shrine on a hill, in a dark forest, or by a riverside.... Some of the shrines were unusual natural phenomena or outstanding landmarks in the landscape. Riba, the premier shrine in Alurland, was physically associated with a cliff on the *Onek-bonyo* (Lake Albert), at a place where from time to time a submarine explosion, *der*, killed shoals of fish. Another shrine at Musongwa was in the form of a small outflow of crude oil, and is associated with the appearance of dead fish "when the mountains have thundered." The hill Abayo, which is the shrine of Labongo chiefdom, is said to have fallen from the sky, burying many dancers underneath. Kilak, Akwang, Labeja are all granite hills that rise suddenly and dominate the surrounding undulation plains. Jok Rukidi at Kal-owang is in a grove of immense forest trees, the only ones of their kind for miles around. (p'Bitek 1980: 59–60)

P'Bitek's comment about Jok Rukidi is perhaps the most striking in terms of implications about shrines and how ritual customs around their use can preserve natural spaces. P'Bitek does not explain why the "immense forest trees" remain. It could be because of unique soils in that area, a microclimate that promotes the growth of these trees, or some other ecological factor. Just as likely an explanation, however, is that much of the land around the grove used to be forested, but because of local land-use practices, much of the forest had been cut down for building and other materials or to clear land for farming and grazing. An unusual natural event could have happened within the grove that influenced its sacredness, but the event was probably coupled with the implicit knowledge that the grove had ecological importance to the region, and if it disappeared the ecological aspects would have also disappeared.

Diedre Bardejo conveys the sense of spirituality and nature-based reverence for Òsun's shrine in the preface of her book *Òsun Seegesi: The Elegant Deity of Wealth, Power, and Femininity*:

> I stood on the banks of the Òsun River breathing the fresh smells of moist earth and rich soil. I crossed the weakened bridge to the other side, and waded in the water. The waters of the Òsun River felt cool, tasted sweet, and I bathed myself, splashing my face repeatedly with her waters. . . . On the several occasions when I aimed towards the Òsun Grove in Òsogbo, I felt drawn, compelled, yes, permitted to enter. Then there were times when I would walk down that same road toward the Grove, reach its gates, only to be turned away. The pervasive presence of Òsun and the other spiritual forces dwelled there, and I followed their guidance. (Badejo 1996)

Òsun Grove in Òsogbo is a good example of the ritualization of the environment, which is commonplace in African traditional religions. Because many important environmental landmarks are ritualized, they are moved into the realm of the sacred. These natural spaces are where deities can be communicated with, and if need be their powers detrimental to humans can be placated, controlled, or appeased. These natural places are not chosen at random—there is usually a historical event that precipitates a place becoming sacred.

It is no exaggeration to argue, as I proposed in my introduction, that we might indeed speak of the ritualization of the environment or nature as the core of African religious practices. Not only does ritual function to control the environment, the environment provides the avenues when the essence of rituals in African religion is realized. Nature necessarily plays an important role in religious ritual and spirituality because of the way that African traditional religion functions in the world and in people's lives. As Zuesse adequately states: "The goal of life, then, is to maintain and join the cosmic web that holds and sustains all things and beings, to be part of the integral mutuality of things. As a result, one does not seek to separate oneself from the world, but to integrate oneself with it" (Zuesse 1991: 173).

One aspect of ritual that is often overlooked is the practical reason behind ritual. What interests us here is how ritual helps to control environmental factors for the mutual benefits of a community. Rappaport (1979: 27–28) points out that a dominant idea in anthropological thought about the relation of ritual the environment is captured in a statement by Homans:

> Ritual actions do not produce a practical result on the external world—that is one of the reasons why we call them ritual. But to make this statement is not to say that ritual has no function. Its function is not related to the world external to the society but to the internal constitution of the society. It gives the members of the society confidence, it dispels their anxieties, it disciplines their social organization. (Homans 1941: 172)

The statement that ritual does not produce a practical result on the world would seem to present a dichotomy between how people view Judeo-Christian ritual and how anthropologists tend to paternalistically view the so-called primitive religions. It is a commonly accepted idea that circumcision and kosher laws presented in the Hebrew Bible are a result of hygienic needs of people at the time. Kosher and circumcision rituals were presumably included in the Bible to prevent outbreaks of disease. The ritual therefore had immediate practical results for the people who followed the rules, at least in part, set out in the Bible.

African traditional religions have similar rituals that are followed to protect people from dangers that exist in the environment, to promote certain ways of interacting with the environment, or to protect resources that are needed by an extended community. Throughout Africa, there are examples of forest groves, rivers, waterfalls, and other natural features that are protected and respected for religious purposes. These sacred spaces are the places where a person can encounter the invisible, and they serve to retain healthy watersheds and medicinal plant populations.

In the highlands in the northwest province of Cameroon, there are large swaths of mountainsides that have been deforested because of increasing human activities such as farming, grazing, and wood gathering. And indeed, where there used to be forest, most of the province is now sweeping grasslands reminiscent of the bare California coastal hills near the city of Davis, where I teach. Much of the remaining forest can still be seen in the northwest province of Cameroon, standing because religious edicts prevented the cutting of trees there. People refrain from entering these spaces except for certain religious ceremonies, rituals, and the gathering of specialized materials. In this way, these necessary natural spaces are set aside for people to carry out ritual activities and, in the process, are preserving natural spaces that may have otherwise been destroyed.

While not an example of an African traditional religion, a celebrated example of traditional religious practices that have a direct impact on people's sustained interaction with the natural world and that have strong implication for rituals in Africa is the work of Rappaport and his study of the Tsembaga and other local territorial groups of Maring-speakers living in the interior of New Guinea. There is a periodic ritual slaughtering of pigs determined by the need to reallocate local resources between different communities. Since Tsembaga are in constant interaction with the environment, the pig slaughter ritual

> maintain[s] the biotic communities existing within their territories, redistributes land among people and people over land, and limits the frequency of fighting (between groups).... It also provides a mechanism for redistributing local pig surpluses in the form of pork throughout a large regional population while helping to assure the local population of a supply of pork when its members are most in need of high quality protein. (Rappaport 1979: 28)

If asked about the meaning and purpose of their rituals, the Tsembaga would reply that they perform their rituals in order to rearrange their relationships with the natural world, rather than the previously mentioned litany of reasons proposed by Rappaport.

Resembling the way that the Tsembaga of New Guinea positively manage natural resources through religious ritual, a similar example comes from Malawi. The Chisumphi cult of the Chewa is headed by a priestess known as the "spirit wife" who guards the land. The priestess officiates in her shrine to appease God and to maintain a proper ecological order. The spirit wife regulates the environment in two main ways: she presides over annual "rain-calling" ceremonies and controls the natural resources found around her shrine. People responsible for rain ceremonies and ritual are well aware of the regularity of seasons, and rains are not brought except in seasons when rain is usually expected. Rainmakers use ritual to bring rain in sufficient quantities, usually to support agricultural production, to end droughts, and to drive rain away in the case of overabundance of water (Parrinder 1976). Among the Lovedu people of South Africa the queen is the appointed person to bring rain to the region. While the ritual surrounding her rain-bringing abilities is a closely guarded secret, it is known that she and a male specialist in the art are the only ones responsible for rainmaking (Krige and Krige 1978).

The rain-calling ritual is filled with symbols and actions that recall the action of rain falling. The priestess wears dark cloth symbolizing dark clouds, young girls' faces are painted with powdered charcoal and dotted with maize-flour paste symbolizing rain drops from the dark clouds, and one girl carries a clean pot containing porridge made from maize flour, while a second girl carries a clay pot of water. The porridge symbolizes food while the water symbolizes the water necessary to grow food that sustains the community throughout the year. By ritualizing natural events and raising the ontological status of ecological features and natural resources to the supernatural realm, implicitly cultural and spiritual value is added to the environment, while believers are called to preserve, maintain, and respect important natural features. Desecration of these same features carries heavy penalties and demands that the perpetrators refurbish and rejuvenate that which they have desecrated (Olupona 1999: 190).

RITES OF PASSAGE AND ENVIRONMENT

Van Gennep's analysis of rites of passage and Victor Turner's elaboration of them have produced some of the most in-depth understanding of the role of rituals in personal and group religious life. What is left out is the central role of the environment in these important ritual activities.

Personal or individual rituals often surround events that happen in everyday life. Birth, transition to adulthood, marriage, and death are four of the most prominent and ubiquitous kinds of life events that are celebrated with religious ritual (Parrinder 1976). Ritual associated with these significant life stages, however, often contains aspects of both communal and personal ritual.

Birth

A personal Fang (central Africa) ritual associated with birth is called *biang ndu* or *biang nzí* (roof medicine). In the case of a difficult delivery, the father of the child climbs onto the roof of the house and finds the precise spot above the mother's belly. After putting a hollow banana stem through the thatched roof, he pours medicinal water through the stem directly onto his pregnant wife's belly. Fernandez's interpretation of this ritual is that the father who has already penetrated the womb to create a child must "penetrate" it again in a symbolic manner through the roof to ensure the child's safe entry into the world (Fernandez 1982: 114–15). This last interpretation should be treated with caution, as the author himself admits that it is not due to explicit explanations by Fang involved in the study. The *biang ndu* (or *biang nzí*) is a personal ritual because it is performed only by a father and witnessed by a small number of family members and neighbors.

Initiation

Transitions into adulthood take many forms in different African cultures, but the change from a child to a community member holding responsibilities that are more adult is a constant. Initiation ceremonies are most commonly held when the youths are going through pubescent transitions that include physical, emotional, and psychological changes (Mbiti 1991: 96). There is abundant ritual involved with initiation. Initiation ties an individual to his or her community, culture, history, and traditions because it is a time for the youth to learn about how to *be* in the world as a member of his or her society. Even though most of a person's initiation is done in seclusion with other individuals of a similar age, separate from the main village or community, initiation is a public recognition that a person is passing from childhood to adulthood by learning the ways and worldview of the community. Participants are taught about the life of their people, their beliefs, history, traditions, how to raise a family, the secrets of marriage, and other practical information. Initiation is a mark of unity with a larger community, and it is a deeply religious affair. Many prayers and sacrifices are offered to God before, during, and after initiation ceremonies to ask for blessings and good luck for the youth going through the arduous process (Mbiti 1991). Female and male circumcisions are often a part, but not the focus, of initiation rites.

As mentioned previously, initiation ceremonies are usually performed separately from the rest of a community, so that people do not know what is happening to the initiates at any particular time. Initiation often takes place over a period of several days or months in a natural space, such as a forest or grassland, where the youth can be in closer contact with the invisible, with spirits, and with God. Nature is a place that does not distract the individual or a person from concentrating on spiritual training. This type of emotional and psychological growth can happen only in a wilder environment than the village and so is essential for the growth of African youth. Unlike the way and manner that many anthropologists theorize the opposition between nature and culture, here initiation shows the dialectical relationship between the two. Culture and nature are intertwined, one lending itself to the other, rather than an opposition that sees culture as superior to nature. The cultural being is refashioned and remade in the bosom of nature.

There are innumerable accounts of initiation rites written by outside observers and anthropologists. One of the best, however, is an account written by Malidoma Patrice Somé as a young adult about his own experience going through an initiation ritual, although he was kidnapped at the age of four from his family by a French Jesuit missionary. His story is about the Dagara of Burkina Faso, and the initiation from the very beginning is filled with connections to nature. Male initiates leave the village and, while still in the presence of family and friends, remove their clothes: "Nakedness is very common in the tribe. It is not a shameful thing; it is an expression of one's relationship with the spirit of nature. To be naked is to be openhearted" (Somé 1994: 193).

The connection between culture, family, and the hugeness of the natural world is captured in the song that children sang as Somé passed from the village into the bush (Somé 1994: 193):

> My little family I leave today.
> My great Family I meet tomorrow.
> Father, don't worry, I shall come back,
> Mother, don't cry, I am a man.
> As the sun rises and the sun sets
> My body into them shall melt,
> And one with you and them
> Forever and ever I shall be.

The entire process is deeply religious, according to Mbiti, and extensive ritual punctuates the spiritual aspects of initiation.

Marriage

Marriage in African traditional religions is an important part of social, spiritual, and cultural life. Mbiti (1991: 104) states that marriage, raising children, and participating in family duties are profound religious duties. Marriage ceremonies are filled with ritual, but economic, social, and religious aspects of marriage ritual overlap to such an extent that the three are difficult to distinguish from one another (Mbiti 1990: 130). Marriage agreements almost always involve both sets of parents of the couple to be married. The binding of a man and woman is often accompanied with an exchange of gifts, which is more a way of thanking the parents of the bride or groom for bringing up their child in a good manner. The gifts do, however, hold some local legal weight, since if a marriage does not hold, it is expected that the value of gifts in cash or kind be returned to the family that gave them (Mbiti 1991: 108).

Death

Long and complex rituals are often associated with death, one of the saddest and most important events that happens in the life of an African community. Again, there is huge variation in tradition and ritual surrounding death. Personal ritual in caring for the deceased may vary widely. Shaving the head or swathing the body with white clay is common. With death comes a permanent physical separation of the deceased and living, and ritual helps to accentuate this fact. Natural objects are often used to wash, clothe, and bury the body. Skins, leather, cotton, bark clothes, leaves, and other objects of nature can be used to cover the body (Mbiti 1991: 119). These objects call attention to the fact that a body is conceived in the earth and is returned to the earth. While the body returns to the earth, the spirit or soul remains as a presence in the lives of individuals and needs to be respected. The world of the ancestors and the abode of the living dead are generally conceived as a sphere beyond the realm of the living. In some societies it is the earth (*ilè*).

In many societies, the departed live in the forests, rivers, riverbanks, hills, or other natural places in the country (Mbiti 1991). These places are generally avoided as the resting place of the departed, and they need to be respected. Places where departed souls reside often are located in, on, or near a remarkable natural phenomenon in the area, and one way these places can be preserved from exploitation is to place certain limits on them. While this reasoning is a Western way of explaining natural resource use, it is an important way for African societies to preserve crucial natural resources from being destroyed. Ritual helps to emphasize how natural resources should be used, and it is just as effective in many cases in inspiring wise resource use as government regulations and written laws are. In many African countries, government employees are not held accountable for their actions and are more willing conspirators in the destruction of their country's natural resources for personal gain. Government regulations often are ineffective in protecting natural spaces. More attention should be paid to local ways of preserving natural resources, particularly on local religious beliefs and ritual in the forefront of any analysis. In this way, natural space can be protected in perpetuity, local support is more forthcoming, and local understanding about why natural resources should be respected is much more likely.

From this discussion of the life stages of birth, passage into adulthood, marriage, and death, we see that ritual is vital to African life and African history. Natural imagery, objects, and places all play an important part of ritual that accompanies every stage of a person's life. Without a natural landscape and reverence for the spirituality and mystery to be found in nature, much of the power of African culture would be greatly diminished. Every African goes through these four life stages. Religious traditions and ritual practices have been passed down from generation to generation for centuries, so a person's current, modern reality is very much living their ancestor's history. History is experienced in every religious ritual that is performed for a particular event. To be sure, ritual has changed over time according to the social, political, environmental, and spiritual needs of individuals, but it is still a very real connection with the past and one that Africans act on very seriously in passing culture from one generation to the next.

WATER AND THE PARADOX OF LIFE AND DEATH

I am very much interested in a major ecological phenomenon whose presence has attracted the attention of the entire world, especially ever since the reality of global warming has been acknowledged. This calls for examination of the place of water in African ecological and religious heritage.

In the book *Water: Life Force*, Maggie Black writes, "Since the earliest times people have honored water, worshipped water, granting [it] a special place in their language, myth and rituals" (Black 2004: 12). Water is an ambivalent matter and resource in the history, culture, geography, and social life of the African people. Although water plays a creative role and is intertwined in African culture and social life, paradoxically it has been the source of problems in the African situation. This duality of water's scarcity and abundance, blessing and destruction, further demonstrates this paradox. This paradoxical image of water will serve as a metaphor for examining the significance of water in Africa's heritage. While I am concerned with the status and history of water, its functions, and its intrinsic value and essence, in my ethnographic portrait I will also show how water reflects some of Africa's contemporaneous problems of environmental sustainability and human survival. The last point will be addressed through theoretical and methodological angles that I often explore in my study of indigenous religious tradition. That is, I will argue that phenomenological and sociological interpretation and the quest for meaning may not be sufficient in understanding any phenomena, in this case water. Scholars must show how the phenomena relate to the concrete lives of individuals, groups, and communities. The modern assault on indigenous culture and knowledge base requires that scholars continue to see how even the tiniest element of phenomena that we study relates to lived realities, to the people who struggle on a daily basis just to stay alive in this hostile global, neoliberal, capitalist world. So theory and practice must relate and connect object with subject.

Water in Historical Perspective

In Ali Mazrui's popular film *The African: The Nature of the Continent*, the director reminds us of the interconnection between African society and water. The natural environment and culture in Africa is marked by the presence of water, and this substance has played a significant historical role in defining Africa's cultural identity for centuries. First, it was the Nile that connected Black Africa with ancient Near Eastern and the Mediterranean civilizations, historically bringing the ancient Judaic tradition and Western civilization in contact with the African continent. European curiosity about the Dark Continent ultimately led to the white race's location of a great many rivers, lakes, oceans, falls, and places that today still bear the hallmark of the colonial legacy. Lake Victoria, Victoria Falls, River Niger, and River Limpopo are examples of such places. By renaming these bodies of water, rivers lost their indigenous identity. For example, the Yoruba people's name for River Niger was Odo Oya, the river belonging to the fiery-looking Yoruba goddess Oya. And the territory beyond this vicinity was called Oke Oya, defining the northern Nigerian cultural and political identity. Beyond the Niger in the Yoruba imagination was the unknown land of the Hausas, Fulani, Kanu, and others.

In traditional modern Africa, the attitude and response to water has been complex, prevalent, and daunting. Water has been the source and origin of towns, villages, and cities. The mythical and historical relationships, the political and social structure, and the development of a large number of places are anchored to forms of water ideology upheld by the people themselves. In western Nigeria, not only was the town of Osogbo founded on a long mythic narrative of a goddess named Osun, indeed one of the states and the inhabitants of the state are also labeled Osun. The reason for this is that the very large Osun River runs through the entire space and place that constitutes the state. Indeed, the two largest bodies of water in Nigeria, the Atlantic Ocean (*okun*) and the Osa (sea), are both depicted as primordial water goddesses in the history and religion of Ile-Ife, the Yoruba cultural center of Nigeria. Constantly in rivalry, Olokun (the goddess of *okun*) and Osara (the goddess of *osa*) are personified as the wives of Odudwa, the legendary founder of Yoruba tradition and civilization. While Olokun is depicted as a wealthy woman with no children, Osara is shown as an indigent woman with many children. Olokun, we could argue is a metaphor for imperial power and mercantilism that characterize the slave trade and the colonial incursion into Nigeria and Africa. But Osara—indigenous, local, and poor—is more acceptable in local parlance because she exemplifies fertility and reproduction, a central value in African cosmology.

In Yoruba mythic history there exists a narrative of an open confrontation that ensued between Olokun and Osara during which Olokun displayed her wealth as a symbol of her authority and power, but was quickly trumped by the appearance of Osara with her many children. The drama of this conflict constitutes the historical and mythical impulse for the Agbon festival in Ile-Ife today, a festival of young children on display. In Ile-Ife today, several festivals begin at the bank of Osara because she bestows fertility on the populace.

The water metaphor invoked in the above narrative portrays two diametrically opposed forces that are constantly at work in society. The metaphor demonstrates the relevance of water in indigenous epistemology and conveys how the indigenous Yoruba society interprets the world. The narrative also conveys the importance of water as the bearer of wealth, prosperity, and fecundity. Money is meaningless in contrast to fertility, and wealth is powerless if it is not tied to human production and people. I should add that the Yoruba goddess of wealth, Aje, is also said to have derived her source from the ocean. I have argued elsewhere that Aje is not only the bearer of money, symbolized in cowries, but it was also Aje who invented the banking system.

Water in Nature and Culture

Ali Mazuri's film clearly shows the intimate relationship of ecology and culture. Bodies of water and rivers ply many functions. Rivers mold the landscape, cutting channels and etching valleys (Black 2004: 34). Rivers also transport particles and cleanse the land; they supply domestic water, irrigation, sources of food, and inspiration to artists, poets, priests, scientists, and philosophers (2004: 34). Rivers also serve as natural boundaries for nations, states, districts, and countries (2004: 38).

Due to the power and reach of water, it is also an element in which life can be celebrated and enjoyed. As Black writes, "Water confers beauty on a landscape," and "the sight and sound are a source of spiritual refreshment" (2004: 114). In cultures across the world there are festivals and celebrations devoted to rain-making and celebration of water. Because water makes the land fertile, as the history of the Nile clearly shows, agricultural production depends on an adequate supply of water. Water provided by rainfall is an essential natural resource in agrarian culture. If rain does not fall at the right place at the right moment, crops will fail (Disanayaka 2000: 51). In addition to the significance and power of water cited in the above examples, for the African religious imagination water carries deep meaning in peoples' spiritual and cultural heritage.

Water, Gods, and Beliefs

Water serves as a tangible manifestation of divine essence (Disanayaka 2000: 138) in all human societies and cultures, and "in many belief systems water is held to have miraculous properties" (Black 2004: 100). For instance, several of Africa's greatest rivers are propitiated as holy, with shrines and temples along the bank, and especially numerous are those built to the rain deities. In some African belief systems, life is viewed as a transitory process of moving from being wet to being dry. A baby is considered "wet" when born, and as the child grows he or she transforms and becomes "drier" (Black 2004: 100).

Because water plays a significant role in human survival, people often appeal to priests and shamans when water does not bring rain or respond to the seasons. Those whose livelihoods depend on the sea or rivers treat the waters with great respect, and they have a whole series of hymns, prayers, and rites dedicated to water (Black 2004: 103). A thrilling body of songs and orature describe the great Osun River:

> She moves majestically in the deep water
> Oh Spirit! Mother from Ijesaland
> The land of the tough and brave people
>
> My mother lives in the deep water
> And yet sends errands to the hinterland
> Aládékojú, my Olódmare [supreme goddess]
> Who turns bad destiny [orí] into a good one.

In African mythology, the earth or the abode of humans, vegetation, animals as we know them today arose from the massive ocean of chaos, when the supreme god ordered that the world be created. From this massive water emerged the primeval hill and the solid earth on which the first human and primordial deities landed to preside over the structuring of the world social order and the creation of cultural land mass. In Edo-Benin mythology, the realm of the ocean (Olokun) is viewed as the exact opposite of the earthly realm and is the domain of the sacred king Oba. Thus, the king is the lord of the sovereign human and earthly realm (he is lord of both the earthly and the water realms).

Water in Ritual Context: Fertility and Purification

Water as a purifier is central to the practice of African religion. A significant aspect of indigenous rituals and practice involves the cleansing of the body, places, and objects with water. Sprinkling water on the body, bathing in running water, entering waterfalls, rinsing the mouth, and the rite of washing the feet in marriage celebrations are several of many purifying rituals. These all point to the cleansing power of water (Fischer 2001: 211), implying that water is in principle a purer element than what it is made to purify. In its ordinary use water is employed to remove dirt, just as in the religious realm of ritual cleansing water has a significant meaning as a ritual purifier.

I am intrigued at how water can also serve as a metaphor for religious conversion, indigenous religious practices, and indeed African's participation in the new global evangelization of the continent. As an example, let me use African Christian reading and appropriation of the Jordan River, which is located in the far distant land of Judeo-Christian faith.

As a Sunday school child, I noticed that part of our Christian indoctrination involved singing about the River Jordan. In this scriptural engagement with Christianity, the story of the baptism of Jesus in the River Jordan was read by the priests over the bowl of water placed at the font of the church during Anglican baptism. There was also the traditional Christian lyric derived from the Bible story of the Aramean general Naaman, a leper who was asked to bathe in the River Jordan. After much hesitation, he did as he was told by the prophet Elijah. The miraculous cleansing of Naaman's impurity was a theme in our Sunday school songs in the 1960s.

Indigenous African Christian churches called the Aladura churches broke from this foreign Jordan model by using African's own indigenous water in its lakes, rivers, and oceans as a place where ritual cleansing could take place. Any flowing water in the vicinity of their churches was commandeered for religious ritual cleansing. One of the beauties of the religious indigenous tradition is that it discovered in its own backyard sacred rivers, sacred lakes, and sacred ocean for spiritual and ritual purposes. But the new evangelical Christianity of the 1990s not only began to return to the Jordan paradigm, but also insisted that there is a need to physically travel in pilgrimage fashion to Jerusalem and the Holy Land of Israel to have a direct Jordan experience so that the pilgrims can see for themselves the marvels of those sacred waters and return home with bottles of water that have become spiritual tangibles for cleansing and ritual purposes. So the Jordan that we read about in the scriptures now becomes collected and stored in physically tangible bottles. Local water that has served for decades as hierophanies of sacred places in the imagination of indigenous African Christians is now assumed to be infected with strange spirits and mammy water spirits, ripe for exorcisms. The Christian evangelicals have succumbed to this new assault on culture and tradition.

The rainmaking ritual best expresses Africa's creativity about water. Rainmakers are found all over the continent. They combine deep indigenous knowledge of the weather with the ritual power to manipulate rainfall in their community. The Nngnga diviner in many Bantu-speaking societies of Malawi, South Africa, Uganda, Gabon, Congo, and other countries is engaged in rainmaking for agricultural and human purposes (Schoffeleers 1997: 75).

Water, Violence, and Liberation

As Black points out in her book, "contrary to the virtues of sweetness and purity" celebrated in prose and poetry, "water is an aggressive substance" with an extraordinary power of solvency that can "secrete poisonous minerals" and contaminate, providing a haven for predatory and parasitic insects (2004: 25). Water has the capacity to be an avenger, and its supernatural powers personified in gods and goddesses can use their powers to devastate humankind (2004: 111).

In many African cultures, creation stories (cosmogonic narratives) relate how the gods and spirits created the universe from the massive water that inhabited the earth. Through ritual actions ordered by the supreme god in the creative process, dry-land vegetations were formed and people began to inhabit the land. The Yoruba deluge myth purports that the newly created world was destroyed after the first creation. There occurred a great flood or deluge that left only a few survivors. Ile-Ife, also called Ooye L'agbo ("the place of those who survived"), may be strange to students of the Bible because of its similarity with the story of Noah and the ark. The missionaries were stunned on arrival to Nigeria to find that the indigenous peoples had stories of creation comparable to the story of the creation of the universe and the flood in the book of Genesis.

While the story of water's destruction is often depicted in negative terms, there are instances when water symbols act as liberative forces. For instance, in one Yoruba narrative the goddess Osun fought with Osanyin, the Yoruba god of herbal medicine. As the story goes, Osanyin left his herbs on a stone grinder when his attention was diverted. Osun then cleverly removed the herbs from the grinder and destroyed them. In the narrative, Osun is praised as the one who liberates her people from evil medicine (*Osun gbologun lo*). Thus Osun's torrential water is powerful enough to destroy evil and rescue innocent worshipers from their enemies. Osun uses her cool water and her torrential water to rescue her worshiper from the evil eye, symbolized in Osanyin, the owner of herbal medicine.

Water and Development

Why should religious scholars be interested in studying water heritage? And why is the study of water heritage significant in the larger study of water by the scientific community whose interests are more focused on water management and the use of water for sustainable development, hydroelectric power generation, underwater maritime exploration, offshore oil exploration? The reasons for our interest are clear. A significant part of the development crisis that many African nations face today relates to the disconnection between development projects and cultural heritage. This has become painfully clear when issues of cultural heritage apparently undermine or get in the way of development projects. As a child growing up in Nigeria, I remember that, with the building of the first major hydroelectric project in the northern city of Kanji, a large part of our national cultural heritage was destroyed. Historical buildings, traditional shrines, ancient vegetation, and many small villages were submerged underwater. Development was simply seen as a superior project that would alter the nation and bring progress and modernity. Traditional values and material cultures would have to give way to heavily financed foreign-sponsored projects.

A major task for African nations is how to incorporate the concept of water as cultural resource and institutionalize the creation of national heritage rivers and to bestow formal state recognition to the important rivers in this vicinity and create monument bodies and agencies that will serve to maintain the historical, recreational, and cultural values of these rivers. The challenge is to provide a link between the perception of water as a cultural heritage and its use for scientific sustainable development and growth.

In Yoruba mythology and imagination, the ocean serves as a good metaphor to provide a connection between the scientific and cultural sphere. The ocean constitutes a separate territory, a domain referred to as Ibu Olokun. This is regarded as the kingdom or realm of influence of Olokun the ocean goddess. D. O. Fagunwa, the celebrated Yoruba author, reminds us in several of his novels that Ibu Olokun is a wealthy kingdom with an abundance of treasures, cowries, gold, silver, and pearls under the control of Olokun. In the scientific world and imagination of our oceanographers, there exists a similar conception of a mysterious underwater world containing a wealth of natural resources. Both cultures share a perception of abundant resources submerged in the water realm.

Contrary to the Western view of development, new forms of development programs are fast discovering that national cultural heritage and dimensions, such as water and river heritage, form an integral part of sustainable development. One cannot function without the other.

REFERENCES

Badejo, D. 1996. *Osun Seegesi: The Elegant Deity of Wealth, Power, and Femininity.* Trenton, NJ: Africa World Press.

Black, Maggie. 2004. *Water: Life Force.* London. New Internationalist.

Crazzolara, J. P. 1950. *The Lwoo.* Verona: [Istituto missioni africane].

Disanayaka, J. B. 2000. Water Heritage of Sri Lanka, Bogota: University of Colombia Press.

Fernandez, J. W. 1982. *Bwiti: Ethnography of the Religious Imagination in Africa.* Princeton: Princeton University Press.

Fischer, Mory Pat. 2001. *Living Religions: A Brief Introduction.* Upper Saddle River, NJ: Prentice-Hall.

Homans, G. C. 1941. "Anxiety and Ritual: The Theories of Malinowski and Radcliffe-Brown." *American Anthropologist* 43:164–72.

Hultkrantz, Å. 1987. "Ecology." In *The Encyclopedia of Religion.* Edited by M. Eliade and C. J. Adams. New York: Macmillan.

Krige, E. J., and J. D. Krige. 1978. *The Realm of a Rain-Queen: A Study of the Pattern of Lovedu Society.* New York: AMS.

Mbiti, J. S. 1990. *African Religions and Philosophy.* Oxford: Heinemann.

———. 1991. *Introduction to African Religion.* Oxford: Heinemann.

Morris, B. 1998. *The Power of Animals: An Ethnography.* Oxford: Berg.

M'Timkulu, D. 1977. "Some Aspects of Zulu Religion." In *African Religions: A Symposium.* Edited by N. S. Booth. New York: NOK.

Olupona, J. K. 1999. "African Religions and the Global Issues of Population, Consumption, and Ecology." In *Visions of a New Earth: Religious Perspectives on Population, Consumption, and Ecology.* Edited by H. G. Coward and D. C. Maguire. Albany: State University of New York Press.

Parrinder, E. G. 1976. *African Traditional Religion.* Westport, CT: Greenwood.

p'Bitek, O. 1980. *Religion of the Central Luo.* Kampala: Uganda Literature Bureau.

Rappaport, R. A. 1968. *Pigs for the Ancestors: Ritual in the Ecology of a New Guinea People.* New Haven: Yale University Press.

———. 1979. *Ecology, Meaning, and Religion.* Richmond, CA: North Atlantic Books.

Ray, B. C. 1976. *African Religions: Symbol, Ritual, and Community.* Englewood Cliffs, NJ: Prentice-Hall.

Schoffeleers, Matthew. 1997. *Religion and the Dramatisation of Life.* East Lansing: Michigan State University Press.

Somé, M. P. 1994. *Of Water and the Spirit: Ritual, Magic, and Initiation in the Life of an African Shaman.* New York: Putnam.

Swantz, M. L. 1970. *Ritual and Symbol in Transitional Zaramo Society with Special Reference to Women.* Lund: Gleerup.

Tucker, M. E., and J. A. Grim. 1993. "Preface." In *Worldviews and Ecology: Religion, Philosophy, and the Environment.* Edited by M. E. Tucker and J. A. Grim. Lewisburg, PA: Bucknell University Press.

Zuesse, E. M. 1987. "Ritual." In *The Encyclopedia of Religion.* Edited by M. Eliade and C. J. Adams. New York: Macmillan.

———. 1991. "Perseverance and Transmutation in African Traditional Religions." In *African Traditional Religions in Contemporary Society.* Edited by J. O. K. Olupona. New York: International Religious Foundation.

CHAPTER 12

INDIGENOUS TRADITIONS

Religion and Ecology

JOHN A. GRIM

No one term in an indigenous language may exactly translate, or even correspond to, the English terms *religion* or *ecology*. Yet something of the depth and variety of life-orienting interactions with bioregions and biodiversity, and the complex human-earth interactions in creation stories and ritual among indigenous peoples, can be communicated through these approaches. No one ritual action and no single mythic cycle among the diverse indigenous peoples provides a "text" or a "language" for understanding their religions. In fact, these genres may mislead an investigation of indigenous religious life that is primarily narrated, danced, sung, heard often in silence, and ritually performed within the community of life.[1] Thus, the experience of sacred relationships forged by indigenous peoples with the world suggests more lived and embodied modes of expression.

Similarly, the term *ecology* is not used here to reference exclusively the scientific study of ecosystems. Rather, this term is used to express indigenous knowledge, traditional ecological knowledge, or traditional environmental knowledge.[2] The term *religious ecologies* can also be used to refer to these different cultural understandings of the interrelationships and interdependencies with natural systems developed by diverse indigenous peoples. These forms of interactive knowledge may not be scientific as that term describes Western modes of empirical, falsifiable, experimental investigation. Indigenous knowledge has it own modes of empirical observation, acquisition through lived experience, and testing in the context of one's community.

Indigenous knowledge is traditional because transmitted from generation to generation. Yet, this knowledge is not static data but a dynamic, lived knowing that grows and adapts to changes. The sources of this knowledge are multiple. Indigenous knowledge might come from close observation of nature, or it might come from visionary and dream sources and endure as traditional environmental knowledge recognized by the wider community. In all settings indigenous knowledge is directly related to the natural world.[3] It is in this sense that it can be said that indigenous peoples live in a universe and not simply in systems of knowledge that characterize such separate academic fields as science, social science, or the humanities.

Despite widespread cultural losses due to colonization and industrialization, many indigenous peoples still hold to their creation stories as the basis of their traditional symbols and rituals of spiritual and ecological intimacy. These creation stories provide the cosmological context for knowing self, society, and world. The values embedded in these stories are manifest in their passages through personal life crises, in community liturgies of renewal, and in their spiritual encounters with the transformations of weather, nightscapes, and landscapes. In other words, the stories are not simply a "narrative literature." They are not merely data for a study that can be called "traditional ecological knowledge." Rather, creation stories are heard as lived, embodied relationships with environments.

Just as indigenous knowledge is embedded in local environments, so also it is at the growing edge of individual and community life. In diverse ways indigenous peoples explore and transmit creative engagements with the natural world and the larger cosmos in a frame that is labeled here as religion and ecology. This work presents an overview of these religious ways of indigenous peoples as embodied life. That is, one interpretive entry into indigenous religions is to explore personal, social, ecological, and cosmological realms as integrated dimensions of a unified body. Embodiment, therefore, is presented here as a phenomenological description found in indigenous religions as well as a helpful metaphor for understanding the relations established between indigenous peoples and the larger world.

Narrations during rituals and statements that describe personal and social experiences provide materials for this investigation. Moreover, since the indigenous traditions of the Americas have been the focus of study for this author, they provide the majority of examples of indigenous religions given here. However, brief references from the indigenous peoples of Africa, Asia, Australia, and the Pacific regions will be included to point the reader toward rich studies of indigenous religion and ecology in those regions.

It is important that these materials should not be understood as simply ethnography or objective narratives and descriptions of personal experiences. That is, the collecting of narratives by anthropologists and other students of culture as well as translated personal statements by indigenous spokespeople themselves can be seen in a broader context of lived embodiments. These statements, then, are not presented as essential, abstract proclamations of a static truth. Rather, these texts are themselves evidence of bodily experiences and interpretive reflections upon the interactive flow of native religious life with the natural world. In that sense, indigenous perspectives on the sacred provide new ways of thinking about the text of all our lives in the context of the world.[4] Indigenous perspectives serve to raise questions about the stories that place dominant societies in the world. What will be the capacity of those stories to draw people forward in meaningful ways? Are there stories that will draw all of us forward into a productive, sustainable future with the whole community of life?

AFFIRMING THE DIVERSITY OF INDIGENOUS PEOPLES

Obviously no one people can be made representative of the thousands of indigenous cultures in Africa, Asia, Australia, the Pacific region, and the Americas. Yet, there are "family resemblances" among indigenous religious experiences that allow an investigation of these plural forms of religious traditions among extremely diverse indigenous peoples. By positing resemblances among indigenous traditions, I do not intend to construct an "essential" indigenous religion. Moreover, the spiritual insights that have guided native peoples through the oppression that they have undergone for the past five centuries cannot be adequately explained by analytical categories. An effort is made here, however, to suggest that these expressions of lived relationship and intimacy with local environments are significant pathways for personal maturity, communal identity, spiritual ecology, and cosmological contemplation among the native peoples who transmit them. Moreover, awareness in dominant societies of the depth and conviction of indigenous religiosity has gradually made indigenous traditions important participants in some cross-cultural, interreligious, and global dialogues of cultures.

Dialogues between peoples may be limited in their grasp of the full range of religious life and practice, but they can lead to deeper levels of empathy and increase understanding of religious differences than many strained sociopolitical exchanges. Dialogues, hopefully, are conducted so that they do not deprive participants of their critical awareness or their particular perspectives. Two points are seminal here. First, discussions of indigenous religion and ecology are typically framed in the language and categories of dominant knowledge systems. What would dialogues look like if they were framed in indigenous ways of knowing? Second, the romanticization of indigenous knowledge can blind dialogue participants from realizing the struggles of native communities themselves to realize their own values and ideals. In that sense, there is no question that indigenous peoples at times also suffer from problematic motivations that blind individuals and communities, driving them toward oppression of one another. Local indigenous communities are certainly not free of their own forms of oppressive authority, gender inequity, and power manipulation.[5]

Yet, for some students the significance of indigenous religions may be their capacity to maintain intimate, meaningful, and sustaining human connections with the local bioregions while struggling to realize their humanity.[6] For others these native traditional religions describe intense religious practices and spiritual experiences that are comparative to those that invigorate the world's religions. Most importantly, for native scholars the recovery and maintenance of indigenous knowledge may be focused on regenerating indigenous communities rather than on implementation and storage in the knowledge systems of dominant societies. What is immediately apparent in traditional native religions is that both narratives of spiritual experiences and religious practices transmit extraordinary encounters with spiritual realms that are not separated from ordinary life lived in relation to local bioregions. Thus, to talk about indigenous religious traditions it is necessary to situate them within their lived communities, or lifeways.

LIFEWAYS IN INDIGENOUS TRADITIONS

The term *lifeway* is used here to suggest the close interaction of worldview and economy in small-scale societies (a cosmology-cum-economy society). Lifeway is not used to suggest a romantic imaging of indigenous peoples as separate from global economic developments. On the contrary, contemporary reserves, reservations, preserves, parks, and the remaining land bases of indigenous peoples are typically set within dominant societies, such as India, Bolivia, Botswana, Malaysia, or Ecuador. This is the case even in the indigenous nation of Papua New Guinea, where highland and island indigenous groups struggle to maintain their land base from mining extraction and logging exploitation.[7] Thus, indigenous lifeways often stand in stark contrast with modern nation-states and multinational corporations, whose mythic values of individualism, unlimited extraction, and global consumerism constantly entice indigenous youth away from traditional values. In critiquing categories of dominance, however, it is important that indigenous modes of survival not be denied.[8] Among these creative realms are mythic expressions of the relationships between individuals and communities, between humans and the fecundity of life, and between humans and the cosmos.

Lifeway can also be used to describe the vital and diverse interactions of urban indigenous communities. Many urban native communities maintain significant relations with their homelands while inextricably involved with, and often dependent on, the global marketplace. These urban indigenous communities still preserve core experiences of lifeways that enable them to construct themselves in traditional ways and to accommodate outside influences. Certainly, the "water wars" in Bolivia provide evidence of coherent communities of urban Aymara and Quecha peoples struggling against privatization of a shared natural resource.[9]

Indigenous peoples as diverse as the Dogon of Mali in Africa and the Bunun of Taiwan in east Asia struggle to preserve traditional ceremonial life in the face of increasing forms of ecotourism based on the exotic other.[10] In some national settings there is no hesitation to encourage the marketing of indigenous ceremonial life. Thus, masked dances, giveaways, birthing, habitat, and life-cycle practices that build on clan and lineage reciprocity to constellate core lifeway experiences are often subject to the intrusive presence of outsiders. The recovery of indigenous voice in determining what is appropriate and what is not appropriate for outsiders to know or to see constitutes significant dimensions of contemporary indigenous religion and ecology. Rather than having been lost or completely subverted by the political, social, and economic values of dominant societies under which indigenous peoples have had to survive, core lifeway experiences continue to be transmitted and reinterpreted by creative indigenous individuals.

Thus, April Bright of the MakMak (Sea Eagle) peoples, of the Marranunggu language group in Northern Territory, Australia, gives expression to the lifeway concept when discussing what "traditional ownership" meant to her mother: "Traditional ownership to country for my Mum was everything—everything. It was the songs, the ceremony, the land, themselves, their family—everything that life was all about. This place here was her heart. That's what she lived for, and that's what she died for."[11] Embedded within ownership, then, for this MakMak elder, as transmitted to her daughter, was not simply individual control of private property or separate knowledge systems of collected data. Rather, ownership appears to be interwoven with diverse forms of indigenous knowledge. The land and knowledge of the land are centered in one's personal body and interwoven with the social body, the ecological body, and the cosmological body.

Similarly, the lifeway context does not reduce religious language, ritual, and roles, such as that of the shaman-healer, to realms of specialization. Rather, the knowledge base from which these activities derive is understood as derived from the ecological and cosmological whole. The descriptions of indigenous elders flow from a lifeway context that is larger than simply social or anthropocentric values. This is evident in the description of indigenous knowledge given by indigenous scholars:

> Perhaps the closest one can get to describing unity in Indigenous knowledge is that knowledge is the expression of the vibrant relationships between people, their ecosystems, and other living beings and spirits that share their lands. . . . All aspects of knowledge are interrelated and cannot be separated from the traditional territories of the people concerned. . . . To the Indigenous ways of knowing, the self exists within a world that is subject to flux. The purpose of these ways of knowing is to reunify the world or at least to reconcile the world to itself. Indigenous knowledge is *the way of living* within contexts of flux, paradox, and tension, respecting the pull of dualism and reconciling opposing forces. . . . Developing these ways of knowing leads to freedom of consciousness and to solidarity with the natural world.[12]

This striking statement presents the insights of Canadian First Nation scholars and is not given here as a panindigenous perspective. Yet it highlights the connections of indigenous knowledge to the way that life is lived by a people in their homelands. The ways in which the flux and tensions of life are embedded and resolved in religion and ecology are ongoing questions for this field of study.

LIFEWAYS AND THE FOURFOLD EMBODIMENTS

Indigenous lifeways express an intimate relatedness between the individual person (or embodied self), the native society, the larger community of life in a region (nature or ecology), and the powerful cosmological beings typically present in ritual actions and mythic narratives. Collectively, this fourfold interpretive context helps to open up interpretations of indigenous religions and ecology rather than reduce them to single explanations. This fourfold interpretive context also seeks to ground native religions in the "text" of local place. A reflexive metaphor of "text" can be helpful insofar as it places indigenous religions in a context that is analogous to, yet strikingly different from, the criteria of literate texts, theologies, or institutional chains of transmission used by many other religions to validate their religious traditions. What is intended is an inscribing onto the reader of indigenous interactive awareness, for example, of sound, verbal expression, and place in indigenous lifeways. This mix of sense knowledge and conceptual thought as interactive process is akin to some notions of text in Western understanding but with the additional nuances that ecologists and cosmologists bring to their understandings of place-as-text. Lifeway points toward the intertextuality of native religions, namely, that which brings together such distinct sacred actions, occasions, and concepts as chants, embodiments, and sites. The concept of lifeway, then, is not simply another "unit of analysis" in a rational study of cultures. Rather, it is an interpretive effort to express indigenous understandings of human-earth relations as an interactive and pervasive context that outsiders might label religion.

ISSUES IN UNDERSTANDING RELIGION AND ECOLOGY IN INDIGENOUS LIFEWAYS

Studies of North American Anishinabe peoples of the Great Lakes region often note references to the natural world and to unmediated experiences when indigenous elders describe sacred encounters. *Ahnishinahbaeótjibway* author Wub-e-ke-niew interweaves person, society, community of life, and spirit forces when he distinguishes the generative character of his peoples' lifeway or *Mide*:

> The *Ahnishinahbaeótjibway Mide* is a way of living in harmony and community; a facilitation of each person's Sovereign relationship with Grandmother Earth, with Grandfather *Mide*, with the Circle of Life which encompasses us, and with the Great Mysteries of the Universe. The *Mide* is experienced, it is directly connected to Grandmother Earth; they are married. This is where we come from.[13]

The multivalent term *Mide* of the Great Lakes woodland people *Ahnishinah-baeótjibway* remains untranslated here because it is presented as a compilation, or synthetic expression, of their knowledge of life. *Mide* as a way of knowing is gendered by this author in English as male and as a holistic presence. His overview is also suggestive of the four embodiments—somatic, social, ecological, and cosmological. *Mide* encircles and pervades all of life. Moreover, the experience of local place transmitted by this term is expressed in intimate kinship terms, namely, as Grandfather "married" to Grandmother Earth, and in cosmological terms, namely, the "Mysteries of the Universe." We can call this an integral lifeway, then, to emphasize the embedded character of sacred teachings and practices among native peoples in ordinary life. Yet, ordinary life is also understood as pervaded by powerful forces. Integral lifeway is marked by personal responsibility to an experienced cosmological balance or harmony directly related to local place and people. This lived community ideal may foster intense visionary experiences of sacred forces in contexts different than the experiences that occur in religions marked by literate scriptures, institutional hierarchies, and theological orthodoxies.

Indigenous communal life has been the subject of intense scrutiny and interpretation by mission-oriented academic scholars and accomplished ethnographers in historical, theological, and anthropological forms. Yet, they often describe the lifeways of native communities in narrow ways using outsider terms and the categories of dominant societies. Stereotypes of the Indian (Americas), Dayak (Indonesia), Adivasi (India), Orang Asli (Malaysia), or Aboriginal (Australia) person have varied widely from pagan savage, primal ancestor, noble hero, or childlike victim. These stereotypes have often been accompanied by views of native communities as void of individual religiosity, civilized ethics, or developed intellectual ideas. Insightful and creative individuals were not expected to arise from indigenous settings, which were perceived as socially limited, superstitious, and so tradition heavy that they disparaged innovative thought. Major obstacles to our understanding of indigenous lifeways are embedded in the ways that statements by native elders and leaders are understood. There is a need to appreciate anew how indigenous thinkers express their own sensual, symbolical, and logical modes of knowing. Dialogue requires that dominant societies cultivate understanding of the wisdom traditions of native religions. Participants in dialogue need to bring participants into their most profound reflections upon themselves, their societies, their natural surroundings, and the universe.

The reversal of some stereotypes of indigenous peoples in dominant societies has been due in part to more insightful ethnographic writings. These empathetic descriptions have played important roles as texts that activate new ways of cultural understanding. However, it is now quite clear that the actual voices of native individuals and communities have been the leading resource for understanding indigenous ways. Indigenous elders have become increasingly more vocal in presenting their positions in national and international settings such as the United Nations.[14] In many cases indigenous oral narrative teachings are now being studied and described by scholars who are themselves indigenous. This collaborative work of native and nonnative scholars has provided sapiential insights into contemporary life by particular indigenous peoples, as well as a deeper understanding of shared human values.

Perhaps the most serious limitation in understanding contemporary indigenous peoples arises from the anxiety of nation-states regarding political sovereignty over native lands. Such anxieties are often manipulated by dominant societies to press for absolute control over indigenous homelands. As a consequence, indigenous peoples in the late twentieth and early twentieth-first centuries have experienced increased pressures on their cultures and homelands from nation-states, multinational corporations, local entrepreneurs, and displaced settlers. In an earlier period of hegemony, for example, the mythic image of Manifest

Destiny in the United States legitimized expansive pressures to exterminate and to remove Native Americans. This symbolic view of a religious destiny fueled an imperialist myth that imagined indigenous peoples as disappearing races.

Indigenous peoples in South America, Indonesia, the Philippines, Africa, and Central Asia now watch in horror as elders, who transmit the storied knowledge of their lifeways, die at the hands of intrusive individuals and institutions or from introduced disease pathogens. These are contemporary tragedies that directly relate to the dismal histories and colonial legacies throughout the world. Many indigenous peoples in the Americas—such as the Macuxi of Brazil and Venezuela, the Nasa peoples of Colombia, the Huarani and Achuar of Ecuador, the Lacandon and other Mayan peoples of Mexico, the Anishinabe and Shoshoni of the United States, and the James Bay and Lubicon Cree of Canada—see their homelands being devoured and their traditions desecrated. In these regions multinational corporations identify as resources what for indigenous peoples has been their homeland. Often the corporate world cooperates with nation-states in asserting national sovereignty over indigenous lands to facilitate extraction industries. In some instances extractive economies have been taken up by tribal governments themselves in the face of massive unemployment on reserves or reservations or as a means of employment when traditional hunting or trapping fail. Indigenous peoples not only search for their political voice, but they struggle to maintain viable lifeways that have at their heart both thick subsistence relations with their homelands and views of reality that engender integral lifeway experiences.

Ironically even the academic discipline of anthropology, which has raised critical awareness about indigenous peoples around the world, provided ethnography in an earlier period that substantiated evolutionary readings of indigenous religions. It is still possible to find forms of "penetration anthropology" solicited by multinational corporations or national governments used to subvert indigenous political activism and cultural self-determination. On the other hand, more recent anthropology, following methods such as "social justice advocacy," "postcolonial," or "postmodern" agendas, critiques the global market economic agendas that exploit the homelands of indigenous peoples.[15]

Through all these exchanges native peoples have ensured their own survival through an amazing variety of assertive and adaptive projects. Their survival, often in desperate material poverty brought about by forced reductions of their homelands and by continuing exploitation of their resources, has been nurtured by profound communal religious sensibilities. Both the threatened survival of indigenous peoples and their reliance on their spiritual worldviews are evident in a statement delivered by a member of the Mapuche political organization, AD-Mapu, in 1984 to the U.N. Special Rapporteur for Chile:

> We would like to point out that the Mapuches have a SACRED and COLLECTIVE concept of the earth and all it produces. There are no concepts like private

property, commercial value, or constantly changing technology that industrial societies have. The religious and sacred dimensions have a global and general quality in Mapuche culture. To alter any aspect of Mapuche culture is to alter the sacred spirituality of Mapuche people.

Traditionally, for the Mapuche the earth is part of life itself and it also has a sacred dimension which encompasses the existence and culture as a whole of Mapuche people.

With this in mind, it is easy to see the vast damage caused to the spirit of the Mapuche people by the division of sacred and collective land. The consequences are unpredictable for the future of the culture of this people.[16]

The Mapuche recognize fundamental differences between themselves and the mainstream society of Chile. Mapuche emphasize collective ownership of land, personal relationships with sacred dimensions of the bioregion, and cultural continuity in transmission of Mapuche traditional environmental knowledge. This is in striking contrast with the national drive for historical and technological change often based on massive public works projects and the commodification of resources. Often deprived of political legitimacy by the judicial processes of the Chilean nation-state, the Mapuche have been subject to the economic rapacity of multinational developers intent on building hydroelectric dams.

From a critical perspective it may appear that the indigenous leader cited above idealistically describes Mapuche "sacred spirituality" as timeless and unchanging. Still, the Mapuche concern is to have a determining voice in making choices affecting their future rather than in having changes forced upon them by outsiders. In this statement, then, Mapuche activists speak from deeper traditional values regarding place and homeland than from valuation of private property or commodity exchange. They speak of sacred cosmology in which their relationships with specific places in their homeland legitimate cultural and subsistence practices transmitted by the people. These insights flow from the Mapuche worldview and reaffirm their lifeway as it is practiced. The holistic wisdom of these indigenous peoples, it is argued here, constitutes an alternative vision of development based on their sustainable relations with their homelands over centuries of habitation.

The rationale here for a study of indigenous religions is not to describe religious activities for re-presentations outside their cultural matrix or to suggest that a discussion of indigenous spiritualities can explain those religions away. Neither is the intention here a subtle form of penetration theology in which indigenous lifeways are studied by evangelical groups with the sole objective of subverting the integrity of those native religious practices through conversion techniques. Nor is this study intended to draw out esoteric, privileged information from native informants so as to store ethnographic data in the computer information highways of mainstream cultures and use it as an intertextual prop in the contemporary assault on indigenous cultures and homelands. What, then, is intended here?

This essay on indigenous religions and ecology seeks to stimulate public intellectuals and interested readers in refining and deepening, in appropriate ways, a broader understanding of these lifeways that will bring about more positive policy decisions for these peoples. This work also attempts to interest readers in further attention to indigenous thought and practice as presented by native scholars in their own studies. Toward that end, this work recommends indigenous perspectives as integral and primary voices in the emerging dialogue of religions and ecology. The writings of native thinkers and the spiritual experiences of native practitioners introduce into this dialogue alternative visions of sustainable development and care for biodiversity. Furthermore, such a dialogue fosters respect, advocacy, and understanding of the diverse cultural ways of native peoples. In many settings indigenous peoples have stood up in legal, political, and social forums on both national and international levels demanding that their rights be heard and their ways respected. Increasingly, the mythologies of indigenous peoples of the Americas, for example, are being presented in regional court systems as bases for legal decisions. Gradually, these forums and dialogues have increased understanding of indigenous religions as complex, interactive processes of self, society, nature, and cosmos.

Person as Somatic Center

Implicit within this category of analysis are different indigenous cultures' views of what constitutes a self. Central to these discussions are different views of what constitutes a person. Foremost among these considerations we might foreground the relationships we call the human body, the senses, and the mind. In much of the nonnative world there is a radical philosophical separation between the body and mind that affirms a pragmatic "use" relationship with the natural world. This is most famously articulated as the Cartesian model and associated with scientific objectivity. The separation of thinking ego from an inanimate world is not a "family resemblance" found in indigenous worldview perspectives. Thus, some attention to indigenous perspectives on personhood brings some insight into the relationships of self and world.

Peter Chipesia, a Dunne-za elder of British Columbia, in describing the vision-quest experience of his people, points toward an embodied knowledge:

> When you are close to something, an animal, you are just like drunk. You don't know anything. As soon as this happens you have trouble thinking straight, like being drunk. Everything is just like when you see this animal it is as if he were a person. If you take water then, everything will get away from you and you will be a person again. You won't see anything. That is why you can't drink water before you go out into the bush. When kids are about the same size as Joe (about five),

when they are just old enough to think, to talk, to walk...when they are older, the animal shows them how to make a medicine bundle.[17]

Here, the young person undergoing the experience was not imaged as an isolated monad so typical in many modes of philosophical individualism. Rather, the quality of "person" is presented as an emergent being in relationship with dreams, ancestors, and animals. Just as medicine bundles are gradually assembled with guidance from the spirit agents, so also personhood is assembled with similar supervision. Moreover, this relationship is not a dogmatic teaching that can be stated or written, nor is it simply the singular religious experience described by Chipesia. This knowledge acquired by individuals directly relates to the mythic stories of the cosmological spirits as well as to specific animals in the environment. The work of implementing the vision is the maturing life path of the individual person in relationship to spiritual forces that have "adopted" him or her. This work of acquiring personhood is an embodied knowledge accomplished over years of interaction with the life of local regions.

An expression of this embodied knowledge is evident in the vision-quest narrative of contemporary Canadian architect Douglas Cardinal. It was Cardinal's design that most significantly influenced the final plan for the Museum of the American Indian on the last public space on the National Mall in Washington, DC. Cardinal opened a 1989 interview with a description of the sweat-lodge experience that marked a ritual moment in his maturing quest. He spoke of his fasting, his awareness of the discipline required, and his "magical" experiences. Near the end of his interview he gave an account of a near-death experience that illustrates how his people, the Canadian Blackfoot, relate personhood to the beauty of life lived in relationship. In this quotation Cardinal responds to a numinous spiritual being who has challenged his self-confidence and pulled him out of his body:

So I said, "I'm ready to go. I feel at peace with myself." He said, "It doesn't matter whether you're ready or not, you're coming anyway. You're still arrogant you know." "Yeah, I know, I'm a human being," I said. So I finally went. It seemed like I was a part of everything, and I felt very, very powerful. I just wasn't there.

The elder came out in the morning and he untied the lodge. He tried to help me come back with sweetgrass and whatever. I could hear him in the distance, "Come back." He was pulling me back. I thought, "I don't want to go back. There's no way I'm coming back. Why would I want to go back? I'm already on the other side and if I come back as a human being, I'm going to have to go through death again. Why should I come back? Then I'd be confined and limited and I would screw up and do all the stupid human being things. I'd be out-of-tune with myself and I'd have to go through all this pain and remorse and suffering. I'm already over here and why do I have to do all that again. Besides, I'm free." The elder said, "You have to come back, just to see this day. You've never seen a day like today. There's dew on the grass, and sun shining on the dew and this golden hue is all over everything. The clouds are all red. The sun is brilliant and the sky is blue. It's the most beautiful day. You have to come back and see this beautiful day. It's wonderful to be alive and walk on this earth."

I thought, "It is wonderful to experience life." I said to that being, whatever it was, "Can I go back for a minute and see that day?" He said, "Well you're a free spirit, you can make your own choice." I said, "I'll just check in for a minute and come right back." I came back in my body and opened my eyes and saw that day. It was a beautiful, fantastic day. I never had seen a day like that. I'd never really looked. The elder said, "See what a beautiful day it is and how wonderful it is to be alive?" I said, "Yes, it's just beautiful."

He said, "Are you afraid of death?" I said, "No. I'm just afraid I ain't gonna live right." He said, "Then you're a fearless warrior."[18]

Here the vision-quest experience brings an individual into a transforming experience that introduces new ways of self-perception as well as an altered vision of the world of space and time. This vision accords with the Blackfoot lifeway in its expression of personal responsibility, its embodied experiences of cosmological harmonies, and its commitment to local place and people. This integral lifeway arises from the possibilities of personhood being reconstructed by means of intense spiritual experiences that occur during traditional rituals in which one's own body becomes closely identified with other embodied realms. Interestingly, ways of knowing are not limited to rational, analytical units of knowledge. In the example above, somatic, sensual, and holistic experiences of being bring the fasting person to an awareness of being "a part of everything." He or she emerges in wonderment with the day's beauty. The esthetic experience described here is different than the contemplative act regarding art that has emerged in Euro-American cultures. That is, in the West in recent centuries the concept of art has arisen in part to fill the void of a materialist scientific worldview emphasizing purposeless, objective reality. Blackfoot thought regarding the individual body and the mutually enhancing relationship of humans to the components of the "day" are strikingly different than art as something that can be marketed, consumed, bought, or sold.[19]

An elder guided the fasting person based perhaps on his or her own somatic knowledge. The elder knew how to summon back the fasting person based on particular cultural expressions of the creative flow of the personal, social, bioregional, and spiritual domains. Embedded within the elder's ability to summon the youth back into ordinary awareness is some reflection of his or her own spiritual journey. Thus, an elder's gentle call across states of being may be seen as implementing a kind of cognitive map of the cosmos arising from his or her own initiating call-experiences. Years of ritual performance may have made that elder acutely aware of the "map" to be followed in vision quests. This path can also be understood as a path of intuitive thought on which a healer arranges symbols according to a personal, imaginative logic. Such intuition-thought processes undertaken by shaman-healers in their rituals provide novel ways of organizing, thinking, and artistically imaging self, society, world, and spirit.

One apparent observation is that these visionary experiences do not stand in isolation but demonstrate a means of social reflection and transformation. They provide insight into lifeways of native peoples in which conscious self-transformation is undertaken for spiritual development. Moreover, the knowledge that flows from such states is gained in ascetic practices undertaken apart from society, but their ultimate fruition is in coming to benefit the social body.

Society

In the religious expressions of many nonindigenous religious traditions the human-divine, or transcendent, relationship may be paramount while the human-to-human social setting of the visionary experience may be secondary. That is not the case in indigenous lifeways. That is, integral lifeways in native communities find their deepest expression, confirmation, and contestation in the community sphere. In *The Spiritual Legacy of the American Indian*, Joseph Brown discussed several aspects of this social context of native lifeways, drawing on the Western concept of mysticism:

> Mysticism, insofar as it is a reality within these native traditions, is not, as the outsider has tended to view it, a vague quality of some supernatural experience that spontaneously comes to individuals whom Providence has allowed to live close to nature. Rather, such mystical experiences are first of all prepared for, and conditioned by, lifelong participation in a particular spoken language that bears sacred power through its vocabulary, structure, and categories of thought, and serves as a vehicle for a large body of orally transmitted traditions, all the themes of which also express elements of the sacred. Secondly, such mystic experiences become more available to those persons who have participated with intensity and sincerity in a large number of exacting rites and ceremonies that have been revealed through time, and that derive ultimately from a transcendent source.[20]

Brown pointedly locates native religion in its social context. He criticizes those who approach indigenous thought as a sentimental nature mysticism that derives spontaneously from proximity to untrammeled wilderness. He also critiques those who project mystical patterns from other religious traditions onto indigenous spiritual life. Especially significant are his emphases on preparation for spiritual experience in the ritual life of indigenous people and in the lifeway context of mystical experiences embedded in languages, arts, and kinship systems. Brown stresses the transcendent orientation in native lifeways, but a different orientation is stressed here. Namely, this essay emphasizes the interdependence of both transcendent and immanent realms of the personal, the communal, the natural, and the cosmological in indigenous religions and ecology.

These fourfold interacting spheres are especially significant for understanding the underlying source and articulation of religious knowledge among indigenous peoples. Rather than separating out religious knowledge from the "fleeting relative reality of the immediate world," this essay explores the implicate world of cosmological relationships folded into the rich array of lifeway activities.[21] In this sense the mysterious unknown, or the divine power, in indigenous religions is neither simply transcendent nor immanent, but a holistic matrix that generates a deeper knowing of the observed world through the interacting spheres of the somatic, the social, the ecological, and the cosmological.

A basic approach in this work is that the encounter and articulation in indigenous communities of a pervasive incomprehensible mystery in the cosmos occurs by means of multivalent experiences and historically changing symbol systems. The formulation of analogies, drawn from the local ecosystem, to express this mystery derives from an indigenous episteme. Episteme, from the Greek for "truth," is used here in conjunction with the concept of embodied knowledge. That is, an indigenous episteme flows out of personal identity and community integrity. Indigenous episteme derives from spiritual forces that reveal themselves in and through local land and animal manifestations.

Participation through bodily experiences such as fasting, visions, or dreams brings one into spheres of the personal, the communal, the natural, and the cosmic that engender an embodied knowing of the mysterious other. This embodied knowing occurs not simply in a contemplative absorption into the divine or a devotional union in religious emotion, but in bodily experiences that are synthetic and synesthetic. That is, encounters in one dimension of the sacred, such as ritual, enrich experiences of other lifeway dimensions, thus forming deeper metapatterns of meaning. This leads inexorably to enhanced ritual and symbolic articulations of cosmological mystery expressed in language that crosses thresholds and boundaries of intellectual knowing and sense feeling. This perceived indeterminacy of the sacred in indigenous lifeways is an active unknowing, or apophatic expression, that arises from effulgent presence in the material world rather than from denial of the inherent spiritual significance and complexity of matter.[22]

Apophatic modes of expression, or articulation by denial, have been used to give some coherence to the experience of the sacred, such as when people say that the sacred is not this or that. So also cataphatic, or positive affirmations and descriptions of the sacred, use metaphorical language from one's experience of the world to define major experiences of the divine. The tendency in academic discussions of indigenous lifeways has been to accentuate relations between humans and nature and to identify these traditions with cataphatic nature mysticism. No doubt this perspective helps to locate many native statements in a comparative context similar to Western religious thought. However, a rich variety of deeply religious experiences are found among indigenous peoples expressed in cosmological symbols rich in both cognitive orientation and sensual, or affective, ecstatic states. Indigenous religious experiences are not simply reducible to the language of the describable, or cataphatic.

An apophatic lineage in any indigenous tradition is suggested, for example, in the powerful description of the *angatkut*, or shaman-healer, Igjugarjuk of the Caribou Inuit near Hudson Bay. He described the somatic deprivation of his arduous fast as well as the foods that purified him and the terrible period of solitude in specific places in the landscape known to be associated with the cosmological beings Pinga and Hila. In contrast to the shamanic performances of coastal Inuit, Igjugarjuk summarized his insight into wisdom:

> For myself, I do not think I know much, but I do not think that wisdom or knowledge about things that are hidden can be sought in that manner [a more cataphatic, performative style of healing]. True wisdom is only to be found far away from people, out in the great solitude, and it is not found in play but only through suffering. Solitude and suffering open the human mind, and therefore a shaman must seek his wisdom there.[23]

Rather than collapsing real wisdom into a mystical retreat or individuated contemplation, the apophatic experience that Igjugarjuk described finds its deepest expression by bringing individual suffering and solitude into the social context. It is as if the shaman-healer stands as a symbolic presence of an experiential, embodied knowledge gained by personal loss or absence that can be brought back for people's welfare. The individual body, as the next example suggests, can embody realms of social meaning by means of both absence and presence.

Language and sound in native religions can be revelatory sources; for example, an experience of sound or song can activate the sacred. A particular bird song or a specific rhythm of the drum can be a sensual trope that activates a range of bioregional, experiential, and lifeway images. Cosmogonic myths among the Hohodene Baniwa of the Brazilian Rio Negro, for example, tell of the primal hero Kuai formed by the song of the supernatural jaguar Yaperikuli. Kuai's body emitted sounds as it was forming and later as it burned in a cosmic conflagration. Baniwa songs are directly related to passages, openings, and transformations. According to the Baniwa, songs can open the human body to invisible worlds and profound experiences.[24]

This overview of Baniwa views regarding song and drum sound manifest "ecological imaginaries," that is, the capacity of local ecological places and lifeforms to activate social imagination. Thus, social identity and orientation often draw on a complex language of sensual symbols such as Baniwa song that integrate the absent—that which was burnt in the cosmological conflagration—with that which is present, namely, the bird song and drum sound associated with places and life-forms. Ecological imaginaries are the deep networks of affective association between bioregions and humans that surface in human imagination as symbols and concepts motivating individuals and communities to action.[25]

Typically the syntaxes of these symbolic languages are cosmological in that they bring one into the whole, while the grammars distinguish spheres of person, community, the local bioregion, and cosmological spirits. The quest for synthetic connectedness is a crucial component of the gradual maturing process in indigenous ways of knowing. In this sense initial visions and mystical experiences open one to familiar images that may be parsed in ways that deepen individual awareness. The native religious vision may remain incomprehensible, hidden, or apophatic, until the other grammars of a fourfold interactive context integrate that knowing into a larger synthetic, cosmological vision.

Later maturing experiences and visions of individuals, revealed to the community at appropriate times and places, expand what has become known of this mysterious incomprehensible presence. Usually this public sharing of private visionary revelations is not presented as a progression toward a teleological goal, but as the unfolding of the holy among the people. Secret shamanic languages among indigenous groups, for example, can be understood in this sense of language as processual and synthetic religious experience, rather than simply esoteric language techniques of individuals promoting arcane knowledge as economic advantage for an elite class.

The lack of a literate canon caused indigenous religions to develop in significantly different directions than traditions that became based, in part, on literate communication systems. Among the speculations that these observations suggest is the development among many indigenous traditions of extremely diverse and complex interior mental states associated with oral modes of cultural transmission. It may even be inappropriate to speak of an "unconscious" among indigenous peoples in the sense that no sharp psychic boundary separated what are called conscious and unconscious realms such as dreams. Moreover, spiritual experiences may have been more readily available to the larger community as the oral traditions suggest rather than being the exclusive practice of isolated individuals.

Such a perspective on indigenous religions is developed in Lee Irwin's book *The Dream Seekers*. He emphasizes the act of indigenous knowing as the unfolding of a whole order in each particular dream and vision. Giving equal weight to individual and community in the cocreative work of elucidating visions, Irwin discerns patterns in the indigenous episteme based on complex relationships. He observes that among indigenous peoples of the Northern Plains of North America

a greater totality of possibility and potential order is conceived as always *implicit* in any particular set of discreet, observable phenomena. Rather than seeking to understand a particular culture through a componential, piece-by-piece analysis strictly determined by mechanistic or intellectual principles of hierarchical order and causal relations between parts, one begins by analyzing the process dynamics of an undivided wholeness from which identifiable, stable, and recurrent patterns of only relative autonomy (rather than strict hierarchy) can be identified. These patterns, as *explicit* manifestations, represent "subtotalities" of meaning that can only be described in terms of their relative autonomy or relationship to other patterns of meaning. The fundamental concept is that rather than a fixed world constructed out of a limited set of known, unchanging laws and relations in a static, deterministic manner, there is a world-process of ongoing explicit manifestations of an implicit, emerging, higher-order dynamics that continues to unfold over the generations through a series of reorganized perceptions coupled with new interpretive perspectives.[26]

Irwin relates the indigenous episteme to patterns of external manifestation, or "subtotalities" similar to the fourfold interpretive context of the personal, the communal, the wild, and the cosmological. The indigenous episteme seeks to understand the cosmological whole implicit in events as a world process in which realizations in one manifestation affect all the other subtotalities. In this essay, subtotalities can be understood as the fourfold embodiments of the individual body, the social body, the ecological body, and the cosmological body that cohere in an integral lifeway. Seemingly autonomous realities are perceived only as separable components so that an aspect of the underlying incomprehensibility might be glimpsed. Thus an indigenous religious experience of indeterminacy can be described not simply as an apophatic incomprehensibility, but as the mutual striving of the knowing visionary and the unknown mystery to come to an awareness of each other in the lived experience of the ecosystem.

What is known by the visionary may arise from a community ritual or explicit ceremonial manifestation. The indigenous episteme serves to open the mythic tradition and the personal formulation of meaning in new analogies put forward by the visionary tradition. Both the mythic tradition and new visions are subject to the challenge of lifeway efficacy; that is, they are challenged to communicate their meaning and benefit for the actual life of the community without loss of their ineffable and incomprehensible mystery. This contested ground between the visionary and the community does not simply depend upon personal narratives or normative social critique. Rather, the voice of local place enters into the emergence of indigenous religions in significant ways. Thus, the natural place—the realm of the more-than-human world—provides the setting for the human community to move beyond itself into the larger community of life. Indigenous modes of prayer often simultaneously empty the individual and society of personal and communal aggrandizement and prepare them for encounter with the cosmological beings in that place.

Ecological Body

One lifeway activity with deeper implications for understanding the close inter-action of religion and ecology in indigenous worldviews is the "giveaway," or "gift-giving." Nuanced in diverse cultural lifeways, this worldview value of com-munal sharing flows out of a deep sense of the collective, interdependent character of life. While no one reading below specifically discusses this ethic of sharing in indigenous communities, nonetheless, this cosmological value is folded into many of the statements. The basic point is that social relationships are established, con-firmed, and renewed through the distribution of food and goods throughout the community. Even among indigenous communities with mixed economies and dif-fering forms of employment there is gift-giving. Urban native peoples often re-ciprocate gifts of meat from hunter-forager relatives by providing places to stay in town, along with meals and rides in city settings. Thus, indigenous relatives in towns and cities reaffirm their membership in indigenous communities by par-ticipating in the reciprocal flow of the life of the people.[27]

This collective reciprocal flow of community life need not be understood simply in an anthropocentric manner. The relationships between any particular indigenous peoples and the diversity of life in their homeland, or bioregion, are expressed in cultural particularity, yet shared characteristics allow us to observe that generally indigenous lifeways extend ideas of "person" into the natural world. In speaking of Anishinabe/Ojibway traditions, ethicist J. Baird Callicott writes:

> Plant and animal species are, as it were, other tribes or nations. Human economic intercourse with other species is not represented as the exploitation of imper-sonal, material natural resources, but as reciprocal gift-giving or bartering, in which both the human and nonhuman parties to the exchange benefit. Game animals give their skins and flesh to human beings, who in return give the animals tobacco and other desirable cultivars and artifacts. The slain animals are reincarnated in the most literal sense of that term—reclothed in flesh and fur— and thus come back to life to enjoy their humanly bestowed benefits.[28]

These reciprocal exchanges of the Anishinabe peoples with the plant, animal, and mineral world of their homeland arise from an integral lifeway. This has been transmitted by native peoples in their subsistence practices, their oral narratives, and their ritual calendars. As the homelands of these indigenous peoples come under increasing pressure, these deep relationships with diverse life in the region are not only stretched to the breaking point, but also activate various forms of political and environmental activism. In a recent study of indigenous thought, an indigenous intellectual observed that

> if non-foragers "gave gifts to the foragers without receiving gifts of food in return, they would shame not only the foragers but also themselves." Is it not also the case that the members of the Whitesands Indian Band would bring shame

upon themselves if they stood by and did nothing while the habitat of those other-than-human persons with whom they exchange gifts is threatened or destroyed? After all it is through the exchange with gifts that one maintains one's membership in Ojibway society. Are not these other-than-human persons with whom they exchange gifts members of that society and entitled to the same respect and help accorded to any other member of the community? There is, we suggest, a moral obligation to protect the habitat of the moose, the beaver, the muskrat, and the lynx; the habitat of geese, ducks, grouse and hare, not just because members of the band wish to continue hunting and trapping, but because these other-than-human persons are also members of Ojibway society.[29]

The authors of this powerful statement interpret the reciprocal exchange in the context of moral obligation required between fellow members of a society. However, the understanding of society obviously extends into the natural world. A further interpretation is that indigenous ethics is always accompanied by cosmology. Thus, this profound religious exchange of indigenous peoples flows from levels of commitment that are deeper and more widespread than simply instrumental alliances between reciprocal beings. In many ways it can be said that indigenous peoples form an ecological body with their bioregion. The ethical and pragmatic orientations of indigenous peoples are a function of life experiences nurtured over generations in particular bioregions described in cosmologies.

For some indigenous peoples, even hearing particular words and phrases of their cosmologies activates the possibility for relatedness with spiritual powers. It is the case that these embodied narrations of the world may be rejected by any given indigenous individual. If, for example, that member was taken at an early age from the matrix of cultural learning, such an integral lifeway may be completely unavailable to that native person. Where indigenous traditions have endured, however, there is often a holistic vision of the world that disposes one to see, experience, and embody a community of life with whom one shares an ethical responsibility.

Moral responsibility for life extends also into ways of imaging local place. For example, the Daur Mongols of Manchuria in Central Asia are described by Caroline Humphrey as implementing and understanding modes of ecological perception that depend on movement. Humphrey draws on the work of James Gibson to refine her sense of this ecological perception. Gibson identified an ecological approach to visual perception as one that "begins with the flowing array of the observer who walks from one vista to another, moves around an object of interest, and can approach it for scrutiny, thus extracting the invariants that underlie the changing perceptive structures and seeing the connections between hidden and unhidden surfaces."[30] In describing the Daur Mongols, Humphrey writes:

> The Mongols recognize this feature of perception, it seems to me, and they construct the sacred object [such as a piling of rocks or *oboo*] in such a way that it requires the subject to become conscious of the relativity of perception and to arrive at knowledge by means of bodily movement. To achieve greater

knowledge, that is, to perceive what Gibson called "invariants," the viewer must be on the move. The viewer must accept the conditionality of his or her perception at any one point in order to understand the true nature of the object. This forces viewers to recognize their own spatial-temporal subordination in relation to the totality of the mountain. In various ways, the Mongols choose the mountain (the rocks, etc.) in such a way as to construct this situation, which emphasizes the qualities all such physical objects actually have, namely, immovability, solidity, and invariance. . . . Circular movement is essential in the ritual cult of mountains. Because the human movement involved in perceiving this invariance happens in time, the mountain's quality of "being" or "standing" is also conceived as a process—the process, if you like, of being the same. This is an intense preoccupation in the ideological concepts of social categories which the cult at the *oboo* seeks to reproduce.[31]

Here not only moral responsibility but embodiment itself flows from a shared understanding of bodies in movement as beings in the world. The movement between different pilings of rock brings the Daur Mongols into perception of mountain as movement also. Thus, the ecological realm shares modes of perception with the human in indigenous perspectives that have amazing ethical implications. These ethical consequences are evident in the following example from the Apache peoples of North America.

The Western Apache of New Mexico are known to name places in the landscape that are associated with established mythic and historical narratives. These named places and historical stories are used to "shoot" those who are irresponsible regarding traditional ways of life. For example, at the sunrise ceremony at which a maturing girl is initiated into womanhood, all the women are expected to wear their hair long. Hair grows from the body of a woman, expressing its inherent vitality and power. This embodiment of power is manifest in individuals but obviously connects them to the larger society and cosmos of power. These somatic and social embodiments also have meaningful ecological references for the Apache evident in this example.

Anthropologist Keith Basso recalls the story of a young woman who inappropriately wore her hair in curlers to a ceremony initiating a girl into womanhood. He remembered how her grandmother later casually reminded her of a place of historical significance to the Apache. It was also the place in which coyote was said to have acted in an inappropriate way. This illustrates a traditional Apache mode of teaching ethics whereby an elder uses a historical story to "shoot" or lodge an ethical teaching in relation to a particular place. That young woman heard that story in a personal way and remembered her grandmother in relation to that place. For her, her grandmother was in that place of the coyote story helping her, shooting her, to live in a traditional way. Because the Western Apache continue to reflect on this capacity of personal experience, narrative and place remain with an individual throughout their life, constantly shooting them, encouraging them to act in responsible ways. Basso notes that geographical features have served the [Western Apache] for centuries as indispensable mnemonic pegs on which to hang the moral teachings of their history.... The Apache landscape is full of named locations where time and space have fused and where, through the agency of historical tales, their intersection is "made visible for human contemplations." It is also apparent that such locations, charged as they are with personal and social significance, work in important ways to shape the images that Apaches have—or should have—of themselves.[32]

Just as breath, sound, and naming orient some indigenous peoples to ecological imaginaries, so also the practices of perception in the piling of rocks and storytelling in other indigenous contexts are uniquely integrated into ecological place. Sensitivities among indigenous peoples to biodiversity in a region have been reported in such diverse practices as hunting languages in which animals are addressed in the language of erotic love used between humans.[33] So also these sensitivities are evident in the types of prayers that elders address to berries on the mountain before they are gathered and to clays in the ground before they are shaped into pottery.[34] A skeptic may respond that these are simply everyday, secular experiences and not necessarily religious experiences. The point is that these acts of reverence and respect open the possibilities for life lived in deep communion between humans and the more-than-human world. Indigenous perspectives on encounters with cosmological power extend into the whole community of life—specific ecosystems and species, as well as nature as a whole. Though these intense religious experiences may not be meaningful or available in the same ways to all contemporary indigenous peoples, the possibility to live in the presence of cosmological beings remains open to those who continue these traditional practices.

Cosmological Body

The diverse cosmological beings of Native American peoples occupy a seminal place in indigenous religions. The pluriform world of spirits has been the subject of intense criticism and scrutiny by practitioners of monotheistic religions. Even the little that might be said opens up questions of religious appropriation. That is, for what ends are such issues being discussed? The unquestioned pursuit of knowledge so central to Western critical method is not generally affirmed in indigenous ways of knowing. Knowledge of spiritual matters is often highly circumscribed in indigenous pedagogy to those who have experiences in such matters. The contemplative paths of the *Ahnishinahbaeótjibway Mide*, for example, are not forbidden to classes or genders of the *Ahnishinahbaeótjibway*, but discussion of *Mide* activity and teachings is prohibited to the uninitiated because the words and practices are believed to be sacred and efficacious. They are themselves cosmological beings capable of numinous transformations.

Another viewpoint regarding the significance of cosmological beings in the fourfold interactive processes of indigenous religions are the observations of Roberta Blackgoat, a Diné/Navajo elder and activist who challenged the U.S. government's decision to remove all Navajo from the Black Mountain area. Blackgoat consistently brought her religious sensibilities to bear on this question of removal. With regard to the removal of her people, she said:

> This is the main thing that I can't set my mind on, . . . I want to have this understood what our religious act is, . . . to take care of this land because when we have a Medicine Man, we need to have him to do an offering, . . . we offer to a tree, or even to a tree that's been struck by lightning, or a rock, or a spring, or a mountain, . . . all these things, it's not only to one place. We offer to a certain place for the rain or either for a Beauty Way, . . . and all these things, that's what has been given us by the Holy People and the Great Spirit. . . . We are strong enough to hold this room which he has surveyed for us, . . . and our Home Song, and this Mountain Song, . . . and now this is not being respected at all. I need to have this be known, . . . it's more important that our ways of being, having a prayer still going, . . . and it's still sacred for us, . . . the way.[35]

The imagery here might be obscure to outsiders, but it is not obscure to the Diné. These images resonate with presence even in English translation. Could it be that the constructs that inform the terms *religion* and *spirituality* stand in the way of outsider understanding? That is, can these terms, as constructed by outside researchers in religion, adequately constitute the felt experience describe by Blackgoat? What does she tell us?

What Blackgoat cannot comprehend is the threat of government agents who tell her that she and her people must leave Black Mountain. Moreover, they warn her that if she stays and dies no one will be around to bury her. She responds by talking about her responsibilities to the holy people as they have manifested themselves to her people on Black Mountain. Being strong is, as she says, "holding the room which [the holy people and the Great Spirit] has surveyed for us." She and her Diné people have responsibilities as cocreators of the cosmological harmony (*hozho*). "Home Songs" maintain the hoganlike habitat of the earth, and "Mountain Songs" ensure the sustaining blessings of long life and happiness. This is the movement of ecological perception on a cosmological scale. Rather than arguing for justice and human rights from her own experiences or from her own life or that of her family, she argues from the responsibilities for the ways that are sacred, harmonious, and beautiful (*hozho*).

This integral lifeway of personal experience and sacred play is evident in Blackgoat's words:

> I had a dream, . . . I dreamed that I was talking to these people, the wildlife people, Tigers and Bears and Lion, . . . they were listening to me, lying down, and they were looking at me and I was talking to them in my prayer, . . . I was still talking

when I woke up, . . . and so I think it's still a way of our sacred ways, so I do need to warn, or teach, our leaders, . . . so the policemen wouldn't handle us here, . . . having us, . . . throwing us around, . . . dragging us, out here on our own ancestral land. This is mighty hard, . . . spoiling our sacred ways, . . . especially our sacred bundles. We have sacred bundles that shouldn't be bounced, . . . they should be taken care of real easy, . . . have a song for it, . . . a prayer for it, with the animals, . . . it holds the animals, and it holds the humans, in this, the whole Indian country.[36]

Rather than a piece-by-piece analysis, an understanding of the process dynamics brings us into a sharpened awareness of the cosmological beings—"holy people" in Diné contexts—whose dream appearance gives Blackgoat the strength to endure the countless humiliations of the oppressed. Her talk brings to bear her own set of symbolic forces—holy people, ancestral land, sacred bundles—to warn her adversaries of powers greater than they realize. The powerful beings form one cosmological body that pervades the world of which Blackgoat speaks.

Blackgoat, like so many indigenous peoples, lives in a universe, not simply in systems of national or corporate statistical identity. She, like so many religious figures of other ages and in other traditions, was vitally active and engaged in the spiritual, political, and social struggles of her people until her death. Unlike the popular stereotype of religious practitioners as withdrawn into passive contemplation, this Diné woman manifested the concerns of a spiritually committed individual working for the well-being of her integrated community. For her the numinous cosmological body was formed by the material bundles, by the ritual performances, and by responsible attention to the vital forces that flow from and through the cosmological depths of life.

Conclusion

Study of religion and ecology among indigenous lifeways accentuates the spontaneities and immediacies of spiritual experiences within the fourfold embodiments. We know that these spiritual intimacies draw together into a holistic unity the human somatic, social, ecological, and cosmological realms. Spiritual kinship and reciprocity reach out into the animal, plant, and mineral domains of local space. Names and mythic narratives give personal and community identity to local places and mark sites of extraordinary encounter with the spiritual presences believed to have created, shaped, and guided the community of life. Indigenous religions provide unparalleled insight into this special knowledge of sacred ecology in such documented expressions as the ecstatic knowing of vision quests and the community knowledge imparted in calendrical rituals. This knowledge is evident in traditional environmental knowledge of plants, animals, and landforms, as well as in resistance to the homogenizing visions of global consumer culture. These are not separate realms but together constitute insight into the regeneration of indigenous religion and ecology.

Indigenous religions are holistic processes that move toward mature integration into a universe of presence. It is in this sense that it is said that indigenous people live in a universe. Individual experiences and their social manifestations in symbols and religious arts are themselves fused expressions inherently related to local ecological and larger cosmological realities. In this sense, indigenous knowing is synthetic, synesthetic, performative, and imagistic—drawing together the fourfold embodiments. Indigenous religions are the cocreative work of the human with the cosmological powers in which affective, cognitive, sensory, and connative faculties give imagistic shape and force to the indeterminacies of the sacred.

This process grows with maturity, resulting not simply in isolated spiritual encounters but in augmented communal ways of knowing. Indigenous communities struggle to maintain their knowledge traditions in the contemporary toxic mix of commercial homogeneity and extractive resource pathology that is often forced on them. The image of the canary in the mine has been likened to their state of well-being, but it is not simply the well-being of indigenous societies that is manifested. Nor is it simply that of human communities. Growing awareness of the creative flow in which indigenous communities socially, spiritually, materially, and intellectually participate gives religious force to the ever-widening consciousness of the plight of our planet.

NOTES

1. Regarding "text" as an interpretive genre, see Paul Ricoeur, "The Model of the Text: Meaningful Action Considered as a Text," in *Interpretive Social Science: A Reader* (ed. P. Rabinow and W. Sullivan; Berkeley: University of California Press, 1979). On the issue of alternative interpretive modes, see Lawrence Sullivan, "Sound and Senses: Toward a Hermeneutics of Performance," *History of Religions* 26.1 (Aug. 1986): 1–33.

2. See George Dei, Budd Hall, and Dorothy Rosenberg, eds., *Indigenous Knowledges in Global Contexts* (Toronto: University of Toronto Press, 2000).

3. For a discussion of these three modes of indigenous knowledge, see Marlene Brant Castellano, "Updating Aboriginal Traditions of Knowledge," in *Indigenous Knowledges in Global Contexts* (ed. George Dei, Budd Hall, and Dorothy Rosenberg; Toronto: University of Toronto Press, 2000), 23–24.

4. It is appropriate to note the special issue of *American Indian Quarterly* 28.3–4 (summer–fall 2004) entitled "The Recovery of Indigenous Knowledge." This issue not only discusses indigenous environmental knowledge but also the position of indigenous knowledge in a decolonized future.

5. See, e.g., David Harvey, *Justice, Nature and the Geography of Difference* (London: Blackwell, 1997).

6. In this capacity see David Suzuki and Peter Knudtson, *Wisdom of the Elders: Sacred Native Stories of Nature* (New York, 1992).

7. See Simeon Namunu, "Melanesian Religion, Ecology, and Modernization in Papua New Guinea," in *Indigenous Traditions and Ecology: The Interbeing of Cosmology and Community* (ed. John Grim; Cambridge: Center for the Study of World Religions, 2001), 249–80.

8. For "survivance," see Gerald Vizenor, *Manifest Manners: Postindian Warriors of Survivance* (Hanover, NH: Wesleyan University Press, 1994).

9. See Vandana Shiva, *Water Wars: Privatization, Pollution, and Profit* (Cambridge, MA: South End, 2002); and an article by Juan Forero in the *New York Times* (24 March 2003), online at www.energybulletin.net/3502.html.

10. Ethnography on the Dogon from the 1930s to the present has generated a lively controversy largely based on the post–World War II fieldwork and publications of Marcel Griaule, especially his *Dieu d'eau: entretiens avec Ogotomméli* (Paris: du Chene, 1948); published as *Conversations with Ogotemmeli: An Introduction to Dogon Religious Ideas* (London: Oxford University Press for the International African Institute, 1965). A critique of Griaule's method and a discussion of the effects of tourism on Dogon ceremonialism can be found in Walter E. A. van Beek, "Dogon Restudied: A Field Evaluation of the Work of Marcel Griaule," *Current Anthropology* 32.2 (April 1991): 139–67.

11. *Country of the Heart: An Indigenous Australian Homeland* (ed. Deborah Bird Rose et al.; Canberra, Australia: Aboriginal Studies Press for the Australian Institute of Aboriginal and Torres Strait Islander Studies, 2002), 15.

12. Marie Battiste and James Henderson, *Protecting Indigenous Knowledge and Heritage* (Saskatoon: Purich, 2000), 35, cited in Deborah McGregor, "Coming Full Circle: Indigenous Knowledge, Environment, and Our Future," *American Indian Quarterly* 28.3–4 (summer–fall 2004): 390.

13. Wub-e-ke-niew, *We Have The Right to Exist: A Translation of Aboriginal Indigenous Thought* (New York: Black Thistle, 1995), 198–99.

14. See, e.g., the roles of indigenous elders in presenting the U.N. Declaration on Indigenous Peoples.

15. It is important to note the work of Cultural Survival and Survival International in promoting the rights of indigenous peoples worldwide. The important work of standing committees of the American Anthropological Association focused on intellectual, cultural, and legal rights of indigenous peoples also needs to be mentioned in this context. So also, mention should be made of the ongoing work of the Indigenous Environmental Network, the Society for the Study of Native American Religions, and the Indigenous Section of the American Academy of Religions. Though they have often differed over the religioeconomic engagement of indigenous peoples in the global market, these organizations have assisted their readership and auditors in confronting a broad range of indigenous cultural issues.

16. Quoted in Roger Moody, ed., *The Indigenous Voice: Visions and Realities* (2nd ed.; Utrecht, The Netherlands: Zed, 1993 [orig. 1988]), 119.

17. Robin Ridington, *Little Bit Know Something: Stories in a Language of Anthropology* (Iowa City: University of Iowa Press, 1990), 127–28.

18. Interview with Douglas Cardinal in *Intervox* 8 (1989/90): 27–31, 44–47; and in Dennis H. McPherson and J. Douglas Rabb, *Indian from the Inside: A Study in Ethno-Metaphysics* (Occasional Paper 14; Lakehead University Centre for Northern Studies, 1993), 73.

19. A related Northern Plains society's view of these spiritual-pragmatic issues can be found in Garter Snake's *Seven Visions of Bull Lodge* (Lincoln: University of Nebraska Press, 1992).

20. Joseph Brown, *The Spiritual Legacy of the American Indian* (New York: Crossroads, 1982), 111–12.

21. Ibid.

22. The penetrating insight of Mayan numerologists into the concept of zero may have developed from reflections on the effulgent faces of time; cf. Miguel Leon-Portilla, *Aztec Thought and Culture* (trans. Jack Davis; Norman: University of Oklahoma Press, 1963), 50–51.

23. Knud Rasmussen, *Observations on the Intellectual Culture of the Caribou Eskimo* (Report of the Fifth Thule Expedition 1921–24; Copenhagen: Gyldendalske Boyhandel, Nordisk Forlag, 1930), 54–55.

24. Robin M. Wright, "History and Religion of the Baniwa Peoples of the Upper Rio Negro Valley" (PhD diss., Stanford University, 1981), 382.

25. See Richard Peet and Michael Watts, *Liberation Ecologies: Environment, Development, and Social Movements* (New York: Routledge, 1996).

26. Lee Irwin, *The Dream Seekers: Native American Visionary Traditions of the Great Plains* (Norman: University of Oklahoma Press, 1994), 23.

27. This point is developed by anthropologist Paul Driben and native lawyer Donald J. Auger, *The Generation of Power and Fear: The Little Jackfish River Hydroelectric Project and the Whitesands Indian Band* (Thunder Bay: Lakehead University Centre for Northern Studies, 1989).

28. J. Baird Callicott, "American Indian Land Wisdom? Sorting Out the Issues," in *In Defense of the Land Ethic: Essays in Environmental Philosophy* (Albany: State University of New York Press, 1989), 214–15. See also the work of A. I. Hallowell.

29. McPherson and Rabb, *Indian from the Inside*, 90.

30. James Gibson, *The Ecological Approach to Visual Perception* (Boston: Houghton Mifflin, 1979), cited in Caroline Humphrey and Urunge Onon, *Shamans and Elders: Experience, Knowledge, and Power among the Daur Mongols* (Oxford: Clarendon, 2003 [orig. 1996]), 88.

31. Ibid.

32. Kicth Basso, "'Stalking with Stories': Names, Places, and Moral Narratives among the Western Apache," in 1983 *Proceedings of the American Ethnological Society* (Washington, DC: American Ethnological Society, 1984), 44.

33. Adrian Tanner, *Bringing Home Animals* (New York: St. Martin's, 1979).

34. Personal experiences of the author with the Yakima in Washington State and at Santa Clara Pueblo.

35. A statement from Roberta Blackgoat, Diné elder and relocation resister, on the occasion of the Beautyway Tour 1999.

36. Ibid.

PART II

RELIGION AND ECOLOGY: CONFLICTS AND CONNECTIONS

CHAPTER 13

POPULATION, RELIGION, AND ECOLOGY

DANIEL C. MAGUIRE

HUMAN life is marked by a special consciousness, and the conclusion that presses itself upon this consciousness is, in the words of Arnold Toynbee, "that human life and its setting are mysteries."[1] We are aware that there is a delicate, perilously perched phenomenon on the surface of the earth. Weight-wise it is almost insignificant since it is less than a billionth of the weight of the planet. It is, of course, not negligible since the name we give to this miracle is *life*. The effort to respond wisely to this precarious and special mystery and its terrestrial matrix in the universe is called *ethics*. Some manifestations of this life event are so precious that we reach for the highest encomium in our language and speak of the "sacredness" and the "sanctity" of this life. The discovery and reaction to the sacred is called *religion*.

Our gifted consciousness recognizes that we have powers of choice—we can choose to enhance life or destroy it. Those choices are the stuff of ethics, making ethics a permanent challenge for humans. Toynbee calls the task of ethics an "intrinsic and universal" characteristic of our human nature.[2] So too is religion the discovery of the sacred—however that sacred is explained, theistically or not.

The burdens, then, on ethics and religion are prodigious and fearsome since our level of consciousness gives us both freedom and power unmatched in the rest of the evolutionary process. We are the best news or the worst news in planet life. Joseph in the Genesis story became powerful in the land of Egypt and was in a position to say to all his brothers that he "could preserve you all . . . and take care of you" (Gen. 45:7, 11). Our "brothers" are the multiple exuberant life-forms of the biosphere into which we are knit. We can preserve and take care of them, but we also have a redoubtable capacity to destroy our biological kin. We are impressively endowed with the talents of creativity and imagination, but we are also flawed—maybe fatally so. Maybe we lack the essential gentleness of Joseph in the use of his power. Can humankind be a caring "steward" of the rest of nature as Judaism and Christianity would see it or, more humbly, as Buddhists would prefer, a "neighbor"? Or are we in spite of these religious protestations more of a threat?

Humanity and the Perils of Power

The Christian teaching on original sin arose from a sense that something is wrong with us human beings. Though the theological dons played with this idea to the point of silliness—Augustine thought it was caused by sexual pleasure—the basic insight was correct. Something is wrong.

What is wrong? While speaking recently to a group of Ford Foundation program officers in Greece, I made reference to "the common good." As we stopped for a break, they asked me to return to the common good and to tell them what it is. For my break I took a walk down a dirt path toward the lovely Aegean Sea. Ahead of me I saw what looked like a black ribbon stretched across the dirt path. As I got closer I saw that it was not a ribbon, but two columns of ants moving back and forth in single file. Those in one row were carrying something; the others were going back for a new load. A real estate change was in process. Every ant was committed to the project. There were no shirkers or apostates from the common effort. There were no special-interest groups. All these insect citizens were bonded to the common good of that community.

How convenient for the insects! The needs of the common good are inscribed on their genes. Human genes have no such inscription. We, like the ants, have need of common good considerations, since the common good is the matrix of minimal livability within which individual good can be pursued. Biblical wisdom would point to that as our potential undoing. Indeed, all the moral traditions of the world religions in their distinct fashions point to this soft center in our makeup. All of them address our tilt toward moral autism. Our genetic impulse seems more directed to egoistic good in opposition to the common good, and since the common good includes the good of all of nature, this momentous flaw in our composition portends planetary ruin.

In a blunt indictment, anthropologist Loren Eiseley says: "It is with the coming of [human beings] that a vast hole seems to open in nature, a vast black whirlpool spinning faster and faster, consuming flesh, stones, soil, minerals, sucking down the lightning, wrenching the power from the atom, until the ancient sounds of nature are drowned in the cacophony of something which is no longer nature, something instead which is loose and knocking at the world's heart, something demonic and no longer planned—escaped, it may be—spewed out of nature contending in a final giant's game against its master."[3] Tragically, we are winning that battle with the rest of nature, self-destructively failing to realize that the economy and human life itself are wholly owned subsidiaries of nature. Edward O. Wilson calls *Homo sapiens* the "serial killer of the biosphere."[4] Nature is mother and matrix, not a competitor to be subdued.

We have lost a fifth of tropical rainforests since 1950. These natural treasures provide oxygen, absorb excess carbon, and supply medicine, not to mention their intrinsic value apart from us (75% of our pharmaceuticals come from plants). We all get hurt when the planetary womb in which we live gets hurt. David Orr records some of the results: male sperm counts worldwide have fallen by 50% since 1938. Human breast milk often contains more toxins than are permissible in milk sold by dairies, signaling that some toxins have to be permitted by the dairies. At death some human bodies contain enough toxins and heavy metals to be classified as hazardous waste.[5] Newborns arrive wounded in their immune systems by the toxins that invaded the womb. One report from India is that "over 80 percent of all hospital patients are the victims of environmental pollution."[6] Human consuming is stressing the oceanic fisheries to their limits, and water tables are falling as there are more of us to share this limited resource.

And there are more of us. Along with our profligate ways, our very numbers are a threat. The question of how many of us is too many has long engaged the human mind. Joel E. Cohen in his monumental book *How Many People Can the Earth Support?* notes that in the seventeenth century a prescient estimate was that planet earth "if fully peopled would sustain" at most thirteen billion. That is close to modern estimates, even though those vary from four billion to sixteen billion depending on variables such as consumption patterns. At any rate, says Cohen: "The Earth has reached, or will reach within half a century, the maximum number the Earth can support in modes of life that we and our children and their children will choose to want."[7] One particularly pessimistic study done at Cornell University estimated that the earth can support a population of only one to two billion people at a level of consumption roughly equivalent to the current per capita standard for Europe.[8] As Harold Dorn says: "No species has ever been able to multiply without limit. There are two biological checks upon a rapid increase in numbers—a high mortality and a low fertility. Unlike other biological organisms [humans] can choose which of these checks shall be applied, but one of them must be."[9] Our species, until recently, was blighted with a sense of an earth that was infinitely supplied with the resources needed to sustain human life, and so we can now increase and multiply without limit. That tragic illusion is no longer sustainable.

ETHICS AND RELIGION TO THE RESCUE?

A chastening look at our history shows that ethics and religion are no match for the efficacy of genetic inscription when it comes to the protection of our species and of our biological and terrestrial neighbors. Yet they are what we have, and in their own fashion they have all addressed the problem, including the problem of family planning and population pressures.

The earliest record we have of the gods being concerned with population is seen in a Babylonian tablet dating back thirty-five hundred years. The gods made humans to serve them, but found that the humans had a perverse tendency to reproduce excessively. Their approach to population management was drastic: they sent plagues to trim the human herd, and then they made it a religious obligation for the surviving humans to limit their births.[10] In the *Cypria*, written in the period 776–580 BCE, Zeus used war to handle the population problem:

> There was a time when the countless tribes of men, though wide-dispersed,
> oppressed the surface of the deep-bosomed earth, and Zeus saw it and had pity
> and in his wise heart resolved to relieve the all nurturing earth of men by causing
> the great struggle of the Ilian war, that the load of death might empty the world.
> And so the heroes were slain in Troy, and the plan of Zeus came to pass.[11]

As Cohen observes, "The notion that gods impose war and plague to prevent the Earth from becoming too full of people persisted at least another two millennia, and survives in the thinking of some people even today."[12]

Aristotle insisted that the number of people should not exceed the resources needed to provide them with moderate prosperity. Somewhat surprisingly, Catholic Saint Thomas Aquinas agreed with Aristotle that the population should not be in excess of the provisions of the community. He even went so far as to say that the limiting of births should be enforced by law as needed. He sidestepped the question of how this would be enforced and did not get into issues of sanctions or incentives.[13]

THE NATALIST THRUST OF RELIGIONS

In the Western world, probably no text has merited the attention that Genesis 1:28 has received: "God blessed them and God said to them, 'Be fertile and increase, fill the earth and master it: and rule the fish of the sea, the birds of the sky, and all the living things that creep on earth.'" In a critique that has been endlessly reproduced, Lynn White indicted Judeo-Christian teaching for Western society's "ruthlessness toward nature."[14] "Be fertile and increase" seems to suggest mindless reproduction. Of course, a text once written enters the free-for-all of historical interpretation and is used and abused in diverse ways. The text in itself, however, is more subtle.

It invites humanity into a special, coprovidential relationship with God. It is the opposite of a mindless writ for maximal reproduction. It puts human sexuality, as Jeremy Cohen writes, into "a singularly human realm." It "charges people to transform their otherwise instinctive sexual drives into subjects of their rational wills, thereby using sexuality to express the distinctively human freedom of choice."[15] It can thus be seen as putting reproduction into the realm of rational and caring choice on the model of God's own wisdom and caring. Overproducing would clearly violate that model, as would heartless domination of the rest of nature since God, in the same passage, pronounced the rest of nature "very good" (Gen. 1:31).

Still, it is clear that religions have contributed to the desire for large families. Small wonder. They were spawned in a world where the problem was depopulation. Emperor Augustus penalized bachelors and rewarded families for their fertility.[16] Widowers and divorcees (of both sexes) were expected to remarry within one month! Only those over fifty were allowed to remain unmarried. Remember that Augustus presided over a Roman society with an average life expectancy of less than twenty-five years. It was a society where, as historian Peter Brown says, "death fell savagely on the young."[17] Only four out of every hundred men—and fewer women—lived beyond their fiftieth birthday. As a species, we formed our fertility habits in worlds that were, in Saint John Chrysostom's words, "grazed thin by death" (De virginitate 14.1). Such instincts are deep rooted. If, as Teilhard de Chardin sagely says, nothing is intelligible outside its history, this reproductive thrust, especially in stressful conditions, is the defining story of our breed. These religions, as we shall see, also contain the cure for this problem, but we concede that for most of history their concern was for more, not for fewer, children.

Chinese Religions and Buddhism

Daoism and Confucianism are the main shapers of Chinese religious culture. Buddhism, which has found expression in many cultures, is also part of the Chinese heritage. Much in these religions is relevant to ecological concerns and the management of human fertility. First is the idea of the fundamental unity of everything in the universe. "All living and nonliving things in the universe are constituted of ch'i. This is the worldview shared by all Chinese religions."[18] Because of the natural unity, harmony is a supreme value in this religious milieu.

There is a "mandate of heaven" which has come to be seen in the sense of "moral destiny, moral nature, or moral order."[19] This is not theism as understood in Western cultures, nor does it have connotations of otherworldliness. Chun-fang Yu says: "There is no God transcendent and separate from the world and there is no heaven outside of the universe to which human beings would want to go for refuge."[20] However, the mandate requires working harmoniously with all of nature with whom we are kith and kin—and not just with humankind. The ecological implications of this are clear.

As with other religions, there is a strong pronatalist thrust in Chinese culture. Confucius said that a noble man would be ashamed of land wasted due to a lack of people. Women were to marry by age fifteen and men by age twenty to maintain the population.[21] Confucian scholars, however, sounded the alarm about too many people and too few resources. Daoists advocated the idea of a "little state with a small population," as promoted by the founder Lao Tzu.[22] From earliest times, the Chinese recognized that excess population would destroy the essential harmony of the universe and so were open to family planning, including abortion. Geling Shang writes: "There has never occurred in the Chinese tradition a ban against abortion: rather, Chinese attitudes toward abortion were mostly tolerant and compassionate."[23]

This was helped by the fact that the Chinese religions did not see sex as essentially reproductive. Sexual pleasure had its own purposes and legitimacy. Chun-fang Yu says: "The indigenous Chinese religions, both Confucianism and Taoism are not ascetic. They do not regard sexuality as a problem, but on the contrary, as natural. In the Book of Mencius, the philosopher Kao Tzu says, 'By nature we enjoy food and sex' (6A.4)."[24]

Regarding Buddhism, Buddhist scholar Rita Gross writes: "Buddhism can in no way be construed or interpreted as pronatalist in its basic values and orientation."[25] Buddhism tends to view sexuality positively, not as something to be justified only by reproduction, much less seeing reproduction as being its primary and dominant purpose. The ecological sensitivity of Buddhism shows in its belief that "the preciousness of human birth is in no way due to human rights over other forms of life, for a human being *was* and could again be other forms of life," this stemming from the traditional Buddhist belief in rebirth.[26] Two principles in Buddhism particularly militate against overreproduction: interdependence (*pratityasamutpada* in Sanskrit and *paticcasamuppada* in Pali) and Walking the Middle Path. Too much fertility as well as too much consumption would indicate selfishness.

Abortion would seem a special problem for Buddhism since "the being about to be born" has a previous record of lives. William LaFleur, professor of Japanese and the Joseph B. Glossberg Term Professor of Humanities at the University of Pennsylvania, addresses this in his *Liquid Life*.[27] He shows in this remarkable study how a contemporary Japanese woman could accept Buddhism with its "first

Precept" against killing, have an abortion, and still consider herself a Buddhist in good standing. LaFleur believes that abortion and infanticide were widely used by Buddhists.

La Fleur discusses how the Japanese Buddhists come to decide that abortion was compatible with their life-affirming religion. The answer is found not so much in texts as in rituals and symbols. There are rituals to remember the aborted fetus which show that they think of them not so much as being terminated as being put on hold, asked, in effect, to bide their time in another world. The "life" that was rejected or that died through miscarriage or early infant death is called a *mizuko*, and parents pray for their well-being in the sacred realms to which they have been "returned." Elaborate rituals are employed to remember these rejected "lives." Little child-size statues of Jizo, a sweet savior-figure associated closely with children, are found in abundance and are visited by parents who lost children or had abortions. In some images, Jizo wraps the *mizukos* under his protective cape and gives them comfort.[28]

Hinduism

Hinduism, properly known as *Sannatan Dharma* ("the eternal tradition"), is more like "a confederation of religions than a dogmatically unified system, as, for example, is Roman Catholicism."[29] Like all the major religious traditions, it is a treasure trove of insights into the moral challenges presented by human power in a finite world. Sandhya Jain writes, "Hinduism is simultaneously a religion and a way of life, and constitutes a unique blend of spirituality and practicality that is inspired by the ideal of universal welfare of all beings, both human and other creatures."[30] Hinduism puts more stress on dharma, a "natural cosmic law" or ethic, than on dogmatic consistency. Thus, "Hinduism recognizes even the atheist as morally valid, and does not deny the atheist space on the religious-spiritual spectrum."[31]

Hinduism is realistic about our capacity to inflict devastation on our earthly home. The epic *Mahabharata* (ca. 500–200 BCE) gives a graphic description of the worst of the periodic destruction of the world: "The population increases, tress do not bear fruit, a drought prevails, people destroy parks and trees, and the lives of the living will be ruined in the world."[32] The concept of karma with its stress on bearing the fruits of our actions has significant ecological importance. The wrecking of the earth, therefore, is not easily viewed as some blind fate that leaves us innocent.

Regarding population, there is strong stress on the blessing of fertility: "Almost all dharmic texts of Hinduism praise the joys of having children, especially sons. The Laws of Vasishta say that nonviolence [*ahimsa*] and procreation are the common duties of members of all castes."[33] Abortion is often seen as a heinous crime. However, one cannot speak of a simple or single Hindu view on abortion. Arvind Sharma writes: "The main body of revealed literature in Hinduism frowns upon abortion, but a body of literature called *Ayur Veda* allows it, and this *Ayur Veda* enjoys the status of a revelation as part of the larger corpus."[34] In Hinduism, the canon of revelation was never closed but grows throughout time. This allows for revelation to respond to new conditions in new ways. Sandhya Jain says that dharma "is not a static notion espousing only the values of a bygone era, but is reflective, contextual, and characterized by movement, change, dynamism, and adaptability."[35]

Abortion is treated with *daya* ("compassion") and this shows in the modern 1971 Indian Medical Termination of Pregnancy act. Abortion is thus available in India, and religious authorities have not objected to current moves to make it available even to minors.

Judaism

Judaism is mindful of the need to care for the earth which God declared with satisfaction to be "very good." There is a strong stress in Jewish history on the human and environmental need for restraint. Laurie Zoloth writes: "We are enjoined always to take less than we could, to wait longer to harvest the first fruits, to let the land rest every seven years, in an entire year of Shabbat, and finally to declare a year of Jubilee every fifty years, when not only the land rests, but the marketplace and social hierarchy itself is restored to its point of origin."[36] Increasingly, holidays like Tu Bishevat and Sukkot are infused with new liturgical stress on habitat preservation.[37]

As seen above in the discussion of the classical and influential text of Genesis 1:28, Judaism does not commend mindless reproduction with no consideration of human and terrestrial needs. Certainly there is a stress on reproduction. Zoloth writes: "The promise that is the basis of the covenant itself is the repeated assurance that the tribe of Abraham will be continued, made numerous, and that the Jewish future and, through it, the human future is safe."[38] However, tractate *Yevamot* in the Mishnah (200 CE) says: "A man may not desist from [the attempt to] procreate unless he already has children." This implies that if he already has children, birth control may be used. When pregnancy involves risks for women, the male obligation to procreate is suspended, and birth control may be used.

The main Jewish tradition has a developmental view of the embryo-fetus. The fetus is not considered an ensouled person: "Not only are the first 40 days of conception considered 'like water' but also even in the last trimester, the fetus has a lesser moral status."[39] "Texts about abortion as the final extreme of reproductive practices are not a matter of deep contention in Jewish thought, since there is wide agreement on textual warrants for abortion under certain circumstances."[40]

There is stress in Judaism on producing not just a human being but "a *humane* being; there is stress on the quality not the quantity of children born.[41] Saadiah Gaon, the rabbinic leader of Diaspora Jewry, said in the early tenth century: "Of what benefit are children to a person if he is unable to provide for their sustenance, covering or shelter? And what is the good of raising them if it will not be productive of wisdom and knowledge on their part?"[42]

Christianity, Protestant and Catholic

Though spoken of in the singular, all religions are plural. Religions are also mutants, constantly changing in the shaping swirl of history's actions and reactions. Thus, there are many Hinduisms, many Judaisms, many Buddhisms, many Christianities, and so on. Even within one religion there are innumerable subdivisions. Thus one cannot speak simply of Protestantism and give "the Protestant" view on population and ecology. The earliest Protestant movement in the German-speaking countries centered on Martin Luther (1483–1546). The Reformed churches were led by Huldrych Zwingli (1484–1531) and John Calvin (1509–64). Anabaptists rose more out of rural communities, and they too spread and took different forms. Today's fundamentalist Protestants are again quite distinct. If anything unites Protestants, says Gloria Albrecht, it is the belief in the "priesthood of all believers" and the responsibilities that that entails.[43]

Some generalizations can be made regarding these richly diverse traditions. The Bible on which the various Christianities leaned was not consumed with ecology so much as with economic justice and peace. It is true, as Catherine Keller writes, that Jesus "did not send us to our jobs on time, but rather—back to the fragrant lilies."[44] Still, as Keller also writes, historical Christendom was not a gentle force in nature. Rather, "it has interpreted its theological sources, its manly mandates to populate, dominate, use, and convert the world. If I do Christian theology, I do it in penance for the effects of Christendom."[45] Still, Protestant theology has led the way in religious ethics in the development of a vibrant ecotheology, witnessed in works by John B. Cobb and Herman Daly, Sallie McFague, and Larry Rasmussen.[46]

Stoic attitudes on sexuality influenced Christian attitudes toward birth control, since reproduction was seen as the only "rational" use of sex. Thus, the Catholic church hierarchy still condemns the use of contraceptives and of abortion. However, changes are taking place. Just as Protestant thought was changed when it moved to a married clergy, so too Catholic theology is no longer a clerical preserve. Men, and for the first time in great numbers, women are doing Catholic theology today. Catholic theologians and Catholic laity largely dissent from Vatican theology, which bans all use of contraceptives and abortion. Even many of the Catholic hierarchy are among the dissenters. When *Humanae vitae*, the 1968 encyclical retaining the ban on artificial contraception, was issued, "the episcopal conferences of 14 different nations issued pastoral letters assuring their laity that those who could not in good faith accept this teaching were not sinners."[47]

Catholic theologian Christine Gudorf predicts that Vatican theology on this point will not stand, for two reasons: first, the command of Genesis to increase and multiply is understood today in terms of "today's fragile biosphere." There is broad recognition among Catholics that we must heed "the divine call for human stewardship over creation also revealed in Genesis." Second, she writes: "The Roman Catholic Church (and Christianity in general) has in the last century drastically rethought the meaning of marriage, the dignity and worth of women, the relationship between the body and the soul, and the role of bodily pleasure in Christian life, all of which together have revolutionary implications for church teaching on sexuality and reproduction. In effect, the foundations of the old bans have been razed and their replacement will not support the walls of the traditional ban."[48]

Signs of the ongoing change are seen in *A Brief, Liberal, Catholic Defense of Abortion* by Catholic professors Daniel A. Dombrowski and Robert Deltete.[49] In this book, relying on Catholic moral traditions, they argue that "the Catholic 'pro-choice' stances is at least as compatible with Catholic tradition as the anti-abortion stance, and may even be more compatible with Catholic tradition than the current anti-abortion stance defended by many Catholics and by most Catholic leaders." They insist that "most twentieth-century Catholic theology on abortion is a caricature on the rich and variegated tradition in Catholicism on this topic."[50]

RELIGION AND EARTH COMMUNITY

Catholic Cardinal John Henry Newman observed that people will die for a dogma who will not stir for a conclusion. There is nothing that so stirs the will as the tincture of the sacred. Thus religions should never be ignored on important social issues. Its deep wells should be tapped since they contain renewable moral energies and poetic power that can animate the will of this species to live at peace with the earth. From the wisdom of indigenous religions to that of the classical major religions, there are insights into the necessity to view our terrestrial setting with wonder and appreciation. Of course, those religions are never pure success stories in the history of ethics, and so critique is part of the service we owe them. Then we can move to reassessment and appropriation of their successes.

NOTES

1. Arnold Toynbee, *Change and Habit: The Challenge of Our Time* (New York: Oxford University Press, 1966), 14.

2. Ibid., 13.

3. Loren Eiseley, *The Firmament of Time* (New York: Atheneum, 1960), 123–24.

4. Edward O. Wilson, *The Future of Life* (New York: Knopf, 2002), 94. Wilson points out (150) that if everyone were to consume as we in the American empire consume, it "would require four more planet Earths."

5. David W. Orr, *Ecological Literacy: Education and the Transition to a Postmodern World* (Albany: State University of New York Press, 1992), 3–5; and *idem, Earth in Mind: On Education, Environment, and the Human Prospect* (Washington, DC: Island Press, 1994), 1–3.

6. Paul Kennedy, *Preparing for the Twenty-First Century* (New York: Vintage, 1994), 191, quoting H. Govind, "Recent Developments in Environmental Protection in India: Pollution Control," *Ambio* 18.8 (1989): 429. Kennedy notes that this use of the term *environmental pollution* is broad and looks to more than respiratory effects.

7. Joel E. Cohen, *How Many People Can the Earth Support?* (New York: Norton, 1995), 367.

8. David Pimentel, Rebecca Harman, Matthew Pacenza, Jason Pecarsky, and Marcia Pimentel, "Natural Resources and an Optimal Human Population," *Population and Environment* 15.3 (1994).

9. Harold F. Dorn, "World Population Growth: An International Dilemma," *Science* (26 Jan. 1962); repr. in *Readings in Conservation Ecology* (ed. George W. Cox; New York: Appleton-Century-Crofts, 1969), 275.

10. Cohen, *How Many People Can the Earth Support?* 6.

11. Anne D. Kilmer, "The Mesopotamian Concept of Overpopulation and Its Solution as Reflected in the Mythology," *Orientalia* 41 (1972): 160–76, at 176.

12. Cohen, *How Many People Can the Earth Support?* 6.

13. Thomas Aquinas, *Sententia libri politicorum* (Omnia opera 48; Rome: Ad Sanctae Sabinae, 1971), A.140–41.

14. Lynn White Jr., "The Historic Roots of Our Ecologic Crisis," *Science* 155 (1967): 1203–7.

15. Jeremy Cohen, *"Be Fertile and Increase, Fill the Earth and Master It": The Ancient and Medieval Career of a Biblical Text* (Ithaca: Cornell University Press, 1989), 15.

16. See Peter Brown, *The Body and Society: Men, Women, and Sexual Renunciation in Early Christianity* (New York: Columbia University Press, 1988), 6.

17. Ibid.

18. Chun-fang Yu, "Chinese Religions on Population, Consumption, and Ecology," in *Visions of a New Earth: Religious Perspectives on Population, Consumption, and Ecology*

(ed. Harold Coward and Daniel C. Maguire; Albany: State University of New York Press, 2000), 163.

19. Ibid., 164.

20. Ibid., 162.

21. Geling Shang, "Excess, Lack, and Harmony," in *Sacred Rights: The Case for Contraception and Abortion in World Religions* (ed. Daniel C. Maguire; New York: Oxford University Press, 2003), 229–30.

22. Ibid., 230.

23. Ibid., 232.

24. Chun-fang Yu, "Chinese Religions on Population," 172.

25. Rita Gross, "Buddhist Resources for Issues of Population, Consumption, and the Environment," in *Population, Consumption, and the Environment: Religious and Secular Responses* (ed. Harold Coward; Albany: State University of New York Press, 1995), 156–57.

26. Ibid., 159–60.

27. William LaFleur, *Liquid Life: Abortion and Buddhism in Japan* (Princeton: Princeton University Press, 1992).

28. See Parichart Suwanbubha, "The Right to Family Planning, Contraception, and Abortion in Thai Buddhism," in *Sacred Rights: The Case for Contraception and Abortion in World Religions* (ed. Daniel C. Maguire; New York: Oxford University Press, 2003), 145–65.

29. Sandhya Jain, "The Right to Family Planning, Contraception, and Abortion," in *Sacred Rights: The Case for Contraception and Abortion in World Religions* (ed. Daniel C. Maguire; New York: Oxford University Press, 2003), 129.

30. Ibid., 129–20.

31. Ibid., 130.

32. Vasudha Narayanan, " 'One Tree Is Equal to Ten Sons': Some Hindu Responses to the Problems of Ecology, Population, and Consumerism," in *Visions of a New Earth: Religious Perspectives on Population, Consumption, and Ecology* (ed. Harold Coward and Daniel C. Maguire; Albany: State University of New York Press, 2000), 111.

33. Ibid., 120.

34. Arvind Sharma, "Conclusion," in *Sacred Rights: The Case for Contraception and Abortion in World Religions* (ed. Daniel C. Maguire; New York: Oxford University Press, 2003), 280.

35. Jain, "Right to Family Planning," 130.

36. Laurie Zoloth, "The Promises of Exiles: A Jewish Theology of Responsibility," in *Visions of a New Earth: Religious Perspectives on Population, Consumption, and Ecology* (ed. Harold Coward and Daniel C. Maguire; Albany: State University of New York Press, 2000), 104.

37. Ibid., 97.

38. Ibid., 31.

39. Ibid., 39.

40. Ibid., 36.

41. Sharon Joseph Levy, "Judaism, Population, and the Environment," in *Population, Consumption, and the Environment: Religious and Secular Responses* (ed. Harold Coward; Albany: State University of New York Press, 1995), 80.

42. Quoted in ibid., 81.

43. Gloria Albrecht, "Contraception and Abortion within Protestant Christianity," in *Sacred Rights: The Case for Contraception and Abortion in World Religions* (ed. Daniel C. Maguire; New York: Oxford University Press, 2003), 81.

44. Catherine Keller, "The Lost Fragrance: Protestantism and the Nature of What Matters," in *Visions of a New Earth: Religious Perspectives on Population, Consumption, and Ecology* (ed. Harold Coward and Daniel C. Maguire; Albany: State University of New York Press, 2000), 90.

45. Ibid.

46. John B. Cobb and Herman Daly, *For the Common Good* (Boston: Beacon, 1989); Dieter Hessel, *Theology for Earth Community* (Maryknoll, NY: Orbis, 1996); Sallie McFague, *The Body of God* (Abingdon, 1991); and Larry Rasmussen's prizewinning *Earth Community, Earth Ethics* (Maryknoll, NY: Orbis, 1996). An early Catholic voice in eco-theology was Rosemary Radford Ruether's *New Woman New Earth* (New York: Seabury, 1975).

47. Christine Gudorf, "Contraception and Abortion in Roman Catholicism," in *Sacred Rights: The Case for Contraception and Abortion in World Religions* (ed. Daniel C. Maguire; New York: Oxford University Press, 2003), 71. See also Charles Curran et al., *Dissent in and for the Church* (Franklin, WI: Sheed & Ward, 1969).

48. Gudorf, "Contraception and Abortion," 70.

49. Daniel A. Dombrowski and Robert Deltete, *A Brief, Liberal, Catholic Defense of Abortion* (Urbana: University of Illinois Press, 2000).

50. Ibid., 1. See also Daniel C. Maguire, *Sacred Choices: The Right to Contraception and Abortion in Ten World Religions* (Minneapolis: Fortress, 2001).

CHAPTER 14

GENETIC ENGINEERING AND NATURE

Human and Otherwise

THOMAS A. SHANNON

THE SHIFTING CONTEXT

ALTHOUGH the *Origin of Species* was published in 1859, the implications of the reality of evolution are still slowly working their way through our ways of thinking. Indeed the situation with respect to the cultural and biological implications is similar to Einstein's comment about the discovery of nuclear power: everything has changed but our way of thinking.

Evolution introduced the notion of change into the world of biology and also culture. Prior to this time the dominant understanding of reality was governed by a static concept of reality. Clearly within this framework, we know that things grew and developed, and thus there were clear signs of change. An acorn was not the oak tree but one knew that was its origin. One knew that a human being emerged from the fetus and went through a series of changes. But what was assumed was the reality of the stability of species, that is, oaks came from acorns, humans from humans, dogs from dogs, corn from corn seeds. While many would not explain this species independence and stability on fundamentalist grounds, the fact of the matter was that the dominant assumption was that like came from like and that species were stable. Thus stasis was the dominant philosophical framework in which all life was understood.

This sense of stasis extended from the biological to the social. Not only were species stable, so too were social classes. While some movement was possible, the dominant assumption was that one stayed within one's social class because that was one's destiny, however one might ground such a vision. Society, as nature, was arranged in a great chain of being, with a place for everything and everything in its place.

The core insight of Darwin—descent through modification—removed the core support of the understanding of nature and society: species stability. Darwin's insight meant that in fact all beings shared a common ancestor and that from that common ancestor all beings arose through a process of a combination of reproductive competition and advantage. Organisms interacted to gain some advantage over their neighbors to that they would survive to pass on this advantage to their descendants. This meant that species as we understand them from our vantage point only gradually emerged out of a long process of change and development. A species is not a boundary marker but a semipermeable membrane that can be penetrated for a reproductive advantage. Species, thus, are a consequence of a process of interaction rather than a beginning point.

Moreover, Darwin argued that this process of descent through modification was a random process. There was no design, no plan, no purpose. Just organisms seeking a reproductive edge to make sure they survived long enough to reproduce. All one had to do to achieve this goal was to be a little bit better than one's neighbor. Perfection was not a goal, nor were organisms designed to survive. They survived by fitting in better, and that achievement was a consequence of how the organism interacted with its neighbors and the environment, not a function of a preexisting plan or design. Extremely important to remember here is that the core concept is process, not perfection or a goal. Evolution is not directional, not teleological, not purposeful. What emerges, organisms a little better off than before, is a function of the process, not the result of an overarching vision directing this process to an end point of perfection.

And there we have the Darwinian challenge: we evolved from ancestors we shared with all other living organisms, and how we got here was the function of a random process. Change and advantage was the name of the evolutionary game, and biological and social stasis were understood as impediments to the process of evolution. This is the third of a set of decenterings of humanity from which we are still reeling. The first was the Copernican revolution that removed our world from the center of the solar system and placed us, as we know now, way to the side. The second decentering was the Freudian revolution that highlighted the unconscious and its influence on our behavior, thus suggesting that the Enlightenment model of rationality might indeed be flawed. Darwin identifies our animal ancestry and links with all creation. Human were clearly no longer unique or the crown of creation. We are one species among many sharing with them a common ancestor.

Genetic engineering is another turning point in the evolutionary story. The how of evolution was not clearly understood, though there was the strong assumption that something in the organism had to be transferred from one generation to another. Scientists came closer and closer to discovering this mechanism until the breakthrough discovery of the structure of the DNA molecule in 1953 by Watson and Crick. Though not perhaps the discovery of the secret of all life that they claimed this to be in the exuberance of their excitement, this discovery certainly opened the door to an explosion of development in the field of the life sciences. For here one did have an understanding into not only the structure of the molecule common to all life, but into how information was transferred from one generation to another. The next critical breakthrough came in the 1970s when Cohen and Berg discovered how to slice the DNA molecule and add in a section of DNA from another organism. A truly transgenic organism was created, and the developmental possibilities expanded exponentially. These developments continued, eventually resulting in the production of maps of the genome of a variety of organisms: C. elegans, mice, rice, humans. Not only do we know how to intervene, we are learning more and more where to intervene to achieve desired results. The engineers can now engineer themselves.

This capacity—to reengineer nature—has profound consequences, not only for nature in general but also for human nature. But it is this capacity, particularly when coupled with the Darwinian insight of change, that is so problematic for us. On the one hand we know that we are the products of an ongoing process of development. On the other hand we seek to plan and shape our future, by implications of our actions if not the direct intent of our interventions. Others continue to reject the Darwinian insight and argue for the integrity and normative status of existing species. Others assume that the randomness of the evolutionary process gives license to intervene at will, seeking the advantage of the moment. Yet the fact remains, as Theodosius Dobzhansky said four decades ago: "Evolution need no longer be a destiny imposed from without; it may conceivably be controlled by man, in accordance with his wisdom and his values."[1] Thus the challenge is raised, and the remainder of this essay will examine several implications of this new situation given us by both nature and our capacities that have emerged from the very evolutionary process we now seek to influence.

TRANSGENICS

One of the clear consequences of the technology of recombinant DNA developed in the 1970s by Cohen and Berg was that DNA could be transferred from one organism to another and bring its particular function with it and continue to operate successfully within the recipient organism. Though first restricted to use in bacteria—after an unprecedented self-imposed moratorium by scientists to evaluate the safety issues—the technology opened the path to a myriad of agricultural and animal applications. Thus we had the transfer of the antifreeze gene from the Artic flounder to tomatoes to prevent their being damaged by frost; the development of a variety of genetically modified grains that were resistant to pesticides; and cross-species organisms such as the geep, a cross between a goat and a sheep. But this was only the beginning. Robert and Baylis note:

> *Human-to-animal embryonic chimeras* are only one sort of novel creature currently being produced or contemplated. Other include: *human-to-animal fetal or adult chimeras* created by grafting human cellular material to late stage nonhuman fetuses or to postnatal nonhuman creatures; *human-to-human embryonic, fetal, or adult chimeras* created by inserting or grafting exogenous human cellular material to human embryos, fetuses, or adults (e.g., the human recipient of a human organ transplant, or human stem cell therapy); *animal-to-human embryonic, fetal, or adult chimeras* created by inserting or grafting nonhuman cellular material to human embryos, fetuses, or adults (e.g., the recipient of a xenotransplant); *animal-to-animal embryonic, fetal or adult chimeras* generated from non human cellular material whether within or between species (excepting human beings); *nuclear-cytoplasmic hybrids*, the offspring of two animals of different species, created by inserting a nucleus into an enucleated ovum (there might be intraspecies, such as sheep-sheep; or interspecies, such as sheep-goat; and if interspecies, might be created with human or nonhuman material); *interspecies hybrids* created by fertilizing an ovum from an animal of one species with a sperm from an animal of another (e.g., a mule, the offspring of a he-ass and a mare); and *transgenic organisms* created by otherwise combining genetic material across species boundaries.[2]

One major goal in this is the development of the possibility of xenotransplantation, the generation of animals whose organs could be transplanted into humans. The key problem in organ transplantation is rejection of the transplanted organ. Pigs with genes from the human immune system are being raised so that when one of its organs is used, the human body will not identify it as a genetic stranger and the organ will be accepted. Clearly this would be a critical breakthrough since the demand for organs outstrips the supply by tens of thousands each year.

But as enticing as this work is clinically, there are substantive anthropological issues raised with respect to and understanding of human nature—as well as the nature of other organisms. At the general level, we have the question of what is a species. At the particular level, we have the question of when is an organism a member of a species—and conversely when is it no longer a member. The general conceptual difficulty with these questions was remotely raised by Darwin, but the practical conceptual difficulty is raised directly by the variety of experiments noted in the quotation above. Genetic information can be moved around from organism to organism, and all manner of chimeras made. For example, cells from a neural cell line were implanted into the brain of a monkey at thirteen weeks gestation. Upon dissection of the brain, these human neural cells had become part of the monkey brain.[3] Additionally Mary Clare King and Allan Wilson showed that "the average human protein was more than 99% identical to its chimpanzee counterpart. . . . This leaves a great paradox: our DNA is almost identical to that of our chimp cousins, but we don't look or act alike."[4] What mutations are responsible for human behavior? Or is it possible for enough new genetic information to be added to one organism so that in effect it is now another organism genetically, even though pheotypically it may well resemble other members of its (previous) kind?

Kevin FitzGerald has identified four philosophical anthropologies as responses to these kinds of scenarios. First are static anthropologies based on "primarily philosophical or theological beliefs about characteristics, including physical characteristics, of human nature which are fundamentally unchanging . . . changing these characteristics would lead to the creation of non-human or deficient human beings."[5] Second are scientistic anthropologies that privileges scientific information—as the philosophical anthropologies privileged philosophical categories. This perspective "leaves no room for data from other branches of knowledge."[6] Third is a "dichotomized interpretation of the human nature by separating human biology from moral characteristics or personhood in order to address ethical issues."[7] Finally is the dynamic anthropology which "attempts to integrate sociohistorically conditioned concepts of human nature with the findings of contemporary scientific research."[8]

The first three models have sharp dividing lines: no data is relevant to the determination of human nature other than that from the respective discipline; personhood is separate from human nature, which is a biological reality, and given the perspective, changes in biology may or may not be relevant to the discussion of the question of human nature. A critical reality remains, however, and that is that humans are embodied realities. We are not pure persons, pure spirits, pure moments of transcendence; we are flesh, and that flesh emerged via evolution from the flesh of other organisms with whom we share DNA. So to speak of human nature is simultaneously to speak of human embodiment and all that that entails. But it is also clear that we do differ from our genetic relatives. We as humans typically have language, transcendence, freedom, symbol-making capacities, a rich emotional life, culture, a history, and a sense of a future. Minimally these capacities are the result of an emergence due to the process of complexification in our own physiology. Evolution reached in us a new level of organization and complexity, and from that new capacities emerged—and will continue to emerge.

Thus the concept of distinct human nature existing as a species absolutely distinct from all others is difficult to maintain. Humans emerged from evolution and are still within that process, though given the time scale of evolution our experience—but not our history—is one of stability of nature. If one accepts that this is no normative human nature and that who we are as humans is yet in a process of development, then an argument for intervention to change or direct that process is in principle easier to make, as suggested in the above quotation from Dobzhansky. Note the phrase *in principle*, however. Much of our culture is predicated on our difference from other organisms, our having a unique nature and a corresponding status in nature based on our uniqueness. Religious traditions also single out the uniqueness of humanity particularly with respect to relation with the deity. Ethical traditions are based on the uniqueness of human nature and its capacity for rule making.

Yet all of these perspectives on human nature were formed in the absence of our knowledge of evolution and how humans fit into that. Such a perspective is critical as we advance further into genetic engineering. One element of this perspective to keep in mind is that evolution is open-ended and adaptive to changing environments. This suggests cautionary notes about attempts to fix a particular genotype. Such fixing of a genotype, though understandable with respect to enhancing crop production in a particular environment, leaves the crop vulnerable to changing environments. Similar attempts to custom-design children through careful selection of traits that are apparently gene linked, though not as potentially devastating on a species level, nonetheless predicate reproduction on momentary desires or aspirations and may well leave the child maladapted for any other opportunities.

Such attempts at intervention need to be characterized by openness, creativity, adaptability, resourcefulness, and capacity for change. We need to recall that frequently in the evolutionary process it is not the organism that has most successfully adapted to an environment that survives but the one that has reserve genetic capacities that make it adaptable. Thus while we must recognize that the concept of human nature is much more fluid and less normative that we have previously thought and that because we share our DNA with many other organisms we may be less distinct as a species than we imagined, insights based on the processes that enable evolution to be successful may help us to be more understanding of who we are and of what qualities are needed to allow the flourishing of that most element of evolution: openness to change.

THE METAPHORS OF INTERVENTION

There are several traditional images of modeling human responsibility.

Steward

The term *stewardship* came out of a particular cosmological and biblical framework. Biblically, the term originates in the Genesis narrative of the garden of Eden in which Adam and Eve are placed. In this garden, they are given certain responsibilities and obligations. Essentially they are to care for the garden that God has created. Imbedded in this narrative is the reality that the garden is not theirs and that their care for it is to be exercised within the ethical framework given them by God. Precisely stated, they are to be stewards, ruling in place of God but on behalf of God. Therefore they had limits, and they answered to a higher authority. From a cosmological perspective, the universe was a static universe designed according to a plan, and this plan specified where everything was to go and how it was to act. This cosmological perspective was shared philosophically and theologically for centuries. It specified a universe of boundaries and limits. And within such a universe, the ethical standards were reasonably clear and well grounded—indeed literally grounded in the universe in which one lived and breathed.

In such a universe, stewardship was not only an ethic easy to validate—it was the only ethic available. Within the worldview of those in this framework, limits were obvious and binding. And actually there was no other choice. There were no other competing frameworks, only very limited interventions were possible, and human powers had very finite and obvious limitations. Stewardship made sense because within the static, hierarchical cosmology and theology of both ancient and medieval Western civilization it was the only option.

Several interlocking events shook this framework to its very foundation and produced a new reality in which the steward and stewardship language are strangers in a very strange land. The first was the Protestant Reformation, which challenged the authority of the ecclesiastical establishment of Rome and replaced it with personal responsibility for one's salvation (faith alone) and private understanding of the scripture (scripture alone). The role of authority in interpreting scripture, imposing limits, and defining responsibility was weakened as responsibility, for that was more and more taken over by private conscience. Second, the Industrial Revolution, fueled by the concurrent rise in capitalism, showed that both human capacity and creativity could be expanded enormously by power sources and money. Limits now seemed to be limits of the imagination— and one's cash reserves or credit line. Third, the theory of evolution forever shattered Aristotle's great hierarchy of being and the social structure implied by it with the insight that the reality of today is the result of a series of random interactions that had no guiding principle. Stability was replaced by process, and one existed at a point in a continuum that had no determined future. Finally, jumping ahead several centuries, we have the reality of genetic engineering that gave us the possibility of engineering or designing the designer.

Thus stewardship, while having an important and useful past, encounters difficulties in light of our contemporary understanding of the cosmos and in light of our increased technical capacities.

Created Cocreator

One response, using somewhat familiar religious language, is suggested by Philip Hefner, who proposes the term *created cocreator*.[9] This term affirms that indeed we are created by God, though that creative process is framed in an evolutionary context. But the "created" dimension of that term suggests some set of limits. We are not on our own; we are not totally independent. We stand in a relation of creaturely dependence to God. The "cocreator" dimension of the term directly acknowledges our capacity for creativity. But again there is a qualification in that we are cocreators, that is, our creativity starts from what is, from the limits that are factually imposed by lack of knowledge and capacity. The lacks are changing all the time, but their reality does remind us that we do have limits. Perhaps the analogy for human creativity that emerges from this phrase is that of the human as a working partner in cooperation with God. We can take initiatives as a partner does, but such initiatives are to be in harmony with our other more senior partner—God. But here again an argument has to be made to justify the harmoniousness of our actions with what we know of God.

Playing God

The term *playing God* signifies a certain mode of acting, an association of human powers with divine powers. To play God is either, from Paul Ramsey's perspective,[10] to abrogate to oneself the powers of God by transgressing God's design or, in Joseph Fletcher's terms,[11] to simply acknowledge reality and realize that the old God no longer exists and to begin acting responsibly in the world. The term is conceived of as a power struggle between two worldviews: one that includes the divine and one that does not.

As such, both Ramsey and Fletcher require us to make many assumptions about God, God's work, and human responsibility and to make these assumptions in a fairly sharp and dramatic way. And because both operate at such an extreme, both are, I think, flawed. Ramsey's perspective makes us almost powerless before nature, even as he acknowledges that interventions are possible. Fletcher assumes there are no limits to intervention as long as we consider carefully the greatest good for the greatest number. Yet both are somewhat uncritical in their orientation and ultimately lead us into dead ends.

Neither of these visions is particularly Christian—or Jewish or Islamic, for that matter. Ramsey's is too biologically deterministic, and Fletcher's is too secular. The vision that both present is too much dependent on the Greek philosophical and mythological tradition. It is certainly much more resonant with the myth of Prometheus, who in stealing fire from the gods and giving it to humans became like the gods and thus played God. However, he suffered the fate of one who usurped the power of the gods. Were the gods suggested by this version of playing God actually this fearful of sharing creation, assumedly that God would never have created in the first place. Why spoil the way things are!

Perhaps a better rendering of playing God is to learn as much about God as one can and then play God by acting as God acts. Minimally this might mean that we are to be creative as well as generous in our creativity and to keep covenant with our God and our creations. To affirm this is to surrender full control because we are not God. But it is also to assert a profound relation between humankind and the rest of created reality. We play God by imitating God—no small task.[12] Of course we still confront the practical problem of responding to radically new situations as we think God would. But we also know there are competing images of God, and, while this is not inherently problematic, the strategy is not as clear as some might think.

The Human as Reader and Editor of the Book of Nature

Bonaventure (1221–74) is one of those figures whose early biography is hidden. Little is known of his early life other than the legend that when he was very young, he became deathly ill and was pledged to Saint Francis of Assisi, whose intercession healed him. Thus began his journey with the Franciscans, leading to his becoming a member of the order, studying at the University of Paris under Alexander of Hales, receiving one of the chairs in theology at that same university, eventually becoming the Minister General of the Friars, guiding them through difficult times, and dying while a participant at the Council of Lyons.

One of the important metaphors that Bonaventure used in his theology was that nature is one of the two books that God has given humans to read, scripture being the other. I think this metaphor is an important one and offers the possibility of another image for us to think about our relation to nature and how we might think about interventions.

Following Bonaventure I would propose the image of the human as the reader of the text of creation. In a very interesting overlap with the perspective developed here, the analogy used to describe the DNA molecule that is common to all living organisms is the book of life. And this book is written with an alphabet composed of four letters, which are the first letters of the four bases that make up this molecule: A (adenine), C (cytosine), G (guanine), and T (thymine). Although adenine can join only with thymine and guanine only with cytosine, these pairs combine and recombine in an almost infinite variety to give us the physical totality of all living creatures that we see around us. The genome of each species can be thought of as an individual book in the library of creation. The Human Genome Project—as do other projects focusing on mapping the genomes of other organisms such as C. elegans, the fruit fly, and the mouse—is the effort to establish the text of the book of the human species so that through a careful reading of it we can understand different elements of the human species.

We need to establish the texts of these various books of the different species so we can come to a greater understanding of how they have come to be, how different themes in these different books might be related to each other, and the composition of the text. Thus we might think of this text as providing us with different kinds of books. One might be an encyclopedia that gives a general overview and a kind of frame of reference. Another book might be a basic science text that provides the specificity of how to understand the rules that make its composition possible. Yet another volume might be a kind of repair manual that shows how various clinical interventions might correct a grammatical or spelling error. Then we can have books of poetry, philosophy, history, anthropology, sociology, or theology. All of these books look at various themes of this core text to give other dimensions, perspectives, and interpretations.

One book is not sufficient to capture the totality of the meaning of the text, as the history of humanity itself reveals. All are necessary to give a more complete or adequate description of the meaning of the text. The alphabet of DNA is very simple, but the books generated from this text are many and complex. This means that the text must be read with an open, but critical mind. While it is true that we will always come to a text with our own context, the text also has some degree of independence from us and in a sense stands over us and critiques our reading of it.

Yet though there is a text in the world of nature, neither the text nor the various story lines in the text are fixed. We know that the text itself, though relying on its alphabet, has developed over time; this is evolution. The different story lines are the different species that have appeared and disappeared over the course of time. This suggests that there is not a normative reading of the grammar but rather that many readings have been available and will continue to be available. Surprise and novelty are the hallmarks of the text, and these elements suggest that we should expect more of the same as we continue to read the various texts. Literalism is not a characteristic of the history of the text, nor should it characterize how we read at present.

We should also recognize that we ourselves are part of the text we are reading. That is, we do not simply hold the text in our hands and read it. Rather we read the various texts through the text of ourselves and through what we already know of ourselves. Our reading is interactive. What we read influences what we have read and changes what we will read. What we read moves us to read other texts—texts from archeology, physics, engineering, and the liberal arts—to further illumine the current text. The process is one of continuous construction and deconstruction of our primary text as we gain further insights into it. Given that the text is in the process of being rewritten or reedited as we read it, no reading is normative or final. What we read is what we read with our best framework, but the reading is provisional. But to say that the reading is provisional is not to say that it is useless or inadequate for meaning or guidance.

This suggests that we must be active in our reading. We must look for relations within the text, for subthemes that can be further developed, for hints and suggestions of themes that were started but not finished, for new starting points. The text of nature is thus a work in progress, a text still being edited. Given that process and our place in it, one consequence that might follow from being an active reader might be that of an editor of the text. We can select different themes, can improvise on the text, perhaps think of shifts of genre within the overall format—adding some poetry, some prose, some narration, some analytical material. What will be important in this editing process is that it must be done in relation to past editions. Thus there may be portions of the text that we do not understand or know how they might have fit into the current edition. That lack of understanding does not mandate immediate deletion, but rather this material could be held in a reserve file for possible future developments. Past editions also suggest that there be some degree of continuity with future editions, though, not a mindless reproduction. For a sequel or trilogy to be successful, for example, the reader needs to have some familiarity with the previous works, but nonetheless the particular volume must be able to stand on its own. Readers who come in at the middle—as we do in reading the book of nature—must be able to engage in the current volume but also be able to see some relation between the plot and characters in past editions. Thus editing calls for a high degree of creativity and freshness as well as the integration of a degree of appreciation for the significance of and remnants of the past in the present.

The act of reading is also an act of the imagination. A text is a catalyst that enables us to create and describe to ourselves and others how we have understood the story. The letters of the text give no representation of the story other than textual, but they are a springboard to the imaginative re-presentation of this text to ourselves and others as we share, compare, and perhaps correct our version of the narrative with that of others. This act of sharing the imaginative reconstruction of the text can range from an informal book-club discussion to a careful reading in a classroom format to an exceptionally close reading in textual criticism as part of a broader formal discussion of the text. Such multileveled discussions continue the creative process of understanding the text but also of developing deeper understandings of it together with further imaginative acts of the significance of the lives of the characters and perhaps even their effect on us. Thus, imaginative engagement with the text allows us to reconstruct the text but also enables us to be changed through this interaction. We imaginatively edit the text but the text also edits us.

Reading and imaginatively editing a text also allows the possibility of suggesting or actualizing a new meaning or even a new text—perhaps a sequel, perhaps the development of a theme or story line from the original, or perhaps striking out in a new direction all together. In all of these instances we use the same alphabet and even the same words, but we also develop new combinations of these words, perhaps a new grammar, syntax, or variations on a genre. In doing this we develop newness and freshness out of the tradition even though we stand within its broad contours. Again we have the reality that out of a finite number of possibilities— the alphabet, whether the genetic or linguistic one—we have the grounds of endless possibilities and creativity.

But the creation of a text is not the end of the story. The text must be read, deciphered, understood, and evaluated. Though one can read a text casually, texts more often than not demand a careful and thoughtful reading. We have a responsibility to engage the text on its own terms but also to recognize that the very act of reading inserts us into a history of the text itself as well as into a history of the reading of the text. If we know the origins of the text, the culture and context in which it was created, and the original language of the text, our reading of a particular text can be much more creative and richly imaginative. Our enriched reading will then positively influence future readings of other texts. This in turn will change how we read because our reading will be more informed, more penetrating, and more imaginative. The present text will guide us, but what we have read in the past will free us to see new interpretative possibilities of the text. We rewrite as we read.

One of the significant dimensions of Bonaventure's thought is the place of nature in his overall scheme. Nature is a locus of revelation, a book given us by the graciousness of the creator that we might learn of the creator. The ground of this is the doctrine of exemplarism through which Bonaventure argues that what is created bears the likeness of its Maker and provides a way of understanding qualities of this Maker. Because the creator has left personal traces in what has been made, we can work our way back to an understanding of who this creator is. This is the basis for a rejection of metaphysical reductionism that assumes that the whole is the sum of the parts. For Bonaventure, the parts can be understood only when seen in relation to the whole, and the whole itself can be understood only in relation to its place in the larger schema of the exemplarity of all created reality. Thus ultimately for Bonaventure, the book of nature by itself is not sufficient to interpret itself. We ultimately need to read the book of nature in conjunction with the book of revelation. Only in that way can be we both appreciate the book of nature and not make the mistake that it will tell us everything we need in order to understand it.

Another significant dimension of Bonaventure's thought is that nature is the means through which the incarnation becomes possible. There are two reasons for this. First, matter is a prerequisite of human existence. Second, the reconciliation of opposites shows the greatest wisdom, and in the incarnation we have the reconciliation of infinite and finite, spirit and matter, divinity and humanity. Matter has a value beyond the instrumental, for it is taken up into a most intimate relation with the Divinity itself. As Delio phrases it: "The very existence of Creation reflects a potency within it for union with the divine because of its exemplary nature. [Everything] in Creation—from stars to protons to humans—bears an expressed relationship to God."[13]

Third, humans are embedded in nature as deeply as any other created entity. Indeed, we are animated flesh. Through our bodiliness we participate in the mineral world and the animal world. Although Bonaventure predates the theory of evolution, this theory captures a real sense of the continuity of humans with nature.

Reading the book of nature at this level calls for both a profound understanding of the text as well as the imaginative interpretation of the text. A close and careful reading of the text through the sciences gives us a much deeper understanding of the text, of its history, and also of its possibilities. If there is anything we should learn from the scientific reading of the text of nature, it is the profundity and the almost extravagant creativity of the plots and themes of the text. The simplest of alphabets has resulted in a text of continuous novelty and surprise.

This openness of the text also leads to its imaginative interpretation. The meaning is not limited to the letters of the text, the words, sentences, or paragraphs that they form. In reading we genuinely play with the text through our imagination. We can improvise on it, imagine the characters in new situations, and imagine paths not taken. This makes for a richer reading of the text, but importantly, not a betrayal of the text. The text itself emerges out of a creative process of the author and the elements out of which the text is created and enters into the creative process of the readers or readers, who then bring their own interpretative powers to the reading of the text. Thus the next discussions of the text or even the next edition of the text is fashioned.

REDUCTIONISM

Scientific Materialism

Although the term *scientific materialism* appears late in Wilson's *On Human Nature,* scientific materialism is a key principle that provides the overarching framework for many of the ideas in sociobiology. Scientific materialism is "the view that all phenomena in the universe, including the human mind, have a material basis, are subject to the same physical laws, and can be most deeply understood by scientific analysis."[14] The core of scientific materialism is the evolutionary epic whose minimum claims are "that the laws of the physical sciences are consistent with those of the biological and social sciences and can be linked in chains of causal explanation; that life and mind have a physical basis; that the world as we know it has evolved from earlier worlds obedient to the same laws; and that the visible universe today is everywhere subject to these materialist explanations."[15]

Scientific materialism is a mythology, and "the evolutionary epic is probably the best myth we will ever have";[16] and it can be "adjusted until it comes as close to truth as the human mind is constructed to judge the truth."[17]

Of critical importance is a discussion of matter, the ultimate grounding—so to speak—of evolution. In Wilson's theory, matter is all that is and all that is needed to account for all activity—insect or animal, private or social. For Wilson, matter is most creatively expressed in the gene, the basic unit of heredity and "a portion of the giant DNA molecule that affects the development of any trait at the most elementary biochemical level."[18] Thus we need to examine human nature through biology and the social sciences. This will lead us to an understanding of the mind as an epiphenomenon of the neuronal machinery of the brain. That machinery is in turn the product of genetic evolution by natural selection acting on human populations for hundreds of thousands of years in their ancient environments.[19]

The Transcendent Potential of Matter

But is matter only matter, inert particles interacting according to the laws of physics and/or chemistry, or is there another level?

One traditional theory explaining the interaction of particles of matter such as electrons and positrons is hylosystemism, which holds that "all bodies, or at least non-living bodies, are composed of elementary particles or hylons which are united to form a dynamic system or functional unit."[20] In this context, system refers to "a functional nature, possessing new powers."[21] When put into various combinations or when actualized under various conditions, these elementary particles form new systems educed from the matter and the properties of this new system and "are not simply the arithmetical sum of the actual properties manifested by these hylons in isolation for the property of any given system such as the nucleus or the hydrogen atom . . . is rooted proximately in the new powers of the respective system, powers which, though ultimately reducible to the two or more hylons that function as essentially ordered causes, exist only virtually in the individual hylons."[22]

Consequently, the properties of individual particles seen in isolation can never tell us the full range of these particles when combined into a system. Therefore, within matter lies a range of possibilities that emerge or are actualized only when these particles are put into a system or when a previous system is restructured.

What are the implications of such a theory? Rahner argues that we are the beings "in whom the basic tendency of matter to find itself in the spirit by self-transcendence arrives at the point where it definitely breaks through."[23] For Rahner, "matter develops out of its inner being in the direction of spirit."[24] This becoming, a becoming more rather than becoming other, must be "effected by what was there before and, on the other hand, must be the inner increase of being proper to the previously existing reality."[25] This notion of becoming more is a genuine self-transcendence, a "transcendence into what is substantially new, that is, the leap to a higher *nature*."[26]

While Rahner does not argue that life, consciousness, matter, and spirit are identical, he does argue that such differences do not exclude development:

> In so far as the self-transcendence always remains present in the particular goal
> of its self-transcendence, and in so far as the higher order always embraces
> the lower as contained in it, it is clear that the lower always precedes the ac-
> tual event of self-transcendence and prepares the way for it by the development
> of its own reality and order; it is clear that the lower always moves slowly
> towards the boundary line in its history which it then crosses in actual self-
> transcendence.[27]

For Rahner, then, the human is the "self-transcendence of living matter."[28] On the one hand, Rahner describes this as the cosmos becoming conscious of itself in the human. Yet on the other hand again, this self-transcendence of the cosmos reaches its final consummation only when the cosmos in the spiritual crea-ture, its goal and its height, is not merely something set apart from its foundation (i.e., something created) by something which receives the ultimate self-commu-nication of its ultimate ground itself, in that moment when this direct self-communication of God is given to the spiritual creature in what we—looking at the historical pattern of this self-communication—call grace and glory.[29]

Eaves and Gross present another view of matter as the ground of new po-tentialities. They argue for a dynamic and holistic conception of matter that emphasizes the "unity of matter, life, and energy and understands nature as a profoundly complex, evolving system of intricately interdependent elements."[30] They suggest a vitality in matter that gives it depth and intensity, value, and the inclination toward organization.

Eaves and Gross operate from a biological and specifically genetic perspective that "seeks a new framework for its comprehension that does justice to all the so-called higher aspects of human consciousness in a phylogenetic and ontogenetic framework."[31] This perspective focuses on the mechanisms of inheritance which "have within themselves the probability of presenting new transcendent possibilities for action within history."[32] Thus they argue that surprise is inherent in nature, and they then develop a view of nature itself as gracious and argue, similarly to Rahner, that "genetics provides *a basis for grace within the structure of life itself.*"[33]

This position serves as the basis for a rejection of crude determinism, for "the material processes of life have produced a person who transcends all conventional definitions of personhood to the point where the term *freedom* is the best we have available."[34]

This gives rise to two consequences: first, "culture creates conditions for completion in community that would otherwise be impossible in a mere aggregation of individuals";[35] second, "recognition that the conditions of life are such that the process that produces pain, in the sense of genetic disease, is also the process that maintains life in the cosmos."[36]

This second point is critical in that it highlights the value of genetic diversity and provides the ground for criticizing simplistic models of genetic waste, unfitness, and disease. Additionally, this point recognized a fundamental ambiguity in the nature of reality. Cancer is a result of the extremely rapid division and growth of cells, the very same process that allows life to continue. In the process of genetic recombination in sexual reproduction, copying errors sometimes occur that result in disease. Yet it is this very same process that allows reproduction to occur at all. These biological processes are the means through which life is transmitted from one generation to another, yet it is through these very same processes that life can be transformed in ways that are sometimes new and helpful and sometimes new and harmful.

A similar point emerges from a consideration of the multiplicity of forms and species: "There are many forms which do not constitute a value or an advantage in the struggle of life; they are useless in this sense, and for that reason they are beautiful. Beauty is a factor that is not necessitated by lower needs, but is something that supposes the liberty of artistic creation."[37]

Considerations such as these regarding the chemical composition of life expressed in the wondrously complex DNA molecule cannot help but also push us in the direction of a radical reconsideration of the nature of matter from both a religious and scientific perspective. For example, theologian Zachary Hayes expresses it this way: "The biblical tradition is a religious tradition that is convinced of the deep religious significance of the material world and of its profound potential for radical transformation into a form so different from its present form in space and time (i.e., the idea of the incarnation and the metaphor of resurrection as the final condition of 'becoming flesh')" (private correspondence).

This is an echo of medieval theologian Saint Bonaventure, who said: "Again, the tendency that exists in matter is ordained toward rational principles. And there would be no perfect generation without the union of the rational soul with the material body" (*On the Reduction of the Arts* §20). Although expressed in what we would consider dualistic language, Bonaventure suggests that matter has within it the potential to transcend itself. Pope John Paul II also articulates this in his address on evolution when he speaks of an ontological leap in which something profoundly different appears within the material reality out of which humans evolve.[38] Such discussions of necessity force us into a more critical dialogue with contemporary physics, particularly quantum mechanics with its take on the nature of matter. While such a discussion is beyond the scope of this essay, I recognize the necessity for such dialogue as articulated in this question by Hayes: "Do we have a spiritual substance such as a 'soul,' or are soul functions such as consciousness, etc. really symptoms of chemical complexification of matter that is still in the process of moving to its final, fulfilling form?" (private correspondence).

Whatever the outcome of such a debate, the view of matter and evolution suggested here is in the tradition of Augustine and his follower Bonaventure, who saw history as a most beautiful song, a "*pulcherrimum carmen* which has been played by the divine Wisdom since the first organisms were called into existence, and of which our present forms are but one scene."[39] Or, as the book of Proverbs says of wisdom: "I was by his side, a master craftsman, / delighting him day after day, ever at play in his presence, / at play everywhere in his world" (8:30–31).

SPECIFIC RELIGIOUS ISSUES

Religion is an interesting test case in an examination of human nature, for what one says about religion also reveals a commitment to a particular ideology and perhaps a methodology. A particular problem is a frequent assumption that a commitment to methodological reductionism also implies a commitment to metaphysical reductionism, which does not necessarily follow. Here, above all, one's prior commitments to particular positions need to be attended to and examined carefully.

For example, Richard Dawkins adopts an explicitly antireligion position. He sees religion as superstition and/or myth (understood as a false statement) whose purpose is to hide scientific truths from the unsuspecting or the naïve. Faith, Dawkins declares, "is such a successful brainwasher in its own favour, especially a brainwasher of children, that it is hard to break its hold."[40] And in addition to faith's being an arbitrary belief—otherwise one could give reasons for one's position— faith leads to fanaticism: "But it is capable of driving people to such dangerous folly that faith seems to me to qualify as a kind of mental illness. It leads people to believe in whatever it is so strongly that in extreme cases they are prepared to kill and die for it without the need for further justification."[41]

Dawkins defines the idea of God as a meme (Dawkins's term for a cultural unit of replication) and part of the meme pool. Thus the meme God gains its survival in this pool through its appeal to our psychology: "It provides a superficially plausible answer to deep and troubling questions about existence."[42] Thus for Dawkins God exists, but only as a meme within the culture.

The Blind Watchmaker, in addition to being a sustained argument for the randomness of evolution, is also an explicit attack on the proof of God based on design in nature. Here Dawkins is "advocating Darwinism not only as a candle in the dark against pseudo scientific beliefs, but also as a direct substitute for personal religion."[43] For Dawkins, evolution has no purpose other than the survival of particular genes, and which ones survive cannot be predicted in advance. Thus there is no design in the process of evolution.

Dawkins also says the following: "You scientists are very good at answering 'How' questions. But you must admit you are powerless when it comes to 'Why' questions. . . . Behind the question there is always an unspoken but never justified implication that since science is unable to answer 'Why' questions there must be some other discipline that *is* qualified to answer them. This implication is, of course, quite illogical."[44]

The question, of course, is on what basis does Dawkins make such a claim. Is this on the basis of scientific methodology? If so, what is it? On what basis does one determine that the differentiation of why and how questions is illogical? Is this a prejudice resulting from a precommitment to a metaphysical reductionism? It is one thing to reject religion—for whatever reason—it is quite another to argue that the rejection of religion follows directly from the acceptance of a scientific or Darwinian perspective.

Dawkins also argues that humans alone among all other species have the capacity to rebel against our genes. The basis on which one might do this is not clearly spelled out. Dawkins recognizes that we do this—the practice of artificial contraception is one of the stock examples of such behavior—but the justification for this is not completely or satisfactorily explained. Here it seems that there is some ambiguity in the nature of reality that escapes a totally scientifically materialistic explanation.

E. O. Wilson comes at the religion question from quite another perspective. First, Wilson was raised as a Southern Baptist and underwent a conversion experience as a youth. But he later underwent another conversion experience, one to evolution and against his own religious upbringing. This led him, according to Segerstråle, to want to "prove the [Christian] theologians wrong. He wanted to make sure that there could not exist a separate realm of meaning and ethics which would allow the theologians to impose arbitrary moral codes that would lead to unnecessary human suffering."[45] Important here is the strong identification of religion and ethics, which is not necessarily the case, as well as the desire to show that religion was not a privileged locus of knowledge for right and wrong. Here Wilson seems quite close to Dawkins in adopting a position of metaphysical reductionism.

On the other hand, Wilson, unlike Dawkins, is sympathetic to the "why" questions that humans ask, for he recognizes that humans have deep emotional needs that must be satisfied. Here Wilson argues that

> our metaphysical quest is an evolutionary one: religious belief can be seen as adaptive. The submission of humans to a perceived higher power, in the case of religion, derives from a more general tendency for submission behavior which has shown itself to be adaptive. By submitting to a stronger force, animals attain a stable situation. In other words, Wilson here used ethological insight to argue that we cannot eliminate our metaphysical quest—it is part of out nature.[46]

For Wilson, the choice is between empiricism and transcendentalism, whether philosophical or religious. His own preference is the empiricist view because it is objective, that is, scientific. It proceeds by "exploring the biological roots of moral behavior, and explaining their material origins and biases."[47] And ultimately, the evolutionary myth of origins will replace the religious one.

Yet, Wilson leans toward deism and states that there could exist a cosmological God whose existence could be proved by astrophysics. On the other hand, "a biological God, one who directs organic evolution and intervenes in human affairs . . . is increasingly contravened by biology and the brain sciences."[48] For all this, though, Wilson says we need our transcendental beliefs: "*We cannot live without them.* People need a sacred narrative. They must have a sense of larger purpose in one form or another, however intellectualized."[49] However, a transcendental form of this narrative will not and cannot endure, for it eventually will not withstand scientific scrutiny. Thus our guiding narrative will need to be taken from "the material history of the universe and the human species."[50] But that is not to our or religion's disadvantage:

> The true evolutionary epic, retold as poetry, is as intrinsically ennobling as any religious epic. Material reality discovered by science already possesses more content and grandeur than all religious cosmologies combined. The continuity of the human line has been traced through a period of deep history a thousand times older than that conceived by the Western religions. Its study has brought new revelations of great moral importance. It has made us realize that Homo sapiens is far more than a congeries of tribes and races. We are a single gene pool from which individuals are drawn in each generation and into which they are dissolved the next generation, forever united as a species by heritage and a common future. Such are the conceptions, based on fact, from which new intimations of immortality can be drawn and a new mythos evolved.[51]

So although disagreeing with Dawkins about the need of raising "why" questions, Wilson essentially lands in the same place: metaphysical reductionism and materialism, for science ultimately will answer all questions. And the answer to the question "Why religion?" is that it is adaptive and leads to social stability. In a concluding perspective on this, Wilson says:

> I just believe, to put it as simply as possible, that science should be able to go in a relatively few decades to the point of producing a humanoid robot which would walk through that door. The first robot would think and talk like a Southern Baptist minister, and the second robot would talk like John Rawls. In other works, *somehow I believe that we can reconstitute, recreate, the most mysterious features of* human *mental activity*. That's an article of faith but it has to do with expansionism. That's expansionism![52]

Thus every element of mental and physical behavior will have a physical basis, and ultimately there will be a materialistic explanation for everything, for science will continue to test every religious assumption and claim about God and humans and ultimately will come to the foundation of all human moral and religious sentiments: "The eventual result of the competition between the two worldviews, I believe, will be the secularization of the human epic and of religion itself."[53] One would then test, in the sociobiological mode, whether the peculiarities of the human brain are inferred to have taken place. If such matching does exist, then the mind harbors a species god, which can be parsimoniously explained as a biological adaptation instead of an independent, transbiological force.[54]

Thus God, and religion, is a product of a brain that is a product of evolution, which leads us to various adaptive behaviors, of which religion is one. And we are back again to biology's being the full explanation of all behavior, Wilson's original point in developing his theory of sociobiology. But is this the whole story?

A problem here might be the lack of distinction between three types of why questions. A scientific why question seeks to answer how one could account for a particular outcome—why bodies fall, for example. A philosophical question tries to seek out inner relations and ultimate principles—Aristotle's seeking out of final causes, for example. Religion pursues its why question in terms of ultimate meaning—for what we may hope, for example. Each of these disciplines has a particular set of rules and a framework in which its particular why question can be answered along with a set of criteria for evaluating the adequacy of the answer. A problem arises when one asks the why question of one discipline from within the perspective of another. Or when one insists on the criteria from one discipline as being the only criteria acceptable for verification. While it is the case that the boundaries of these disciplines more frequently resemble semipermeable membranes rather than fixed borders, one continually needs to be sensitive to what kind of question one is asking and what are the acceptable criteria for evaluating an answer. Border crossings are to be expected in our interdisciplinary world, but one must also remember to respect the customs and culture of the territory we visit.[55]

NOTES

....

1. Cited in Kevin FitzGerald, "Philosophical Anthropologies and the Human Genome Project," in *Controlling Our Destinies: The Human Genome Project from Historical, Philosophical, Social, and Ethical Perspectives* (ed. Philip R. Slaon; Notre Dame: University of Notre Dame Press, 2003), 401.

2. Jason Scott Robert and Françoise Baylis, "Crossing Species Boundaries," *American Journal of Bioethics* 3 (summer 2000): 1–2 (emphasis original).

3. Kevin FitzGerald, "Embryonic Stem Cell Research," in *Stem Cell Research: New Frontiers in Science and Ethics* (ed. Nancy E. Snow; Notre Dame: University of Notre Dame Press, 2003), 25.

4. Ann Gibbons, "Which of Our Genes Makes Us Human?" *Science* 281 (4 Sept. 1998): 1432.

5. FitzGerald, "Embryonic Stem Cell Research," 398–99.

6. Ibid., 399.

7. Ibid.

8. Ibid., 401.

9. See Philip Hefner, "Biocultural Evolution and the Created Co-creator," in *Science and Theology: The New Consonance* (ed. Ted Peters; Westview, 1998).

10. See Paul Ramsey, *Fabricated Man* (New Haven: Yale University Press, 1970).

11. See Joseph Fletcher, *The Ethics of Genetic Engineering: Ending Reproductive Roulette* (Garden City, NY: Anchor, 1974).

12. Allen D. Verhey. " 'Playing God' and Invoking a Perspective," *Journal of Medicine and Philosophy* 20 (1995): 347–64.

13. Ilia Delio, "Revisiting the Franciscan Doctrine of Christ," *Theological Studies* 64 (2003).

14. E. O. Wilson, *On Human Nature* (Cambridge: Harvard University Press, 1978), 221.

15. Ibid., 201.

16. Ibid.

17. Ibid.

18. Ibid., 216.

19. Ibid., 195.

20. Allan B. Wolter, "Chemical Substance," *Philosophy of Science* (Jamaica, NY: St. John's University Press, 1960), 98.

21. Ibid., 105.

22. Ibid.

23. Karl Rahner, "Christology within an Evolutionary View," in *Theological Investigations* (trans. Harl-H. Kruger; New York: Crossroad, 1983), 5.160.

24. Ibid., 164.

25. Ibid.

26. Ibid., 165.

27. Ibid., 167.

28. Ibid., 168.

29. Ibid., 171.

30. Lyndon Eaves and Linda Gross, "Explaining the Concept of Spirit as a Model for the God-World Relation in an Age of Genetics," *Zygon* 27 (1992): 226.

31. Ibid., 274.

32. Ibid., 278.

33. Ibid., 274.

34. Ibid., 275.

35. Ibid., 277.

36. Ibid., 278.

37. Ibid., 257.

38. Pope John Paul II, message to Pontifical Academy, 22 Oct. 1996, §6.

39. Philotheus Boehner, "The Teaching of the Sciences in Catholic Colleges," *Franciscan Educational Conference* (1955): 150–59.

40. Richard Dawkins, *The Selfish Gene* (2nd ed.; New York: Oxford University Press, 1989), 330.

41. Ibid.

42. Ibid., 193.

43. Cited in Ullica Segerstråle, *Defenders of the Truth: The Battle for Science in the Sociology Debate and Beyond* (New York: Oxford University Press, 2000), 401.

44. Ibid.

45. Ibid., 38.

46. Ibid., 402.

47. E. O. Wilson, *Consilience: The Unity of Knowledge* (New York: Knopf, 1998), 240.

48. Ibid., 241.

49. Ibid., 264.

50. Ibid., 265.

51. Ibid.

52. Segerstråle, *Defenders of the Truth*, 160.

53. Wilson, *Consilience*, 265.

54. Segerstråle, *Defenders of the Truth*, 160.

55. Zachary Hayes, *A Window to the Divine: A Study of Christian Creation Theology* (Quincy, IL: Franciscan Press, 1997), 11–13.

....

CHAPTER 15

SO NEAR AND
YET SO FAR

Animal Theology and Ecological Theology

....

ANDREW LINZEY

WHAT do we see when we look at nature? As Max Weber famously said: "All knowledge comes from a point of view." What, then, are the points of view that make animal theology and ecological theology such, apparently, uncomfortable bedfellows? Of course, every reforming movement is notoriously hostile to those who see, but do not yet quite see what they have seen, and so it is with animal and ecological theologies. On paper, the agreements appear so considerable that many cannot quite see that there *is* an issue of difference at all. Take, for example, some of the following points of religious agreement in environmental ethics:

- The natural world has value in itself and does not exist solely for human needs.
- There is a significant continuity of being, between human and non-human living beings, even though human beings do have a distinctive role. This continuity can be felt and experienced.
- Non-human living beings are morally significant, in the eyes of God and/or the cosmic order. They have their own unique relations to God, and their own places in the cosmic order.

- Moral norms such as justice, compassion and reciprocity apply (in appropriate ways) both to humans and to non-human beings. The well-being of humans and the well-being of non-human beings are inseparably connected.[1]

No animal theologian would dissent from these key propositions, and so it appears that there is unanimity. But the world, and especially the world of nature, is not quite that straightforward. Animal theologians and ecological theologians still do not *see* the same things when they peer into nature, or even if they see them, they count them in different ways.

Perhaps the best way of seeing the difference is to see through the eyes of Annie Dillard. Her *Pilgrim at Tinker Creek* is widely regarded as a testimony to ecological wisdom, and the author has been praised as "a writer of genuinely original vision to teach us anew 'To see a World in a grain of sand.'"[2] Dillard spent a year in the Roanoke Valley in the Blue Ridge Mountains of Virginia, at a place called Tinker Creek. There, throughout the seasons of the year, she watches and waits. She encounters what may be one of the last remnants of a genuinely undisturbed ecosystem, at least for now. She observes and notes and then philosophizes on the myriad of strange life-forms that she encounters and how they interrelate in ways that appear (to us) unaccountable and mysterious. But Dillard is not just a "nature writer," commonly understood as someone who wants to point out unusual or marvelous things, still less a "travel writer," who aims to bring new worlds to our attention; her underlying concern is to *make sense* of the world, or worlds, that lie around us. Her religious background, never specifically spelled out, but always close to the surface, helps provide a kind of spiritual autobiography.

As the narrative moves on, the darkness of nature, as well as its delights, presses itself upon her. One form of life really does live at the expense of another. There is pain, apparent waste, futility, and premature destruction. The world that presages paradise becomes, on closer inspection, a hellish one. And she does not hide from our gaze the awfulness of what she discovers, but how is she meant to make sense of it?

The clues mount up. "It's a chancy out there," we are told. "Dog whelks eat rock barnacles, worms invade their shells, shore ice razes them from the rocks and grinds them to a powder. Can you lay aphid eggs faster than chickadees can eat them? Can you find a caterpillar, can you beat the killing frost?"[3] "Evolution," it is suggested, "loves death more than it loves you or me."[4] Again, "We value the individual supremely, and nature values him not a whit." Perhaps she will have to extricate herself from the creek in order to *remain* human, to prevent herself from becoming "utterly brutalized." And here the religious impulses to "make sense" become overwhelming:

Either this world, my mother, is a monster, or I myself am a freak.
Consider the former: the world is a monster. Any three-year-old can see how unsatisfactory and clumsy is this whole business of reproducing and dying by the

millions. Yet we have not encountered any god who is as merciful as a man who flicks a beetle over on its feet. There is not a people in the world who behaves as badly as praying mantises. But wait, you say, there is no right and wrong in nature; right and wrong is a human concept. Precisely: we are moral creatures, then, in an amoral world. The universe that suckled us is a monster that does not care if we live or die—does not care if it itself grinds to a halt. It is fixed and blind, a robot programmed to kill. We are free and seeing; we can only try to outwit it at every turn to save our skins.[5]

Although Dillard's precise philosophy is difficult to discern, the various recognitions of parasitism force her to conclude that the universe is designed to be a self-sacrificing system. The references to priests and sacrifice increase. Can it really be that the creator has so willed it to the advantage of humankind? "God *look* at what you've done to this creature [the sacrificial ram], look at the sorrow, the cruelty, the long damned waste!" she expostulates. "Can it possibly, ludicrously be for *this* that on this unconscious planet with my innocent kind I play softball all spring, to develop my throwing arm?"[6] Difficult though it is to summarize, the conclusion (at least) betokens resignation, even (paradoxically) "thanksgiving." As in Emerson's dream, she "ate the world":

> All of it. All of it intricate, speckled, gnawed, fringed and free. Israel's priests offered the wave breast and the heave shoulder together, freely, in full knowledge, for thanksgiving. . . . And like Billy Bragg I go my way; and my left foot shouts "Glory," and my right foot says "Amen": in and out of Shadow Creek, upstream and down, exultant, in a daze, dancing, to the twin silver trumpets of praise.[7]

Time spent with Dillard is always rewarding because, excellent writer that she is, she helps us to see in an anguished and poetic way the currently fatalistic view of the world that is common among ecological theologians. She sees the pain, the waste, and the futility—according to Richard Adams, she is even "obsessed with Nature's apparently futile waste and indifference to suffering, to a degree no British writer (as far as I know) has been."[8] But the vital point to grasp is that there is no other world for which we should strive, no other world is morally or theologically available. We just have to bear it. The world is a self-murdering system of survival. Like Israel's priests we have to accept it and offer thanksgiving.

Ecological theologians reinforce Dillard's line almost to a tee, except that they usually go further. Not only is the world there to be eaten, it is manifestly God's will that it *should* be. I choose just three examples.

The first is from Richard Cartwright Austin, a prolific ecotheologian with more than four books to his credit, who extols the "beauty" of predation. In the context of a meditation on a fish eagle taking its prey, he writes: "Now I think that death may be part of the goodness of God's creation, so long as death and life remain in balance with each. To eat, and finally to be eaten, are part of the blessing of God."[9]

The second is from a well-known exponent of creation theology, Matthew Fox, whose *Original Blessing*[10] did so much to encourage an alternative theological view of creation. In a conversation with Jonathon Porritt, he makes clear that he ascribes equal value to all parts of creation, but also endorses predation as God's will:

> One of the laws of the Universe is that we all eat and get eaten. In fact, I call this the Eucharistic law of the Universe, even Divinity gets eaten in this world. And so the key is not whether we are going to be doing some dying in the process of being here, but whether we kill reverently. And that, of course, means with gratitude. You know, in the Christian tradition, it's interesting that the sacrifice of Divinity is called eucharist, that is, "thank you," gratitude. Gratitude is, I think, the test of whether we are living reverently this dance of the equality of being on the one hand, but also the need to sacrifice and be sacrificed on the other.[11]

In both writers, killing, even of sentient creatures, is unambiguously God's will. We might even say that nature *is* God's will.[12] That is not just how the world *is*, but how it *should be*. The task is not to reject or question such a world or be appalled by it, but to live reverently within it—with thanksgiving.

A third and similar, but rather more nuanced, view is provided by Jay B. McDaniel, another prolific writer on ecotheology. McDaniel has led the way in imaginatively combining process theology and ecological awareness. He sets his readers a poser drawn from the writings of influential nature writer Gary Snyder. The scenario takes place in the Chukchi Sea, just north of western Alaska, where a friend of his had watched "with fascinated horror" as orcas methodologically battered a grey whale to death. McDaniel asks us to envision God in relation to that spectacle. It is understandable, he notes, that we should feel empathy with the whale and wonder why an all-powerful God could allow it to happen, but he suggests we ask a different question: "As the orca was chasing the grey whale in the Chukchi Sea, whose side was God on?"

"The answer must be," writes McDaniel, that "God was 'on the side' of both creatures, to the exclusion of neither."[13] Since God wills the survival of both creatures, we must so envision God as immanent within both. From a process perspective, "things happen in the universe which God does not will, but which are nevertheless part of God's life." Process thinkers, says McDaniel, may discern a "fallen" dimension in predator-prey relations, "but they also see God as partly responsible for their existence in the first place." In short, the "fall" into carnivorous existence was lured by God, with cooperation from creatures. It was a "fall upward."[14]

It should be clear that whatever the individual sentiments of the writers we have considered (and McDaniel, for one, has been in the forefront of encouraging Christian concern for animals), it is difficult to see how their perspectives can provide a robust theological case against the human killing of other sentient creatures. In the end, not just death, but killing is wholly or partly God's will. Again, that is not just how the world *is*, it is how it *should be*. We have to resign ourselves to it. What counts is the "system," "nature as a whole," "creation as God made it"; individual creatures are just pawns in the game or, rather, means to another's end. There is no dilemma save that provided by our own misguided moral senses.

Now, I do not want to underplay the significance of what may be termed the "realist" perception of nature, which these writers, variously, express. Indeed, ironically, thanks to ecological awareness such a perception of nature is now widespread. Nature films on television have vastly reinforced it by regularly focusing on nature as an untrammeled struggle for survival. Every schoolboy or schoolgirl now knows that it is a jungle out there where human notions of right and wrong really do not apply—at least to individuals. Animal protectionists, like me, are accused of simply not facing up to the world as it is.[15]

But I do want to challenge the notion that this perception is the *only* one that the study of nature can afford us and that it can be sustained solo without obscuring other insights arguably much more fundamental, one in particular. So, let me return to my starting point: what do we see when we look at nature? The rival perception bequeathed by historic theology is to see nature not as a whole, but rather as unwhole, as tragic, incomplete, divided against itself, even fallen. Such a perception does not deny pain, suffering, and apparent waste and futility in nature, rather it begins with it and seeks to relate it to the wider Christian themes of creation and redemption. Saint Paul classically gives this expression when he writes of the creation in bondage, suffering in travail, awaiting its deliverance by the redeemed children of God (Rom. 8:14–24).

This idea has found wide resonance within Western culture. Even Nietzsche—surely one of the most unromantic of thinkers as well as one who resolutely resists theologizing of any kind, writes lyrically of how the "sight of blind suffering is the spring of the deepest emotion": "The deeper minds of all ages have had pity for animals, because they suffer from life and have not the power to turn the sting of the suffering against themselves, and understand their being metaphysically."[16]

That is why Nietzsche, quite remarkably, postulates that nature needs human interpreters, specifically the artist, the philosopher, and also, most revealingly, the saint:

> In him [the saint] the ego is melted away, and the suffering of his life is, practically, no longer felt as individual, but as the spring of the deepest sympathy and intimacy with all living creatures: he sees the wonderful transformation scene that the comedy of "becoming" never reaches, the attainment, at length, of the high state of man after which all nature is striving, that she may be delivered from herself.[17]

Although he would scarce have wanted it described this way, Nietzsche is here articulating the fundamental Christian hope for creation's redemption. From this perspective what is significant about creation is not what it currently is—but what it can become by grace. What is most telling is that we can hear the sighs and groans of inarticulate (to us) fellow creatures. Martin Luther, in his commentary on Romans 8, says that "anyone who searches into the essences and the functioning of the creatures rather [than] into their sighings and earnest expectations is certainly foolish and blind" because "he does not know that also the creatures are created for an end."[18]

Although largely unrepresented in modern theology, Paul Tillich well grasped the point in his sermon "Nature, Also, Mourns for a Lost Good," inspired by Schelling's poignant line: "A veil of sadness is spread over all nature, a deep unappeasable melancholy over all life," a sadness "manifest through the traces of suffering in the face of a nature, especially in the faces of animals."[19] The upshot is that nature "is not only glorious; it is also tragic. It is suffering and sighing with us. No one who has ever listened to the sounds of nature with sympathy can forget their tragic melodies."[20] But because the tragedy of nature is bound to the tragedy of human beings, so we can hope that, as humans are redeemed by grace, there is also hope for nature itself. Significantly, the Eucharist, far from being a legitimization of predation, as Fox supposes, becomes a symbol for Tillich of what E. L. Mascall once called the "Christification"[21] of nature itself: "Bread and wine, water and light, and all the great elements of nature become the bearers of spiritual meaning and saving power."[22]

The result, then, of attending to this other perception is to affirm the ambiguity of creaturely existence. Creation is both good, even "very good," yet it is also incomplete and unfinished. We perceive a symmetry in nature, but one that points beyond itself. Karl Barth says there is both a yes and a no in creation. A yes because God loves it, and values it, and creates it for an eternal relationship with itself. But there is also a no because it is not divine, and like all earthly stuff it is, in its own way, tragically incomplete without divine grace.[23] More poignantly still, Albert Schweitzer speaks of creation as the "ghastly drama of the will-to-live divided against itself."[24]

Ecotheologians commonly resist this conclusion and maintain that seeing nature as less than what it should be devalues God's gift of creation. What we need, they say, is a robust acceptance that all nature is *unambiguously* good as it is (it should follow, of course, that the same should be said of human nature—but that logical step is never taken), even that it is sacred. In the blunt words of Anne Primavesi: "If Nature is seen as 'not God,' then this licenses human control over it."[25] But this is a non sequitur; deifying all nature does not by itself save it from exploitation. On the contrary, one of the weaknesses of McDaniel's approach, based on process theology, is that God is eclipsed in the most vital work of all, namely, redemption. For how can God, logically, redeem herself? The attempt to sacralize nature is understandable as a protest against ruthless exploitation, but as a theological position it simply undercuts the whole human-responsibility-for-ecology program.

The bottom line is that the transcendence of God constitutes the hope of a *this-worldly* redemption. A God that is so compromised by immanence that she cannot escape the eternal earthly round is not one that can offer redemption for any creature, let alone command responsible stewardship.

It is now appropriate to consider the divergent practical implications that flow from these different perspectives. Some words of caution are required. Not all ecologists are antianimals and vice versa. One of the remarkable things about Dillard's work is that, although she seemingly resolves the dilemma through resignation, she is also acutely conscious of the pains that animals have to undergo. So it is with many ecologists, and, it should be noted, many animal advocates are often at the forefront of ecological campaigns. Nevertheless, there is typically a divide when it comes to practical issues, which is often acutely felt by those who campaign in these areas. I select just three.

The first relates to the ethics of killing, and vegetarianism in particular. If God wills a self-sacrificing system of survival, as many ecotheologians suppose, then there cannot be any theological basis for desisting from killing, not just plants, but also sentient animals (humans, too, if one is to be strictly logical). This is why so many ecotheologians are not vegetarians or, if they are, why they appeal not to straightforward ethical considerations about killing, but to environmental factors, like the harmful effects of intensive farming and the deforestation commonly involved in rearing cattle in Third World countries. Although some ecotheologians will adopt vegetarianism on these grounds, or because of the unnatural methods of factory farming, on the general issue of killing, they will most likely be found on the side of those who do not see a moral issue at all.[26]

Animal theologians, on the other hand, see vegetarianism as the first step toward the creation of a violence-free world, and even though absolute consistency is hard to embody, they will recommend it as a symbolic and practical renunciation. In the trenchant words of Stephen R. L. Clark: "Honourable men may honourably disagree about some details of human treatment of the non-human, but vegetarianism is now as necessary a pledge of moral devotion as was the refusal of emperor-worship in the early Church.... Those who still eat flesh when they could do otherwise have no claim to be serious moralists."[27]

It is important to make explicit the *theological* basis of this divergence. For animal theologians, the issue of not killing except when essential is not just about ethical precepts. It is about a contrasting vision of what humans should be in creation. Animal theologians take seriously the idea that humans are made in God's image—in the image of a just, holy, and loving God—and therefore expect humans to acknowledge duties to animals who cannot acknowledge them in return. *Human* killing in creation is attended (in most cases) by free will and moral choice. Therefore animal killing cannot provide a basis for justifying human acts. C. S. Lewis rightly observed: "It is our business to live by our own law not by hers [nature's]."[28] Moreover, theological vegetarianism has an anticipatory character: "By refusing to kill and eat meat, we are witnessing to a higher order of existence, implicit in the Logos, which is struggling to be born in us. By refusing to go the way of our 'natural nature' or our 'psychological nature,' by standing against the order of unredeemed nature we become signs of the order of existence for which all creatures long."[29]

The second concerns suffering. One might think that, even if animal and ecotheologians were divided about killing, then at least they would be united in opposing the infliction of suffering on animals, but not so. Greenpeace and other environmental organizations do not campaign against hunting for sport or even trapping for fur. The great campaigns on these issues have been waged—not with the tacit support of ecologists—but rather in the face of their opposition. In England, even supposedly green bishops like Hugh Montefiore, John Oliver, and Jim Thompson have lined up in favor of the continuance of fox and deer hunting. The appeal is invariably not to justice for individual sentients or the undesirability of humans obtaining pleasure from chasing and killing animals, but rather to general environmental considerations and most especially to predation in nature. Foxes, we are told, are not "kindly in their ways," and "nature is not a kindly place." Bishop Thompson opined that those "who believe in God must come to terms with a creation of mutual hunting and eating," as if nature were a moral textbook or capable of relieving us of our obligations as moral agents.[30] Even worse was the pronouncement of former Archbishop of York that hunting with dogs could be justified because of the "fascination" which people have "in the kind of competitive encounter that one has with a wild animal."[31] No less than twelve Anglican bishops have supported the continuance of hunting, with only one or two against.

Again, it is important to grasp that the issue is not just about morality. If one begins with the perception of the tragic character of nature as locked in, even against itself, to bondage, then intensifying that bondage by exploiting the natural antipathy that one creature has for another is a sub-Christian pursuit. Former Dean of Westminster Edward Carpenter argued that hunting is "in the strictest sense of the word, deplorable . . . it is to fall back into that bondage, into that predatory system of nature, from which the Christian hope has always been that not only man but the whole natural order itself is to be released and redeemed." Hunting does "violence to Christian faith and witness[es] to a lower order than that redeemed creation to which Christ leads us."[32]

The third concerns human management of the environment and of animals in particular. One might think, given the high doctrine of creation as God's will, that ecologists would be content to "let it be" according to its own perceived God-given law. On the contrary, ecologists typically espouse an *active* form of management, bolstered by the notion of dominion, in which humans have a right and a duty to intervene. If this management was confined to helping ailing species or individuals within species, then animal theologians could only applaud, but conservationists go much further. In the interests of the whole (for they often claim to know what are the interests of the whole), they appear only too eager to sacrifice one species for another, even if this means ruthless and indiscriminate killing and the infliction of considerable suffering. I give two examples, drawn from the United Kingdom.

The first is the case of the ruddy duck. First brought to Britain in the 1940s for ornamental purposes, some escaped and bred in the wild. Their numbers increased, and there is now a population of around five thousand in Britain. The British government is currently planning their systematic eradication. The putative justification for this killing ("culling" is the word used by so-called conservationists) is that the ruddy mates with another nonnative species called the white-headed duck, and hybridization occurs. As a contracting party to the Convention on Biological Diversity, the British government is required to control or eliminate alien species which "threaten ecosystems, habitats or species."[33] The upshot is that around one thousand ruddies have already been killed (mainly through shooting), and another five thousand killings are planned. In the words of the Department for Environment, Food, and Rural Affairs (DEFRA), the "ruddy duck control trial final report (2002) concluded that there is an 80% certainty that the population can be reduced to fewer than 175 birds in between four and six years, at a cost of between £3.6m and £5.4m."[34]

There is a range of objections to this policy. In the first place, it is not obvious that hybridization (interspecies mating) is undesirable in itself. The statement that "hybridization is recognized as the most significant threat to the species' [white-headed duck's] long-term survival"[35] is an odd kind of claim since in human terms we would hardly regard interracial marriage as a threat to the survival of one or both races. But, accepting for the sake of argument that hybridization is undesirable, and that control is therefore necessary, it still does not follow that *lethal* control is justifiable. Moreover, while the Convention on Biological Diversity may require the British government to act, it is by no means clear that a variety of nonlethal controls could not be utilized.[36]

Consider: through no fault of its own (since it was introduced by human agencies), the ruddy duck having adapted successfully to its new environment is now going to be ruthlessly exterminated over a period of six years, involving the infliction of incalculable suffering, at the expense of millions of pounds to the British taxpayer. And all this to prevent what is little more than a kind of species purity. But the most alarming fact of all is this: while we often expect governments to be morally regressive when it comes to environmental issues, in this case it was putative conservationists and ecologists, including the Royal Society for the Protection of Birds (RSPB), who were unanimous in support of the killing, while opposition came principally from animal-protection organizations.[37]

The second example is similar and concerns the attempt to control the numbers of grey squirrels. This is proposed, inter alia, in order to protect the red squirrel. DEFRA says that "it is widely (though not universally) accepted by the scientific and ecological communities that the presence of grey squirrels is inimical to reds and that if numbers of greys go unchecked reds will die out."[38] When asked what evidence there is that control by killing is effective, DEFRA responds by stating that "removal by killing is bound to be effective, if it is successful."[39] The reply is an oxymoron. Of course "removal by killing" is bound to be effective, but the fact is that killing does not always, if ever, totally remove a species unless the species is entirely eliminated by killing. We know that all animals breed in relation to the food and environment available. Killing may dent a population, but unless the entire population (in a definable area) is disposed of, the animals frequently breed to compensate for their loss. Hence, killing is not always effective and can even result over time in an *increased* population. What is true for squirrels is equally true for ruddy ducks. It is by no means clear that killing—even regular kills over a prolonged period of time—will always result in a substantial reduction, let alone elimination.

Consider: the British government through public funds continues to deploy a policy of killing grey squirrels, supposedly (in part) to "protect" red squirrels through programs of "targeted control" leading to "local eradication." But when asked how many need to be killed, by whom, and over what timescale, DEFRA informs us that the "Forestry Commission [a government department responsible for the administration of forests and woodlands] ceased collation of data on the number of squirrels killed in woodland about 10 years ago so we have no current data on numbers actually killed."[40] This admission is remarkable and undermines all claims to control squirrels by killing. Calculating squirrel populations is itself problematic, but not even counting the numbers killed means that all justifications for control are devoid of a scientific basis.

No wonder that DEFRA says that "unless grey squirrel populations are permanently and naturally reduced by some other factor (such as disease) or our research uncovers another method of reducing populations, we believe that control by killing is an *indefinite* commitment where a landowner or manager decides it is merited."[41] Note the word *indefinite* here. This is killing without end and, it seems, without evidence as to its effectiveness. Again, the most alarming fact of all is that conservationists and ecologists are among the first to defend such killing, even, as in this case, when there is insufficient empirical evidence to make a reasoned judgment.

These two cases alone expose how ready ecologists are to countenance the killing of wild animals and the insubstantial nature of their putative justifications. But counterquestions may be asked: Do animal people not neglect or overlook the whole? Are they so concerned with the welfare of individuals that they fail to see the common (ecological) good? Does not an obsessive concern for individuals undercut the working out of ecological responsibility? In the two cases I have cited (and many others I could mention), the undercutting is all one way. I know of no animal advocate who opposes ecological schemes save those that harm animals—and that is the rub. Injury to individual sentients should, and could in most cases, be avoided. Even in utilitarian terms, it is doubtful whether the purity of one species of duck could justify the destruction of six thousand other ducks, or whether the protection of the red squirrel could merit the indefinite killing of thousands of grey ones. It is not self-evident that we maintain the good of the whole by the destruction of individual parts.

It is important to see, however, that the apparent disagreements over killing and management are not just about the value of individual sentients, significant though that is. They concern much deeper issues about how humans should understand themselves in the world of creation. While animal theologians are keen for humans to be proactive in defense of threatened individual sentients, their overwhelming concern is to let creation be. They look askance at human attempts to control, manipulate, and dominate more and more of the natural world, as if salvation consisted in human management of creation. Rather the hubris involved in such management is itself in need of salvation. Animal theologians do not deny that being made in the image of God means that sometimes humans have to intervene (normally to rectify—as in the case of ruddy ducks—previous human mismanagement) or that sometimes difficult choices have to be faced about the killing of individual sentients (despite what is said in the media, very few animal advocates believe that either human or animal rights are absolute), but such intervention requires explicit justification and a much greater sense of humility. The *imago dei* does not confer infallibility.

In the cases cited, what characterizes human involvement is the unquestioned assumption that humans always know better, that we alone can properly judge what is in the interests of creation or know how nature really ought to work. Despite years of disastrous mismanagement, humans still go on—now increasingly under the guise of ecological responsibility—wanting to subject more and more of the natural world to our designs of how it should be. Acknowledging past mistakes only serves the impetus to manage more, instead of acting as a warning against all meddling. It is paradoxical, to say the least, that a philosophy which emerged in a newly found sensibility to nature should issue in a more and more tyrannous attempt to control it.

In conclusion, I have tried—from my admittedly partial perspective—to provide an account of the divergence of perception which undergirds the current tension between animal and ecological theologians and how this tension results in sharp divergences over practical issues. There is, I believe, no easy way to harmonize these perspectives. It is not essential that they should be in conflict. Theoretically it could, surely must, be possible to care for the whole as well as (or at least not to the detriment of) its individual parts. But any attempt at harmonization really must begin by recognizing the deep theological cleavage that separates these two perceptions. Some really do look at nature, perhaps sigh a little, but then offer thanksgiving, while others see only suffering that any decent creator God must long to redeem.

But I have one emollient thought to offer. The enduring value of theology to thinking about animals and ecology consists in a recognition that God relativizes all our human perceptions. If we have not seen this then all our visions will be hopelessly partial. Theology promises a God-centered rather than a human-centered view of the world. We cannot simply assume that our perspective is God's perspective. That recognition alone should give us all pause, even hope that there is yet more to be seen.

NOTES

1. See "Points of Religious Agreement in Environmental Ethics," online at http://www.wildbirds.org/info/religion.htm. The summary is the result of research by Kusumita P. Pedersen, "Environmental Ethics in Interreligious Perspective," in *Global Ethics: Comparative Religious Ethics and Interreligious Dialogue* (ed. Sumner B. Twiss and Bruce Grelle; Boulder, CO: Westview, 1998).

2. Annie Dillard, *Pilgrim at Tinker Creek* (London: Pan, 1976 [orig. 1975]), unattributed cover blurb.

3. Ibid., 154.

4. Ibid., 157.

5. Ibid., 158.

6. Ibid., 231 (emphasis original).

7. Ibid., 237.

8. Richard Adams, "Introduction," in Annie Dillard's *Pilgrim at Tinker Creek* (London: Pan, 1976 [orig. 1975]), 9.

9. Richard Cartwright Austin, *Beauty of the Lord: Awakening the Senses* (Atlanta: John Knox, 1988), 196–97. Both this passage and the following one from Matthew Fox are cited and discussed in Andrew Linzey, *Animal Theology* (Urbana: University of Illinois Press, 1994), 119.

10. Matthew Fox, *Original Blessing* (Santa Fe: Bear, 1983).

11. Interview with Matthew Fox in Matthew Fox and Jonathon Porritt, "Green Spirituality," *Creation Spirituality* 7.3 (May–June 1991): 14–15; and in Linzey, *Animal Theology*, 119–22.

12. For an important discussion of this question, see Stephen R. L. Clark, "Is Nature God's Will?" in *Animals on the Agenda: Questions about Animals for Theology and Ethics* (ed. Andrew Linzey and Dorothy Yamamoto; SCM/University of Illinois Press, 1998), 123–36.

13. Jay B. McDaniel, "Can Animal Suffering Be Reconciled with Belief in an All-Loving God?" in *Animals on the Agenda: Questions about Animals for Theology and Ethics* (ed. Andrew Linzey and Dorothy Yamamoto; SCM/University of Illinois Press, 1998), 163.

14. Ibid., 168.

15. "There was also a failure to face up to the realities of the natural world as revealed by biologists and others"; from a review of my *Animal Theology* by C. S. Rodd in *Expository Times* 106.1 (Oct. 1994): 4. It is a charge, of course, to which I gladly plead guilty. All Christian theology has to some degree, as Karl Barth once noted, to be logically inconsequent in the face of the facts, or rather *some* facts.

16. Friedrich Wilhelm Nietzsche, "Schopenhauer as Educator" (orig. 1874), in *Thoughts Out of Season* (trans. Adrian Collins; Edinburgh: Foulis, 1909), 2.149; extract in Andrew Linzey and Paul Barry Clarke, eds., *Animal Rights: An Historical Anthology* (New York: Columbia University Press, 2005), 148.

17. Nietzsche, *Thoughts out of Season*, 54; and in Linzey and Clarke, *Animal Rights*, 151–52.

18. Martin Luther, *Lectures on Romans* (ed. William Pauck; Library of Christian Classics 15; London: SCM, 1961), 237.

19. Cited in Paul Tillich, "Nature, Also, Mourns for a Lost Good," in Tillich's *The Shaking of the Foundations* (New York: Scribner, 1962), 83.

20. Ibid., 81.

21. E. L. Mascall, *The Christian Universe* (London: Darton, Longman & Todd, 1966), 163–64. Mascall's thesis of the progressive "Christification" of the natural order has been overlooked by theologians.

22. Tillich, *Shaking of the Foundations*, 86.

23. For my work on Barth, see Andrew Linzey, "The Neglected Creature: The Doctrine of the Human and Its Relationship to the Non-Human in the Thought of Karl Barth" (PhD diss., University of London, 1986). Barthian themes also emerge in Andrew Linzey, *Christianity and the Rights of Animals* (London: SPCK/New York: Crossroad, 1987).

24. Albert Schweitzer, *Civilization and Ethics* (trans. C. T. Campion; London: Allen & Unwin, 1967 [orig. 1923]), 216. For a full-length treatment, see Ara Barsam, *On the Ogowe River: The Quest for Reverence for Life in the Mystical Theology of Albert Schweitzer* (forthcoming).

25. Anne Primavesi, *From Apocalypse to Genesis: Ecology, Feminism, and Christianity* (Minneapolis: Fortress, 1991), 146. For my critical review, see *Scottish Journal of Theology* 45.2 (1992): 265–66.

26. See, e.g., my debate with Hugh Fearnley-Whittingstall, "Should We All Be Vegetarians?" *The Ecologist* 34.8 (Oct. 2004): 23–27. Although Fearnley-Whittingstall is not a theologian, he reflects much of ecothought by not seeing a moral problem about killing.

27. Stephen R. L. Clark, *The Moral Status of Animals* (Oxford: Clarendon, 1977), 183.

28. C. S. Lewis, *Present Concerns* (ed. Walter Hooper; New York: Harcourt Brace Jovanovich, 1986), 79, cited and discussed in Wesley A. Kort, *C. S. Lewis: Then and Now* (New York: Oxford University Press, 2001), 156. See also my extensive discussion on the value of Lewis's thought: Andrew Linzey, "C. S. Lewis's Theology of Animals," *Anglican Theological Review* 80.1 (winter 1998): 60–81.

29. Linzey, *Animal Theology*, 90–91.

30. Bishop Jim Thompson of Bath and Wells, *Hansard* 623.45 (12 March 2001): 537. See my response: "An Open Letter to the Bishops on Hunting," *Church Times* 20.27 (Dec. 2002): 10.

31. Archbishop John Habgood, *Hansard* 623.45 (12 March 2001): 611.

32.   Edward Carpenter, "Christian Faith and the Moral Aspect of Hunting" in *Against Hunting* (ed. Patrick Moore; London: Gollancz, 1965), 137. For my discussion, see Andrew Linzey, *Christian Theology and the Ethics of Hunting with Dogs* (London: Christian Socialist Movement, 2003), 1–24.

33.   Communication from DEFRA to the author, 18 Oct. 2004. For my critique of the concept of biodiversity and the way in which it is used to justify killing, see Andrew Linzey, "Against Biodiversity," *The Animals' Agenda* 21.2 (March–April 2001): 21.

34.   Communication from DEFRA, 18 Oct. 2004.

35.   Communication from DEFRA to the author, 10 Aug. 2004. The scientific work which forms the basis of the government's case can be found, inter alia, in J. M. Rhymer and D. D. Simberloff, "Extinction by Hybridization and Introgression," *Annual Review of Ecology and Systematics* 17 (1996): 83–109. But, like most scientific work, it fails to question underlying assumptions about the relative value of genetic purity, extinction, and welfare.

36.   The answer from DEFRA is that "a range of control measures have been tested and the most effective identified [i.e., shooting]"; communication from DEFRA, 18 Oct. 2004. But this rush to the gun must be questioned. Why should not taxpayers' money be spent on morally preferable options, even if they yield slower results, especially since total eradication is a near impossibility?

37.   In particular, Animal Aid has led a valiant fight against the slaughter; see http://www.animalaid.org.uk/campaign/wildlife/ruddycull.htm.

38.   Communication from DEFRA, 18 Oct. 2004.

39.   Ibid.

40.   Ibid.

41.   Ibid.

CHAPTER 16

RELIGIOUS ECOFEMINISM

Healing the Ecological Crisis

ROSEMARY RADFORD RUETHER

IN this essay I explore the religious roots of the ecological crisis and how these religious patterns that have promoted exploitation and negation of nature are interconnected with the religiously mandated patterns of exploitation and negation of women. In other words, I wish to confirm the basic ecofeminist thesis; namely, that there is an interconnection between the domination of women and the domination of nature. Although this thesis can be explored in terms of various ideologies, philosophical, psychological, and scientific, in this essay I will concentrate on the religious mandates for this interconnected pattern of domination of women and of nature. These religious mandates are older in human cultures and set the pattern for philosophical, psychological, and pseudoscientific mandates.

I will explore this interconnection of domination of nature and of women primarily in terms of the Christian tradition. This is not because the Christian patterns are theoretically more negative for ecology than other religions, but because, as the religious basis of Western civilization that has been a primary agent of creating and spreading the ecological crisis throughout the world, it has had the most influence. I focus on Christianity also because it is my own tradition and therefore the only one on which I can speak authoritatively and for which I can and must take responsibility.

But this story would be only a hopeless one if these mandates for domination were the only one available in the Christian tradition. Then the conclusion to this argument would simply be the necessity of rejecting Christianity in order to overcome the ecological crisis. But I will argue that there are powerful patterns of Christian thought in its scriptural roots that can help overcome and heal the ecological crisis. This healing cannot be only theoretical and cannot concentrate on only a generic "man-nature" relation. Rather this "man-nature" relation itself has always carried in it a hidden or overt social hierarchy, male over female, ruling class over subjugated classes, racial or ruling race over subjugated races and ethnicities. Thus the healing of the ecological crisis must encompass ecojustice; it must embrace a healed relation between the ruling sex, class, and race and the subjugated sex, classes, and races. It also must be practical, a way of life. It must overcome the split between public and private, between theory and daily life, and hence between the realms traditionally allotted to men and to women.

In 1967 historian of science Lynn White published an article that became the foundation for the debate about the negative impact of Christianity on ecology: "The Historical Roots of Our Ecologic Crisis."[1] In this article White argued that Christianity had played a major role in causing the ecological crisis by promoting a notion of "man's" unlimited mastery over nature. He pointed particularly to Genesis 1:26: "Let us create man in our image, after our likeness, and let them have dominion over the fish of the sea and over the birds of the air and over the cattle and over all the earth and over every creeping thing that creeps upon the earth." The passage became the basis for the Christian claim that humans are mandated by God to use the earth as their possession for whatever use they desire. White dubbed Christianity "the most anthropocentric religion the world has seen."[2]

This article sent Christian theologians and especially biblical scholars scrambling to defend their tradition from what seemed like an unequivocal condemnation. Many biblical scholars argued that the term *dominion* in the Genesis passage had been misinterpreted. God gives humanity "care" for the earth, not the right to abuse it. Humans are accountable to God for its well-being. They do not have unlimited mastery over it. The true biblical view of the human relation to nature is one of stewardship, not ownership. God is and remains the lord of creation to whom humans must account for their caretaking role.[3] This argument, however, did not explain why the Christian tradition had interpreted this mandate from Genesis as one of unlimited mastery, if indeed they had always done so, as White claimed. Nor did White unpack the term *man* in the text to ask if this referred to all humans or only males of the ruling class and race, either in the text itself or in later Christian interpretations of it.

In my view the problem of the historic impact of the Christian worldview on both the domination of nature and the domination of women is broader than simply the (mis)interpretation of one text from one book of the Bible. The Christian tradition was forged by the marriage of two major patriarchal cultures, Hebrew and Greek. It brought together a Hebrew tradition of religious law that defined the male head of family as representing divine rule over nature, as well as over the subjugated people within the family, women, children, and slaves, with a Greek philosophical tradition that split reality into a hierarchy of spirit over matter, mind over body, as male over female. It sees God as Father uniquely and definitively incarnated in one human male, "his son." Although women too can be saved through this one male, son of the Father (until recently, for some denominations), their defective nature as females was seen as making them incapable of representing this "son" as priests.[4]

Christianity early (by the second century) developed an ascetic understanding of spirituality that continued to define its ideals until modern times. Christian asceticism not only sees the body and the material world as ephemeral, destined to pass away very soon, but as inferior to the soul, as a temptation that draws the "manly mind down from its heavenly heights to wallow in the flesh,"[5] to quote a famous phase by Saint Augustine. Although not evil in itself, fixing one's affections on the corporeal world is the primary cause of sin or human alienation from God. The spiritual journey of redemption is directed toward an escape from material life into communion with a transcendental God to be perfected only after death.

Thus Christianity fused together two different patterns of human superiority to nature: domination over nature and negation or flight from nature. The first tradition of domination suggested unbridled mastery and exploitation of the material world, while the second tradition promoted a disconnection from the material world that denied one's (ruling-class males) actual relation to it and dependence upon it. This means one can use the world as one wills, but one need not take responsibility for it or even be aware of how one's use of it is harming it and harming those humans exploited in the process of exploiting the material world. This is a fatal combination and helps explain the destructive combination of abusive use and yet obliviousness to the results of this abuse, even on the bodies and daily lives of those supposedly benefiting from this use, typical of Western consciousness.

This combination of attitudes of mastery over the body and nature and disconnection from the body and nature that is used/abused is always deeply intertwined with domination of subjugated and despised groups of human beings who both do the work, that is, mediate the wealth extracted from nature to the ruling class/race/gender, and are also inferiorized by being identified with those roles. Domination of males over females plays a central role in this pattern of domination of subjugated others/nature, both ideologically and practically. That is, women are the primary group whose bodies are subjugated both for reproduction and for material labor, and sexual domination forms a central psychological/cultural lens through which all dominations of inferiorized others are culturally constructed.

The expression of this sexual symbolism of domination is blatant in both the biblical and Greek philosophical roots of Christianity. Plato in his creation treatise the *Timaeus* sees the soul as originating in the heavens, where it has a preincarnate vision of the good, the true, and the beautiful in its spiritual form. But the soul is then placed in bodies and given the task of dominating the inferior passions that arise from the body and thus preparing itself for "mortification," that is, the ultimate separation of the soul from the body so it can return to its true home in the heavens. Thus for Plato the soul neither originates from the earth nor has its true home on earth. The embodied state is only a brief interlude and testing place between an original and a final disincarnate life in the heavens.

If, however, the soul fails to dominate the body and its passions and instead succumbs to them, it will be reincarnated in a woman or a brute that resembles the low estate to which it has fallen. It must then go through a continual cycle of reincarnations until it rises to its highest embodiment as a male ruling-class person (Greek) and is then able to doff the body to return to its "native star" (Plato, *Timaeus* 42). Plato's view of reincarnation seems to have belonged to a worldview of rising asceticism that spanned the ancient world from India to Greece and is found in a similar form in Hinduism and Buddhism. In these traditions, too, incarnation as a female is seen as punishment for past sin and itself merits the abusive treatment accorded to women in the society.[6]

Aristotle added to this Platonic denigration of women a second perspective that would also be deeply incorporated into Christian theological anthropology. For Aristotle, women were innately defective, lacking the fullness of humanity in mind, will, and body. They lack the fullness of rationality or mental capacity, of will or capacity for self-control, and of strength of body. Thus they are inherently nonnormative. Only the male represents normative humanness.

This defectiveness of women Aristotle traces to the very biological processes by which the male and the female are conceived. For Aristotle the male seed provides the form of humanness, normatively male. The female is merely the gestator who provides the material stuff formed by the male seed. But when the female matter is insufficiently formed by the male seed, a defective human or female is born. Aristotle, in effect, denies the existence of the female ovum as part of the genetic basis of the fetus. The male alone is the active source of human life. The mother is merely its passive incubator (Aristotle, *Generation of Animals* 2.4.20–21).

Aristotle also makes slavery a central metaphor for all relations of domination by which the ruling-class (Greek) male makes instrumental use of subjugated people and things. All relations of domination are seen as ones where a male ruling class, that alone possesses autonomous mind and will, appropriates and uses the bodies of others as "tools." Women, barbarians, slaves, and animals are thus reduced to mindless instruments incapable of ruling themselves and fit primarily to be ruled over and used by this male ruling class (Aristotle, *Politics* 1.5). In the renewal of slavery in sixteenth- and seventeenth-century European colonialism and its appropriation of the lands and bodies of African and indigenous peoples as slaves and serfs to work the land, Aristotle's views played a major role. Class, race, and gender domination and the domination of body and nature are thus integrally linked within the ruling metaphor of mastery and reduction to servile instrumentality.

Christianity not only appropriated these Platonic and Aristotelian perspectives as integral to its view of women, but it also married these with a misogynist interpretation of the Genesis story of Adam and Eve's creation and fall. In the New Testament letter of 1 Timothy, attributed to Paul, the Genesis 2–3 story of the creation of Eve, her seduction of Adam into sin, and their expulsion from the garden is interpreted as a woman-blaming myth of the origin of original sin. This myth of Eve, further elaborated by Saint Augustine, would form the basis of Western Christian teaching about women until the present time. Augustine argued that woman, although possessing a soul and hence a capacity for spiritual salvation, was in her female and bodily nature created by God to be subjugated to the male. In God's original and intended order of things, woman would have accepted her subjugated status and thus there would have been harmony in the human family and in relation to God.

But the woman disobeyed, thus inverting the proper relation of male and female as ruling head and subjugated body. The male also sinned in accepting this improper initiative by his wife, thus causing humanity as a whole to fall into sin, which Augustine defines as both "carnality" (the rule of the passions over the mind) and loss of free will (the ability to love God before Self). The restoration of "order" thus demands that woman be coercively put back in her "place" as a subjugated being or, better, converted to accept her subjugation as both her proper place in the order of creation and her due punishment for having caused the fall into sin. Although for Augustine the women who accepts her subjugation may also be saved and will become spiritually "equal" to the male when sexuality and gender roles are overcome in heaven, she attains salvation only through the path of repentant submission to her own servitude.[7]

The lens of sexual domination as the rule of mind over body, sovereign will over inferior groups, seen as innately inferior yet also rebellious and to be reduced by force to submission, deeply shapes the Western view of a variety of others. This double take on the inferiorized others, as both naturally subordinate and naturally insubordinate, shaped a tripartite ideology of these others. That is, the myths of the other are split into three views: the other as natural brute, the other as demonic adversary, and the other as romanticized auxiliary or helper. I will briefly summarize these three mythic patterns in relation to five classes of negated others in Christian Western culture: women, Jews, African Americans, "Indians" or native

Americans, and homosexuals. Finally I will show how this tripartite pattern is played out in relation to nature.

These three mythic identities are evident in Christian views of woman. First of all woman is Eve, the woman created to be subjugated to the male from the beginning, lacking the fullness of humanity in mind, will, and body, unable to be a autonomous human being, much less one who can lead others, shaped to be the submissive wife, helpmeet of the male. But Eve is also the rebellious wife who overturned her proper place. When she persists in her rebelliousness and refusal of submission, she becomes the witch, the helpmeet of the devil, the channel of the demonic whose proper fate is to be destroyed by fire.

On the other hand, as totally submissive wife and mother, one who gives life to the male redemptive child while somehow remaining miraculously untouched by the pollution of sexuality, she is Mary, the virgin Mother, the ideal embodiment of the spiritual feminine whom women are called to emulate, but, of course, can never literally imitate.

The Christian myth of the Jew is compounded of a similar tripartite pattern. First, Jews are people of the "old law" which cannot redeem, who have been superseded by the people of the New Testament, the dispensation of grace that frees humanity from the rule of the law. The relation of Christianity and Christians to Judaism and Jews is construed in the church fathers as a variant on the dualisms of body/spirit and materiality/spirituality. Jews understand only the exterior, the bodily meaning. Their religion is one of materiality, in contrast to Christians who have been given the redeeming spirit.[8]

But Jews are also crafty and demonic aliens, Christ-killers who continually conspire against Christianity and Christians, members of the synagogue of Satan who plot to install the devil as world ruler. Yet, they are also the chosen people of God whose heritage the church appropriates and makes its own, the source of the prophets who predict the coming Christ, and finally Christ himself through whom the world is redeemed from evil.

The African peoples brought to the New World as slaves were also divided into these three forms of the "mythic gaze" by their white Christian masters. First, it was assumed that these Africans were simply childlike beings that could never grow up into autonomous adults. Lacking adequate intellect and self-control, they are primarily suited to be servile labor. This assumption was also extended to various other "brown" people whom the United States sought to colonize in its history, such as Mexicans and Filipinos.

But African Americans, especially males, were also viewed as dangerously rebellious, desiring not only to throw off white control, but also to sexually desecrate the white man's most precious possession, the white woman. Thus the African American male in rebellion is the monstrous rapist, the most heinous of criminals appropriately dealt with by castration and lynching. But, on the other hand, when properly submissive, purged of all resistance to white rule, the elderly

African American male becomes gentle Uncle Tom, who along with his wife Mammy, loves and nurtures the white child in preference to their own children.

The native Americans whom the Christian settlers of the Americas sought to displace from the land (which they claimed to have been given to them by God as their promised land) were likewise split into three different perspectives. First, they were seen as primitives, people incapable of full humanity, worshiping idols, lacking authentic culture. Perhaps they might be redeemed if they became totally submissive children under the tutelage of their new Christian masters.

But the Indian in rebellion became the "devil Indian," the incarnation of an alien demonic world that must be totally purged in order to redeem the land for God's true people. On the other hand, the Indian could also be seen as the "noble savage," remnants of an innocent humanity in paradise before the fall, from whom the white "man" might learn a certain nobility and skills in hunting and tracking the wilderness, but destined to vanish before the triumphant expansion of God elect.[9]

Homosexuals or gays as others have been subjugated to the mythic gaze more recently in American thought, but have emerged in the last thirty years as major ideological adversaries of the Christian fundamentalist agenda of a redeemed America. First, gays can be looked at as simply "disordered" persons who have failed to mature, to rightly develop to normative heterosexuality. Thus the Christian fundamentalist can look on them as people on whom to exercise compassion, extending to them the possibility of conversion from immature homosexuality to mature heterosexuality.

But as persons who persist in their error, who even desire to legitimize and sanctify their relationships as committed love and marriage rivaling heterosexuals, they become demonic aliens, so-called perverts whose very existence debilitates our army and threatens our children. Such people should be purged from the army and the schools, where they might contaminate the innocent and debilitate the virile.[10] No full-blown romantic myth of the gay exists in Christian culture, although the making of it might be found in the idea of the gay as the "sensitive, feminine" man, the hairdresser and fashion designer for wealthy women, the artist and poet who lacks real manliness, but can offer some gifts of creativity and esthetic adornment. (These myths focus only on the gay male, ignoring lesbians.)

Finally how does the tripartite mythic gaze shape the views of nonhuman nature? Here too we find a similar splitting into three perspectives. First of all nonhuman nature is simply inferior to the human, lacking rationality or subjectivity, not made in the image of God, which is reserved for the human, not even a place where God is present (contra false nature religions). Nonhuman nature, animals, plants, land are given to us for our use. They are resources to be eaten, to be exploited, but not a part of a living community into which we can enter in I-Thou relationship. Yet, as fallen nature the devil lurks in the dark corners of the natural world, tempting us into brutishness and carnality. But nonhuman nature is also Mother Nature, the original paradise that still beacons us in beautiful scenery, a place of momentary memories of a lost past of blissful nurture, comfort, and happiness.

Having explored these pathological patterns of the relationships of domination toward various human others and toward the ultimate other, the nonhuman world, what are the chances for a healthy alternative, a genuine mutuality and community between people and between humans and the earth? White, in his famous article referred to earlier, insisted that we need to find an alternative religious worldview to heal the ecological crisis: "Since the roots of our trouble are so largely religious, the remedy must also be essentially religious."[11] Since White was a historian of science, this is an interesting assertion. I take it to mean that the solution cannot be simply secular/scientific. The reason for this is that secular scientific thinking in the West has been modeled on mathematics and claims to be value-free and objective, rejecting esthetic, poetic, and ethical thinking. Such language is good at description but impoverished at prescription or motivation.

What is needed is a holistic vision that integrates accurate descriptions of cosmological and earth-based ecological relations with the feelings and ethical mandates of love, concern, and commitment that can call us to a new way of life. These are the traditional elements of religious language. This does not mean that it must or even can arise from the reform or renewal of one particular religion. Rather, humans of many religions need to grapple with the resources in their traditions for the needed vision. Perhaps what must eventually emerge is a new planetary consciousness that draws humans together across cultures to a life-giving way of being with each other and the planet, although one that would not negate but renew the values of their diversities.

As a contribution toward recovering positive ecological principles for a dialogue toward a new planetary consciousness, I will briefly explore two patterns of thought in the Bible (Hebrew scripture and New Testament): a covenantal view and a sacramental view. As the outset it is important to say that the Bible has been interpreted as resolutely antinature through a reading popularized in German biblical scholarship in the nineteenth and twentieth centuries. This scholarship read into the Bible their own sharp dualism between nature and history, the (false) gods of nature and the (true) God of history. This dualism distorts the biblical worldview.[12]

Although Hebrew scripture develops the historical dimensions of God's relation to creation, there is also a lively sense of God's interrelation with and presence in nature. God is seen as creating nature and not as an expression of it, but the nature that God creates is alive (animate) and enters into lively relation with God. God delights in God's creation, and the creatures return this rejoicing in joy and praise. Divine blessing inundates the earth as rain, and the mountains skip like a calf, the hills grid themselves with joy, the valleys deck themselves with grain; they shout and sing together for joy (Ps. 29:6; 65:9–13).

The modern Western nature-history split does not exist in the biblical worldview. Rather, all things, whether human wars between cities or whether rain that brings abundant harvest or drought that brings disaster to the people who depend on rain for their crops, are seen as events. In all these events the Hebrews saw the presence and work of God as blessing or as judgment. All such events have moral meaning. When enemies conquer a city and lead its people into exile, this might be read as divine judgment on human infidelity. When the people repent and return to fidelity to God, then justice and peace will reign, not only between humans but also between humans and animals, between humans and the well-being of the earth (Isa. 65:17–25).

Hebrew legislators developed a body of law to express God's commandments to maintain and restore right relations between humans and with the rest of creation. These laws mandate periodic rest and restoration of right relations between humans, animals, land, and God. This Hebrew understanding of God and creation rejects the aristocratic split between a leisure-class groups of divinities who mirror the ruling class and a humanity who serves the gods through slave labor, a view typical of ancient Near Eastern mythology.[13] In Genesis God is described as both working and resting, thereby setting the pattern for all humans, as well as for animals and the land, who together form the covenant of creation.

This pattern of work and rest was developed as a series of concentric circles, seven days, seven years, and seven times seven years. On the seventh day, not only the farmer, but "his" human and animal work force are to rest: "On the seventh day you shall rest, so that your ox and your donkey may have relief, and your home-born slave and your resident alien shall be refreshed" (Exod. 23:12). In the seventh year attention is given to the rights of the poor and to wild animals, as well as to the renewal of the land itself: "For six years you shall sow your land and gather its yield; but the seventh year let it lie fallow, so that the poor of your people may eat and what they leave the wild animals may eat. You shall do the same with your vineyard and your olive orchard" (Exod. 23:10–11). Slaves are to be set free and laborers should rest.

Finally in the Jubilee year, the fiftieth year, there is to be a great restoration of all relationships. Those who have lost their land through debt are to be restored to their former property. Debts are to be forgiven and captives set free. The earth is to lie fallow; animals and humans should rest (Lev. 25:8–55). All the accumulated inequities of the past two generations between humans, some of whom fell into debt, lost their land, and then fell into slavery to others, are overcome, putting the former debtors and slaves into a position of equal opportunity with the others who have benefited from their misfortune. Every household should have "their own vine and fig tree" or, as one might say in the American tradition, "forty acres and a mule." Exploitation of land and animals is to be overcome through a period of rest and renewal. Right balance between humans and with the land is to be restored.

This vision of periodic redemption, of restoration of right relations, underlies Jesus's language about "kingdom come" in the Lord's Prayer (Luke 11:2–4; Matt.

6:9–13), as well as his proclamation of the "acceptable year of the Lord, as "good news to the poor," "release to the captives," the setting "at liberty of those who are oppressed," in his inaugural sermon in his home synagogue in Nazareth (Luke 4:18–19). This vision of periodic cycles of redemption is more compatible with finitude and ecological realities than the radical notions of the millennium and a once-for-all apocalyptic end of history through which modern scholarship has read Jesus's concept of the kingdom of God.

The sacramental tradition of the Hebrew wisdom tradition and New Testament Christology and cosmology complements the covenantal tradition of sabbatical legislation and prophetic judgments and hopes. Not just the human community but the entire cosmos is the starting point for this tradition. The human mirrors the cosmic community as micro to macro, but also communes with the cosmic body. God bodies forth and is immanent within it. In some sense the cosmic community is God's sacramental body. God is incarnate in it and is the body of the cosmos, although not reduced to it. Strikingly the wisdom tradition saw this immanent manifestation of God in the cosmos as female:

> Wisdom . . . pervades and permeates all things.
> . . . .
> She rises from the power of God,
> the pure effluence of the glory of the Almighty.
> . . . .
> She is the brightness that streams from everlasting light,
> the flawless mirror of the active power of God
> and the image of his goodness.
> She is one but can do all things.
> Herself unchanging, she makes all things new;
> age after age she enters into holy souls
> and makes them God's friends and prophets.
> . . . .
> She spans the world from end to end
> and orders all things in Blessing. (Wisdom of Solomon 7:23–8:1)

In the New Testament this cosmogonic wisdom is identified with Christ. Jesus as the Christ not only embodies, in crucified form, the future king and redeemer (Messiah), but also incarnates and manifests the cosmogonic principles through which the universe was created, is sustained, redeemed, and reconciled with God. In the cosmological Christology found in the preface to the Gospel of John, the first chapter of Hebrews, and in some Pauline letters, Christ is understood as the Alpha and the Omega, the beginning and end of all things. As John puts it, "In the beginning was the Word [logos] and the Word was with God and the Word was God. He was in the beginning with God. All things were made through him and without him was not made anything that was made. In him was life, and the life was the light of men" (John 1:1–4).

In the letter to the Colossians, the divine word that dwelled in Christ is the same Logos which founded and sustains the world from the beginning: "All things have been created through him . . . and in him all things hold together." This same Logos manifest in Christ is now bringing the whole cosmos back into union with God: "In him all the fullness of God was pleased to dwell and through him to reconcile to himself all things whether on earth or in heaven" (1:16–20). Saint Paul in his sermon to the Athenians on the Areopagus gives the most pithy expression of this view of God as pervading and sustaining all things: "In him we live and move and have our being" (Acts 17:28).

These two views of the relation of God, humanity, and nature should not be seen as alternatives, but rather as dynamically complementing one another. On the one hand, all humans in their concrete locations in relation to one another as workers to produce the means of life and together with the land and animals on whom they depend must be seen as members of one covenant with God. The well-being of the whole and all members in it are interdependent. All must be taken into account; all must rest and be refreshed. Unjust relations between them must be periodically rectified to maintain right relations so that all have the means of life and none exploit and impoverish others for long.

Yet the God with whom this community of creation lives in covenant should not be seen as a tribal war god with whom a group of warriors enters into a military pact, but rather as the creativity and sustaining life of the cosmos itself, "in whom we live and move and have our being." We are bound together in community because this common divine life underlies the life of all in interdependence; none can exploit or destroy others without disrupting the foundations of life that sustains us all.

As theologian Gordon Kaufmann argues in his 2004 book *In the Beginning . . . Creativity*,[14] we can no longer make sense of an idea of God as a superhuman agent who stands outside the universe and intervenes in it. Rather we need to think of the divine as the creativity of the cosmos itself, a creativity which is not random, but extraordinarily nuanced to sustain possibility of the unfolding of the cosmos and the creation of life on earth. But the rejection of an anthropocentric and agental concept of the divine does not undermine, but indeed can broaden and strengthen an ecojustice ethic.

In a universe created by its own creativity, love, justice, and service to others remain the ethic by which we enable life to flourish, while their absence threatens the continuation of life on earth. If humans are to make the creative leaps to heal the ecological crisis, it will be by deepening the ethic of love and justice and widening its application, not just to all humans without exception, but also to the "wide orders of life," that is, plants, animals, air, water, and soil.

What are some of the ways forward toward a healing of the ecological crisis? I suggest three important steps or aspects. First, the ecological crisis needs to be seen not just as a crisis in the health of nonhuman ecosystems, polluted water, contaminated skies, threatened climate change, deforestation, extinction of species, important as all these realities are. Rather one needs to see the interconnections between the impoverishment of the earth and the impoverishment of human groups, even as others are enriching themselves to excess. The phenomenon in which 20 percent of humans own and benefit from 83 percent of the resources of the earth while 80 percent share the remaining 17 percent, with the poorest 20 percent living in deepest misery, is an expression of a human world crisis interconnected with a severely damaged ecosystem. Healing the ecosystem must be a holistic healing of humans to one another in the context of their ecosystems as one community of life. One must think of ecology and justice as integral parts of one system.

Second, a healed ecosystem—humans, animals, land, air, and water together—needs to be understood as requiring a new way of life, not just a few adjustments here and there. Traditional religions—Judaism, Islam, Daoism, Confucianism, and Hinduism—never saw themselves as a few prayers or rituals to be performed on holidays, but as a total way of life that defined all relationships between different groups of humans and in relation to the environment. Modern individualism might see such a way of life as oppressive, valuing as we do personal choice and the split of public and private. Yet clearly it should be recognized that we cannot adequately recycle our waste, compost our garbage to renew the soil, overcome the contamination of the air by ceasing to burn fossil fuels, and convert to sustainable energy, without a different system in which new technology and new social relations are knit together.

One needs pioneering ecological communities that demonstrate a new way of life in which the community as a whole lives in an ecologically sustainable way. This means a way of life that is integrated into the buildings and transportation they use, the way they produce their food and dispose of their wastes and also the myths, stories, rituals, and visions by which they live. But fairly quickly such demonstration communities need to move from the context of exceptional communes to local municipalities and then to bioregions and global systems. Shared religion vision and ethical mandates will be crucial for creating the spirit that would underlie the shaping of such ways of living together in community. Overcoming the split between public and private, between the sphere of woman and family in the home, and between the public systems of politics and economics would be an integral part of such a vision of community. Men and women together need to recognize that recycling garbage, conserving water, and not polluting the air are personal and political. They cannot be split into private domestic tasks and public responsibilities.

Finally one needs to nurture the emergence of a new planetary vision and communal ethic that can knit together people across religions and cultures. There is rightly much dismay at the role that religions are playing in right-wing politics and even internecine violence today. But we need also to recognize the emergence of new configurations of interreligious relations. What we find in the late twentieth and early twenty-first centuries is not so much groups split from each other by denominations and religions, as denominations and religions split within themselves. Progressive and Opus Dei Catholics have little in common. Feminist social-justice Methodists are at war with "Good News" Methodists, and so on across Christianity.

Likewise within Judaism and Islam and other religions there is emerging a sector of the religious community concerned with ecology and social justice, open to the equality of women and to ecumenical dialogue. This progressive interreligious community is evident at the meeting of the Parliament of the World's Religion which gathered in Chicago in 1993; in Cape Town, South Africa, in 1999; and in Barcelona, Spain, in 2004.[15] But the sort of Christians, Jews, Muslims, Buddhists, Hindus, Sikhs, and others meeting in a venue such as the Parliament of the World's Religions are different from and often at odds with conservative sectors of each of their religions that would reject such dialogue.

Perhaps what this situation reflects is the birth pangs of a new planetary consciousness that is manifest today as sharp divisions that split groups within the same cultures and religions. Some sectors of each culture and religion cling to old certainties and patterns of separation and even seek to reinforce them, while others see themselves as entering into a new vision, not against their traditions, but as a valid development of its true vision. What is striking is that while the conservatives in each camp have little in common with each other, as well as with the progressives of their own tradition, the progressives largely share a common vision on such matters as mutual ecumenical respect for different religions, feminism, and ecology, although they may use somewhat different language to express these common visions.

I suggest that this emerging commonality among progressives of many denominations, religions, and cultural contexts represents the emerging new planetary consciousness that is needed to rebuild a healed world. These commonalities need to be nurtured (and are being nurtured) with vast ramifications of networking and interconnected means of communication between all these different contexts of progressive thought, found in such gatherings as the World Social Forum.[16] In so doing we help to birth the new vision and way of life that is needed to heal life on earth.

NOTES

1. Lynn White, "The Historical Roots of Our Ecologic Crisis," *Science* 155 (10 March 1967): 1203–7; repr. in *This Sacred Earth: Religion, Nature, and Environment* (ed. Roger S. Gottlieb; New York: Routledge, 1996), 184–93.

2. Ibid., 189.

3. See, e.g., James Barr, "Man and Nature: The Ecological Controversy and the Old Testament," in *Ecology and Religion in History* (ed. David Spring and Eileen Spring; San Francisco: Harper & Row, 1974), 48–75.

4. This is the classic view of Thomas Aquinas, drawing on Aristotle's definition of women as biologically defective; see Kari Borreson, *Subordination and Equivalence: The Nature and Role of Women in Augustine and Thomas Aquinas* (Washington, DC: University Press of America, 1981), 141–334.

5. Augustine, *De sermone domini in monte* 15.41; see Kim Power, *Veiled Desire: Augustine on Women* (New York: Continuum, 1996), 160.

6. See Arvind Sharma, ed., *Women in World Religions* (Albany: State University of New York Press, 1987), 69, 113.

7. See Borreson, *Subordination and Equivalence*, 1–87.

8. On the dualistic patterns in the church fathers' views of the Jews, see Rosemary Ruether, *Faith and Fratricide: The Theological Roots of Anti-Semitism* (New York: Seabury, 1974), 117–81.

9. For American mythology toward the Indian, see Richard Slotkin, *Regeneration through Violence: The Mythology of the American Frontier, 1600–1860* (New York: Harper, 1996).

10. For the Christian right's ideology and campaign against homosexuals, see Rosemary Ruether, *Christianity and the Making of the Modern Family: Ruling Ideologies, Diverse Realities* (Boston: Beacon, 2000), 171–74.

11. White, "Historical Roots of Our Ecologic Crisis," 192–93.

12. These two biblical views of covenant and sacrament are elaborated in Rosemary Ruether, *Gaia and God: An Ecofeminist Theology of Earth Healing* (New York: Harper, 1992), 205–53.

13. On the master-slave analogy for gods and humans in the Babylonian creation myth, see ibid., 16–19.

14. Gordon Kaufmann, *In the Beginning . . . Creativity* (Minneapolis: Fortress, 2004).

15. The history and documents of the successive Parliaments of the World's Religions can be accessed at www.cpwr.org.

16. For a brief account of the meetings of the World Social Forum, see Rosemary Ruether, *Integrating Ecofeminism, Globalization, and World Religions* (Lanham, MD: Rowman & Littlefield, 2005), 139–41. Their website is www.forumsocialmundial.org.

CHAPTER 17

SCIENCE AND RELIGION IN THE FACE OF THE ENVIRONMENTAL CRISIS

HOLMES ROLSTON III

BOTH science and religion are challenged by the environmental crisis, both to reevaluate the natural world and to reevaluate their dialogue with each other. Both are thrown into researching fundamental theory and practice in the face of an upheaval unprecedented in human history, indeed in planetary history. Life on Earth is in jeopardy owing to the behavior of one species, the only species that is either scientific or religious, the only species claiming privilege as the "wise species," *Homo sapiens*.

Nature and the human relation to nature must be evaluated within cultures, classically by their religions, currently also by the sciences so eminent in Western culture. Ample numbers of theologians and ethicists have become persuaded that religion needs to pay more attention to ecology, and many ecologists recognize religious dimensions to caring for nature and to addressing the ecological crisis. Somewhat ironically, just when humans, with their increasing industry and technology, seemed further and further from nature, having more knowledge about natural processes and more power to manage them, just when humans were more and more rebuilding their environments, thinking perhaps to escape nature, the natural world has emerged as a focus of concern. Nature remains the milieu of culture—so both science and religion have discovered. In a currently popular vocabulary, humans need to get themselves "naturalized." Using another metaphor, nature is the "womb" of culture, but a womb that humans never entirely leave. Almost like God—to adopt classical theological language—nature is "in, with, and under us."

Religious persons regularly find something "beyond" such naturalizing, holding that nature is not self-explanatory. Believers point to deeper forces, such as divine presence or Brahman or emptiness (*sunyata*) or the Dao underlying. Religions may detect supernature immanent in or transcendent to nature, perhaps even more so in human culture, though some religions prefer to think of a deeper account of Nature, perhaps enchanted, perhaps sacred.

Scientists search for natural explanations; at least a methodological naturalism may be required as scientific method. But scientists are of mixed minds about whether, when some so-called natural explanation is discovered, explanation is over. At cosmological scales there is deep space and time; at evolutionary scales Earth is a marvelous planet, a wonderland lost in this deep space-time. Humans can seem minuscule at astronomical levels; they can seem ephemeral on evolutionary scales. Humans do not live at the range of the infinitely small or at that of the infinitely large, but humans on e\Earth do seem, at ecological ranges, to live at the range of the infinitely complex, evidenced both in the biodiversity made possible by genetics and in the cultural history made possible by the human mind. In humans, there is the most genesis—so far as we yet know.

Contemporary biologists have not only described but come to celebrate the diverse array of forms of life on Earth and then to lament that humans are placing so many of them in jeopardy. What to make of Earth, the home planet? What to make of who we are, where we are, what we ought to do? The "deep" thoughts about this "deep" nature are right here, in religious and scientific minds. What seems always to remain after science are the deeper value questions. After four centuries of Enlightenment and Western science, and with due admiration for impressive successes, the value questions in today's world are as urgent, sharp, and painful as ever—evidenced by the environmental crisis. There is no scientific guidance of life.

This analysis raises issues surrounding (1) value in nature, then asks about (2) the connections between science, conscience, and conservation. (3) Puzzling over attitudes toward evolutionary natural history, we view that life struggle as renewal and regeneration and, in more religious terms, as life's redemption, leading to reverence for life. Western monotheist traditions affirm the goodness of creation. (4) Is there something to be learned encountering Eastern and indigenous faiths? (5) We look at ecology as a science and its joining with human ecology, where the religious dimension is more evident. (6) That leads to questions about nature and human nature, the human place and possibilities. Humans are part of and yet apart from nature; we evolved out of nature, yet require a sustainable biosphere.

(7) Humans are moral agents, and, facing current development and a sustainable future, questions of environmental justice have become urgent. (8) Indeed, facing the next century Earth, the planet of promise, is a planet in peril. Science and religion will both be required for our salvation.

VALUE IN NATURE

Value is a frequently encountered term in evolutionary biology and ecosystem science—and this despite the "value-free" aspect of science about which we will worry below. "An ability to ascribe value to events in the world, a product of evolutionary selective processes, is evident across phylogeny. Value in this sense refers to an organism's facility to sense whether events in its environment are more or less desirable" (Dolan 2002: 1191). Adaptive value, survival value, is the basic matrix of Darwinian theory. An organism is the loci of values defended; life is otherwise unthinkable. Such organismic values are individually defended; but, ecologists insist, organisms occupy niches and are networked into biotic communities. At this point ethicists wonder whether there may be goods (values) in nature which humans ought to consider and care for. Animals, plants, and species, integrated into ecosystems, may embody values that, though nonmoral, count morally when moral agents encounter these. Perhaps also religious convictions can illuminate such values, those (as Judeo-Christian monotheists will say) of a good creation.

Environmental science informs any environmental evaluation in subtle ways. Consider some of the descriptive categories used of ecosystems: the *order*, *stability*, and *diversity* in these biotic *communities*. Ecologists describe their *interdependence* or speak of their *health* or *integrity*, perhaps of their *resilience* or *efficiency*. Biologists describe the *adapted fit* that organisms have in their niches, the roles they play. Biologists may describe an ecosystem as *flourishing*, as *self-organizing*. Strictly interpreted, these are just descriptive terms; and yet often they are already quasi-evaluative terms. Order, stability, diversity, interdependence, fitness, health, integrity are values too—perhaps not always so but often enough that by the time the descriptions of ecosystems are in, some values are already there and putting constraints on what we think might be appropriate human development of such areas (Keller and Golley 2000; Golley 1998).

At the same time, however much ecology reframes nature for our reevaluation, the deeper evaluative questions are still left open. In that sense, science cannot teach us what we most need to know about nature, that is, how to value it. At this point one may need to turn to religion or something like it. Bible writers, for example, have an intense sense of the worth of creation, what we today would call its value. Nature is a wonderland: "Praise the LORD from the earth / you sea monsters and all deeps, / fire and hail, snow and frost, / stormy wind fulfilling his command! / Mountains and all hills, / fruit trees and all cedars! / Beasts and all cattle, / creeping things and flying birds!" (Ps. 148:7–10). "You crown the year with your bounty; / the tracks of your chariot drip with fatness. / The pastures of the wilderness drip, / the hills gird themselves with joy, / the meadows clothe themselves with flocks, / the valleys deck themselves with grain, / they shout and sing for joy" (Ps. 65:11–13). Encountering the vitality on their landscapes, the Hebrews formed a vision of creation, cast in a Genesis parable about a series of divine imperatives that empower earth with vitality.

The brooding Spirit of God animates the earth, and earth gives birth: "The earth was without form and void, and darkness was upon the face of the deep; and the Spirit of God was moving over the face of the waters. And God said, 'Let there be'" (Gen. 1:2–3). "Let the earth put forth vegetation. . . . Let the earth bring forth living things according to their kinds" (1:11, 24). "Let the waters bring forth swarms of living creatures" (1:20). Earth speciates. God, say the Hebrews, reviews this display of life, finds it "very good," and bids it continue: "Be fruitful and multiply and fill the waters in the seas, and let birds multiply on the earth" (1:22). In current scientific vocabulary, there is dispersal, conservation by survival over generations, and niche saturation up to carrying capacity. After that, one has to go beyond science to say, "Amen, and so be it!"

Value in nature is recognized again when the fauna is included within the Hebrew covenant: "Behold I establish my covenant with you and your descendants after you, and with every living creature that is with you, the birds, the cattle, and every beast of the earth with you" (Gen. 9:9–10). The fallow fields and vineyards, for example, were to be open to the birds and beasts. In modern terms, the covenant was both ecumenical and ecological. It was theocentric, theologians might insist, but if so it was also less anthropocentric and more biocentric than traditional Jews or Christians realized. Noah with his ark was the first Endangered Species Project, despite the disruptions introduced by human evil. The science is rather archaic, but the environmental policy ("keep them alive with you"; Gen. 6:19) is something the U.S. Congress reached only with the Endangered Species Act in 1973.

The values in nature found by biologists can couple with the values in nature detected by prophets, sages, saviors. To continue with the monotheist tradition, although nature is an incomplete revelation of God's presence, it remains a mysterious sign of divine power. In the teachings of Jesus, "the birds of the air neither sow nor reap yet are fed by the heavenly Father, who notices the sparrows that fall. Not even Solomon is arrayed with the glory of the lilies, though the grass of the field, today alive, perishes tomorrow" (Matt. 6). There is in every seed and root a promise. Sowers sow, the seed grows secretly, and sowers return to reap their harvests. God sends rain on the just and unjust. "A generation goes, and a generation comes, / but the earth remains forever" (Eccles. 1:4).

True, Jesus says, "My kingdom is not of this world." Teaching as he did in the imperial Roman world, his reference, however, in "this world" is to the fallen world of the culture he came to redeem, to false trust in politics and economics (or science, we add), in armies and kings (or scientists). God loves the world, and in the landscape surrounding him Jesus found ample evidence of the presence of God. He teaches that the power organically manifest in the wildflowers of the field is continuous with the power spiritually manifest in the kingdom he announces. Nature, ever dynamic and changing, is valued for its glorious, divine creativity, in contrast to the political world with its misplaced values, in need of redemption.

Science, Conscience, and Conservation

Humans are a quite late and minor part of the world in evolutionary and ecological senses. Resulting from processes in natural history, they are one more primate among hundreds, one more vertebrate among tens of thousands, one more species among many millions. But there is also a way in which this last comes to be first. *Homo sapiens* is the first and only part of the world free to orient itself with a view of the whole, to seek wisdom about who we are, where we are, where we are going, what we ought to do. Such inquiries have classically been thought of as philosophical or religious. But, facing an ecological crisis, the question arises whether and how far human environmental ethics can be drawn from the scientific study of nature.

Scientists and ethicists alike have traditionally divided their disciplines into the realm of the *is* and the realm of the *ought*. By this division, no study of nature can tell humans what ought to happen, on pain of committing the naturalistic fallacy. Ecologists who claim to know what we ought to do, or theologians who claim to base ethics on ecology, may be violating the long-established taboo against mixing facts and values. This neat division has been challenged by ecologists and their philosophical interpreters. There is ambiguity: ecology, as we noticed above, reframes ways that we think about nature, but leaves deeper questions unanswered.

Though biologists (in their philosophical moments) are typically uncertain whether life arrived on Earth by divine intention, they are almost unanimous in their respect for life and seek biological conservation on an endangered planet.

Biologists find biological creativity indisputable, whether or not there is a creator. Biologists have no wish to talk theologians out of genesis. Whatever one makes of God, biological creativity is indisputable. There is creation, whether or not there is creator, just as there is law, whether or not there is a lawgiver. Biologists are not inclined, nor should they be as biologists, to look for explanations in supernature, but biologists nevertheless find a nature that is super! Superb! Biologists are taught to eliminate from nature any suggestions of teleology, but no biologist can doubt genesis.

Earth's impressive and unique biodiversity, evolved and created, warrants wonder and care. Anciently, the Hebrews marveled over the "swarms" (= biodiversity) of creatures that earth brings forth in Genesis 1, brought before man to name them (a taxonomy project!). Classically, theologians spoke of "plenitude of being." Contemporary biologists concur that Earth speciates with marvelous fecundity; the systematists have named, catalogued a far more vast genesis of life than any available to ancient or medieval minds. There is but one species aware of this panorama of life, a species at the same time jeopardizing this garden Earth.

While this seems congenially to couple the concerns of biology and religion, others are more cautious. They worry about that naturalistic fallacy and warn that the values surrounding the pursuit of science, as well as those that govern the uses to which science is put, are not generated out of the science. Science can, and often does, serve noble interests. Science can, and often does, become self-serving, a means of perpetuating injustice, of violating human rights, of making war, of degrading the environment. Where science seeks to control, dominate, manipulate either persons or nature or both, it blinds quite as much as it guides. Nothing in science ensures against philosophical confusions, against rationalizing, against mistaking evil for good, against loving the wrong gods. The whole scientific enterprise of the last four centuries could yet prove demonic, a Faustian bargain, and as good an indication as any of that is our ecological crisis (Rasmussen 1996; Ruether 1992).

How do persons rise from the *facts* of natural history, Earth's biodiversity as described by biology, to what *ought to be*, human caring for a valuable creation, as urged by religious faith? Whatever the uses to which science is put, better or worse, there are the quite commanding discoveries that science has made about the history of life on Earth. There is something awesome about an Earth that begins with zero and runs up toward five million to ten million species in several billion years, setbacks notwithstanding. The long evolutionary history, fact of the matter, seems valuable; it commands respect, as biologists recognize, even reverence, as theologians claim. When one celebrates the biodiversity and wonders whether there is a systemic tendency to produce it, biology and theology become natural allies. Perhaps this alliance can help humans to correct the misuses to which science has been put—with more respect and reverence for life.

THE LIFE STRUGGLE: REGENERATION, REDEMPTION, REVERENCE

Critics at this point caution again against too easy an alliance between biology and faith. There has been, and continues to be, conflict between Darwinian evolutionary biology and monotheistic faith. Living organisms must compete, pressed for adapted fit, struggling to hold a place against other lives. Survival is the name of the game. Nature is "red in tooth and claw." The process is prolific, but evolutionary history can seem tinkering and makeshift. One might respect the survivors for their achievement and adapted fit, but this blind struggle for survival is not a process one can reverence.

This dimension of struggle, however, can be fitted into a more inclusive perspective, which ecology adds to evolution. The community of life is continually regenerated, as well as creatively advanced, and this requires value capture as nutrients, energy, and skills are shuttled round the trophic pyramids. From a systemic point of view, this is the conversion of a resource from one life stream to another—the anastomosing of life threads that characterizes an ecosystem. Ecology traces the systematic interconversion of life materials—how nature re-cycles. Death *in vivo* is death ultimately; death *in communitatis* is death penul-timately but life regenerated over the millennia of species lines and dynamic biotic communities, continuing almost forever.

The idea of adapted fit also requires a niche, a place to be, and includes a life-support system. An ecology is a home. The currents of life flow in the interplay of environmental conductance and environmental resistance. An environment that was entirely hostile would slay all; life could never have appeared within it. An environment that was entirely irenic would stagnate life. The vital process is conflict and resolution. Take away the friction, the stress, and would the struc-tures stand? Would they move? The organism is tested for how much information it can contribute to the next generation. Survival of the fittest is survival of the senders. Suffering? Struggle? Yes. But if we may borrow a word from the Socratic philosophers, life is in "dialectic."

The evolutionary picture is of nature laboring in travail. The root idea in the English word *nature*, going back to Latin and Greek origins, is that of "giving birth." Birthing is creative genesis, which certainly characterizes evolutionary nature. Birthing (as every mother knows) involves struggle. Earth slays her chil-dren, a seeming evil, but bears an annual crop in their stead. Birthing is nature's orderly self-assembling of new creatures amidst this perpetual perishing.

In ecosystems, organisms are both challenged and supported. Every organism is what it is where it is, the "skin-out" environment as vital as "skin-in" me-tabolisms. Early ecologists favored ideas such as homeostasis and equilibrium.

Contemporary ecologists emphasize more role for contingency or even chaos. Others incline to emphasize self-organizing systems (autopoiesis), also an ancient idea: "The earth produces of itself [Greek 'automatically']" (Mark 4:28). Some find that natural selection operating on the edge of chaos offers the greatest possibility for self-organization and self-transformation.

Within the structural constraints and mutations available, the process optimizes adapted fit. There is much openness, emergence, surprise, struggle, loss, gain, or wandering. Natural selection is thought to be blind, initially in the genetic variations bubbling up without regard to the needs of the organism, some few of which by chance are beneficial, and also in the evolutionary selective forces, which select for survival, without regard to advance. Many evolutionary theorists insist that nothing in natural selection theory guarantees progress; most doubt that the theory predicts, or even makes probable, the long-term historical innovations that have occurred. Others think that the creative results are inherent in the system.

Whether biologists can find such selective principles, it seems that something is at work making the system fertile, prolific, sustaining development, combining both innovations and novelties with stabilities and regularities so as to perpetuate a swelling wave. This portrays a loose teleology, a softer concept of creation than that in classical monotheism; and yet one that permits genuine integrity and autonomy in the self-developing creatures. The classical theology of design will need reforming. What theologians once termed an established order of creation is rather a natural order that dynamically creates, an order for creating.

But any biology of randomness and bloody struggle will equally need reforming. The system historically uses pain for creative advance. Ecologists subsume struggle under the notion of a comprehensive situated fitness. With this, one begins to get a new picture painted over the old, although some of the old picture still shows through. Theologically speaking, this position is not inconsistent with a theistic belief about God's providence; rather, it is in many respects remarkably like it. There is grace sufficient to cope with thorns in the flesh (2 Cor. 12:7–9).

The better biological categories are those of values achieved, actualized, shared, and conserved in a natural history of dramatic creativity. Such a reinterpreted biology will be reasonably congenial to theology. The facts of the matter give sufficient cause to wonder about reverence for creation. Where there is creativity, the religiously minded have cause to wonder whether there may lurk a creator.

Even the secular naturalists will find that a prolife principle is overseeing this earthbound history of matter and energy; they may be moved toward respect, even reverence. Stephen Jay Gould, for example, found on Earth "wonderful life," if also "chance riches" (1989; 1980), and he was moved, among the last words he wrote, to call the earthen drama "almost unspeakably holy" (2002: 1342). Edward O. Wilson, a secular humanist, ever insistent that he can find no divinity in, with, or under nature, still exclaims: "The biospheric membrane that covers the Earth, and you and me, . . . is the miracle we have been given" (2002: 21).

Asking about respect for creation, critics of Western monotheism may reply that the problem is the other way round. Judeo-Christian religion has not adequately cared for nature because it saw nature as the object of human dominion. Famously, historian Lynn White laid much of the blame for the ecological crisis on Christianity, an attack published in *Science*, the leading journal of the American Association for the Advancement of Science (White 1967). God's command in Genesis 1 for humans to "have dominion" over nature flowered in medieval Europe, licensed the exploitation of nature, and produced science and technology that have resulted in an ecological crisis. Equally, of course, White was attacking science for buying into a secular form of the dominion hypothesis, but the original authorization, so he claimed, was religious. After the fall and the disruption of the garden earth, nature too is corrupted and life is even more of a struggle than before. Nature needs to be redeemed by human labor.

Christians, it may further be complained, are headed for heaven; they have little use for Earth. One would first assume that religious people, like everyone else, will protect the environment when they realize its importance to their own health and the health of their children. But religious conservatives of an apocalyptic bent believe that the end days are near. Why worry about conserving Earth if you are shortly to be taken up in rapture? Indeed the famine (from desertification, escalating populations), floods (from melting polar caps), and pestilence (rapidly spreading exotic diseases, drug-resistant pathogens) are signs of the tribulation foretold in the Bible. But such apocalyptic views hardly seem characteristic of mainstream Christianity.

Theologians have replied that dominion requires stewardship and care (Birch, Eakin, and McDaniel 1990; Cobb 1972; DeWitt 1998; Nash 1991; Northcott 1996; Fern 2002). Adam and Eve are also commanded to "till the garden and keep it" (Gen. 2:15). There are more positive senses of dominion. Adapting biblical metaphors for an environmental ethic, humans on Earth are and ought to be prophets, priests, and kings—roles unavailable to nonhumans. Humans should speak for God in natural history, should reverence the sacred on Earth, and should rule creation in freedom and in love. As we have already seen, ample biblical passages celebrate the goodness of nature and urge its respect. In fact, religious persons can bring a perspective of depth on nature conservation. They will see in forest, sky, mountain, and sea the presence and symbol of forces in natural systems that transcend human powers and human utility. They will find in encounter with nature forces that awful fill one with awe and are overpowering, the signature of time and eternity. In the midst of its struggles, life is ever "conserved," as biologists might say; life is perpetually "redeemed," as theologians might say. Or, to adapt a biblical metaphor: the light shines in the darkness, and the darkness has not overcome it. That natural history does command our respect and our reverence.

EASTERN AND INDIGENOUS FAITHS

Given that the monotheistic faiths, characteristic of Western peoples, are implicated in the ecological crisis, perhaps a turn to Asian faiths or to those of indigenous peoples will result in a better joining of religion and ecological science, with better proposals for addressing the environmental crisis.

In Daoism, for example, might the yang and yin suggest ecological harmony and cooperation? Ecosystems undergo periodic successions, cycles and rhythms, returnings. Everything results from the negentropic yang and the entropic yin, dialectically entwined. Might this account bring into better balance the human drive to dominate nature? Huston Smith diagnoses how the West has been on a wild "yang trip," evidenced in science and resulting in the ecological crisis, against which Daoism "throws its ounces on the side of the yin, but to recover the original wholeness" (Smith 1972: 80). The ecological crisis results from too much muscle, macho; the West needs a recovery of the feminine; we need to flow with nature, properly to attune ourselves to its rhythms in counterpoint. The Dao that is descriptive of nature becomes prescriptive for human behavior.

Some excess in Western religion (the dominion of humans) is driving the scientific view in both theory and practice (yielding analytic science and technological science) which can be corrected by an Eastern metaphysics (binary complementarity), moderating the arrogance operating in science. Then again, a Daoist ecology might be more muddle than model, like a Buddhist biochemistry, a Hindu meteorology—or like the Christian ecology we were just exploring. Used as a comprehensive explanation, the Dao conflates many processes that, outside of the persuasive influence of this paradigm, have no discoverable connection with each other in nature that the sciences said to be congenial with it have yet revealed. Male and female have little to do with wet and dry, with mountains and valleys, or with eating foods that grow above and below ground. The waxing of youths and the waning of the elderly is a different phenomenon from the sweet and the sour. The near omnipresence of sexuality in biology has nothing to do with succession in ecology. The yang and yin, some mystic force bonding complements, is only superficially congenial with ecology, and advice such as "More yin!" can only superficially orient humans how they ought to behave in encountering nature.

The first Buddhist commandment is the injunction to *ahimsa* ("noninjury"), reverence for life. The Buddhist analysis of the phenomenal world is of karma and rebirth; all things are forever born and reborn; they fit together like gems in a net. Nothing is despised, however lowly it might be; the *bodhisattvas* vow to enlighten the last blade of grass. Phenomenal things are intimately identified with the divine. All is one; one is all. Continuing its analysis, the phenomenal world is like a cycling wheel (*samsara*) driven by desire (*tanha*, "thirst"). Buddhism urges the control of desires, surely a requisite for any solution to environmental problems. Humans should stop their thirsts (abetted by Western dominance and consumerism) and find a more meaningful life of balance. The Dalai Lama has made repeated appeals for caring for nature and living in harmony with nature.

But again there are misgivings. The Buddhist natural history, like human history, is appearance (*maya*, "illusion") spun over emptiness (*sunyata*). As with the monotheists troubled by the character of the evolutionary process, the Buddhist tradition too may have difficulties knowing how much of the phenomenal world to embrace and how much to see through or transcend. The core ideas of emptiness and of extinction of desires (*nirvana*) do not initially seem promising for conserving the phenomenal world.

Moving from theory to practice, as the West has encountered an environmental crisis, the East has imported Western science and technology. The East now faces on its own soil the task of applying its religions to these sciences for an effective valuing of nature. But China has, officially at least, largely repudiated its classical past and turned to Western (if also Maoist) outlooks. Perhaps Buddhism can help resolve environmental problems in industrialized Eastern nations—Japan, Taiwan, or Korea, for example. So far, however, the results are not impressive.

Aboriginal worldviews are plural and diverse: Native Americans differ from Australian indigenous peoples, who differ from African peoples. Many such accounts are animist; nature is animated with spirits. Humans are part of such inspirited nature. Such peoples live in a sacred, almost an enchanted world, which they encounter with reverence. This combines with an intimate knowledge of local environments, since such peoples lived immediately in nature for centuries, a knowledge sometimes superior to that of Western ecologists. By contrast Western scientific views of nature have been mechanistic, although biologists and especially ecologists now prefer to think of earthen nature as organic system. Toward such nature, contemporary religious thinkers—so we have been arguing—advocate reverence also, but the character of such reference must be postscientific, and this cannot be the same as the aboriginal reverence, which is (to use a pejorative word) superstitious.

Conflicts have arisen over sacred sites (such as Devil's Tower in Wyoming or Uluru in Australia) or wildlife (sacrificing eagles or cougars in religious ceremonies). In practice, critics argue that native peoples were seldom ecologically noble savages; they did what they had to do to survive and were quick to adopt guns and horses and to modernize when offered Western technology. Their aboriginal accounts worked well enough over millennia, but they served as mythologies for local coping. They deserve respect. But these views are not likely to supply useful models to address contemporary environmental issues, such as loss of biodiversity, global warming, or global capitalism.

ECOLOGY AND HUMAN ECOLOGY

The term *ecology* is, etymologically, the logic of living creatures in their homes, a word suggestively related to "ecumenical," with common roots in the Greek *oikos*, the inhabited world. Human culture is entwined with nature; humans must have their ecology. An evident form of value in ecosystemic nature arises because humans, since they must be at home on Earth, have a great deal at stake in the condition of their ecosystems. To use a word that has come to center stage since the U.N. Conference on Environment and Development, humans require "sustainability" in their relations to natural systems (World Commission on Environment and Development 1987). Over 150 nations have endorsed sustainable development.

But whereas the U.N. Conference on Environment and Development focus was on "sustainable development," ecologists have insisted that the ultimate criterion is a "sustainable biosphere." The Ecological Society of America, for example, has made a sustainable biosphere a priority in its research: "Achieving a sustainable biosphere is the single most important task facing humankind today" (Risser, Lubchenco, and Levin 1991). Any sustainable human development must come within those more fundamental parameters. "Maximizing human development," even sustainably, returns to a dominion that exploits nature solely for human purposes. Humans ought rather to fit themselves into a sustainable biosphere, as members of a larger community of life on e\Earth. That is a better logic of our being at home on Earth.

Scientists turning to environmental policy often advocate ecosystem management. This promises to combine what ecosystems are, scientifically, with what we humans wish to do, employing them in our cultural stories. This appeals alike to scientists, who see the need for understanding ecosystems objectively and for applied technologies; to landscape architects and environmental engineers, who see nature as redesigned home; to developers, who like the idea of management; and to humanists, who desire benefits for people. Further, this seems balanced to politicians and environmental policymakers, since the combined ecosystem/management principle promises to operate at the systemwide level, presumably to manage for indefinite sustainability, alike of ecosystems and their outputs for human benefit. Such management sees nature as "natural resources" at the same time that it has a "respect-nature" dimension. Christian ethicists note that the secular word *manage* is a stand-in for the earlier theological word *steward* and also note the connections with biblical *dominion* as caring for a garden earth.

Ethicists have frequently thought of ethics as a social contract; religions teach people how to get along with people. Ecologists add that ethics needs also to be a natural contract, human responsibility for this ecological/ecumenical planet on which we reside. The U.S. Congress, for example, in the National Environmental

Policy Act expects that the ecological sciences can help the nation "to create and maintain conditions under which man and nature can exist in productive harmony" (U.S. Congress 1969: §101(a)). Can religious faith enter into this effort?

There are both problems and opportunities when religious ethicists look toward ecological science and wonder what (use) to make of it. An environmental ethic is foolish not to be informed by the best such science available. The success of an environmental policy does not depend merely on the cultural and religious values, the policy preferences, or the social institutions that drive the human actors. Success depends on coupling such prescriptive values with an environmental science that is descriptively accurate and operationally competent. On the other hand, there are many pitfalls and one has to proceed cautiously.

Ecology is a natural science, and one might make a mistake to think that the classical faiths knew anything about such science, any more than they knew astronomy or molecular biology. The British Ecological Society surveyed their members to rank the fifty most important concepts in ecology (Cherrett 1989). On that list one finds concepts such as the ecosystem, succession, the niche, habitats, food webs and trophic levels, carrying capacity, territoriality, keystone species, energy flow, life history strategies. None of these appears as such in the Bible. There is nothing about nutrient cycles or the Lotka-Volterra equations, which relate population size, the number of organisms that the environment will support, to time, growth rate, and carrying capacity. At this point, religious accounts and scientific accounts can seem quite far apart.

At the same time, ecologists do not have much grand theory, laws that are always and everywhere true all over the e\Earth, seemingly because of the complexity, dynamism, change, and openness in ecosystems. Ecology is a piecemeal science that can, despite any general principles, when it comes to specifics, be good only at generalizations of regional or local scope—what will happen in the Sonoran Desert in drought or to the mussels in the Tennessee River at certain pollution levels. But if this is true, then we might have dismissed the ecological wisdom embodied in classical and indigenous faiths too quickly. There is an important difference between ecology and astrophysics or cellular biology. Ecology is a science at native range. Perhaps the Bible writers did not know the Lotka-Volterra equations or the nitrogen cycle in the soil. But they did live at the pragmatic ranges of sower who sows, waits for the seed to grow, and reaps the harvest.

The Hebrews knew how to grow vineyards and olive trees; they knew how to prepare "wine to gladden the heart of man, / and oil to make his face shine" (Ps. 104:15), although they did not know the bacteria of fermentation, much less had they any knowledge of unsaturated fats in the olive oil. They knew to let land lie fallow, on Sabbath in the seventh year, to restore its productivity. Abraham and Lot, and later Jacob and Esau, dispersed their flocks and herds because "the land could not support both of them dwelling together" (Gen. 13:2–13; 36:6–8). The

Hebrews worried about livestock trampling and polluting riparian zones (Gen. 29:1–8; Ezek. 34:17–19). Residents on landscapes live immersed in their native range ecology. Academic ecologists might be too quick to think that the Hebrews knew no ecology.

Any science is an abstraction; that is, it achieves its successes by a "pulling away" (*abs-traction*) from concrete reality. The scientist detects generalities in the particulars. The Lotka-Volterra equations (which formalize Abraham and Lot's problem) take a part out of the whole, the lawlike or repetitively patterned aspect isolated out for the science, while there is in real nature law mingled with the particulars of the local environment: the pastures "from the Negev as far as Bethel" and on to where Abraham's tent was pitched "between Bethel and Ai" (Gen. 13:3) on which these nomads realized they were trying to keep too many sheep and goats. As we earlier noted, the textbook ecologist is likely to be able to learn something from any people indigenous to a landscape for centuries. Rural peoples in ancient Palestine might have been better in contact with nature than biological academics, who sit in offices and laboratories.

Passing from such human ecology to ethics, religious ethicists can with considerable plausibility make the claim that neither technological development, nor conservation, nor a sustainable biosphere, nor sustainable development, nor any other harmony between humans and nature can be gained until persons learn to use the earth both justly and charitably. Those twin concepts are not found either in wild nature or in any science that studies nature. They must be grounded in some ethical authority, and this has classically been religious.

One needs human ecology, humane ecology, and this requires insight more into human nature than into wild nature. True, humans cannot know the right way to act if they are ignorant of the causal outcomes in the natural systems they modify—for example, the carrying capacity of the Bethel-Ai rangeland. But there must be more. The Hebrews were convinced that they were given a blessing with a mandate. The land flows with milk and honey (assuming good land husbandry) if and only if there is obedience to Torah. Abraham said to Lot, "Let there be no strife between me and you, and between your herdsmen and my herdsmen" (Gen. 13:8), and they partitioned the common good equitably among themselves.

The big problem is not figuring out what intrinsic values are out there in wild nature and how much they count; the problem is rather dealing with the disvalues in humans—their irrationality, their greed, their shortsightedness, in short, their sinfulness. Sacred scriptures are books about how to live justly, not about how natural history works. The righteous life depicted in the Hebrew Bible is about a long life on earth, sustainable until the third and fourth generations. Whatever it has to say about heaven or life after death, the Bible is also about keeping this earthly life divine, godly, or at least human, humane, or righteous and loving. Any people who cope on a landscape for centuries will have some store of ecological wisdom, but that is not what we really turn to classical religious faiths to learn.

How *nature* works is the province of the physical and biological sciences. How *human nature* works is the province of religion, both how human nature does and how it should work. Religions emphasize the *human,* not the *ecology* side of the relationship.

NATURE AND HUMAN NATURE

Religions are about a gap between what *is* and what *ought to be* and how to close that gap. This often requires revealing how human nature functions and dysfunctions, works and fails to work, and how to reform or redeem this "fallen" nature. Nature can take care of itself; God will perennially regenerate, redeem the creation, as has happened for millennia. But in religion God is redeeming humans, and that is redemption of a different order. One does not turn to ancient scriptures to learn modern ecological science, but there might be classical insights into human character there, as vital today as they ever were. Humans face a perennial challenge. We must get oriented with regard to values. We have to face ultimate questions. On these issues, science is not so knowledgeable, and religion comes to the fore. What it means to be blessed, what it means to be wicked: these are theological questions. Humans must repair their broken wills, discipline innate self-interest, and curb corrupt social forces. One is not going to get much help here from ecology, any more than from astrophysics or soil chemistry. Science could be part of the problem, not part of the solution; we met earlier those claims that science is used for Western dominion over nature, and science is equally used for Western domination of other nations.

Biologists appear at this point with a further concern, deeply troubling. If the theologians have found in humans a tendency to self-interest, to selfishness, to sin, biologists concur. Now the biologists may indeed claim to know something about human nature; humans are innately selfish by Darwinian natural selection. The nature inherited in human nature is self-interested, and this, in an environmental crisis, may prove self-defeating. Theologians and biologists alike find too much in human nature that is irrational, blind. Although the conservation biologists celebrate earth's biodiversity, the sociobiologists (or, later, evolutionary psychologists) worry that the human disposition to survive, a legacy of our evolutionary heritage, has left humans too shortsighted to deal with the environmental crisis at the global level.

Such biologists hold that we are naturally selected to look out after ourselves and our families, perhaps also to cooperate in tribes or for reciprocal benefits. Beyond that, humans are not capable of more comprehensive altruism, considering the interests of others in foreign nations or in future generations, if this is at the expense to our own interests (Sober and Wilson 1998; D. Wilson 2002). Humans are not rational in any absolute or even global sense; they bend their reason to serve their interests, competitively against others—other nations, tribes, or neighbors—when push comes to shove. Hence the escalating violence and terrorism in today's world, often as not claiming their cause in the name of some faith. Humans inherit Pleistocene urges, such as an insatiable taste for sugar, salt, fats, traits once adaptive, but which today make obesity a leading health problem. Our global environmental problems—escalating populations, escalating consumerism, escalating capitalism, escalating nationalism, the rich getting richer, the poor poorer—result from such Pleistocene urges.

A further trouble is that many environmental problems result from the incremental aggregation of actions that are individually beneficial. Coupled with a long lag time for environmental problems to become manifest, this masks the problem in both nature and human nature. A person may be doing what would be, taken individually, a perfectly good thing, a thing he or she has a right to do, were he or she alone, but which, taken in collection with thousands of others doing the same thing, becomes a harmful thing. A good thing escalates into a bad thing. This is Garrett Hardin's tragedy of the commons (1968). Pursuit of individual advantage destroys the commons. Biologists may claim that humans are not genetically or psychologically equipped to deal with collective issues that upset individual goals (Ehrlich and Ehrlich 2004).

Religious believers may welcome enlightened self-interests, caring for family, patriotism, and mutually beneficial reciprocity. But they go on to insist that, while indigenous faiths may remain local or locals may bend a more comprehensive faith to their denominational interests, the major world faiths have been quite trans-tribal, transnational. Their ethics is quite ecumenical, for example, the Golden Rule or the Good Samaritan. They have spread widely around the globe; Christianity began in the Middle East but there are now more Christians in South America than in the Middle East. Reformed Christianity was launched in Switzerland, spread to Scotland and North America, and today there are more Presbyterians in Korea than in any other nation. These Korean Presbyterians have themselves sent out forty thousand missionaries to over one hundred countries. This is good evidence that religious faith can transcend narrowly self-calculating family, tribal, or national interests with comprehensive global concern for the salvation of others, genetically unrelated. If so, and if this salvation now comes to be seen to require caring for the Earth, sustaining the biosphere, the home for us all, religious faiths may already have on hand the commitment and resources for addressing global problems that require this larger, transgenetic, sense of community.

Religions do not teach that humans simply follow either nature or human nature. Nature does not teach us how we ought to behave toward each other. Compassion and charity, justice and honesty, are not virtues found in wild nature.

There is no way to derive any of the familiar moral maxims from nature: "One ought to keep promises." "Do to others as you would have them do to you." "Do not cause needless suffering." No natural decalogue endorses the Ten Commandments. Humans, if uniquely the wise species, are also uniquely the species that needs redemption. Religions may celebrate creation or struggle what to make of evolutionary history. But the real business of religion is salvation, mending the perennial brokenness in human nature.

Humans sin, unlike the fauna and flora. Religion is for people and not for nature, nor does salvation come naturally; even the earthly good life is elusive. Ultimately such salvation is beyond the natural, perhaps supernatural by the grace of the monotheist God, perhaps in some realization of depths underlying the natural, such as Brahman or *sunyata*. Meanwhile, whatever the noumenal ultimate, humans reside in a phenomenal world, which they must evaluate and in which they must live, hopefully redeemed or enlightened by their faiths.

ENVIRONMENTAL JUSTICE

Much concern has come to be focused on environmental justice; the way people treat each other is related to the way they treat nature. If humans have a tendency by nature to exploit, they will as soon exploit other people as nature. These are the underlying theological and ethical issues underlying global capitalism, consumerism, nationalism. The combination of escalating populations, escalating consumption, global capitalism, struggles for power between and within nations, and militarism results in environmental degradation that seriously threatens the welfare of the poor today and will increasingly threaten the rich in the future. The four critical items on our human agenda are population, development, peace, and environment. All are global; all are local; all are intertwined; in none have we modern humans anywhere yet achieved a sustainable relationship with our Earth. Our human capacities to alter and reshape our planet are already more profound than our capacities to recognize the consequences of our activity and deal with it collectively and internationally.

Today 80 percent of the world's production is consumed by 20 percent of its peoples, and 80 percent of the world's population is reduced to living on the remaining 20 percent; there is strife about this, and environmental degradation in result. What is the equitable thing to do? That is human ecology with a focus on the ethics, not the science. No ecological science can supply answers. Answers are not easy to come by from the Hebrew, Christian, Buddhist, Daoist, or Hindu scriptures, but in consensus these religions proclaim that justice and love are necessary parts of the answer.

Many of Earth's natural resources, unevenly and inequitably distributed, have to flow across national lines, if there is to be a stable community of nations. People have a right to water; that seems plausible and just. But consider the nations in relation to the hydrology of the planet: At least 214 river basins are multinational. About fifty countries have 75 percent or more of their total area falling within international river basins. An estimated 35 percent–40 percent of the global population lives in multinational river basins. In Africa and Europe most river basins are multinational. The word *rival* comes from the Latin word for river, *rivus*, those who share flowing waters. With escalating population and pollution levels, sharing water has become increasingly an international issue.

In an ethics that provides for a shared commons, the international fabric will have to be stable and dynamic enough that a nation which is not self-contained can contain itself within the network of international commerce. This involves living in a tension within a community of nations where there is access that redistributes resources across national lines sufficiently for nations to repair their own resource deficiencies in international trade. Unless such commerce can be arranged, the environment will suffer. Human rights to a decent environment, to a fair share of the world's resources and goods, will be denied. Insecurity, hunger, and a sense of injustice will breed despair and outrage that will find a voice in violence, war, terrorism. The classical religions claim that they alone among the human institutions have a deep-seated universal ethical concern adequate to address such transnational issues.

But demanding one's rights and fair share is only half the truth. If pushed to the whole, this pushing becomes as much part of the problem as part of the answer. Perhaps the deepest trouble is this forever putting ourselves first, never putting ourselves in place in the fundamental biosphere community in which we reside. If we ask "What is the matter?" the deepest problem may be the conviction that nothing matters unless it matters to us. That returns to the original sin, to the beast within us not yet elevated to humanity. That disrupts, first, our nations and our cultures; it disrupts, second, and with equal damage our life support systems on Earth. Our welfare, our well-being, is a matter of living in sustainable communities, human and natural; this flourishing requires policies and behavior that keep population and development in harmony with landscapes. It is going to be difficult to keep peace with each other, until we are at peace with our environment.

What we want is not just riches, but a rich life, and appropriate respect for the biodiversity on Earth enriches human life. Humans belong on the planet; they will increasingly dominate the planet. But we humans, dominant though we are, want to be a part of something bigger. Contemporary ethics has been concerned to be inclusive. Environmental ethics is even more inclusive. It is not simply what a society does to its poor, its blacks, slaves, children, minorities, women, handicapped, or future generations that reveals the character of that society, but also what it does to its fauna, flora, species, ecosystems, landscapes. Environmental justice needs to be ecojustice, as with the World Council of Churches' emphasis on "justice, peace, and the integrity of creation."

PLANET OF PROMISE, PLANET IN PERIL

Since the coming of science with its technology, since the invention of motors and gears in the mid-nineteenth century, giving humans orders of magnitude more power to transform the landscape, since the coming of modern medicine, there have been unprecedented changes in world population, in agricultural production, in industrial production, in transportation and communication, in economic systems with global capitalization, in military commitments—all these literally altering the face of the planet and threatening the future of the planet.

Earth is the only planet with this display of life, so far as we yet know. This natural history warrants respect, reverence. Managing a landscape that has reared up such a spectacle of life becomes a matter of ethics and religion as well as of science. Entering a new millennium, we face a crisis of the human spirit. In the twentieth century science flourished as never before, but left us with deep misgivings about the human relation to nature. In other centuries, critics complained that humans were alienated from God. In the twenty-first century, critics complain that humans are alienated from their planet. One may set aside cosmological questions, but we cannot set aside global issues, except at our peril. We face an identity crisis in our own home territory, trying to get the human spirit put in its natural place.

The geophysical laws, the evolutionary and ecological history, the creativity within the natural system we inherit, and the values these generate are, at least phenomenally, the ground of our being, not just the ground under our feet. Theologians may wish to demur that, noumenally, God is the ground of being, but "ground" is an earthy enough word to symbolize this dimension of depth where nature becomes charged with the numinous. Life persists because it is provided for in the ecological Earth system. Earth is a kind of providing ground, where the life epic is lived on in the midst of its perpetual perishing, life arriving and struggling through to something higher. Ultimately, there is a kind of creativity in nature demanding either that we spell nature with a capital *N* or pass beyond nature to nature's God. When Earth's most complex product, *Homo sapiens*, becomes intelligent enough to reflect over this earthy wonderland, everyone is left stuttering about the mixtures of accident and necessity out of which we have evolved. But one does not undermine presently encountered values by discovering that it had uncertain origins. We can remain puzzled about origins while we greatly respect what we now find on Earth. Nobody has much doubt that this is a precious place.

Earth could be the ultimate object of duty, short of God. And if one cannot get clear about God, there is ample and urgent call to reverence the Earth. Whether or not one detects here the brooding Spirit of God, nature has been brooding spirits; we ourselves are the proof of that. And that sets us brooding over our place and our responsibility in this place. In this sense evolution and ecology urge us on a spiritual quest. If there is any holy ground, any land of promise, this promising Earth is it.

Theologians claim that humans are made in the image of God. Biologists find that, out of primate lineages, nature has equipped *Homo sapiens*, the wise species, with a conscience. Ethicists, theologians, and biologists in dialogue wonder if conscience is not less wisely used than it ought to be when, as in classical Enlightenment ethics, it excludes the global community of life from consideration, with the resulting paradox that the self-consciously moral species acts only in its collective self-interest toward all the rest. Biologists may find such self-interest in our evolutionary legacy; but now, superposing ethics on biology, an *is* has been transformed into an *ought*. Ecologists and religious believers join to claim that we humans are not so enlightened as once supposed, not until we reach a more inclusive ethic.

Perhaps ecology is something of a piecemeal science, but that is testimony to how complex, diverse, intricate, open to possibilities the earthen ecosystem network really is; the evolutionary natural history on Earth is quite a "grand narrative," dislike this idea though the postmodernists may. Several billion years worth of creative toil, several million species of teeming life, have been handed over to the care of this late-coming species in which mind has flowered and morals have emerged. Ought not those of this sole moral species do something less self-interested than count all the produce of an evolutionary ecosystem resources to be valued only for the benefits they bring? Such an attitude today hardly seems biologically informed (even if it claims such tendency as our inherited Pleistocene urge) much less ethically adequate for an environmental crisis where humans jeopardize the global community of life.

Ecologists and theologians agree: humans need a land ethic. Anciently Palestine was a promised land. Today and for the century hence, the call is to see Earth as a planet with promise.

REFERENCES

Birch, Charles, William Eakin, Jay McDaniel, eds. 1990. *Liberating Life: Contemporary Approaches to Ecological Theology.* Maryknoll, NY: Orbis.

Callicott, J. Baird. 1994. *Earth's Insights: A Survey of Ecological Ethics from the Mediterranean Basis to the Australian Outback.* Berkeley: University of California Press.

Chapple, Christopher Key, ed. 2003. *Jainism and Ecology: Nonviolence in the Web of Life.* Cambridge: Center for the Study of World Religions, Harvard Divinity School.

Chapple, Christopher Key, and Mary Evelyn Tucker, eds. 2000. *Hinduism and Ecology: The Intersection of Earth, Sky, and Water.* Cambridge: Center for the Study of World Religions, Harvard Divinity School.

Cherrett, J. M. 1989. "Key Concepts: The Results of a Survey of Our Members' Opinions." Pages 1–16 in *Ecological Concepts: The Contribution of Ecology to an Understanding of the Natural World.* Edited by J. M. Cherrett. London: Blackwell Scientific.

Cobb, John B., Jr. 1972. *Is It Too Late? A Theology of Ecology.* Beverly Hills, CA: Bruce.

DeWitt, Calvin B. 1998. *Caring for Creation: Responsible Stewardship of God's Handiwork.* Grand Rapids: Baker/Washington, DC: Center for Public Justice.

Dolan, R. J. 2002. "Emotion, Cognition, and Behavior." *Science* 298:1191–94.

Ehrlich, Paul R., and Anne H. Ehrlich. 2004. *One with Nineveh: Politics, Consumption, and the Human Future.* Washington, DC: Island Press.

Fern, Richard L. 2002. *Nature, God, and Humanity: Envisioning an Ethics of Nature.* Cambridge: Cambridge University Press.

Foltz, Richard C., ed. 2003. *Worldviews, Religion, and the Environment: A Global Anthology.* Belmont, CA: Wadsworth/Thomson Learning.

Foltz, Richard C., Frederick M. Denny, and Azizan Baharuddin, eds. 2003. *Islam and Ecology: A Bestowed Trust.* Cambridge: Center for the Study of World Religions, Harvard Divinity School.

Giradot, N. J., James Miller, and Liu Xiaogan, eds. 2001. *Daoism and Ecology: Ways within a Cosmic Landscape.* Cambridge: Center for the Study of World Religions, Harvard Divinity School.

Golley, Frank B. 1993. *A History of the Ecosystem Concept in Ecology: More Than the Sum of the Parts.* New Haven: Yale University Press.

———. 1998. *A Primer for Environmental Literacy.* New Haven: Yale University Press.

Gould, Stephen Jay. 1980. "Chance Riches." *Natural History* 89.11 (Nov.): 36–44.

———. 1989. *Wonderful Life: The Burgess Shale and the Nature of History.* New York: Norton.

———. 2002. *The Structure of Evolutionary Theory.* Cambridge: Harvard University Press.

Grim, John A., ed. 2001. *Indigenous Traditions and Ecology: The Interbeing of Cosmology and Community.* Cambridge: Center for the Study of World Religions, Harvard Divinity School.

Hardin, Garrett. 1968. "The Tragedy of the Commons." *Science* 169:1243–48.

Hessel, Dieter, and Rosemary Radford Ruether, eds. 2000. *Christianity and Ecology: Seeking the Well-Being of Earth and Humans.* Cambridge: Center for the Study of World Religions, Harvard Divinity School.

Keller, David R., and Frank B. Golley, eds. 2000. *The Philosophy of Ecology: From Science to Synthesis.* Athens: University of Georgia Press.

McFague, Sallie. 1993. *The Body of God: An Ecological Theology.* Minneapolis: Fortress.

———. 1997. *Super, Natural Christians: How We Should Love Nature.* Minneapolis: Augsburg Fortress.

Nash, James A. 1991. *Loving Nature: Ecological Integrity and Christian Responsibility.* Nashville: Abingdon Cokesbury.

Northcott, Michael S. 1996. *The Environment and Christian Ethics*. Cambridge: Cambridge University Press.

Rasmussen, Larry L. 1996. *Earth Community, Earth Ethics*. Maryknoll, NY: Orbis.

Risser, Paul G., Jane Lubchenco, and Samuel A. Levin. 1991. "Biological Research Priorities—A Sustainable Biosphere." *BioScience* 47:625–27.

Rolston, Holmes, III. 1988. *Environmental Ethics*. Philadelphia: Temple University Press.

———. 1996. "The Bible and Ecology." *Interpretation: Journal of Bible and Theology* 50:16–26.

Ruether, Rosemary Radford. 1992. *Gaia and God: An Ecofeminist Theology of Earth Healing*. San Francisco: Harper.

Smith, Huston. 1972. "Tao Now." Pp. 62–81 in *Earth Might Be Fair*. Edited by Ian Barbour. Englewood Cliffs, NJ: Prentice-Hall.

Sober, Elliott, and David Sloan Wilson. 1998. *Unto Others: The Evolution and Psychology of Unselfish Behavior*. Cambridge: Harvard University Press.

Taylor, Bron, ed. 2005. *Encyclopedia of Religion and Nature*. London: Continuum.

Tirosh-Samuelson, Hava, ed. 2003. *Judaism and Ecology: Created World and Revealed Word*. Cambridge: Center for the Study of World Religions, Harvard Divinity School.

Tucker, Mary Evelyn, and John Berthrong, eds. 1998. *Confucianism and Ecology: The Interrelation of Heaven, Earth, and Humans*. Cambridge: Center for the Study of World Religions, Harvard Divinity School.

Tucker, Mary Evelyn, and Duncan Ryuken Williams, eds. 1997. *Buddhism and Ecology: The Interconnection of Dharma and Deeds*. Cambridge: Center for the Study of World Religions, Harvard Divinity School.

Ulanowicz, Robert E. 2004. "Ecosystem Dynamics: A Natural Middle." *Theology and Science* 2.2:231–53.

United States Congress. 1969. *National Environmental Policy Act*. 83 Stat. 852. Public Law 91-190.

White, Lynn, Jr. 1967. "The Historical Roots of Our Ecological Crisis." *Science* 155 (March 10): 1203–7.

Wilkinson, Loren, ed. 1991. *Earthkeeping in the Nineties: Stewardship and the Renewal of Creation*. Grand Rapids: Eerdmans.

Wilson, David Sloan. 2002. *Darwin's Cathedral: Evolution, Religion, and the Nature of Society*. Chicago: University of Chicago Press.

Wilson, Edward O. 2002. *The Future of Life*. New York: Knopf.

World Commission on Environment and Development. 1987. *Our Common Future*. Oxford: Oxford University Press.

....

CHAPTER 18

RELIGION AND ECOLOGY

Survey of the Field

....

MARY EVELYN TUCKER

THE environmental crisis has been well documented in its various interconnected aspects of resource depletion and species extinction, pollution growth and climate change, population explosion and overconsumption. Each of these areas has been subject to extensive analysis by scientists, recommendations by policymakers, and regulations by lawyers. While comprehensive solutions have remained elusive, there is a mounting consensus that the environmental crisis is both global in proportions and local in impact and that the health of humans and ecosystems is being severely affected.

Moreover, there is a dawning realization that the changes that humans are making on planetary systems are comparable to the changes of a major geological era. Indeed, scientists have observed that we are damaging life systems on the planet and causing species extinction at such a rate as to bring about the end of our current period, the Cenozoic era. No such mass extinction has occurred since the dinosaurs were eliminated sixty-five million years ago by an asteroid. Our period is considered to be the sixth such major extinction in earth's 4.7-billion-year history, and in this case humans are the primary cause. Having grown from two billion to six billion people in the twentieth century we are now a planetary presence devouring resources and destroying ecosystems and biodiversity at an unsustainable rate. This increasing damage to ecosystems was made abundantly clear in the Millennium Ecosystems Assessment Report issued in March 2005. We are making macrophase changes to the planet with microphase wisdom. We are not fully aware of the scale of the damage we are doing and are not yet capable of stemming the tide of destruction.

While this stark picture of mass extinction and its effects on the environment has created pessimism among many and denial among others, it is increasingly evident that human attitudes and decisions, values and behavior will be crucial for the survival and flourishing of numerous life-forms on earth. Indeed, the formulation of viable human-earth relations is of central concern for a sustainable future for the planet. Along with such fields as the natural sciences and social sciences, and in concert with ecological design and technology, religion, ethics, and spirituality are contributing to the shaping of such viable relations. Moreover, a more comprehensive cosmological worldview of the interdependence of life is being articulated along with an ethical responsiveness to care for life for future generations.

The Role of Religions: Cosmologies, Symbols, Rituals, and Ethics

The emerging field of religion and ecology is playing a role in this. That is because world religions are being recognized in their great variety as more than simply a belief in a transcendent deity or a means to an afterlife. Rather, religions are seen as providing a broad orientation to the cosmos and human roles in it. Attitudes toward nature thus have been significantly, although not exclusively, shaped by religious views for millennia in cultures around the globe.

In this context, then, religions can be understood in their largest sense as a means whereby humans, recognizing the limitations of phenomenal reality, undertake specific practices to effect self-transformation and community cohesion within a cosmological context. Religions thus refer to those cosmological stories, symbol systems, ritual practices, ethical norms, historical processes, and institutional structures that transmit a view of the human as embedded in a world of meaning and responsibility, transformation and celebration. Religions connect humans with a divine presence or numinous force. They bond human communities, and they assist in forging intimate relations with the broader earth community. In summary, religions link humans to the larger matrix of indeterminacy and mystery from which life arises, unfolds, and flourishes.

Certain distinctions need to be made here between the particularized expressions of religion identified with institutional or denominational forms of religion and those broader worldviews that animate such expressions. By worldviews we mean those ways of knowing, embedded in symbols and stories, which find lived expressions, consciously and unconsciously in the life of particular cultures. In this sense, worldviews arise from and are formed by human interactions with natural systems or ecologies. Consequently, one of the principal concerns of religions in many communities is to describe in story form the emergence of the local geography as a realm of the sacred. Worldviews generate rituals and ethics, ways of acting, which guide human behavior in personal, communal, and ecological exchanges. The exploration of worldviews as they are both constructed and lived by religious communities is critical because it is here that we discover formative attitudes regarding nature, habitat, and our place in the world. In the contemporary period, to resituate human-earth relations in a more balanced mode will require both a reevaluation of sustainable worldviews and a formulation of viable environmental ethics.

A culture's worldviews are contained in religious cosmologies and expressed through rituals and symbols. Religious cosmologies describe the experience of origination and change in relation to the natural world. Religious rituals and symbols arise out of cosmologies and are grounded in the dynamics of nature. They provide rich resources for encouraging spiritual and ethical transformation in human life. This is true, for example, in Buddhism, which sees change in nature and the cosmos as a potential source of suffering for the human. Confucianism and Daoism, on the other hand, affirm nature's changes as the source of the Dao. In addition, the death-rebirth cycle of nature serves as an inspiring mirror for human life, especially in the Western monotheistic traditions of Judaism, Christianity, and Islam. All religions translate natural cycles into rich tapestries of interpretive meanings that encourage humans to move beyond tragedy, suffering, and despair. Human struggles expressed in religious symbolism find their way into a culture's art, music, and literature. By linking human life and patterns of nature, religions have provided a meaningful orientation to life's continuity as well as to human diminishment and death. In addition, religions have helped to celebrate the gifts of nature such as air, water, and food that sustain life.

In short, religions have been significant catalysts for humans in coping with change and transcending suffering, while at the same time grounding humans in nature's rhythms and earth's abundance. The creative tensions between humans seeking to transcend this world and yearning to be embedded in this world are part of the dynamics of world religions. Christianity, for example, holds the promise of salvation in the next life as well as celebration of the incarnation of Christ as a human in the world. Similarly, Hinduism holds up a goal of *moksha*, of liberation from the world of *samsara*, while also highlighting the ideal of Krishna acting in the world.

This realization of creative tensions leads to a more balanced understanding of the possibilities and limitations of religions regarding environmental concerns. Many religions retain other-worldly orientations toward personal salvation outside this world; at the same time they can and have fostered commitments to social justice, peace, and ecological integrity in the world. A key component that has been missing in much environmental discourse is how to identify and tap into the cosmologies, symbols, rituals, and ethics that inspire changes of attitudes and actions for creating a sustainable future within this world. Historically, religions have contributed to social change in areas such as the abolitionist and civil-rights movements. There are new alliances emerging now that are joining social justice with environmental justice.

In alignment with these ecojustice concerns, religions can encourage values and ethics of reverence, respect, redistribution, and responsibility for formulating a broader environmental ethics that includes humans, ecosystems, and other species. With the help of religions humans are now advocating for a reverence for the earth and its long evolutionary unfolding, respect for the myriad species who share the planet with us, restraint in the use of natural resources on which all life depends, equitable distribution of wealth, and recognition of responsibility of humans for the continuity of life into future generations.

Clearly religions have a central role in the formulation of worldviews that orient humans to the natural world and the articulation of rituals and ethics that guide human behavior. In addition, they have institutional capacity to affect millions of people around the world. Religions of the world, however, cannot act alone with regard to new attitudes toward environmental protection and sustainability. The size and complexity of the problems we face require collaborative efforts both among the religions and in dialogue with other key domains of human endeavor, such as science, economics, and public policy.

RELIGIONS IN ENVIRONMENTAL
THOUGHT AND PRACTICE

The potential and actual contribution of religions to environmental discussions and action is being recognized by a variety of persons and institutions within academia and beyond. Within the academic community, scientists such as E. O. Wilson at Harvard and Paul Ehrlich at Stanford, as well as social scientists such as economist Richard Norgaard at Berkeley and anthropologist David Maybury-Lewis at Harvard, are acknowledging the important role of religion and values in environmental studies and policymaking. Environmental-studies programs at major research universities including Harvard, Yale, Columbia, Stanford, and Duke are seeking to complement their science and social-science orientation with a humanities approach that draws on religious values and environmental ethics.

Beyond academia, international civil servants and national policymakers are calling for religions to become involved with environmental protection and conservation. This effort ranges from U.N. leaders such as Klaus Toepfer, director of the U.N. Environment Program, and Mikhail Gorbachev, director of Green Cross International, to ministers of the environment such as Masoomeh Ebtekar in Iran and Juan Mayr in Colombia.

In addition, in many parts of the world, grassroots projects, such as tree-planting and river restoration, are illustrating the effectiveness of religiously based groups to initiate and support local environmental efforts. Some examples of this include clean-up of the Ganges and Yamuna rivers in northern India, the ordaining of trees to protect them by monks in Thailand, and the efforts to plant trees by an alliance in Zimbabwe of the traditional Shona people along with the African Zionist churches.[1] These grassroots efforts to insure the integrity of ecosystems are being recognized as essential to long-term peace and security of nations and local communities. This was especially highlighted when Wangari Maathai in Kenya was awarded the Nobel Peace Prize for her Greenbelt Movement of tree-planting that affirms the sacredness of creation.

Call for the Participation
of Religions

Religions were acknowledged by scientists in the early 1990s as having an important role to play in revisioning a sustainable future. They recognized the importance of religion as key repositories of deep civilizational values and indispensable motivators in moral transformation around consumption, energy use, and environmental protection. Two important documents were issued by scientists calling for collaboration with religious leaders, laypersons, and institutions.

One is the statement of scientists titled "Preserving the Earth: An Appeal for Joint Commitment in Science and Religion," which was signed at the Global Forum meeting in Moscow in January 1990. It states: "The environmental crisis requires radical changes not only in public policy, but in individual behavior. The historical record makes clear that religious teaching, example, and leadership are powerfully able to influence personal conduct and commitment. As scientists, many of us have had profound experiences of awe and reverence before the universe. We understand that what is regarded as sacred is more likely to be treated with care and respect. Our planetary home should be so regarded. Efforts to safeguard and cherish the environment need to be infused with a vision of the sacred."

A second document is called "World Scientists' Warning to Humanity." This was produced by the Union of Concerned Scientists in 1992 and signed by over two thousand scientists, including more than two hundred Nobel Laureates. The document also suggests that the planet is facing a severe environmental crisis and will require the assistance and commitment of those in the religious community. It states: "A new ethic is required—a new attitude toward discharging our responsibilities for caring for ourselves and for the earth. We must recognize the earth's limited capacity to provide for us. We must recognize its fragility. We must no longer allow it to be ravaged. This ethic must motivate a great movement, convincing reluctant leaders and reluctant governments and reluctant peoples themselves to effect the needed changes."

RESPONSE OF RELIGIOUS LEADERS AND COMMUNITIES

The response to these appeals was slow at first but is rapidly growing. It might be noted that some strong voices advocated a religious response over half a century ago. These included Walter Lowdermilk, who in 1940 called for an eleventh commandment of land stewardship, and Joseph Sittler, who in 1954 wrote an essay titled "A Theology for the Earth." Likewise, Islamic scholar Seyyed Hossein Nasr has been calling since the late 1960s for a renewed sense of the sacred in nature drawing on perennial philosophy. Lynn White's 1967 essay entitled "The Historical Roots of Our Ecologic Crisis" sparked controversy over his assertion that the Judeo-Christian tradition has contributed to the environmental crisis by devaluating nature. In 1972 theologian John Cobb published a prescient book titled *Is It Too Late?*

Over the last two decades some key movements have taken place among religious communities that have shown growing levels of concern and commitment regarding alleviating the environmental crisis. Some of these include the interreligious gatherings on the environment in Assisi under the sponsorship of the World Wildlife Fund in 1984 and under the auspices of the Vatican in 1986. The Parliament of World Religions held in Chicago in 1993 and in Cape Town,

South Africa, in 1999 issued major statements on global ethics embracing human rights and environmental issues. The 1993 statement on global ethics was formulated by Catholic theologian Hans Küng, who continues to pursue efforts in this regard through his institute in Germany. The 1999 Parliament in Cape Town issued a challenge to lead institutions (educational, economic, political) to participate in the transformation toward a sustainable future.

The Global Forum of Spiritual and Parliamentary Leaders held international meetings in Oxford in 1988, Moscow in 1990, Rio de Janeiro in 1992, and Kyoto in 1993, which had the environment as a major focus. Since 1995 a critical Alliance of Religion and Conservation has been active in England and in Asia for environmental protection and restoration. Similarly, the National Religious Partnership for the Environment has organized Jewish and Christian groups on this issue in the United States. The Coalition on Environment and Jewish Life (COEJL) has activated American Jewish participation in environmental issues. In August 2000 a historic gathering of more than two thousand religious leaders took place at the United Nations during the Millennium World Peace Summit of Religious and Spiritual Leaders, where discussion of the environment was one of four major themes.

Several major international religious leaders have emerged as strong spokespersons for the importance of care for the environment. Tibetan Buddhist leader, the Dalai Lama, and Vietnamese Buddhist monk Thich Nhat Hanh have spoken out for many years about the universal responsibility the human community has toward the environment and toward all sentient species. Church leaders, such as Anglican Archbishop Rowan Williams and Robert Edgar, president of the United States National Council of Churches, are pointing to environmental problems, such as resource use and climate change, as major ethical challenges. Greek Orthodox Patriarch Bartholomew has sponsored a series of symposia at sea that have brought together scientists, religious leaders, civil servants, and journalists to highlight the problems of marine pollution and fisheries depletion. These have included symposia on the Mediterranean, the Black Sea, the Adriatic, the Baltic, and the Danube River. The symposium on the Adriatic concluded in Venice with a joint statement signed by both the patriarch and Pope John Paul II on the urgent need for environmental protection and care for nature's resources.

It is now the case that most of the world's religions have issued statements on the need to care for the earth and to take responsibility for future generations. These statements range from various positions within the Western monotheistic traditions to the different sectors within Asian traditions of Buddhism and Daoism. By no means monolithic, they draw on different theological perspectives and ethical concerns across a wide spectrum. They reflect originality of thought in bringing religious traditions into conversation with modern environmental problems, such as climate change, pollution, and loss of biodiversity. Within the various denominations of Christianity, for example, the Protestant-based World Council of Churches has published treatises on "justice, peace, and the integrity of creation"; Greek Orthodox Patriarch Bartholomew has issued statements on destruction of the environment as "ecological sin"; the evangelical community has published letters and position papers calling for care for creation; the Catholic bishops of the Philippines issued a pastoral letter on the environment; and the American Catholic bishops have published several statements on ecology, including a letter on the Columbia River bioregion. These statements are being used as a moral call to engage in further action on behalf of the environment.[2]

INTELLECTUAL INFLUENCES ON RELIGION AND ECOLOGY

It is within this global context that the field of religion and ecology has emerged within academia over the last decade. While it is still a relatively new field that is defining its scope, the academic study of religion and ecology is drawing on other disciplines and thinkers to develop theoretical, historical, ethical, cultural, and engaged dimensions. Among many thinkers, some of the theoretical and historical foundations have been laid by key philosophers. These include Clarence Glacken, who developed a study of nature in Western culture, and Arne Naess, who drew on Baruch Spinoza and south Asian thought to elaborate a theory of deep ecology emphasizing the primacy of the natural world over human prerogatives. Other philosophers and ethicists such as Baird Callicott and Holmes Rolston have helped to develop the field of environmental ethics. The cultural dimensions are influenced by the work of anthropologists such as Julian Steward, who coined the term *cultural ecology* to describe the relations between the environment and the economic and technological aspects of society. Furthermore, anthropologist Roy Rappaport extended cultural understanding of the ways in which ritual sustains social life in specific bioregions. Geographers David Soper and Yi Fu Tuan have investigated the spatial and ecological characteristics of religion.

Historians such as Thomas Berry and William McNeill provided a perspective from world history for understanding the mutual influences involved in human interactions with ecosystems. Theologians such as John Cobb and Gordon Kaufman have brought together theoretical and engaged perspectives by suggesting ways in which Christian beliefs can be more effectively expressed theologically and in environmental action. Ecofeminists such as Rosemary Ruether, Sallie McFague, and Heather Eaton have illustrated the contested nature of the treatment of the earth and the exploitation of women. Ecojustice writers such as Robert Bullard, Dieter Hessel, and Roger Gottlieb have also made important contributions to understanding the linkages between social injustice and environmental pollution. In many of these thinkers the theoretical, historical, ethical, cultural, and engaged perspectives are not separate but mutually inclusive. It is, however, appropriate to distinguish these approaches as they are currently informing the emerging field of religion and ecology.

These approaches are animated by several key questions. Theoretically, how has the interpretation and use of religious texts and traditions contributed to human attitudes regarding the environment? Ethically, how do humans value nature and thus create moral grounds for protecting the earth for future generations? Historically, how have human relations with nature changed over time and how has this been shaped by religions? Culturally, how has nature been perceived and constructed by humans, and conversely how has the natural world affected the formation of human culture? From an engaged perspective, in what ways do the values and practices of a particular religion activate mutually enhancing human-earth relations? What are the contributions of ecofeminist or ecojustice perspectives to a sustainable future? These questions and others have been raised by individuals and groups as the field began to take shape over the last decade.

A comprehensive theoretical and historical examination of religious traditions and ecology was undertaken by historians of religions, theologians, and environmentalists in the Harvard conference series on World Religions and Ecology. At the same time constructive theologians such as Jay McDaniel and Catherine Keller and engaged ethicists such as Larry Rasmussen and Ronald Engel have been working to formulate an expansion of religious and ethical sensibilities across time (intergenerational) and across space (to include other species and the planet as a whole). In addition, there has emerged a lively exploration of responses to nature experienced as sacred outside the religious traditions in so-called nature religions by Catherine Albanese, Bron Taylor, and Graham Harvey.

The Emerging Academic Field
of Religion and Ecology

The emergence of an academic field of religion and ecology over the last decade is marked by a number of key efforts of individuals and groups. This includes conferences organized, forums created, websites constructed, books published, courses taught, and undergraduate and graduate programs created. All of this can be seen within the larger context of the humanities, which are now making significant contributions to environmental studies.

Harvard Conference Series

A three-year international conference series took place at Harvard Divinity School's Center for the Study of World Religions from 1996 to 1998. The goal was to examine the varied ways in which human-earth relations have been conceived in the world's religions. The project was launched to provide a broad survey that would help to ground a new field of study in religion and ecology. It was not intended to be exhaustive but rather suggestive of the wide variety of resources—intellectual and engaged—to draw on from the world's religious traditions. Recognizing that religions are key shapers of people's worldviews and formulators of their most cherished values, this research project uncovered a wealth of attitudes and practices toward nature sanctioned by religious traditions.

Acknowledging the gap between ancient texts and traditions and modern environmental challenges, it drew on a broad method of retrieval, reevaluation, and reconstruction. The intention was to avoid simplistic ecofriendly or apologetic readings of scriptures written in vastly different times and circumstances. The scholars were engaged in critically retrieving aspects of the religious traditions for reexamination and reevaluation in the contemporary context. This has been part of the dynamic unfolding of religions historically as they have struggled to balance orthodoxy with the urgencies of adapting to new circumstances or cultures. Religious traditions have never been monolithic, but rather have embraced a broad range of interpretive positions ranging from Orthodox to Reform. Discerning appropriate change and the abiding value of tradition has been an important part of the life of religious teachers for centuries. Jewish rabbis, Christian theologians, and Islamic imams in the West and Hindu pundits, Buddhist monks, and Confucian scholars in Asia have all been involved with interpretation of their respective traditions over time. The Harvard project drew on that ongoing process of discernment so as to move toward a constructive phase. In the constructive phase the scholars of the various religions could point toward actual or potential sources of ecological awareness and action from within the particular traditions.

The Harvard conferences were also designed to foster interdisciplinary conversations that drew on the synergy of historians, theologians, ethicists, and scientists as well as on the work of grassroots environmentalists. This synergy proved to be indispensable as it provided a dynamic open space for fresh conversations. An awareness emerged that religion and ecology was a new field of study that was being created in both dialogue and in an ongoing network of exchange. The openness of the discussions was also enhanced by the fact that there were no "experts," as participants were discovering new approaches together. A spirit of collaborative scholarship rather than individualistic research emerged naturally in the conferences. This was in part because participants realized there was not one way forward, but multiple possibilities that each of the religions might contribute. Moreover, there arose a remarkable sense that cooperative efforts for the future of the planet were more valuable than the claims to a superior perspective from one tradition or from one scholar. Individual traditions, scholars, and projects were seen as part of larger and long-term efforts for the flourishing of life on the planet for future generations.

Thus, with this sprit of engagement, from 1996 to 1998 over eight hundred scholars participated in a series of ten conferences examining the traditions of Judaism, Christianity, Islam, Hinduism, Jainism, Buddhism, Daoism, Confucianism, Shinto, and indigenous religions. The conferences were organized by Mary Evelyn Tucker and John Grim with a team of area specialists in the world's religions. Each of the conferences was designed to include a spectrum of positions ranging, for example, from Orthodox, Conservative, and Reform in Judaism to Catholic, Protestant, Orthodox, and evangelical in Christianity to Theravada, Mahayana, and Vajrayana in Buddhism. The conferences were also intended to embrace both historians and scholars of the traditions along with religious spokespersons for the traditions. Moreover, scientists, environmentalists, and activists as well as graduate and undergraduate students were invited. Each conference included plenary sessions for a broader public. A wide range of funders insured that participants could be brought from around the world. This attempt at breadth and inclusivity resulted in some remarkable gatherings and some inevitable challenges. The indigenous conference had representatives from every continent and from numerous ethnic groups. The Shinto conference was the largest gathering of Shinto priests and practioners ever to occur outside of Japan. The Islam conference, with representatives across the Islamic world, fostered lively discussions over differences between Sunni and Shi'ite interpretations of jurisprudence.

The edited papers from these conferences have been published in ten volumes by the Harvard Divinity School's Center for the Study of World Religions. The purposes of the conferences and books were as follows:

- To examine varied attitudes toward nature from the religions of the world with attention to the complexity of history and culture.
- To contribute to the articulation of functional environmental ethics grounded in religious traditions and inspired by broad ecological perspectives.
- To stimulate the interest and concern of religious leaders and laypeople as well as students and professors of religion in seminaries, colleges, and universities.

This research project assumed that religions could contribute toward a more sustainable future, but that multidisciplinary approaches were needed. With this assumption in mind, three culminating interdisciplinary conferences were held in the fall of 1998 at the American Academy of Arts and Science in Cambridge, at the United Nations, and at the American Museum of Natural History in New York. These conferences included scientists, economists, educators, and policy-makers as well as scholars from the various world religions. Journalist Bill Moyers interviewed the religion scholars to highlight the insights from their particu-lar perspectives for a sustainable future. Maurice Strong (secretary general of the Stockholm and Rio de Janeiro U.N. environmental conference) and Timothy Wirth (director of the U.N. Foundation) participated in the conferences. Other participants included Jane Lubchenco (past president of the American Academy of Arts and Science) from the field of science, Ismail Serageldin (of the World Bank) from economics and policy, and George Rupp (president of Columbia University) from higher education. Cultural historian Thomas Berry and cos-mologist Brian Swimme spoke from the perspective of the evolutionary story of the universe and our current environmental crisis.

Forums

At the October 1998 U.N. conference, the formation of the Forum on Religion and Ecology was announced. It had three objectives: to continue research in the area of religion and ecology, to foster the development of teaching in this area, and to encourage outreach within academia to interdisciplinary environmental-studies programs and outside of academia to religious and policy groups. It has since grown into a global network of some five thousand people and is coordinated by Mary Evelyn Tucker and John Grim. Since the initial Harvard conferences, the forum has continued its research agenda by organizing several other Harvard conferences: World Religions and Animals, which resulted in a volume from Columbia Uni-versity Press titled A Communion of Subjects;Ecological Imagination with leading nature writers, which resulted in collaborative work with Orion magazine; and World Religions and Climate Change, which resulted in a Daedalus volume in 2001.[3] In addition, the forum convened a multiyear seminar on cosmology and religion with scientists Brian Swimme, Ursula Goodenough, Barbara Smuts, George Fisher, and Terry Deacon. Moreover, to promote teaching religion and ecology the forum has sponsored workshops for college and high school teachers.

Internationally, the forum has encouraged outreach by organizing panels at the Parliament of World Religions in Capetown and Barcelona as well as at environmental conferences in Asia, Europe, and the Middle East. It has partici-pated in the symposiums on the Aegean and the Baltic seas convened by Greek

Orthodox Patriarch Bartholomew. It has worked with the U.N. Environmental Program on various projects, including participation in two symposia in Iran in 2001 and 2005. It has also been involved in the Earth Charter movement through workshops in North America and international conferences in South America and in Europe. Forum co-director, Mary Evelyn Tucker, was a member of the Earth Charter International Drafting Committee and is also a member of the Earth Charter International Council.

In 2003 a group of Canadian scholars, including Heather Eaton, James Miller, Anne Marie Dalton, and Stephen Scharper, formed a Canadian Forum on Religion and Ecology. They have been active in Canada in developing the field of study as well as in sponsoring talks and workshops and participating in public forums and radio discussions.[4] They are sponsoring a series of books on religion and ecology from the University of Toronto Press.

Websites

A website was created by the Forum on Religion and Ecology under the Harvard Center for the Environment to assist in fostering research, education, and outreach in the area of religion and ecology.[5] To encourage research there are annotated bibliographies of the literature on the world religions along with selections from sacred texts and environmental statements from the world's religious communities. There are posted examples of some one hundred grassroots religiously inspired environmental movements around the world that illustrate the engaged practices in this area.

To enhance teaching, the website contains introductory essays about each of the world's religious traditions and their environmental contributions. It posts syllabi and lists audiovisual resources. It links to the high school teachers' website in this area: religious studies in the secondary schools.[6]

To illustrate the importance of interdisciplinary dialogue in partnership with science, economics, and policy the website contains introductory sections on each of these areas. An annotated bibliography of the evolutionary and ecological sciences is posted, along with bibliographies of ecological economics and ecological ethics.

Publications

The academic literature has been growing rapidly, and interest among students at both the secondary and collegiate level has been robust. The ten-volume Harvard series World Religions and Ecology edited by Mary Evelyn Tucker and John Grim was published between 1997 and 2003. This involved key area specialists in the world religions and hundreds of scholars and environmentalists. Two years later another major multiyear project was completed with the publication of the two-volume *Encyclopedia of Religion and Nature* edited by Bron Taylor and Jeffrey Kaplan. This has been years in production, has involved hundreds of scholars, and makes an invaluable contribution in identifying the many approaches, topics, and movements included in religion and ecology.[7] A Society for the Study of Religion, Nature, and Culture has emerged out of this effort and the first conference was held at the University of Florida in April 2006.

In addition, two significant anthologies have been edited by Roger Gottlieb, *This Sacred Earth* (1995), and Richard Foltz, *Worldviews, Religion, and the Environment* (2003). Two journals were established: *Worldviews: Culture, Religion, and the Environment* and *Ecotheology*. *Worldviews* was edited for ten years by the environmental ethicist, Clare Palmer, and is now edited by the historian of religion, Christopher Key Chapple. It has published a wide range of articles ranging from world religions, environmental ethics, deep ecology, and ecojustice issues. Moreover, the published literature in each of the world's religions on this topic—both books and journal articles—has increased exponentially over the last decade. An important indication that the field of religion and ecology is now established within religious studies is the fact that the revised Macmillan *Encyclopedia of Religion* edited by Linsey Jones (2005) includes a new section on this area with a dozen articles on the topic.

Courses

Courses in religion and ecology are spreading across North America, and there are now several graduate programs offering concentrations in this area of study. These include the University of Florida at Gainesville, the University of Toronto, Drew University, and the University of Hawaii. Within the American Academy of Religion the Religion and Ecology Group has been active in sponsoring panels and convening planning meetings since 1993. High school teachers in private schools have taken up the field with enthusiasm and helped to sponsor workshops on the topic, develop courses, review books, and create a website.

Humanities Contributions

The emergence of the field of religion and ecology is part of a broader movement within other branches of the humanities over the last thirty years to contribute in various ways to environmental studies. Environmental philosophy and ethics have grown in size and significance. Nature writing—both prose and poetry—has developed so that an academic society and various journals (such as *Orion*) have emerged in this area. Environmental historians are researching topics ranging from studies of cities or bioregions (William Cronin) or countries (Vaclav Smil) to wider surveys of periods (John McNeill) or civilizations (Clive Ponting). The arts have also made significant contributions to reflection on the environment and human interaction with it. Many scholars involved in religion and ecology are in dialogue with other humanists who are exploring the contributions of the humanities to environmental studies and policymaking.

CHALLENGES TO RELIGION AND ECOLOGY

As the field of religion and ecology emerges within academia and beyond, it is clear that religions offer both promise and problems for ameliorating environmental issues. Religions have sustained human aspirations and energies for centuries, but they have also contributed to intolerance, violence, and fundamentalist views of various kinds. The world's religions may thus be seen as necessary but not sufficient for ecological solutions. Religions have their problematic and dogmatic tendencies and have been late in coming to recognize the scale and scope of the global environmental crisis.

There are limits, then, to what religions may contribute to solving environmental problems. One example of these limits is the issue of population. Some of the religious traditions have presented recurring obstacles to open discussion of certain kinds of birth control at U.N. population conferences. These religious groups are associated largely with Islam, Roman Catholicism, and evangelical Christianity. However, there is an alternative research project identifying a more plural approach among world religions to population control. Led by Daniel Maguire at Marquette University, this project is called the "Religious Consultation on Population, Reproductive Health, and Ethics."

Hence while noting that religions may at times be problematic there is also recognition that religions may bring a broadened ethical perspective to environmental issues. There is a felt need for creative humanistic and religious initiatives so as to formulate more interdisciplinary approaches to environmental science, policies, law, and economics. The directors of many environmental-studies programs at leading universities in the United States are understanding this and exploring ways to integrate religion and ethics into traditional science- and policy-based programs. At the same time these directors of environmental programs are trying to define the parameters of scholarship and public service.

Should environmental programs be simply centers of research? To what extent should they be arenas for debating public policy or even advocating certain environmental policy approaches?

Analogous questions are arising in the field of religion and ecology as it begins to define itself and seeks to be in dialogue with science and policy. Should religion and ecology simply be a scholarly field of historical or theoretical research apart from contemporary issues? How should it relate to science and policy concerns? Should it pursue engaged scholarship such as ecojustice? What, if any, is the role of advocacy within academia? Can academics be engaged scholars or public intellectuals in the environmental field within academia and beyond? These are potentially creative and healthy tensions that have emerged in the field of religion and ecology.

The pressing nature of the environmental crisis is urging some scholars within academia to become public intellectuals who are contributing to the understanding of environmental problems and pointing toward possible solutions. This debate on the role of academics engaged in environmental studies and policymaking crosses the disciplines from the sciences and social sciences to the humanities. Many people are calling on higher education and research universities to make a larger contribution to the solution of environmental problems. It is at this lively intersection between theoretical, historical, and cultural research and engaged scholarship that the field of religion and ecology is growing.

THE LIMITS OF SCIENCE AND POLICY

The field of religion and ecology is becoming well situated to make a contribution to interdisciplinary environmental studies within academia as well as to be in conversation with scientists and policymakers outside of academia. In analyzing the current global environmental situation, leaders from both science and policy fields are wondering why we have not made more progress in solving environmental problems. Over the last fifty years, the enormous contributions of science to our understanding of many aspects of environmental problems, both global and local, are being fully recognized.

However, while thousands of scientific studies have been published and then translated into policy reports, many experts have concluded that we have not made sufficient progress in stemming the loses of ecosystems and species. We are stymied by a range of obstacles, from lack of political will to unchanging human habits. For many environmentalists there is a growing realization that a broader sense of vision and values is missing.

Scientists are noting that dire facts about environmental problems, as overwhelming as they may be, have not altered the kinds of human behavior that are rapaciously exploiting nature. Nor have such facts affected human habits of addictive consumption, especially in the richer nations. Moreover, policy experts are realizing that legislative or managerial approaches to nature are proving insufficient to the complex environmental challenges at hand. One cannot simply legislate change or manage human nature.

In short, environmentalists are observing that while science and policy approaches are necessary, they are not sufficient to assist in transforming human consciousness and behavior for a sustainable future. They are suggesting instead that values and ethics, religion and spirituality are important factors in this transformation. This is being articulated in conferences, in books and articles, and in policy institutes like Worldwatch. Here is where the field of religion and ecology is beginning to make an important contribution both to environmental studies within the academy and to policy initiatives outside the academy.

RESPONSE OF POLICY GROUPS
AND SCIENTISTS

One such initiative has been promoted by the U.N. Environment Program, which has established an Interfaith Partnership for the Environment that for some twenty years has distributed thousands of booklets entitled *Earth and Faith* for use in local congregations and communities. It has encouraged the participation of religious leaders, scholars, and laypeople in conferences that it has organized. Klaus Toepfer, until 2006 the executive director, has called for environmental ethics and spiritual values to be more actively integrated into environmental protection. He draws on Hans Jonas's principle of responsibility as crucial for future generations. He notes that legal and compliance mechanisms are indispensable, but a more holistic approach to environmental issues is needed. He has suggested, for example, that resources such as water should not be seen as simply economically important for human use but also spiritually valuable. In this light, he cites the need to develop indicators for assessing not just market values but spiritual and ethical values as well. Toepfer has been instrumental in encouraging this broader ethical approach in many international conferences. These include two conferences that the U.N. Environment Program organized in Tehran in cooperation with the Islamic Republic of Iran in June 2001 and May 2005.

A conference in Lyon in 2001 chaired by Mikhail Gorbachev also reflected this search for broader ethical approaches to environmental problems. Its title was

"Earth Dialogues: Is Ethics the Missing Link?" This Earth Dialogue conference was followed by another in Barcelona in 2004 where religious and ethical issues were also prominent. While not looking for quick solutions or easy answers, many thoughtful people are observing that human motivation, values, and action are critical in making the transition to a sustainable future.

Think tanks such as the Worldwatch Institute in Washington, DC, are also realizing that statistics and alarming reports are not enough to help initiate the changes for an ecologically sustainable world. In the final chapter of the Worldwatch *State of the World* 2003 report, senior researcher Gary Gardner wrote of the growing role of religions in shaping attitudes and action for a broader commitment to environmental protection and restoration. His essay received significant attention, and the larger version of the chapter was published in a separate Worldwatch paper (no. 164) titled "Invoking the Spirit: Religion and Spirituality in the Quest for a Sustainable World."

Several prominent scientists and policymakers are recognizing that human values and ethical perspectives need to be part of the equation in environmental discussions. They are noting that arguments from "sound science" and computer models that draw on reams of data and statistics do not necessarily move people to action. Harvard biologist E. O. Wilson, in his book *On the Future of Life* (2003), observes the potential power of religious beliefs and institutions to mobilize large numbers of people for ecological protection. In this vein, James Gustave Speth, dean of Yale's School of Forestry, in his book *Red Sky in the Morning* (2004), acknowledges that ethics and values will need to play a larger role in environmental discussions.

Stanford biologist Paul Ehrlich voiced similar concerns in an address to the Ecological Society of America in August 2004. He observed that "for the first time in human history, global civilization is threatened with collapse." Thus, he suggests, "the world therefore needs an ongoing discussion of key ethical issues related to the human predicament in order to help generate the urgently required response." He observed that the Millennium Ecosystem Assessment Report was undertaking an evaluation of the conditions of the world's ecosystems. He noted, however, that "there is no parallel effort to examine and air what is known about how human cultures, and especially ethics, change, and what kinds of changes might be instigated to lessen the chances of a catastrophic global collapse." He called for the establishment of a Millennium Assessment of Human Behavior to address these problems.

In the thirty-year anniversary edition of *Limits to Growth* in 2004 Dennis Meadows and his colleagues observe that we need new "tools for the transition to sustainability." The authors admit, "In our search for ways to encourage the peaceful restructuring of a system that naturally resists its own transformation we have tried many tools. The obvious ones are rational analysis, data systems thinking, computer modeling, and the clearest words we can find. Those are tools that anyone trained in science and economics would automatically grasp. Like recycling, they are useful, necessary, and they are not enough" (Meadows, Randers, and Meadows 2004: 269). Instead, they suggest qualities beyond the usual frame of environmental science and policy in mapping the road toward sustainability, namely, in planning for a future that will sustain the life needs of humans and other species. The qualities for ensuring such a future include visioning, networking, truth telling, learning, and loving. These qualities indicate a major shift for social planners and policy-oriented environmentalists. The authors identified the importance of such "soft tools" in 1992 but now feel that they are not simply optional but rather essential for the transition to sustainability.

CONCLUSION

It is becoming increasingly clear that environmental changes will be assisted by a variety of disciplines in very specific ways: namely, scientific analysis will be critical to understanding nature's ecology; educational awareness will be indispensable to creating modes of sustainable life; economic incentives will be central to adequate distribution of resources; public policy recommendations will be invaluable in shaping national and international priorities; and moral and spiritual values will be crucial for the transformations, both personal and communal, required for the flourishing of earth's many ecosystems. All of these disciplines and approaches are needed. In this way, the various values, incentives, and knowledge that motivate human activity can be more effectively channeled toward long-term sustainable life on the planet. It is in this nexus that the field of religion and ecology is making important contributions, within academia and beyond.

NOTES

1. One hundred such projects are documented on the Harvard website: www.environment.harvard.edu/religion.

2. Many of these can be viewed on the Harvard website: www.environment.harvard.edu/religion.

3. See www.amacad.org/publications/fall2001/fall2001.aspx.

4. See www.cfore.org.

5. See www.environment.harvard.edu/religion.

6. See www.rsiss.net.

7. See www.religionandnature.com.

REFERENCES

Anderson, E. N. 1996. *Ecologies of the Heart: Emotion, Belief, and the Environment.* New York: Oxford University Press.

Barbour, Ian, ed. 1973. *Western Man and Environmental Ethic.* Reading, MA: Addison-Wesley.

Barnhill, David, and Roger Gottlieb, eds. 2001. *Deep Ecology and World Religions: New Essays on Sacred Ground.* Albany: State University of New York Press.

Berry, Thomas. 1988. *The Dream of the Earth.* San Francisco: Sierra Club.

———. 1999. *The Great Work.* New York: Bell Towers/Random.

———. 2006. *Evening Thoughts: Reflecting on Earth as Sacred Community.* (Sierra Club Books and University of California Press.

Boff, Leonardo. 1995. *Ecology and Liberation: A New Paradigm.* Translated by John Cumming. Maryknoll, NY: Orbis.

Callicott, J. Baird. 1994. *Earth's Insights: A Survey of Ecological Insights from the Mediterranean Basin to the Australian Outback.* Berkeley: University of California Press.

Callicott, J. Baird, and Roger Ames. 1989. *Nature in Asian Traditions of Thought: Essays in Environmental Philosophy.* Albany: State University of New York Press.

Chapple, Christopher Key, ed. 2003. *Jainism and Ecology: Nonviolence in the Web of Life.* Cambridge: Center for the Study of World Religions, Harvard Divinity School.

Chapple, Christopher Key, and Mary Evelyn Tucker, eds. 2000. *Hinduism and Ecology: The Intersection of Earth, Sky, and Water.* Cambridge: Center for the Study of World Religions, Harvard Divinity School.

Chryssavgis, John, ed. 2000. *Cosmic Grace, Humble Prayer: The Ecological Vision of the Green Patriarch Bartholomew I.* Grand Rapids: Eerdmans.

Cobb, John. 1995. *Is it Too Late? A Theology of Ecology.* Denton, TX: Environmental Ethics (orig. 1972).

Devall, Bill, and George Sessions. 1985. *Deep Ecology.* Salt Lake City: Smith.

Ehrlich, Paul. 2001. *Human Natures.* Washington, DC: Island Press.

Foltz, Richard C. 2003. *Worldviews, Religion, and the Environment: A Global Anthology.* Belmont, CA: Wadsworth.

Foltz, Richard C., Frederick M. Denny, and Azizan Baharuddin, eds. 2003. *Islam and Ecology: A Bestowed Trust.* Cambridge: Center for the Study of World Religions, Harvard Divinity School.

Giradot, N. J., James Miller, and Liu Xiaogan, eds. 2001. *Daoism and Ecology: Ways within a Cosmic Landscape.* Cambridge: Center for the Study of World Religions, Harvard Divinity School.

Glacken, Clarence. 1969. *Traces on the Rhodian Shore: Nature and Culture in Western Thought from Ancient Times to the End of the Eighteenth Century.* Berkeley: University of California Press.

Gottlieb, Roger S. 1995. *This Sacred Earth: Religion, Nature, Environment.* 2nd ed. New York: Routledge.

———, ed. 2003. *Liberating Faith: Religious Values for Justice, Peace, and Ecological Wisdom.* Lanham, MD: Rowman & Littlefield.

Grim, John A. 1983. *The Shaman: Patterns of Religious Healing among the Ojibway Indians.* Norman: Oklahoma University Press.

———, ed. 2001. *Indigenous Traditions and Ecology: The Interbeing of Cosmology and Community*. Cambridge: Center for the Study of World Religions, Harvard Divinity School.

Hessel, Dieter, and Rosemary Radford Ruether, eds. 2000. *Christianity and Ecology: Seeking the Well-Being of Earth and Humans*. Cambridge: Center for the Study of World Religions, Harvard Divinity School.

Jones, Lindsay, ed. 2005. *Encyclopedia of Religion*. 2nd ed. New York: Macmillan.

Kaufman, Gordon, D. 1993. *In Face of Mystery: A Constructive Theology*. Cambridge: Harvard University Press.

McDaniel, Jay. 1989. *Of Gods and Pelicans: A Theology of Reverence for Life*. Louisville: Westminster/John Knox.

McFague, Sallie. 1987. *Models of God: Theology for an Ecological Nuclear Age*. Philadelphia: Fortress.

Meadows, Donella, Jørgen Randers, and Dennis Meadows. 2004. *The Limits to Growth: The 30-Year Update*. White River Junction, VT: Chelsea Green.

Millennium Ecosystems Assessment Report. March 2005.

Naess, Arne. 1989. *Ecology, Community, and Lifestyle*. Cambridge: Cambridge University Press.

Nasr, Seyyed Hossein. 1996. *Religion and the Order of Nature*. Oxford: Oxford University Press.

———. 1997. *Man and Nature: The Spiritual Crisis in Modern Man*. Dunstable, England: ABC/Chicago: Kazi.

Peet, Richard, and Michael Watts. 1996. *Liberation Ecologies: Environment, Development, and Social Movements*. New York: Routledge.

Rapport, Roy A. 1969. *Pigs for the Ancestors: Ritual in the Ecology of a New Guinea People*. Oxford: Oxford University Press.

Rasmussen, Larry. 1997. *Earth Community, Earth Ethics*. Maryknoll, NY: Orbis.

Ruether, Rosemary. 1992. *Gaia and God: An Ecofeminist Theology of Earth Healing*. San Francisco: Harper.

Soper, David E. 1967. *The Geography of Religions*. Englewood Cliffs, NJ: Prentice-Hall.

Spretnak, Chalene. 1986. *The Spiritual Dimension of Green Politics*. Santa Fe: Bear Press.

Tirosh-Samuelson, Hava, ed. 2003. *Judaism and Ecology: Created World and Revealed Word*. Cambridge: Center for the Study of World Religions, Harvard Divinity School.

Tuan, Yi-Fu. 1974. *Topophilia: A Study of Environmental Perception, Attitudes, and Values*. Englewood Cliffs, NJ: Prentice-Hall.

Tucker, Mary Evelyn. 2003. *Worldly Wonder: Religions Enter Their Ecological Phase*. Chicago: Open Court.

Tucker, Mary Evelyn, and John Berthrong, eds. 1998. *Confucianism and Ecology: The Interrelation of Heaven, Earth, and Humans*. Cambridge: Center for the Study of World Religions, Harvard Divinity School.

Tucker, Mary Evelyn, and John Grim, eds. 2001. *Religion and Ecology: Can the Climate Change?* Special issue of *Daedalus* 130.4. Cambridge, MA: American Academy of Arts and Sciences.

Tucker, Mary Evelyn, and Duncan Ryuken Williams, eds. 1997. *Buddhism and Ecology: The Interconnection of Dharma and Deeds*. Cambridge: Center for the Study of World Religions, Harvard Divinity School.

White, Lynn, Jr. 1967. "The Historical Roots of Our Ecologic Crisis." *Science* 155 (March).

CHAPTER 19

THE SPIRITUAL DIMENSION OF NATURE WRITING

DAVID LANDIS BARNHILL

This world as a glorious apartment of the boundless palace of the
sovereign Creator, is furnished with an infinite variety of animated
scenes, inexpressibly beautiful and pleasing, equally free to the inspection and enjoyment of all his creatures. (Bartram 15)

I believe that spiritual resistance—the ability to stand firm at the
center of our convictions when everything around us asks us to
concede—that our capacity to face the harsh measures of a life,
comes from the deep quiet of listening to the land, the river, the
rocks. (Williams *Red* 17)

FROM 1773 to 1778, William Bartram, a trained botanist like his father John, explored the southeastern areas of North America, looking for biotic specimens. It
was a long and at times risky journey through areas that few Europeans had ever
seen at that time. Bartram's account of the expedition epitomizes nature writing's
combination of passionately pursued scientific knowledge about nature and a
personal intimacy and emotional involvement with the land.[1] As the quotation
above shows, this work, like so many in nature writing, overflows with an earth-
focused spirituality. And as is so often the case among nature writers, Bartram's
spirituality is eclectic. He gushes with romantic awe at the sublime, seeing nature as

an "earthly paradise" and "Elysian fields." At other places he exhibits the Enlightenment's vision of nature as manifesting the rationality of God, which supported his cataloguing of biota in Linnaeus's strict taxonomic categories.

The quotation from Terry Tempest Williams is also rooted in earth spirituality, although of a very different kind. Her spirituality is sensual and involves what she calls an "erotics of place." It also leads her to a "spirituality of resistance"[2] against the political hegemony of our time, exemplifying the religiously based political critique and activism that have become a common feature of contemporary nature writing.

A RANGE OF SOURCES

These two quotations hint at the rich complexity of the religious dimension of North American nature writing. Certainly the range of traditions that North American nature writers have drawn from has been highly diverse. Christianity has been one source, although in widely different ways. Susan Fenimore Cooper was in many ways orthodox in her nineteenth-century Christianity (especially in the claim of it being the one true religion), but she also articulated a twenty-first-century theology of stewardship. The works of Kenneth Rexroth and Annie Dillard are influenced heavily by the contemplative tradition of Christianity. Wendell Berry's writings present a radical Protestantism, while Terry Tempest Williams offers an unorthodox Mormon perspective.

Another common source for nature spirituality has been Asian culture. Starting in the 1940s, Rexroth's nature poetry showed an increasing influence of Hua-Yen and Tantric Buddhism as well as Chinese and Japanese literature. Gary Snyder, Peter Matthiessen, and Gretel Ehrlich are all practicing Buddhists, and Ursula Le Guin's science fiction and anarchist politics are informed by Chinese Daoism.

The other principal source for the religious dimension of nature writing has been Native American cultures. Some of our finest nature writers, such as N. Scott Momaday, Leslie Marmon Silko, and Linda Hogan, draw on their Native American religious traditions. In addition, several European American writers, such as Mary Austin, Gary Snyder, Terry Tempest Williams, and Richard Nelson, have incorporated a sophisticated understanding of native traditions in their works.

But not all nature writers have a specific tradition they identify with. Some draw on a wide variety of traditions. Thoreau considered himself to be an heir of all religious traditions. More recently Scott Russell Sanders has searched the sacred texts of world religions in order to explore the ground of being: "After combing through the Bible," he says, "I scoured the holy texts of other religions. Everywhere I looked, I came across rumors of the primal power" (*Staying Put* 130).

Others have what we might call a generic earth-centered spirituality. Rick Bass, for instance, speaks in general terms of holiness, sanctity, sacred place, and mystery: "The utter holiness of being alive and part of such a system," he exclaims, "the holiness of being allowed to be a lichen within the system" (219). Edward Abbey comes up with his own term for earth-centered spirituality. He raises the issue of the existence of a deity and replies: "Why confuse the issue by dragging in a superfluous entity? Occam's razor. Beyond atheism, nontheism. I am not an atheist but an earthiest. Be true to the earth" (*Desert Solitaire* 208).

This brief overview of sources for spirituality of nature writers suggests something of the diversity involved, but it fails to give us a sense of the specific content of their religious perspective. A number of themes are common in nature writing, and an exploration of them helps to show both the similarities and divergences in the spirituality of nature writing, what they share and where they travel distinct paths.

What Nature?

In one way or another, spiritual nature writers present nature as sacred. This simple statement leads immediately to two crucial questions: what is the nature that is considered sacred; and sacred in what way? In fact, "what is nature?" has become a thorny issue in both nature writing and environmental philosophy. While the question is quite complex, we can highlight three major trajectories in nature writing, three different modes of viewing nature as sacred. The first we might call "nature as sacred other." This term involves two different but interrelated qualities. The first is extraordinary: the distinctive, exceptional nature of a place, that which differentiates it from the world of normal, conventional life. The second is pristine: areas where human presence is absent or nearly so. Not all exceptional places are pristine in this sense, but it has been a major element in much nature writing to feel the presence of the sacred where the presence of humans is absent.

This idea of sacred other follows the perspective proposed by religious studies scholar Mircea Eliade, who argued that sacred space "constitutes a break in the homogeneity of space" (39), with certain areas ritually sanctified. Such an orientation toward nature's spirituality responds to what Henry David Thoreau called our need "to witness our own limits transgressed, and some life pasturing freely" (*Walden* 306).[3] Mountains, whether they are the Sierras of Muir and Snyder or the Himalayas of Matthiessen in *The Snow Leopard* and Ehrlich in *Questions of Heaven*, are often considered to have spiritual power, to be places where religious vision is cultivated. We are not talking here of Yosemite overrun with cars and cameras but places apart from all that. They move us with their sweeping majesty and with being both physically and psychologically above our normal lives. The desert has also served as sacred space, for Austin in *The Land of Little Rain*, Williams in *Red*, and, of course, Abbey in *Desert Solitaire*: "A weird, lovely, fantastic object out of nature like Delicate Arch has the curious ability to remind us—like rock and sunlight and wind and wilderness—that *out there* is a different world, older and greater and deeper by far than ours, a world which surrounds and sustains the little world of men as sea and sky surround and sustain a ship" (41–42).[4] Travel to foreign lands has also been an occasion of encountering the sacred other of nature. When visiting Japan, Gretel Ehrlich "wanted to see how and where holiness revealed itself, to search for those 'thin spots' on the ground where divinity rises as if religion were a function of geology itself: the molten mantle of sacredness cutting through earth like an acetylene torch, erupting as temple sites, sacred mountains, plains, and seas, places where inward power is spawned" (*Islands* 90).

These areas are not places to live in but to journey to and return from. Perhaps one can sojourn there a short while (Thoreau's two-year experiment at Walden, Henry Beston's year in his "outermost house" on the Atlantic coast, Abbey's "season in the wilderness"), but it is not a permanent home where human business is transacted—at least not a home for the nature writers. In some cases, however, indigenous peoples do live there—Tibetans, Paiute, or Inuits—and often nature writers present them, admiringly, as part of the other. In our postmodern and postcolonial age, this is problematic approach to take, for it could involve a modern-day notion of the noble savage or the idea, common among early European settlers, that the natives are just a part of wild nature, rather than being a complex people who inhabit a homeland that we are invading. However, some of the best nature writing, such as Lopez's *Arctic Dreams*, is characterized by serious and subtle reflections on cross-cultural encounters.

This perspective of nature's sacrality has much in common with the pilgrimage tradition. In a pilgrimage, one leaves one's normal life and social structure to venture to a sacred space far removed, and one then returns home changed and deepened.[5] But the tradition that is most relevant to nature as sacred other is wilderness: "Where the earth and its community of life are untrammeled by man, where man himself is a visitor who does not remain."[6] "God's wilderness," Muir called it, a place were creation works its beauty and mystery undiminished by humanity's many limitations and machinations. Again, Muir: "Thousands of tired, nerve-shaken, over-civilized people are beginning to find out that going to the mountains is going home; that wildness is a necessity; and that mountain parks and reservations are useful not only as fountains of timber and irrigating rivers, but as fountains of life."[7]

But there are a number of problems with this perspective on sacred nature.[8] One is the very distinctiveness of the area. *There* nature is sacred and should be preserved—as opposed to *over here*. John Elder has analyzed this issue in an intercultural context, comparing esthetically refined Japanese gardens walled off from a polluted outer world and American wilderness areas metaphorically and legally fenced off from the rest of the world. Both, he notes, can lead to a devaluing and abuse of nature outside of the privileged areas.[9]

Another problem is seen in the very definition of the term found in the Wilderness Act. If wilderness is "the earth and its community of life . . . untrammeled by man," humanity is positioned outside of nature's community of life, a foreign and dubious visitor. Interestingly, Aldo Leopold, one the proponents of wilderness protection and one of the forces in the founding of the Wilderness Society, argued that we should view ourselves as plain members and citizens of the land community.[10] There is a clear tension between such an ideal and locating the sacred in a land community untrammeled by man or woman.

What would that ideal of citizenship in the land community involve? And what is the alternative to nature as sacred other? Here we come to the second principal mode of experiencing sacred nature: the sacrality of place. Increasingly over the last fifty years, a "sense of place" has joined wilderness as one of the central ideas—even basic orientation—in nature writing. For all the commonality of the term *place*, in this context the idea is extraordinarily complex, the subject of studies by geographers, philosophers, religious-studies scholars, historians, and scholars of literature.[11] Place is where people are invested in the land, where human presence is well-established and an intimacy with nature is achieved. Nature is not other but home. "Familiarity has begun," Berry has put it; "one has made a relationship with the landscape."[12] He often speaks of this relationship as one of marriage, a chosen and sacramental commitment. Sanders has used the same imagery: "Loyalty to place arises from sources deeper than narcissism. It arises from our need to be at home on the earth. We marry ourselves to the creation by knowing and cherishing a particular place, just as we join ourselves to the human family by marrying a particular man or woman" (*Staying Put* 13).

Involved here is not distinctiveness *from* one's home place but the distinctiveness *of* it. Wilderness is everyone's; place is personalized. And not awe but intimacy is the prevailing condition. Such a place could be anywhere,[13] but it must be a special place one is tied closely to. The most common examples are rural. Susan Fenimore Cooper, for example, did not scale Muir's peaks but paid detailed, loving attention to the earth in her *Rural Hours*. Berry has continued that tradition, whether it is an essay on walking on "A Native Hill" (*Recollected Essays*) or the poetry of a "Man Born to Farming," who has become so intimate with the earth that his "hands reach into the ground and sprout, / to him the soil is a divine drug" (*Collected Poems* 103). However, such a life in place is not always a divine drug. If you are trying to raise organic peaches, like David Mas Masumoto, the appearance of a few worms can bring a less peaceful state of mind: "But how can I live with nature? By learning to live with five worms and my stress? I realize that for the rest of the season . . . I'll stand on my farmhouse porch thinking about five worms" (62). Particularly with a sense of place, spirituality coexists with practical difficulties.

One of the most important sources for the notion of a sacred sense of place is in Native American culture. Both Leslie Marmon Silko and N. Scott Momaday, for instance, have discussed native culture as an alternative to the landscape idea mentioned above. Silko notes that in native culture "human consciousness remains within the hills, canyons, cliffs, and the plants, clouds, and sky," and as a result, "the term landscape . . . is misleading. . . . [It] assumes the viewer is somehow outside or separate from the territory he or she surveys." In native culture, on the other hand, "Viewers are as much a part of the landscape as the boulders they stand on" (32). Momaday summarizes the integration of culture and nature involved in a sense of place: "In the natural order man invests himself in the landscape and at the same time incorporates the landscape into his own must fundamental experience. This trust is sacred." He emphasizes that nature is something *used* by native peoples rather than merely looked at. But at the same time, "You say that I use the land, and I reply, yes, it is true; but it is not the first truth. The first truth is that I love the land; I see that it is beautiful; I delight in it; I am alive in it" ("First American" 27–28).

While the notion of wilderness is linked to our urge to go beyond the human and encounter nature on its own terms, this sacred sense of place is associated with a similarly vital need to feel fully at home in nature, to be an integral part of the biotic community, and to become native to a place. This drive is associated with a contemporary movement that is important to spiritual nature writing: bioregionalism. The central thrust of bioregionalism is the focus on becoming a conscious, harmonious part of one's unique local place. To return to Leopold: what is it to truly live, as an individual and as a community, as a citizen of the land? Bioregionalism's answer to that question involves numerous aspects: knowledge of the local flora, fauna, and geography; a "deep time" awareness of the geological and human history; a psychological and spiritual sense of identity with the land and the other beings; a responsibility to the holistic well-being of the local land community now and into the future; and decentralized, locally based agriculture, economy, and politics.[14] The bioregional ideal is to cultivate the complex art of "reinhabitation." "I aspire to become an inhabitant," Sanders has said, "one who knows and honors the land" (*Staying Put* xiii)

What our academia has neatly divided into disciplines and divisions—natural sciences, social sciences, and humanities—here blend together. Snyder, for instance, asserts that the ecological sciences "have been laying out (implicitly) a spiritual dimension. We must find our way to seeing the mineral cycles, the water cycles, air cycles, nutrient cycles as sacramental—and we must incorporate that insight into our own personal spiritual quest" (*Place in Space* 188).

While nature as sacred other and nature as sacred place have been the dominant modes of perceiving nature's sacrality, there is a third and very important mode. As the sacred-other approach is highly dualistic, this one is totally monistic: everywhere is sacred. The whole of creation is holy, be it mountaintop, farm—or an abandoned city lot. If God made the world, it would seem that all of it must be intrinsically good. Buddhism has articulated this perspective most fully. Snyder, commenting on the nondualistic philosophy of Japanese Zen master Dogen, argues that the East Asian term *mountains and waters* "is a way to refer to the totality of the process of nature. . . . The whole, with its rivers and valleys, obviously includes farms, fields, villages, cities, and the (once comparatively small) dusty world of human affairs." Contra Eliade, space *is* homogenous: all of nature is sacred. Hence, Snyder has spoken on numerous occasions of "the preciousness of mice and weeds." In such a perspective, if sacred means some special place set apart—as other or as one's place—it is "a delusion and an obstruction" (*Practice of the Wild* 102–3) because it keeps us from recognizing the ultimate, unlimited value of every place.

These three modes are ideal types only. While the model of sacred other and model of sacred place can sometimes seem to be adversaries,[15] they also can be and are combined in various ways. Berry has distinguished in a subtle way between the human and the natural, the domestic and the wild, culture and nature; they are "indivisible, and yet are different." He thus advocates preserving wilderness which we take brief journeys into and also cultivating a sense of place where one lives and works: "In the recovery of culture *and* nature is the knowledge of how to farm well, how to preserve, harvest, and replenish the forests. . . . In this *double* recovery, which is the recovery of our humanity, is the hope that the domestic and the wild can exist together in lasting harmony" (*Home Economics* 139, 142).

Sanders affirms not only the sacrality of every place but also the need to focus on particular places: "If *all* of creation is holy . . . then why do we identify certain groves, mountains, or springs as sacred? Because they concentrate our experience of the land. We cannot hold the entire earth or even a forest or river in our minds at once; we need smaller places to apprehend and visit" (*Staying Put* 154). In particular he points to the need to cultivate the art of fully inhabiting your own place: "If you stay put, your place may become a holy center, not because it gives you special access to the divine, but because in your stillness you hear what might be heard anywhere. All there is to see can be seen from anywhere in the universe, if you know how to look; and the influence of the entire universe converges on every spot" (*Staying Put* 115–16).

Snyder often considered *the* wilderness, and in his essays and poetry he advocates strenuously for wilderness protection. But he is also a leading figure in bioregionalism, articulating a philosophy of living in and with a place you become intimate with. Indeed, he has criticized a pure preservationist stance that separates humans from the wild. And as we have seen, he experiences the nondualistic comprehensive sacrality of the entire world. Looking at the corpus of spiritual nature writing, we can say that some combination of all three perspectives is necessary to enter into the full complexity of what we call nature and our multifaceted relationship to it.

NATURE AND THE SACRED

The answer to the question "what nature is sacred?" responds in part to the second question: sacred in what way? It can be sacred as a place set apart, as one's home place, or as part of an unbroken field of being. But there is more to this second question. It can be put another way: what is the relationship between nature and ultimate reality or the divine?

Nature can be sacred in various ways. Many early settlers, for instance, saw the New World in a biblical framework. For some, like Thomas Tillam in 1638, it was positive: "Hayle holy-land wherein our holy lord / Hath planted his most true and holy world." For others, such as Governor William Bradford, nature was "a hideous and desolate wilderness, full of wild beasts and wild men." In such an anti-Christian land, "What could now sustain them but the Spirit of God and His Grace?" (quoted in Gatta 15, 18, 19). For Bradford and many others, it was a religious duty to turn the wilderness into a garden. In one sense or another, New England was seen as a new Israel where the Christian drama had a fresh start.

Slowly, the natural world became valued as more than merely a religious context. Anne Bradstreet's poem "Contemplations" (probably 1660s) was a meditation on the wonder of creation in the new land, making her "the first poet to record a sustained, appreciative response to outdoor experience in British North America" (Gatta 40). In her paeans for the land, however, creation was not the sacred per se. The creator remained the true locus of ultimate reality. Still, the world of nature was clearly of spiritual value as the gift of God, worthy of our spiritual attention. When we get to the *Travels* of William Bartram a century later, the natural world has clearly become the principal focus. The land overflows with a richness of species new to Europeans, and it shimmers with spiritual value. That value came from the creator, but creation had become the center of interest.

The tension between nature and the divine took a different turn with transcendentalism. Ralph Waldo Emerson's philosophy presents a fascinating paradox. On the one hand, nature is clearly and enthusiastically prized. He speaks of "the spiritual element" of nature, its "high and divine beauty." Emerson experiences not simply awe at distant splendor but intimacy with the more-than-human world: "The greatest delight which the fields and woods minister is the suggestion of an occult relation between man and the vegetable. I am not alone and unacknowledged. They nod to me, and I to them." And yet ultimately, for Emerson, it was not nature itself that had spiritual value: "All the facts in natural history taken by themselves, have no value, but are barren, like a single sex. But marry it to human history, and it is full of life." What he truly valued was the "radical correspondence between visible things and human thoughts" and the similarly radical correspondence between the visible world and the invisible. It is not so much that nature itself is sacred: "the world is emblematic" of the spiritual world (11, 7, 16, 18).[16] Despite Emerson's glorification of nature, it was the human mind and a metaphysical reality that had fundamental value.

Emerson's disciple Thoreau made the crucial turn to make nature itself truly sacred. John Gatta notes that Thoreau used the word *sacred* with regularity "to evoke not a distinct supernatural order, but a transcendent dimension of this physical world antithetical to the 'profane'" (129). Thoroughly eclectic, he spoke of ultimate reality in various ways. But we might say his most fundamental term for sacred reality was *the wild*. Concerning our spiritual cultivation, "we need the tonic of the wild" (*Walden* 306); concerning the natural world, "in Wildness is the preservation of the World" (*Essays* 162). It is this very world, shot through with wildness beyond our imaginings, that is sacred. So too for John Muir, who was even less interested in any transcendental realm. It is "God's wilderness," but it is wilderness that was the primary spiritual reality.

One of the most explicit stances affirming the natural world as being the locus of the sacred comes from Edward Abbey, a self-avowed "earthiest": "If a man's imagination were not so weak, so easily tired, if his capacity for wonder not so limited, he would abandon forever such fantasies of the supernal. He would learn to perceive in water, leaves and silence more than sufficient of the absolute and marvelous, more than enough to console him for the loss of the ancient dreams" (*Desert Solitaire* 200). Uncharacteristically invoking Buddhism, Abbey adds: "I stare long at the beautiful, dimming lights in the sky but can find there no meaning other than the lights' intrinsic beauty. As far as I can perceive, the plants signify nothing but themselves. 'Such suchness,' as my Zen friends say. And that is all. And that is enough. And that is more than we can make head or tail of" (*Down the River* 19–20).

But for a number of nature writers, the tension between a transcendental dimension of reality and the sacrality of nature generates a vital complexity. Annie Dillard reflects on two different types of mystical experiences, one in which she saw a tree "with the lights in it" shining from eternity and one when she "patted the puppy" in a state of total concentration of the present moment she was in:

> I had thought, because I had seen the tree with the lights in it, that the great door, by definition, opens on eternity. Now that I have "patted the puppy"—now that I have experienced the present purely though my senses—I discover that, although the door to the tree with the lights in it was opened *from* eternity, as it were, and shone on that tree eternal lights, it nevertheless opened on the real and present cedar. It opened on time: Where else? That Christ's incarnation occurred improbably, ridiculously, at such-and-such a time, into such-and-such a place, is referred to—with great sincerity even among believers—as "the scandal of particularity." Well, the "scandal of particularity" is the only world that I, in particular, know. (81)

For Dillard there is an eternal beyond this world, and yet she is wholly enmeshed in this incarnational world.

Sanders presents a different dynamic between the phenomenal world and a metaphysical one. The "ground of being" is "the whole thrust of the world that heaves us into existence and then draws us back to the source" (*Staying Put* 89). Occasionally, he uses theological language: "All that we perceive, think, and feel is gathered up in the mind of the Creator, and the Creator, in turn, ponders and probes the universe through us. Even these sentences, even your thoughts as you read them, are filaments that flicker in the great Mind" (*Hunting for Hope* 165). But ultimately his faith is "in the healing energy of wildness, in the holiness of Creation" (*Hunting for Hope* 39), and his fundamental stance is a spirituality of the earth: "What a seductive phrase: the ground of being. But if you waited to plant your beans or build your house until you had dug down to that elusive ground, you would go hungry and cold. I'll make do with the dirt I can touch" (*Staying Put* 141).

Sanders, in fact, has said he hesitates to speak of religious meanings. Similarly, Berry speaks of his discomfort in speaking of his deepest questions as "religious" because of how abstract religions usually are, with the result that a religious person, "while pursuing Heaven..., [can] turn his heart against his neighbors and his hands against the world." "But when I ask [these questions] my aim is not primarily to get to heaven. Though heaven is certainly more important than the earth if all they say about it is true, it is still morally incidental to it and dependent on it, and I can only imagine it and desire it in terms of what I know of the earth" (*Recollected Essays* 102). With Dillard, Sanders, and Berry we can speak of an earth-centered spirituality with a transcendental dimension.

Looking at these and other nature writers who address the spiritual dimension of our relationship with the earth, one can conclude that what we need, and what they as a group offer, is a broad and inclusive sense of the sacred. Our relationship with the holy cannot be exhausted by one perspective or approach. To fulfill our relationship with the earth, we need to experience the sacred otherness of it and also the spirituality of place, feel life beyond us living free of us and work with and as part of the land in a harmonious way. We need to experience the distinctively religious dimension of special places and the holiness of the whole of nature. To feel the full spirituality of the earth we are called to open to the divine quality of this phenomenal, material world and be open as well to a sense of a sacred dimension that somehow, in some way, goes beyond but remains linked with the natural world. Part of the power and significance of nature writing is that it invites us, it impels us, to such an expansive and multilayered relationship with this sacred earth.

QUALITIES OF A SACRED EARTH

While nature writers usually give us detailed, perceptive accounts of the natural world about them, they also reflect more generally on nature. How does it work? How is it put together? Perhaps the most common insight in regard to these questions is that of organic interrelatedness. Rather than being a collection of independent objects, nature is a web of relationships. "When we try to pick out anything by itself," Muir famously said, "we find it hitched to everything else in the universe" (*My First Summer* 157). This awareness of a profound interrelatedness gives a sense of the wholeness of the world. For Robinson Jeffers, "Integrity is wholeness, the greatest beauty is / Organic wholeness, the wholeness of life and things, the divine beauty of the universe" (594).

Biologists recognize the interconnections in the natural world, but nature writers experience this as a metaphysical quality of the physical world. While the idea of interrelationship is one of the most common themes in nature writing, writers influenced by east Asian religions have offered particularly complex views of the profound interpenetration of things. Le Guin, drawing on Daoism, speaks of "the universe of power. It [is] the network, field, and lines of the energies of all the beings, stars and galaxies of stars, worlds, animals, minds, nerves, dust, the lace and foam of vibration that is being itself, all interconnected, every part part of another part and the whole part of each part, so comprehensible to itself only as a whole, boundless and unclosed" (*Always Coming Home* 308). Rexroth and Snyder have drawn on Hua-Yen Buddhism and its image of Indra's jewel net. In this image, the universe is considered to be like a vast web of many-sided and highly polished jewels, each one acting as a multiple mirror. In one sense each jewel is a single entity. But when we look at a jewel, we see nothing but the reflections of other jewels, which themselves are reflections of other jewels, and so on in an endless system of mirroring. Thus in each jewel is the image of the entire net.

Snyder notes that Hua-Yen "Buddhist philosophy sees the world as a vast interrelated network" in which "the universe and all creatures in it are . . . acting in natural response and mutual interdependence." As a result, "any single thing or complex of things *literally* [is] as great as the whole" (*Earth House Hold* 91–92, 90, 31).[17] Rexroth has presented this perspective in various poems, often emphasizing the dynamism involved in nature, with "all / The webs, and nets, of relationships / Changing" (*Complete Poems* 610).

It is not merely that all things "out there" in nature are interrelated. We are a part of the web as well. In her essay "All My Relations," Linda Hogan describes in detail some of the symbolism of the sweat-lodge ceremony, in which "the entire world is brought inside the enclosure"—plants, animals, water, wind, earth— reminding humans of their connection to all things. As in all ceremonies, the sweat-lodge experience is meant to reshape a person "by restructuring the human mind." "By the end of the ceremony, it is as if skin contains land and birds" and all life; and people, reminded of their rightful status as part of everything, are reconnected to the natural world (39, 40, 41).

Another common feature of spiritual nature writing is a sense of the vastness of the universe. From the perspective of the immensity of nature, human life takes on a humbler position and brings a feeling of awe. Henry Beston, for instance, gazing from his outermost house, writes: "For a moment of night we have a glimpse of ourselves and of our world islanded in its stream of stars—pilgrims of mortality, voyaging between horizons across eternal seas of space and time. Fugitive though the instant be, the spirit of man is, during it, ennobled by a genuine moment of emotional dignity, and poetry makes it own both the human spirit and experience" (173–74).

Although humans have a diminished stature in this view of the world's vastness, the feeling of the smallness can also involve a sense of integration into a boundless universe. In the brilliant one-page prologue to his Pulitzer Prize– winning novel *House Made of Dawn*, N. Scott Momaday describes a desert morning: "The land was still and strong. It was beautiful all around." He concludes his description with a depiction of a Native American running within and as a part of the landscape: "The road curved out and lay into the bank of rain beyond, and Abel was running. Against the winter sky and the long, light landscape of the valley at dawn, he seemed almost to be standing still, very little and alone" (7).

In a vast and interrelated world, time also can take on a different character. A number of writers have explored "deep time," a spiritual sense of long reaches of prehistorical and even geological time, which can yield a sense of timelessness. For some writers this involves a strong sense of connectedness with the past. Berry writes of a farmer putting his hand into the earth: "It has reached into the dark like a root / and begun to wake, quick and mortal, in timelessness, / a flickering sap coursing upward into his head / so that he sees the old tribespeople bend in the sun, digging with sticks, the forest opening / to receive their hills of corn, squash, and beans, / their lodges and graves, and closing again. / He is made their descendant" (*Collected Poems* 119). For Sigurd Olson, a wilderness explorer senses "ancient rhythms" of earlier human life, sees in human intuition a link with other species in the evolutionary process, and believes that "it is easy to slip back into the ancient grooves of experience" (30, 21, 34). John Haines, homesteading in Alaska, speaks of viewing "a sheer sense of the land in its original presence," as he "easily slipped back a thousand years into a twilight approaching winter" ("Shadows and Vistas" 58). He finds himself able to enter "the original mystery of things, the great past out of which we came" (*Living Off the Country* 9). John Hanson Mitchell has a similar experience as he enters into what he calls "ceremonial time." In his book by that name, an Indian friend describes the ability to "actually see events that took place in the past. You can see people and animals who have been dead for a thousand years; you can walk in their place, see and touch the plants of their world" (12). Mitchell, exploring his "one square mile" in the town of Scratch Flat, is caught in a snowstorm: "I witnessed then, briefly, the essence of timelessness. I saw Scratch Flat as it must have been fifteen thousand years ago, saw the fields of ice, the heartless whip of blowing snow, the endless winters, and, at the base of it all, the insignificance of the human experiment" (25).

As nature writers have explored a sense of a space and time different than that of common sense, they have frequently felt a vitality that diverges from the ancient Greek and Enlightenment view that matter is inert. Silko states that animism has been a central part of Native American cultures and links humans with the rest of creation: "A rock shares this fate with us and with animals and plants as well. A rock has being or spirit, although we may not understand it. . . . In the end we all originate from the depths of the earth. Perhaps this is how all beings share the spirit of the Creator" (31).

Richard Nelson, formerly an anthropologist, learned of this view when he lived with the Koyukon people of Alaska: "Koyukon elders . . . teach that everything in the natural world has its own spirit and awareness, and they give themselves to that other world." Nelson has now integrated that view in his own life and hunting practice: animals have spirits, which calls for deep gratitude: "When we eat the deer, its flesh becomes our flesh. . . . And each time, we should carry a thought like a prayer inside: 'Thanks to the animal and all that made it' " (275–76, 268, 137).

One of the ways that nature writers have experienced the earth's vitality is in nature's ongoing creation. Instead of creation being a one-time event long, long ago, it is a continual process. If we are attuned to this fact, we recognize that every moment of every day we live within the act of creation. Berry finds that act most fully realized in the woods, "where the creation is yet fully alive and continuous and self-enriching, where whatever dies enters directly into the life of the living" (*Recollected Essays* 240).

Ultimately, however, ongoing creation is happening in every place. "None of us lives at the point where the Creation began," writes Sanders; "but every one of us lives at a point where the Creation *continues*" (*Writing from the Center* 166 [emphasis original]). Sanders experiences this as "the issuing-forth," "the perennial thrusting-forth of shapeliness out of the void" (*Staying Put* 139, 140). Sanders, however, emphasizes that there is another side to the generosity of creation: destruction. "It is a prodigal, awful, magnificent power, forever casting new forms into existence, then tearing them apart and starting over" (*Hunting for Hope* 34).

Despite all of the metaphysical reflections by nature writers, the notion of mystery remains central. Or perhaps we could say that the abiding sense of mystery is the root of those metaphysical reflections. There is far more to reality and there is far more meaning than we can ever behold. In *Walden*, Thoreau declared, "we require that all things be mysterious and unexplorable, that land and sea be infinitely wild, unsurveyed and unfathomed by us because unfathomable" (chap. 17). Sharon Butala finds she can enter more deeply into a sense of place by means of a sense of mystery: "I'd begun to approach the unbroken prairie in the way people approach a church or a great work of art—with a sense of awe and reverence at entering a powerful mystery" (126). For Rick Bass, mystery is not a fuzziness or lack of awareness, it "is as real and necessary a component of these mountains as are the tangible elements of bears, rock, sun, trees, water, ferns" (124–25). Williams states that she chooses to "court the mysteries" (quoted in Pearlman 132), in part because mystery is a quality of connectedness: "My connection to the natural world is my connection to self—erotic, mysterious, and whole" (*Unspoken Hunger* 56). But mystery is not always beneficent. If you are trying to make a living off the land—especially if you eschew the chemical solution to uncertainty—one can, as Masumoto learned, encounter "a grand mystery that can alternately frustrate and torment or amaze and initiate" (59).

A SPIRITUAL AWARENESS

Nature writers do not simply reflect on nature and its sacrality. "They are students of the human mind, literary psychologists," Scott Slovic has claimed; "and their chief preoccupation, I would argue, is with the psychological phenomenon of 'awareness'" (3). At least we can say that *one* of their chief preoccupations is a spiritual consciousness of the natural world. Whether they draw on Christian or Buddhist, Native American or neopagan traditions, nature writers often depict types of awareness that include some commonly shared qualities. One is focused attention and energy. Nelson has stated that "most of hunting is like this—an exercise in patient, isometric endurance and keen, hypnotic concentration" (259). Dillard has termed such a state "innocence": "What I call innocence is the spirit's unself-conscious state at any moment of pure devotion to any object. It is at once a receptiveness and total concentration" (83).

As Dillard notes, a state of receptivity needs to be part of such awareness: "Experiencing the present purely is being emptied and hollow; you catch grace as a man fills his cup under a waterfall" (81–82). Part of the process of cultivating such a state of mind is subtractive. In the words of Snyder, "All the junk that goes with being human / Drops away" (*Gary Snyder Reader* 400). One of the things that is emptied is the will. Rexroth opens the poem "Empty Mirror" with these words: "As long as we are lost / In the world of purpose / We are not free" (*Complete Poems* 321). Dillard puts it this way: "You don't run down the present, pursue it with baited hooks and nets. You wait for it, empty-handed, and you are filled. You'll have fish left over" (104). This empty waiting allows the moment to open to you.

One of the most famous statements about this type of pure perception comes from Emerson: "Standing on the bare ground,—my head bathed by the blithe air, and uplifted into infinite space,—all mean egotism vanishes. I become a transparent eye-ball; I am nothing; I see all; the currents of the Universal Being circulate through me; I am part or particle of God" (6). That image has been repeated by other writers. In reflecting on his consciousness during a hunt, Nelson has found that "my existence is reduced to a pair of eyes" (274). Rexroth has a similar experience as he peers into his telescope aimed at the night sky: "My body is asleep. Only / My eyes and brain are awake" (*Complete Poems* 535–36).

In such a state, the distinction between subject and object, between consciousness and reality can break down. The sensation of being cut off from the world has dissolved. "I fight against the optical illusion of separateness," Ehrlich has said. And when she succeeds, she can ask, "Where do I break off and where does water begin?" (*Islands* 165, 196). Rexroth had a similar experience when he was viewing the night sky: "The stars stand around me / Like gold eyes. I can no longer / Tell where I begin and leave off" (*Complete Poems* 535).

If we empty ourselves to the present, we may be able to achieve what the Buddhists call "direct perception," in which we see things "as they are." Such an ideal is not limited to Buddhists, however. Abbey the "earthiest" has said, "I want to be able to look at and into a juniper tree, a piece of quartz, a vulture, a spider, and see it as it is in itself, devoid of all humanly ascribed qualities, anti-Kantian, even the categories of scientific description" (*Desert Solitaire* 6).

In an emptied and concentrated state, one experiences the "present only," in which the moment seems timeless. Nelson depicts such an experience while hunting. As a doe approaches him, "time expands and I am suspended in the clear reality of the moment" (275). Abbey puts this in terms of "a suspension of time, a continuous present" (*Desert Solitaire* 13). Often the observer experiences the paradox of an "eternal" moment that is at the same time fleeting. "Catch it if you can," Dillard says; "the present is an invisible electron; its lightning path traced faintly on a blackened screen is fleet, and fleeing, and gone" (80). In a powerful political poem criticizing the powerful military and economic henchmen of nation-states, Rexroth concludes that "we have our own / Eternity, so fleeting that they / Can never touch it, or even / Know that it has passed them by" (*Complete Poems* 618).

One of the most interesting themes associated with this type of awareness is meaninglessness. This is not the despairing absence of meaning where it should be. With a pure absorption in the present, there is no interpretation or analysis, that is, no "meaning," only direct perception. Abbey refers to this a number of times in *Desert Solitaire*: "The desert reveals itself nakedly and cruelly, with no meaning but its own existence" (155).[18] About the desert he asks, "What does it mean? It means nothing. It is as it is and has no need for meaning. The desert lies beneath and soars beyond any possible human qualification. Therefore, sublime" (219). This theme is found most often among Buddhist nature writers. Matthiessen speaks of this idea in *The Snow Leopard*, presenting the Buddhist ideal strikingly while noting his own failure to fully realize it: "The secret of the mountains is that the mountains simply exist, as I do myself: the mountains exist simply, which I do not. The mountains have no 'meaning,' they are meaning; the mountains are" (217–18). Rexroth begins one of his poems with a startling statement for a poet: "There are no images here / In the solitude, only / The night and its start which are / Relationships rather than / Images." He repeats the claim that there are no images, then later in the poem states that the night or a tree "doesn't mean / Anything. It isn't an image of / Something. It isn't a symbol of / Something else. . . . It is just an / Almond tree, in the night, by / The house, in the woods, by / A vineyard, under the setting / Half moon, in Provence, in the / Beginning of another Spring" (*Complete Poems* 610–11). The tree, in its "suchness," is a node of relationships within the limitless wholeness of the night.

Slovic has noted an important tension in this ideal spiritual awareness. On the one hand there is a profound sense of unity with the world, a direct correspondence between the natural world out there and the perceiving mind. At the same time, as we have seen, many nature writers affirm the otherness of nature, its intrinsic, independent life. "This dialectical tension between correspondence and otherness," Slovic says, "is especially noticeable in Thoreau, Dillard, and Abbey" (5). Abbey puts the tension this way:

> We must beware of a danger well known to explorers of both the micro- and the macrocosmic—that of confusing the thing observed with the mind of the observer, of constructing not a picture of external reality but simply a mirror of the thinker. Can this danger be avoided without falling into an opposite but related error, that of separating too deeply the observer and the thing observed,

subject and object, and again falsifying our view of the world? There is no way out of these difficulties. (*Desert Solitaire* 270)

In fact, Abbey puts this tension in terms of retaining the sense of his own individual self: "I dream of a hard and brutal mysticism in which the naked self merges with a nonhuman world and yet somehow survives still intact, individual, separate. Paradox and bedrock" (*Desert Solitaire* 6).

SOCIAL SPIRITUALITY IN NATURE WRITING

Nature writing before World War II was dominated by accounts of the solitary pilgrimage in wilderness or at least the lone stroll in field and wood.[19] Since 1945, solitude has continued to be a major part of spiritual nature writing, but the *social* dimension of nature writing—the context of family and community, economics and politics—has also become a central part of the genre. Going out into nature with a daughter or son, seeing a bird sanctuary with a grandmother, working the fields with neighbors, facing down what Snyder calls the "Growth-Monster" (*Practice of the Wild* 5), these also have become common contexts for nature writing.

This social dimension is part of the spirituality of nature writing. In our largely Protestant culture, we have tended to associate religion with the individual's encounter with the sacred. And the term *spirituality* is often used to designate one's unique personal religion outside the bounds of institutional religions with their hierarchy, dogma, and social conformity. But in recent nature writing, there has been an increasing interest in what we can call social spirituality: religious beliefs, values, and practices that are not confined to conventional religious institutions (and thus I use the term *spirituality*), but at the same time are fundamentally social. These nature writers explore questions such as, What is the spiritual significance of nature for me as a member of a family or community or culture? How does my family or my community or my culture relate spiritually to nature?

We can distinguish different types of this spiritually based social nature writing: family and community, cross-cultural, environmental, and political. An example of the first type is Williams's *Refuge*, her moving account of both the flooding of a bird refuge and her mother's fight with cancer. The birds lose their refuge, and Williams and her mother lose that same quiet place as their personal and shared refuge from the city. And as the birds are driven from their home, Williams's mother passes away. Williams's experiences of nature and of family are indivisible. The same is true, with less tragic circumstances, when

Sanders canoes the boundary waters with his daughter or hunts for hope in response to his son's despair.[20] Similarly, we read in a number of Snyder's poems about his experiencing nature with his wife and sons and experiencing his family in nature. Berry's agrarian fiction offers a rich tapestry of narratives about several generations of people in the small town of Port William. Their relationship with each other and their relationship with rural Kentucky are indissoluble. In addition, one of the reasons Berry writes these novels is to articulate an alternative to the industrial, consumer society that is devouring the earth. Le Guin's science fiction presents different imaginary cultures with a variety of relationships with nature, some distant and destructive, others intimate and harmonious. In novels such as *The Dispossessed*, *Eye of the Heron*, and *Always Coming Home*, she presents a compelling interwoven philosophy of nature and of society.

Another form of spiritually based social nature writing is *cross-cultural*, with particular emphasis on Native American cultures. Austin's *Land of Little Rain*, published in 1903, was the first significant reflection on native cultures by a nature writer, and over the last fifty years, such accounts have become numerous. Native writers such as Momaday, Silko, and Hogan have described Indian cultures as having an exemplary relationship with nature. The focus is not the individual experience but how society as a whole and the cultural values, beliefs, and behaviors are harmonious with the natural world. European American writers such as Snyder, Williams, and Nelson have also discussed the significance of nature in Indian culture and explored the ways that white culture can learn from the native inhabitants of "Turtle Island," an Indian term for North American that Snyder likes to use.[21] In this cross-cultural type of nature writing, the individual's relationship with nature cannot be separated from his or her relationship with the society and its relationship with nature. As such, nature as a social experience is inextricably social. In addition, these descriptions are significant as a stark contrast with the way that the dominant culture today relates to nature. Learning about native cultures reinforces social questions: What kind of society should we have if we are as a society to live harmoniously with nature? What kind of social spirituality should inform our relationship with nature?

Those questions have been intensified with the increasing awareness of the degradation of the planet and have lead to another kind of social nature writing: *environmental*. Nature writing does not only sing of the beauties of the earth, it also witnesses to the despoiling of the world we all depend on. Since Rachel Carson's *Silent Spring*, environmental concerns have become prominent in nature writing. Issues such as the polluting of the earth and the crippling of its ecological systems are far more inclusive than wilderness preservation, because the poisons affect our neighborhoods and our bodies. The human community, not just pristine nature, is under attack. Patrick Murphy has distinguished these two types of nature-oriented literature. He uses the terms *nature writing* and *nature literature* for works that describe natural history and the individual's experience in nature. He uses the term *environmental writing* for works that discuss "the ways in which pollution, urbanization, and other forms of human intervention have altered the land and environment." This mode of writing, he says, is distinguished by "authorial self-consciousness about environmental issues and problems" (5). These concerns are not just personal; the social dimension is fundamental. And spirituality has special power here. It is a powerful motivation to help preserve that which has ultimate value. It also engenders deep compassion for both people and nonhuman nature, an abiding sense of responsibility for the well-being of the entire community of life, a drive to fight what is so wrong, and an ability to endure all the suffering and loss.

Human society is not only the victim but also the cause. The environment is degraded by the workings of economic systems, social ideologies, and political structures—all of these rooted in fundamental worldviews. Spiritual nature writers have engaged the *political* dimension of nature writing as they have critically analyzed the cause of the pillaging of the planet and explored alternative social ideals. Spirituality sharpens the critique of the dominant culture and the analysis of its causes; it gives a deeper sense of responsibility to the well-being of the earth; and it shapes the alternative communal ideals that are considered. Politics have been associated with nature writers since Emerson, a strong critic of slavery. But while Emerson criticized one social institution, Thoreau criticized the entire political system with his version of anarchism. He opened his famous essay "Civil Disobedience" with the claim: "That government is best which governs not at all" (*Essays* 125).

Radical political views have become a fairly common feature in contemporary nature writing. In these cases nature writers do not call simply for a change in environmental policy, the establishment of an environmental advocacy organization, or a shift from one established political party to another. The grave situation of environmental degradation and the social suffering and injustice that is intertwined with it point to a pervasive problem with our fundamental worldview and with the social institutions that put it so ruthlessly into practice. With nature and culture intimately interconnected, with social problems and environmental destruction co-arising (to use a Buddhist term), affirming the spiritual value of nature involves rejecting the basic fabric of political life.

Williams, for instance, has taken up the call for an "open space for democracy" to counter the Bush administration's response to the 9-11 terrorist strike and its invasion of Iraq. For her and other writers, the spiritually based love and defense of the earth cannot be isolated from a love and defense of social justice and true democracy; a commitment to the biotic community blends into a "new patriotism."[22] For Williams and others, the spiritual values that make the despoiling of wilderness intolerable make social injustice and imperialism a source of outrage. Her feminist call for a politics of place based on an "erotics of place" challenges the basic assumptions of American politics.

Other writers have developed a critique of modern society and a vision of a radical alternative. Rexroth's nature poetry is infused not only with Catholic/ Buddhist contemplation but also with his commitment to the anarchist tradition. He quotes the anarchist aphorism "Liberty is the mother / Not the daughter of order" in one his various poems that make the contemplation of nature's beauty the context for recollections of Sacco and Vanzetti and commiseration with the anarchists in the Paris Commune, the Kronstadt Rebellion, and the Spanish Civil War.[23] He said in his preface to *The Art of Worldly Wisdom*, "My poems are acts of force and violence directed against the evil which murders us all. If you like, they are designed not just to overthrow the present State, economic system, and Church, but all prevailing systems of human collectivity altogether." In contrast to collectivity, Rexroth articulates a communal ideal based on a spirituality of mutual love and responsibility that is grounded in his vision of nature's order and his tragic sense of human history. He locates himself in an apostolic community that runs through time, seeking to be one who has "turned from the world and embarked on the spiritual pilgrimage toward divinization in company with the beloved community" (*Communalism* 98).

Abbey's masters thesis analyzed arguments for revolutionary violence in anarchism, and in one of his essays he articulates his "Theory of Anarchy" (*One Life at a Time* 25–28). Like Rexroth, he claims that there is special community of free individuals. He contrasts the stultifying effect of culture to his ideal of civilization, by which he means

> the brotherhood of great souls and the comradeship of intellect, a *corpus mysticum*. The Invisible Republic open to all who wish to participate, a democratic aristocracy based not on power or institutions but on isolated men—Lao-Tse, Chaung-Tse, Gautama, Diogenes, . . . Socrates, Jesus, . . . Paine and Jefferson . . . and ten thousand other poets, revolutionaries and independent spirits, . . . whose heroism gives to human life on earth its adventure, glory and significance. (*Desert Solitaire* 276)

Abbey has commented that "the traditional conflict between our instinctive urge toward fraternity, community, and freedom, and the opposing demands of discipline and the state . . . runs through everything I have written, binding it together into whatever unity it may have" (*Slumgullion Stew* ix).

Snyder has blended the anarchist Industrial Workers of the World legacy that was part of his upbringing with the spiritual liberation and compassion of Buddhism and the tribal social organizations of Native American cultures.[24] For Snyder, "Buddhism holds that the universe and all creatures in it are intrinsically in a state of complete wisdom, love and compassion, acting in natural response and mutual interdependence." As a result, "in the Buddhist view, that which obstructs the effortless manifestation of this is Ignorance, which projects into fear and needless craving." Modern nation-states foster that craving and fear and are, in effect, "monstrous protection rackets." For Snyder, Buddhist wisdom leads "to a deep concern for the need for radical social change through a variety of non-violent means" (*Gary Snyder Reader* 41–42).

Le Guin, informed by Chinese Daoism, has portrayed in her science fiction a number of anarchistic communities living in relative harmony with nature (see especially *Dispossessed, Always Coming Home,* and *Eye of the Heron*). "Odonian-ism," the social theory depicted in *The Dispossessed,*

> is anarchism. Not the bomb-in-the-pocket stuff, which is terrorism...; not the social-Darwinist economic "libertarianism" of the far right; but anarchism, as pre-figured in early Taoist thought, and expounded by Shelley and Kropotkin, Goldman and Goodman. Anarchism's principal target is the authoritarian State (capitalist or socialist); its principal moral-practical theme is cooperation (soli-darity, mutual aid). It is the most idealistic, and to me the most interesting, of all political theories. (*Wind's Twelve Quarters* 260)

In *The Eye of the Heron,* she depicts another culture whose bioregional anarchism is informed by Quaker practice.

Berry's radical Christianity takes the form of a rather anarchistic agrarianism. His criticism of our government, economy, and religious institutions is withering: "In this state of total consumerism...all meaningful contact between ourselves and the earth is broken.... [This] is madness, mass produced." He describes the current economy as "firmly founded upon the seven deadly sins and the breaking of all ten of the Ten Commandments." Christianity, "in its de facto alliance with Caesar,...connives directly in the murder of Creation." "It has, for the most part, stood silently by, while a predatory economy has ravaged the world, destroyed its natural beauty and health, divided and plundered its human communities and households." True Christianity is radical: "The most important religion in the Bible is unorganized, and is sometimes profoundly disruptive of organization." In particular, "Christ's life, from the manger to the cross, was an affront to the established powers of his time, as it is to the established powers of our time" (*Art of the Commonplace* 85, 309, 319, 311). He is also quick to criticize the abstractness of social movements, including the environmental movement itself.[25]

CONCLUSION

Nature writing thus draws from a wide range of spiritual traditions, and it offers differing views of the sacrality of nature. In various ways it presents a dynamic, interwoven natural world that one can access in special states of consciousness. It also involves a political and a social spirituality, as well as a personal one. The richness of nature writing as a means of exploring the religious dimension of the natural world continues to deepen, as does its capacity to nourish a spirituality of resistance that Williams addresses in the quotation at the beginning of this essay. The power of nature writing to combine a deep quiet with an ability to stand firm in the face of the contemporary assault on the earth gives it both a timely and timeless significance.

NOTES

1. The original title consisted of thirty-one words, which Dover retitled simply as *The Travels of William Bartram.*

2. For an extended discussion of the term *spirituality of resistance* and its relevance to environmental issues, see Gottlieb's brilliant *Spirituality of Resistance.*

3. Many critics point out that Walden Pond was hardly isolated in a distant wilderness; it was close to Concord and Emerson's friendly house. But psychologically, Thoreau experienced his place as one of both deep solitude from conventional culture and connected to far off cultures. In the chapter entitled "Solitude," he says, "I have my horizon bounded by woods all to myself; a distant view of the railroad where it touches the pond on the one hand, and of the fence which skirts the woodland road on the other. But for the most part it is as solitary where I live as on the prairies. It is as much Asia or Africa as New England" (*Walden* 126).

4. Abbey ends the book reflecting on nature's indifference to the coming or going of people.

5. See, for instance, the many studies of pilgrimage by Victor Turner. For discussions of Japanese nature writer Matsuo Bashō, who presents an alternative to pilgrimage—unending wayfaring—see Barnhill's "Bashō as Bat" and "Impermanence, Fate, and the Journey." There are interesting similarities between Bashō's wayfaring ideal and Thoreau's notion of sauntering ("at home everywhere"), propounded in his essay "Walking."

6. The definition in the 1964 Wilderness Act, in Callicott and Nelson 121.

7. "The Wild Parks and Forest Reservations of the West," in Cronon, *John Muir* 721. Note that "home" here is spiritual and metaphorical: Muir was advocating for parks, not for housing development, in the mountains. Wilderness has been a recurrent theme in ecocriticism as well; see, e.g., Oelschlaeger and O'Grady.

8. Two important collections that consider the problems with the wilderness idea are Cronon, *Uncommon Ground* and Callicott and Nelson.

9. See "Wildness and Walls," an essay in Elder's fascinating *Following the Brush,* his reflections on the year he lived in Japan.

10. See his "Land Ethic" in *Sand County Almanac.*

11. Among the numerous studies of place, see Casey, Gallagher, Hiss, Jackson, Kemmis, Seamon and Murgerauer, and Vitek and Jackson.

12. Berry, "Native Hill," in *Recollected Essays* 90. Slovic has stated that "what we experience reading Wendell Berry's essays about Kentucky is the steady intensification of *Hiemlichkeit,* or the feeling of being at home" (118).

13. In *At Home on the Earth*, I have collected essays on the sense of place arranged by different areas and lifestyles: homesteading, ranching, farming, living between city and country, and urban living.

14. Among the many books on bioregionalism, see Carr, McGinnis, Sale, and Thayer.

15. Lueders 56–60 contains an interesting exchange between Terry Tempest Williams, who calls for greater protection of wilderness because it is "sacred ground," and Robert Finch, who calls government-regulated wilderness preservation a dead-end in terms of saving land.

16. A modern example of deeply religious—but not ultimate—reverence for nature is John Cobb, who argues from a process Protestant perspective that "the right way to speak [of the spiritual value of nature] is incarnational, immanental, or sacramental. God is present in the world—in every creature. But no creature is divine. Every creature has intrinsic value, but to call it sacred is in danger of attributing to it absolute value" (223).

17. For an analysis of Snyder's use of Indra's jewel net image in his understanding of nature, see my "Indra's Net as Food Chain."

18. Abbey may have had in mind a similar statement made by James Agee: "For in the immediate world, everything is to be discerned, for him who can discern it, and centrally and simply, without either dissection into science, or digestion into art, but with the whole of consciousness, seeking to perceive it as it stands . . . and all of consciousness is shifted from the imagined, the revisive, to the effort to perceive simply the cruel radiance of what is" (11).

19. See Roorda's important analysis of solitary nature writing in his *Dramas of Solitude*.

20. See Sanders's "Voyageurs" in *Writing from the Center* and *Hunting for Hope*.

21. Snyder has said that one of the reasons for his interest in Native American cultures is that while the Buddhism tradition is monastic—separating serious religiosity from family, community, and culture—native societies incorporated spirituality throughout the culture to create a "total integrated life style," which Buddhist monasteries cannot have (*Real Work* 15).

22. See, for instance, *Patriotism and the American Land*, which includes essays by Nelson, Lopez, and Williams.

23. See, for instance, "Climbing Milestone Mountain, August 22, 1937," "From the Paris Commune to the Kronstadt Rebellion," and "Requiem for the Spanish Dead" (*Complete Poems* 151, 143, 148).

24. For example, see his "Buddhism and the Possibility of a Planetary Culture" in *Gary Snyder Reader* 41–43.

25. See *Citizenship Papers*, including "In Distrust of Movements."

REFERENCES

Abbey, Edward. *Desert Solitaire: A Season in the Wilderness*. New York: Ballantine, 1968.
———. *Down the River*. New York: Plume, 1991.
———. *Good News*. New York: Plume, 1991.

———. *Let Us Now Praise Famous Men.* Boston: Houghton Mifflin, 1941.

———. *Monkey Wrench Gang.* New York: Avon, 1975.

———. *One Life at a Time, Please.* New York: Holt, 1987.

———. *Slumgullion Stew: An Edward Abbey Reader.* New York: Dutton, 1984.

Agee, James, and Walker Evans. *Let Us Now Praise Famous Men.* Boston: Houghton Mifflin, 1941.

Austin, Mary Hunter. *The Land of Little Rain.* New York: Dover, 1996 (orig. 1903).

Barnhill, David Landis, ed. *At Home on the Earth: Becoming Native to Our Place.* Berkeley: University of California Press, 1999.

———. "Bashō as Bat: Wayfaring and Anti-Structure in the Journals of Matsuo Bashō (1644–1694)." *Journal of Asian Studies* 49 (May 1990): 274–90.

———. "Impermanence, Fate, and the Journey: Bashō and the Problem of Meaning." *Religion* 16 (1986): 323–41.

———. "Indra's Net as Food Chain: Gary Snyder's Ecological Vision." *Ten Directions* 11.2 (spring/summer 1990): 20–28.

Barnhill, David Landis, and Roger S. Gottlieb, eds. *Deep Ecology and World Religions: New Essays on Sacred Ground.* Albany: State University of New York Press, 2001.

Bartram, William. *Travels of William Bartram.* Edited by Mark Van Doren. New York: Dover, 1928.

Bass, Rick. *The Lost Grizzlies: A Search for Survivors in the Wilderness of Colorado.* Boston: Houghton Mifflin, 1995.

Berry, Wendell. *Art of the Commonplace.* Washington, D.C.: Counterpoint, 2002.

———. *Citizenship Papers.* Washington, DC: Shoemaker & Hoard, 2003.

———. *Collected Poems, 1957–1982.* San Francisco: North Point, 1984.

———. *Home Economics.* San Francisco: North Point, 1987.

———. *Recollected Essays, 1965–1980.* San Francisco: North Point, 1981.

Beston, Henry. *The Outermost House: A Year of Life on the Great Beach of Cape Cod.* New York: Holt, 1992.

Butala, Sharon. *Perfection of the Morning: A Woman's Awakening in Nature.* St. Paul: Hungry Mind, 1994.

Callicott, J. Baird, and Michael P. Nelson, eds. *The Great New Wilderness Debate.* Athens: University of Georgia Press, 1998.

Carr, Mike. *Bioregionalism and Civil Society: Democratic Challenges to Corporate Globalism.* Vancouver: UBC, 2004.

Carson, Rachel. *Silent Spring.* Boston: Houghton Mifflin, 1994.

Casey, Edward S. *The Fate of Place: A Philosophical History.* Berkeley: University of California Press, 1997.

Cobb, John B., Jr. "Protestant Theology and Deep Ecology." Pp. 213–28 in *Deep Ecology and World Religions: New Essays on Sacred Ground.* Edited by David Landis Barnhill and Roger S. Gottlieb. Albany: State University of New York Press, 2001.

Cooper, Susan Fenimore. *Rural Hours.* Edited by Rochelle Johnson and Daniel Patterson. Athens: University of Georgia Press, 1998.

Cronon, William, ed. *John Muir: Nature Essays.* New York: Library of America, 1997.

———. *Uncommon Ground: Rethinking the Human Place in Nature.* New York: Norton, 1996.

Dillard, Annie. *Pilgrim at Tinker Creek.* Toronto: Bantam, 1973.

Ehrlich, Gretel. *Islands, the Universe, Home.* New York: Penguin, 1991.

———. *Questions of Heaven.* Boston: Beacon, 1997.

———. *The Solace of Open Spaces.* New York: Penguin, 1986.

Elder, John. "Wildness and Walls." Pp. 141–53 in *Following the Brush: An American Encounter with Classical Japanese Culture.* Boston: Beacon, 1993.

Eliade, Mircea. *The Sacred and the Profane: The Nature of Religion.* New York: Harcourt, Brace, 1959.

Emerson, Ralph Waldo. *The Selected Writings of Ralph Waldo Emerson.* New York: Random, 1950.

Gallagher, Winifred. *The Power of Place: How Our Surroundings Shape Our Thoughts, Emotions, and Actions.* New York: Poseidon, 1993.

Gatta, John. *Making Nature Sacred: Literature, Religion, and Environment in America from the Puritans to the Present.* New York: Oxford University Press, 2004.

Gottlieb, Roger S. *A Spirituality of Resistance: Finding a Peaceful Heart and Protecting the Earth.* Lanham, MD: Rowman & Littlefield, 2003.

Haines, John. *Living Off the Country: Essays on Poetry and Place.* Ann Arbor: University of Michigan Press, 1981.

———. "Shadows and Vistas." Pp. 57–62 in *At Home on the Earth: Becoming Native to Our Place.* Edited by David Landis Barnhill. Berkeley: University of California Press, 1999.

Hiss, Tony. *The Experience of Place: A New Way of Looking at and Dealing with Our Radically Changing Cities and Countryside.* New York: Knopf, 1990.

Hogan, Linda. *Dwellings: A Spiritual History of the Living World.* New York: Touchstone, 1995.

Jackson, John Brinckerhoff. *A Sense of Place, a Sense of Time.* New Haven: Yale University Press, 1994.

Jeffers, Robinson. *The Selected Poetry of Robinson Jeffers.* New York: Random, 1938.

Kemmis, Daniel. *Community and the Politics of Place.* Norman: University of Oklahoma Press, 1990.

Le Guin, Ursula. *Always Coming Home.* New York: Harper & Row, 1985.

———. *The Dispossessed.* New York: Harper & Row, 1984.

———. *The Eye of the Heron.* New York: Harper & Row, 1978.

———. *The Wind's Twelve Quarters.* New York: Harper & Row, 1975.

Leopold, Aldo. *A Sand County Almanac and Sketches Here and There.* New York: Oxford University Press, 1949.

Lopez, Barry. *Arctic Dreams: Imagination and Desire in a Northern Landscape.* New York: Scribner, 1986.

Lueders, Edward, ed. *Writing Natural History: Dialogues with Authors.* Salt Lake City: University of Utah Press, 1989.

Masumoto, David Mas. *Epitaph for a Peach: Four Seasons on My Family Farm.* San Francisco: Harper, 1996.

Matthiessen, Peter. *The Snow Leopard.* New York: Bantam, 1979.

McGinnis, Michael Vincent, ed. *Bioregionalism.* New York: Routledge, 1999.

Mitchell, John Hanson. *Ceremonial Time: Fifteen Thousand Years on One Square Mile.* Readings, MA: Addison-Wesley, 1984.

Momaday, N. Scott. "A First American Views His Land." Pp. 19–29 in *At Home on the Earth: Becoming Native to Our Place.* Edited by David Landis Barnhill. Berkeley: University of California Press, 1999.

———. *House Made of Dawn.* New York: New American Library, 1969.

————. *My First Summer in the Sierra.* Boston: Houghton Mifflin, 1916.

Murphy, Patrick D. *Farther Afield in the Study of Nature-Oriented Literature.* Charlottesville: University Press of Virginia, 2000.

Nelson, Richard. *The Island Within.* New York: Vintage, 1991.

Nelson, Richard, Barry Lopez, and Terry Tempest Williams. *Patriotism and the American Land.* Great Barrington, MA: Orion Society, 2002.

Oelschlaeger, Max. *The Idea of Wilderness: From Prehistory to the Age of Ecology.* New Haven: Yale University Press, 1991.

O'Grady, John P. *Pilgrims to the Wild: Everett Ruess, Henry David Thoreau, John Muir, Clarence King, Mary Austin.* Salt Lake City: University of Utah Press, 1993.

Olson, Sigurd. *Reflections from the North Country.* Minneapolis: University of Minnesota Press, 1998.

Payne, Daniel G. *Voices in the Wilderness: American Nature Writing and Environmental Politics.* Hanover: University Press of New England, 1996.

Pearlman, Mickey. *Listen to Their Voices: Twenty Interviews with Women Who Write.* New York: Norton, 1993.

Philippon, Daniel J. *Conserving Words: How American Nature Writers Shaped the Environmental Movement.* Athens: University of Georgia Press, 2004.

Rexroth, Kenneth. *The Art of Worldly Wisdom.* 2nd ed. Sausalito, CA: Golden Goose, 1953.

————. *Communalism: From Its Origins to the Twentieth Century.* New York: Seabury, 1974.

————. *Complete Poems.* Edited by Sam Hamill and Bradford Morrow. Port Townsend, WA: Copper Canyon, 2003.

Roorda, Randall. *Dramas of Solitude: Narratives of Retreat in American Nature Writing.* Albany, State University of New York Press, 1998.

Sale, Kirkpatrick. *Dwellers in the Land: The Bioregional Vision.* Athens: University of Georgia Press, 2000.

Sanders, Scott Russell. *Hunting for Hope: A Father's Journey.* Boston: Beacon, 1998.

————. *Staying Put: Making a Home in a Restless World.* Boston: Beacon, 1993.

————. *Writing from the Center.* Bloomington: Indiana University Press, 1995.

Seamon, David, and Robert Murgerauer, eds. *Dwelling, Place, and Environment: Towards a Phenomenology of Person and World.* New York: Columbia University Press, 1985.

Silko, Leslie Marmon. "Landscape, History, and the Pueblo Imagination." Pp. 30–42 in *At Home on the Earth: Becoming Native to Our Place.* Edited by David Landis Barnhill. Berkeley: University of California Press, 1999.

Slovic, Scott. *Seeking Awareness in American Nature Writing: Henry Thoreau, Annie Dillard, Edward Abbey, Wendell Berry, Barry Lopez.* Salt Lake City: University of Utah Press, 1992.

Snyder, Gary. *Earth House Hold: Technical Notes and Queries to Fellow Dharma Revolutionaries.* New York: New Directions, 1969.

————. *The Gary Snyder Reader: Prose, Poetry, and Translations, 1952–1998.* Washington, DC: Counterpoint, 1999.

————. *A Place in Space: Ethics, Aesthetics, and Watersheds.* Washington, DC: Counterpoint, 1995.

————. *The Practice of the Wild.* San Francisco: North Point, 1990.

————. *The Real Work: Interviews and Talks, 1964–1979.* New York: New Directions, 1980.

Thayer, Robert L., Jr. *LifePlace: Bioregional Thought and Practice.* Berkeley: University of California Press, 2003.

Thoreau, Henry David. *The Essays of Henry D. Thoreau.* Edited by Lewis Hyde. San Francisco: North Point, 2002.

———. *Walden: A Fully Annotated Edition.* Edited by Jeffrey S. Cramer. New Haven: Yale University Press, 2004.

Turner, Victor, and Edith Turner. *Image and Pilgrimage in Christian Culture: Anthropological Perspectives.* New York: Columbia University Press, 1978.

Vitek, William, and Wes Jackson, eds. *Rooted in the Land: Essays on Community and Place.* New Haven: Yale University Press, 1996.

Williams, Terry Tempest. "A Peach in the Wilderness." *Nature Conservancy* (Jan.–Feb. 2001).

———. *Pieces of White Shell: A Journey to Navajoland.* Albuquerque: University of New Mexico Press, 1984.

———. *Refuge: An Unnatural History of Family and Place.* New York: Vintage, 1992.

———. *Red: Passion and Patience in the Desert.* New York: Pantheon, 2001.

———. *An Unspoken Hunger: Stories from the Field.* New York: Vintage, 1994.

CHAPTER 20

RELIGION, ENVIRONMENTALISM, AND THE MEANING OF ECOLOGY

LISA H. SIDERIS

In recent decades, the rise of religious environmental ethics and ecological theology has engendered a number of positive and fruitful connections between the study of religion and other disciplines. At the same time, ecotheologians have scrutinized their own traditions in search of ethical resources that can be mined for environmental content. Probably no tradition has worked harder than the Christian tradition in the quest to locate—or create—positive environmental teachings. In part, the Christian response was generated in the aftermath of Lynn White's now famous critique of the tradition (or, more broadly, the Judeo-Christian tradition) decades ago. White conferred to Christianity the dubious distinction of being the world's most anthropocentric religion. In contrast both to the animistic paganism that preceded Christianity and to Asia's religions, White argued, Christianity "established a dualism of man and nature but also insisted that it is God's will that man exploit nature for his proper ends."[1] At best, nature confronts the Christian as mere "physical fact" and, at worst, as an adversary to be mastered and conquered.[2] Within ecotheological circles, there is also a general sense—often promoted by

Christian ecotheology as much as by its critics—that the Eastern religions have always been more or less on the right environmental track.

Thus, in the process of making the tradition more environmentally friendly, Christian ecotheologians have sometimes groomed it to look more like the Eastern religions or, at least, more like what they assume the Eastern religions look like. Since dualisms between heaven and earth, spirit and matter, are assumed to be Western patterns of thought, one often finds "wistful glancings eastward and intimations that there are ways of thinking in Asia based on interconnection rather than dichotomy."[3] Some Western theologians have called for a resacralization of nature and a reenvisioning of a less transcendent deity, as ways of mitigating the supposedly disenchanted material world and unearthly God of Christianity. Taking cues from Eastern religions, many have promoted a relational, radically inter-connected model of humankind, the natural world, and our animal kin. Eco-theologians have emphasized that Christianity, like Buddhism or Jainism, understands, or ought to understand, the suffering of all beings as inextricably linked to our own and part of our moral universe. After all, as some have stressed, the bedrock of Christian ethics is care for the suffering and oppressed neighbor, and why should this be limited to human neighbors? A religion that has so often made social justice and liberation for the oppressed a moral priority can easily extend that agenda to the whole community of life.

If one approach to developing a greener form of Christianity has been through dialogue and comparison with Eastern religions, another approach involves a turn to modern (or, frequently, postmodern) science in search of concepts that reso-nate with these enlightened religious perceptions of the God-human-nature in-terconnection. Along these lines, a number of ecotheologians have begun to locate, and celebrate, a community-based, interconnected, relational model that they see as demonstrably central to cutting-edge scientific accounts of nature. Having now incorporated some of the best ethical teachings of greener religions, while cultivating environmental aspects inherent in Christianity, *and* having now bolstered all these teachings with findings from the latest science, Christian ecotheology is in an excellent position to respond to our global environmental crisis.

Or is it? Certainly, Christian ecotheologians have made a number of prom-ising connections with other religions, with other disciplines, and with larger social and environmental movements. And yet, ironically, as one critic observes, despite its best efforts and the emphasis on an ecological model, "the intersection that ecotheology has had difficulty making has been with the science of ecology."[4] In particular, ecotheology has downplayed or ignored the existence of negative processes of the natural world—processes involving predation, competition, disease, and premature and painful death of organisms. When they are acknowl-edged, these features are commonly assumed to be a consequence of the fall, not part of the original plan for nature. In short, ecotheologians have ignored Darwinian evolutionary processes that are central to biological and ecological science. Moreover, what passes as ecology or the ecological model in much of this literature is not necessarily representative of current ecological concepts but is rather a pastiche of only those scientific concepts that ecotheologians find most palatable and most amenable to religious ethics.

The failure to incorporate evolutionary science into religious descriptions of, and ethical prescriptions for, nature creates a number of problems. I will mention just a few of these here. The first and most obvious is that an inaccurate understanding of the natural world may generate ethical imperatives toward nature that are inappropriate for, and even disruptive of, natural processes. This is the case with much of ecotheological ethics, owing in part to the way in which the problem of nonhuman suffering is interpreted in this literature. The ethics proposed (or implied) for remedying suffering are oriented toward the establishment or perhaps *re-creation* of a world that is, as far as we know, biologically impossible. Put differently, nearly four decades after White's critique, Christian environmental ethics still remains too otherworldly, too preoccupied with a prefall ideal of the natural world.

Second, neglecting Darwinism means that religious environmentalists cut themselves off not only from an important source of information about nature but also from a vital source of environmental and animal ethics. By this I do not mean that Darwinism is unambiguously or straightforwardly normative or that there exists a single monolithic form of evolutionary ethics. But talk about animals as our kin in a literal, biological sense and not just a metaphorical or sentimental sense owes everything to the advent of Darwin's theory. Ecotheologians who enthusiastically embrace interrelationship and interconnection must learn to embrace Darwin, even if doing so results in some dampening of their enthusiasm for "interrelationship" as a positive normative concept, as I suspect it might. Yet, historically, Darwinism has contributed much to elevating the moral status of animals, both as individuals and as species, and Darwin's work also provided a foundation for an important body of holistic or ecocentric ethical deliberation, such as that exemplified by land ethics (more on this below). Darwin's theory is arguably the most important development in the history of science in terms of clarifying the place of humans in nature and our relationships, past and present, to all other forms of life. Ecotheologians wishing to steer Christianity away from an overtly anthropocentric worldview should find a valuable reference point in Darwinism.

A third potential problem stemming from a neglect of Darwinian evolution is one that I shall not pursue in detail in this essay but it bears mentioning. Some ecotheological accounts of nature so little resemble an evolutionary account that they could virtually pass for creationism or other arguments from design. At a time when a new and more sophisticated generation of creationists is attempting to wedge themselves, unnoticed, into the academy, ecotheology needs to affirm a Darwinian stance. If it cannot do so, it should not attempt to weigh in on issues pertaining to ecological science and ethics.

The Ecological Model and the Mechanical Model

It is necessary to get a sense of what ecotheologians have in mind when they speak of "current" science. The ecological model is a dominant concept in much of contemporary ecotheology. While not all ecotheologians explicitly invoke it, many assume the same basic account of the natural world and derive from that account a similar set of normative guidelines. First and foremost, the ecological model is consistently offered as an alternative to mechanistic (or, variously, atomistic, Cartesian, Baconian, or Newtonian) models believed to perpetuate neglect and exploitation of nature. We can discern, in frequent juxtapositions of the ecological model to its inferior, disenchanted, mechanical predecessor, the legacy not only of White's critique of Christianity but also, it seems, the lasting impact of Carolyn Merchant's account of the "death of nature" at the hands of mechanistic science and philosophy.[5] White himself, it might be noted, actually discerns a type of reverential natural theology even in the work of Newton and other thinkers of the scientific revolution, and he holds that it is not so much modern mechanical science that is taking its toll on nature but the more recent and nefarious fusion of science with *technology* (rooted, of course, in Genesis "dominionism"). He and Merchant are in agreement that the animistic paganism that once envisioned powerful spirits inhabiting every realm of nature has gone by the wayside. But in White's account, Christianity systematically destroyed animism and its value system—thus effecting "the greatest psychic revolution in the history of our culture"—long before mechanical science emerged.[6]

Merchant's claim is that seventeenth-century mechanical science, particularly exemplified by the work of Francis Bacon and René Descartes, disenchanted nature, stripping it of its vital animism—and, thereby, its value—because the scientific revolution conceptualized a universe of mere matter, a kind of lifeless collection of atoms set in motion by a deity external to the system. Since nature is nothing but dead, inert particles and since God is remote, humans (particularly male scientists) have attempted to reclaim a godlike role of manipulating and rearranging those parts as they see fit, with an attitude of complete detachment and ready-made "justification for power and dominion over nature."[7]

Mechanistic models thus legitimize abuse, domination, and objectification of nature and its inhabitants.

Merchant's book-length argument is, of course, far more complex than this, building upon environmental history, feminist theory, rival interpretations (both religious and scientific) of the fall narrative, as well as the rise of technology and the increasing dominance of the capitalist worldview. But her basic claim, namely, that modern science (i.e., that heralded by the scientific revolution and taught in classrooms to this day) and its mechanistic framework put nature in peril remains a crucial point of departure for ecotheology, as well as for other forms of environmentalism. As a result, ecotheologians have championed their model as an alternative to the hegemony of the mechanistic model. But another lingering after-effect of Merchant's work and similar works touting the virtues of "postmodern science" vis-à-vis the old paradigm has been a deep ambivalence about science and scientific method; many ecotheologians share a vague sense that the dominant, Western mode of science is inherently aggressive and repressive. This ambivalence is unfortunate, and it leads some to borrow selectively from scientific concepts and data or to neglect science altogether.

Merchant's account and others based on it assume that prior to seventeenth-century science a different model of nature held sway, one in which nature was seen as a sort of nurturing mother, a living, animate organism, an entity with subjecthood.[8] This assumption is shared by some ecotheologians who present their model as one that reaches backward to premodern sensibilities and forward to postmodern theories. So, for example, Sallie McFague argues that the modern ecological model resembles medieval notions of nature's subjecthood, relationality, and interconnectedness, though the medieval paradigm was somewhat more hierarchical. The current ecological model is said to be informed by the account of the physical world "coming to us from postmodern science and ecology," wherein all beings have equal worth and membership in the larger system, community, or organism that is earth.[9] An understanding of our planet as a single integrated entity, a living organism, is upheld as an important feature of both the premodern and postmodern views. (For this reason, many ecotheologians also gravitate toward the Gaia hypothesis of James Lovelock, which describes our planet in similar terms and also evokes a sense of earth as mother or goddess.)

The main features of the ecological model, then, are as follows. The model champions interdependence and interrelationality, stressing the importance of the relations and links between all living things rather than interpreting them atomistically, in isolation from one another. Interdependence and interrelationality, in turn, suggest an ethic built around mutuality, care, liberation, and perhaps love for all other beings. The model embraces community, a common good for all beings, while celebrating difference and highlighting the value of the individual distinct from communal value. "Interdependence" in this model is simultaneously descriptive (of nature and the nature-human relationship) and prescriptive. That is, interdependence is understood to be something inherently and self-evidently good, and, it is often argued, if we would but take nature's lessons to heart, deriving from nature a blueprint for our own societies, all beings could live a more peaceful and fulfilled existence and our impacts on earth would be lessened significantly. In this model, notes of cooperation, symbiosis, and solidarity are sounded far more frequently and clearly than are discordant notes of competition, predation, and strife. Rosemary Radford Ruether, for example, argues that contemporary ecology is not only a science of biotic communities but also "suggests guidelines for how humans must learn to live as a sustaining, rather than destructive, member of such biotic communities."[10] The normative import of ecology emerges from the fact that "cooperation and interdependency are the primary principles of ecosystems."[11] Human communities which operate on principles of competition—"social hostility and competition for resources"—are in violation of natural patterns and guidelines.[12]

From statements such as these, one would assume that the model in question is derived from nature's functioning and applied as an ideal to human communities. But it is not always clear that the ecological model has much at all to do with sciences such as ecology, ethology, or biology, nor is it clear whether this blueprint is derived from nature, imposed on it, or (paradoxically) both. McFague, for example, sometimes urges us to apply or extend the ecological model *to* nature, as if it originated elsewhere. Similarly, process thinkers Charles Birch and John Cobb pose the question of whether the ecological model, which seems so well suited to human communities, might also be applicable "to other animals."[13]

Indeed, closer scrutiny of this model suggests that it has little to do with nature as we know it. The ecological model, McFague writes, assumes that the "well-being of the whole is the final goal, but that this is reached through attending to the needs and desires of the many subjects that make up the community."[14] If so, then the model cannot be derived *from* nature as it really is, because in nature the needs and desires of each and every individual are rarely if ever met simultaneously. Nature's interdependence is constructed from strands of conflict, and organisms often relate to one another in and through competition for resources, as Darwin knew. Nature's interdependence, in other words, is not a guarantee that needs will be met, nor is it an antidote to want and strife, as ecotheologians sometimes assume. But if the model does not derive from nature, what is its source? In McFague's case, the model seems to be generated primarily from Christian sensibilities regarding the needs of the oppressed and the claims of the suffering individual. The ecological model is inspired by the ministry of Jesus, as McFague readily concedes. We ought to care for *nature's* marginalized, despised, and oppressed beings, "healing the wounds of nature and feeding its starving creatures" just as a Christian community would focus on "feeding and healing its needy human beings."[15] Thus the ecological model is something we extend to the natural world as a model that is "highly compatible with the spirituality of Jesus's ministry, for it sets us in a world of radical relationality at all levels."[16] It appears that this model is termed "ecological" because it is *applied* to natural systems, not because it is based on them. What role, then, does science play in this model, and why do many ecotheologians claim to find support for it in natural science?

In fact, the science of particular interest to many ecotheologians seems to be physics rather than biology. Process thinkers like Birch and Cobb draw inferences from modern (or, again, postmodern) physics, such as quantum theory and chaos theory, which reveal a physical universe of complex, interconnected events, or *processes*, rather than an atomistic array of discrete objects. Contemporary physics underscores continuity among all entities and undermines dualistic categories, particularly any absolute distinctions between subject and object or between the observer and the observed. In these laudatory allusions to postmodern science, it is assumed that the old mechanistic perspective and its presumption of objectivity is on its way out. In postmodern sciences such as physics and (according to these thinkers) ecology, "alienation from the matter being studied is overcome," as Frederick Ferre argues, and the "esprit, from the sense of being involved, from participating in what is of scientific interest rather than being merely the disinterested observer, places the science of ecology quite apart from other sciences in obvious ways."[17]

Yet this eagerness to dispense with objectivity and disinterestedness may have pernicious results. Paul R. Gross and Norman Levitt argue in *Higher Superstition* that the old Baconian paradigm is routinely characterized as "giving way to an holistic, nonexploitative way of knowing, a new form of science that expresses the spiritual pantheism of the new order."[18] In the spirit of postmodernism, they note with exasperation, "subjectivity is not only not suspect: it is demanded. Objectivity, on the other hand, is dismissed, curtly, as the delusion of a Western consciousness obsessed by domination, exploitation, and profit."[19] As they suggest, objectivity is somehow understood to be the same as *objectification*, disinterestedness the equivalent of injurious neglect. Thus, objectivity must be jettisoned in order to liberate life from this ancient, oppressive stranglehold.

Gross and Levitt attribute these views primarily to what they consider anti-science "ecoradicals," but in fact such sentiments are commonplace in ecological theology. Note, for example, the similarities in the following accounts of the old mechanistic paradigm and its welcome displacement by postmodern science and the ecological model:

> The Cartesian objectification of the world destroys the natural environments of living things in order to incorporate them in the environment of the dominating human subject.[20]

> Western sensibility has traditionally been nurtured by an atomistic, reductionistic perspective that separates human beings from other beings and reduces all that is not human to objects for human use.... In the early years of the

twentieth century, there was a movement toward a model more aptly described as organic, even for the constituents with which physics deals, for there occurred a profound realization of the deep relations between space, time, and matter, which relativized them all.[21]

Events in an electromagnetic field exemplify the ecological model as much as does animal behaviour in the wild. . . . The ecological model counts against any dualism between the inorganic and the organic, the non-living and the living.[22] The distinctively "modern" view of the world, which through successful propaganda became virtually equated with the "scientific" worldview, had at its basis the rejection of "animism." Included under animism were all doctrines that saw the basic constituents of nature as having perception or experience of any sort.[23]

The shell of biology today is still an old one from the past. . . . To break down the old shell and to begin to grow a new one is to begin the process of liberating the concept of life. . . . If we can liberate the concept of life we might be able to liberate life itself. . . . The ecological model . . . is a liberated and liberating alternative.[24]

The postmodern sciences deconstruct and transcend the Cartesian metaphysical distinctions between humankind and Nature, observer and observed, Subject and Object[,] overthrow the static ontological categories and hierarchies characteristic of modernist science. In place of atomism and reductionism, [they] stress the dynamic web of relationships between the whole and the part; . . . the postmodern sciences appear to be converging on a new epistemological paradigm, one that may be termed an *ecological* perspective. . . . A liberatory science may arise from interdisciplinary sharing of epistemologies.[25]

Clearly, there is a great deal of consensus among these thinkers regarding the achievements and novel insights of postmodern science. The first five quotations are from prominent ecotheologians. The final quotation is from the text of that famous parody of postmodern thought—the well-known "Sokal hoax"—published in the journal *Social Text* nearly a decade ago, entitled "Transgressing the Boundaries: Toward a Transformative Hermeneutics of Quantum Gravity." There Alan Sokal, a physicist, suggested a link between quantum physics and postmodern thought (like many ecotheologians, he cites Merchant freely in making his case). The article was published despite numerous, intentional factual errors regarding physics and the author's deliberate use of obfuscating jargon and, in places, utterly nonsensical prose. Recounting the lessons to be learned from his hoax and its aftermath, Sokal makes the reasonable observation that "anyone who insists on speaking about the natural sciences—and nobody is forced to do so—needs to be well-informed and to avoid making arbitrary statements about the sciences or their epistemology."[26]

Whether or not these devotees of postmodern physics are well informed, the fact remains that wildlife managers, conservation biologists, and other environmental scientists do not make management decisions or set their priorities according to the findings of quantum physics. To say that the ecological perspective and, say, quantum science are essentially one because both recognize the importance of interdependence is absurdly simplistic; it is like drawing the conclusion from relativity theory that "everything is relative" (a point that McFague comes close to making in the second quotation above). In any case, the ethical import of quantum interdependence is ambiguous if not entirely vacuous. For example, postmodern epistemology (according to its devotees) tells us essentially that what is known is not objective or real but is always conditioned by the perspective of the knower. Moreover, reality, as one postmodernist puts it, consists of "one unbroken whole" implying that "things could apparently be connected with other things any distance away."[27] It is difficult to see how these insights, considered so crucial by postmodern environmentalists, provide in themselves much useful guidance for the environmental problems we face. Certainly, for both ethical and scientific reasons, wildlife biologists conducting experiments in the field want to be conscious of impacts they may be having on the organisms under study and how those impacts may spread from one area to another. But this is not a new form of consciousness gleaned from postmodern physics. Rather, this awareness merely underscores the need for attaining greater objectivity in research; it is not grounds for *abandoning* measures aimed at ensuring that data are not skewed by the method of collection or observation, nor does it constitute an admission that organisms being studied have no objective reality independent of the scientist observing them.

However, my concern is not so much with whether these pronouncements about postmodern science accurately reflect current physics (I leave that to Sokal and Gross and Levitt, among others, who argue persuasively that they do not). Rather, I am troubled by the fact that ecotheologians largely ignore more pertinent ecological and evolutionary thought in favor of these rather abstruse theories and concepts. It is also, I believe, a sign of stagnation in any discipline when there is so much repetition of the same ideas, as is the case within ecotheology, where there is little dissent from the now decades-old assumption that mechanistic science is being replaced by some new, organic, and liberatory paradigm.

But why are so many environmentalists more enamored of the Heisenberg Uncertainty Principle than, say, the theory of natural selection or Lotka-Volterra equations for predator-prey and parasite-host interactions which so perfectly illustrate intimate interconnections between predator and prey? Is Darwinism also to be junked as a modern, mechanistic, manipulative science? Perhaps the enthusiastic turn to postmodern physics has something to do with the fact that it is much harder to blend ethics of liberation with frank acknowledgement of predation and parasitism as fundamental. Darwinism seems to connote something blind, aggressive, antagonistic, and hierarchical compared with the unbroken whole of quantum ecology. "The overall impact of [Darwin's] thought," we are told by the author of a widely used textbook in religion and ecology, "supported a view of nature as hostile to human beings and mandated the human species' attempts to subdue the natural world through scientific and technological means."[28] Here we see that it is *Darwinism*—cast as the currently reigning mechanistic science—and not Genesis that provides ongoing justification for dominionistic attitudes.

THE FALL AND THE PEACEABLE KINGDOM

A large part of the reason that ecotheologians do not dwell on the details of evolutionary biology, I believe, is because they seem in conflict with the vision of nature they most cherish and toward which their ethics are aimed. Newcomers to the field might be surprised to discover just how rampant the belief in nature's inadequacy—its fallenness—is in ecotheology.[29] The motif of the peaceable kingdom, in which predator and prey exist in harmony and nature is once more infused with tranquil abundance, recurs throughout this literature. As we have seen, ecotheologians want to join the good of the individual and the good of the natural "community"—the parts and the wholes—in a way that dissolves all conflicts between them. The problem of predation is the most obvious, though not the only, form of conflict between the individual and the natural system in which it is embedded.

For some of the most prominent ecotheologians, biblical stories of a lost paradise serve as a touchstone for environmental ethics. Ruether hopes for a final restoration of nature that will usher in "right relations" among all creatures, thus "healing nature's enmity." The book of Isaiah, she reminds us, promises that "even the carnivorous conflict between animals will be overcome in the Peaceable Kingdom."[30] McFague's "subject-subjects model" draws inspiration from biblical stories "in which the lion and the lamb, the child and the snake, lie down together; where there is food for all; where neither people nor animals are destroying one another."[31] Jürgen Moltmann awaits the time of *creatio nova* when the spirit of God "drives out the forces of the negative, and therefore also banishes fear and the struggle for existence from creation."[32] A "peaceable kingdom of shalom and ecological harmony" in which "predatorial behavior will no longer characterize human and non-human relations" comprises part of Michael Northcott's vision of God's will for creation.[33] Birch also anticipates a time when "paradise is regained, and everyone not only goes back to a nonmeat diet, but the friendliest relations subsist between all species."[34] Even secular ethicists such as Tom Regan have at times advocated animal rights as an important step in the "journey back (or forward) to Eden [and] God's original hopes for and plans in creation," though here the issue is more one of humans eating other animals than nonhuman animals eating each other.[35]

As ecotheologian Larry Rasmussen explicitly acknowledges, predation is "not a pattern of morality we praise and advocate" for our own communities—at least, he quips, "not on our better days."[36] Yet the sort of natural community envisioned by many ecotheologians could come about only if natural selection were somehow to cease. Much of this literature seems aimed at wholesale recreation of nature rather than working with natural patterns and processes to repair the damage we have done. This preoccupation with the idea that natural processes are inherently sinful or the result of sinful corruption is a major obstacle not only to a more appropriate environmental ethic but also to more meaningful dialogue with the natural sciences.

APPLYING ECOTHEOLOGICAL ETHICS TO NATURE

In order to appreciate the potential mismatch between certain ecological ethics and nature's *true* nature, let us linger for a moment over some of the ethical imperatives being offered and the potential results of their application to nature.

Among ecotheologians, McFague has argued most forcefully for the extension of a Christian ethic of love and care for the oppressed to the natural world. In *Super, Natural Christians*, McFague's "subject-subjects" ethic is offered, not surprisingly, as an alternative to the (old) subject-*object* model which is "dualistic, hierarchical, individualistic, and utilitarian," regarding animal others as *its* rather than *thous*.[37] Her own approach emphasizes love and care for all subjects as subjects and calls on us to reenvision ourselves and all selves as "constituted by relationships and exist[ing] only in relationships."[38] This expanded vision is essentially an ecological vision. What makes the subject-subjects model distinctly Christian is its previously noted affinity with the ministry of Jesus and his preferential love for the oppressed, despised, marginalized, hungry, suffering neighbor. In today's world, nature is the most oppressed class, the "new poor" as McFague terms it, and "nature" here includes both wild and nonwild nature, pristine forests and urban landscapes.[39]

McFague's apparent endorsement of intervention in nature to feed its starving creatures and heal their wounds does not sit well with biological realities, given that natural selection is fueled by competition for habitat, food, and mates. That every organism has bodily needs is beyond dispute, but the claim that humans ought to attend to those needs is highly suspect. But McFague may object that I am taking too literal a reading of her metaphorical theology and that she does not really intend that Christians directly minister to animal bodies in this way. Yet, she repeatedly stresses the literal importance of bodily health in the Christian tradition, noting that Jesus did not just minister to people spiritually but physi-cally, healing their wounds, curing their ailments and feeding those who hun-gered. Moreover, Jesus, as the embodied form of God, calls attention to the value of bodies. McFague even concedes at times that her ethic is "counterbiological" and that natural selection does not operate "on the themes of sharing and in-clusion" found in Jesus's ministry. Rather, her ethic is meant to herald a "new stage of evolution, the stage of our solidarity with other life-forms, especially with the needy and outcast forms."[40] In other words, rather than make her ethic more consistent with biological reality, she proposes that nature's reality be *changed*, bringing about new, post-Darwinian forms of human-animal coexistence. It is hard to see how this approach constitutes a less manipulative attitude toward nature than the controlling, mechanistic attitudes that McFague wants to displace.

Compare this account to that endorsed by Birch and Cobb. Their process view of nature certainly adheres more closely to a Darwinian account than many other ecotheological descriptions, despite their penchant for turning to postmodern physics more often than to evolutionary biology. In fact, one of the stated inten-tions of Birch and Cobb's project is to think more critically about notions such as balance, harmony, and order in the natural world, to recognize the role of com-petition and strife in nature. In their collaborative work, *The Liberation of Life*, for example, Birch and Cobb endeavor to "take evolutionary history seriously in the effort to understand humanity, its future, its relations to other living things and to the life that is inwardly experienced."[41] They note that struggle and suffering are inherent features of nature. Process metaphysics interprets the fall not as original sin in the biblical garden of Eden but as a past evolutionary stage that introduced both greater complexity and greater discord into the human psyche. "The emer-gence of conscious experience was the crossing of a great new threshold," Birch and Cobb argue, "which we associate with the development of a central nervous system."[42] That is, as evolutionary processes led to increasing complexity, and particularly to the advent of higher consciousness, they also created possibilities for evil. In humans, rich experience was and is inextricably linked to possibilities for disharmony and discord. We may choose to live in such a way as to minimize discord in our lives and our experience, but not without losing much of what makes life rich and enjoyable, rather than merely trivial. In the absence of some measure of suffering and discord, intense enjoyment is not possible.

The same can be said, more or less, of the nonhuman world—and this "more or less" approach is significant: process thought embraces a form of panexper-ientialism or panpsychism that sees all life as potentially capable of both discord and enjoyment. (Here is where postmodern science comes in, undermining and breaking down traditional boundaries between subjects and objects and sharp categorizations of different forms of life and modes of experience.) Like us, nonhuman animals may learn transcending responses in the face of the strife they encounter, and novelty is continuously introduced into the evolutionary process via those responses. The removal of all suffering would also shut down these creative, transcending responses in both humans and animals. And so suffering is the flip side of intensity of experience, enjoyment. Birch and Cobb conclude: "Contentedness and discontentedness belong together" and suffering is "necessarily entailed in the creation of beings capable of high grades of enjoyment."[43]

However, two problems immediately arise from this argument about suffering and richness. The first has to do with the way in which animals are "more or less" like humans; the second stems from the coupling of an ethic of liberation with the process view of suffering.

Process thought attributes richness of experience to all life-forms but in varying degrees. In this "tiered" system of value, those organisms capable of great richness—generally signaled by the presence of a central nervous system, large brains, complex mental activity, and even possession of language—have a greater capacity to suffer as well, and thus our greater obligations are to them. We give moral consideration to animals "in proportion to their capacity for rich experience."[44] Lacking a central nervous system, plants, on the other hand, are more like aggregates or "societies" than like true individuals, and they have only instrumental value. Unlike higher animals, they can appropriately be used as means to human and other animal ends. Of course, there is a gradation of richness and thus a gradation of value *within* the animal kingdom as well, as some animals are more highly developed and more complex than others. So, we find Birch and Cobb affirming that when it comes to assessing the relative value—and relative moral claims—of different animals, a porpoise, for example, has greater richness than a tuna or shark. "Accordingly," they continue, "porpoises make claims upon us beyond those made by tuna or sharks."[45]

The problem with this hierarchy of value based on richness is that it cannot operate as a *biodiversity* principle. While it has the advantage of casting the net widely in terms of valuing a range of life-forms, such an ethic would seem to give priority and preferential protection to organisms according to their status as experiencers—regardless of whether such organisms are members of an endangered or overabundant species, for example, or whether they are native or introduced, or generally how they fit into the natural system as a whole. Birch emphasizes that "greater obligation is entailed toward those creatures that have more significant experience."[46] Ironically, in process thought, individual qualities, such as self-awareness, capacity for pain, and so on, are valued in relative isolation from the systems and *processes* in which organisms are embedded. In this sense, as Clare Palmer has observed, process thought is closer to animal rights/liberation positions in its *individualist* thrust than it is to holistic or systemic approaches (such as those I allude to at the end of this chapter).[47] We are left with no obvious grounds for valuing organisms with little or no richness, particularly when conflicts arise between rich organisms and those with little richness, as often happens in wildlife management. A native but nonsentient or minimally sentient species might be threatened by an introduced but highly sentient ("rich") one. Following Birch and Cobb, the former species has little or no moral claim. Culling the sentient species, as wildlife managers must sometimes do, would violate the basic ethical presumption in favor of promoting rich experience.[48]

Consider some further implications of the imperative to promote richness of experience. Since the experience of nonhuman animals is largely a function of their bodily enjoyment (or discord), we are obligated to "reduce the amount of suffering *we inflict*."[49] That is, since nonhuman animals' enjoyment is largely mediated through their physical experiences, physical suffering greatly reduces the richness of those experiences. Like McFague's, this account treats nonhuman life as an oppressed class deserving of an ethic of liberation, where liberation includes freedom from painful conditions. But why do we have an obligation to liberate life from suffering when, as process metaphysics has already established, suffering is intertwined with and inextricable from richness and creative transcendence? Would not the elimination of suffering impede the development of transcending responses that propel the evolution of consciousness onward? Perhaps Birch and Cobb are merely asking us to reduce the amount of unnecessary or excessive suffering that *humans* introduce ("we inflict") in animals lives by cultural practices such as experimentation on animals, factory farms, and the like. But this remains unclear, especially in light of their celebration of postmodernism's destruction of categories and distinctions between forms of life, and thus Birch and Cobb call for liberating all life from "the cell to the community" (as the subtitle of their book phrases it). These arguments seem to drift back to the assumption that suffering is not natural or inherent but symptomatic of nature's fallenness, as when Birch claims that our duty consists in reestablishing a peaceable kingdom wherein "the friendliest relations subsist between all species."[50]

Like McFague, Birch and Cobb move further and further away from an ecological, evolutionary account of nature as their ethic unfolds, leaving us with vague assertions of interdependence and an obligation to liberate life that is supposedly consistent with and promoted by a recognition of interdependence.

CONCLUSION: VALUING NATURAL PROCESSES

Despite the emphasis repeatedly laid upon holism, interdependence, and process in ecotheology, far too little attention is given to the possibility that great value resides in the patterns and processes—including natural selection—evident in the world we actually inhabit. Ironically, in much of this literature, nonhuman animals are valued according to similarities to humans, evidenced by their possession of certain (evolutionarily produced) traits such as sentience or richness, even while the processes that produced these traits are ignored or even considered in some sense evil. What would it mean for ecological models and environmental ethics to pay closer attention to these natural patterns and processes? Let me suggest the basic contours of what I see as a preferable, and more realistic, environmental ethic.

In the first place, the fall, and the account of evil that it implies, makes little sense in light of natural history since struggle and suffering were features of nature long before humans evolved. Holmes Rolston argues that a "peaceable natural kingdom, where the lion lies down with the lamb...is a cultural metaphor and cannot be interpreted in censure of natural history."[51] This reminder also casts doubt on the proposal that we minister to nature with Jesus's ethics in mind, healing and feeding those who need it. Such ethics is not only too interventionist but also too skewed toward the interests of the most complex and sentient beings (those most capable of suffering). Second, Christian environmental ethics needs to shift some of its focus away from the needs of the individual and to recognize that in nature the larger system—the ecosystem, the species as a whole—has properties and values that may count over and above the value of each individual. Much of what is valuable about nature resides in the processes that propel the system forward, not merely in the products (i.e., the individual organisms). Natural processes often involve pain or death that is instrumental for species survival, even while not directly beneficial for the individual whose life and genes are selected against by evolution. Saving individual organisms from natural conditions of suffering only undermines evolution's way of adapting species to changing environments. A particular organism may or may not be a good evolutionary fit, but a species line has options, a collective gene pool, and thus a kind of collective creativity and value. Our ethics toward nature entails responsibilities not primarily to individuals but to species—as well as to processes such as *speciation*. Interventions in natural processes are warranted primarily when humans' past interventions have disrupted natural function: damage to nature (and attendant suffering) caused by humans should be addressed, but with the somewhat paradoxical goal of managing nature in order that it can manage itself. This may involve us in a complex task of mimicking holistic, natural processes and, at times, treating the parts as relatively expendable vis-à-vis the whole, as nature appears to. The parts and the wholes cannot always be harmonized and preserved simultaneously, contrary to the wishes of ecotheologians, without seriously contravening natural processes. Postmodern theorizing about the unbroken wholeness, the radical continuity of reality, and the need to recognize subjectivity in all life-forms will not change these facts or resolve these conflicts.

In ethics, distinctions are important. Admittedly, an ethic of this sort introduces a fundamental distinction between nature and culture that some might find problematic: an interventionist ethic of compassion for the suffering individual, regardless of the source or cause of that suffering, may be appropriate for humans and for domesticated animals, but not for nature. Humans are no longer buffeted by the forces of natural selection as our wild relatives are; for better or for worse, we have had considerable success in removing ourselves (and our domesticated animals) from these refining fires. But in dealing with our wild neighbors, we have an obligation to respect their wildness, to refrain from intervening even when an impulse of love and compassion might dictate otherwise. ·

The ethic I have briefly outlined here is roughly consistent with the ecocentric land ethics of American conservationist Aldo Leopold and modern-day proponents of his work, such as Holmes Rolston and J. Baird Callicott. For Leopold, an appropriate ethic toward nature was captured in the golden rule of land ethics, namely, that an action is right when it tends to preserve the integrity, stability, and beauty of a biotic community and wrong when it tends otherwise. In practice, this means that management of nature should be kept to a minimum and that natural processes that shaped the character of a biotic community should be permitted to operate, regardless of whether their operation meets with human approval, values, or preferences. Clearly, this means permitting, and even facilitating when necessary, some of the very functions of nature that ecotheologians find troubling (Leopold was particularly interested in protecting trophic relations, e.g., between predator and prey).

One question that remains is whether an ethic that at times subordinates the individual to the group and leaves that individual's suffering unattended can ever be compatible with theism, and more particularly with Christianity. I would argue that a failure to value nature as it really is—in its *given* ordering rather than its prefall or eschatologically anticipated perfection—constitutes a lack of piety, a failure to show respect and gratitude for the deity who created and sustains this ordering. Theologian James Gustafson argues that the desire to alter or perfect this basic, natural ordering implies a denial of God as God, a refusal to "consent" to the patterns and processes that manifest the deity in the realm of nature.[52] An ecocentric ethic, by which I mean a general presumption in favor of respecting natural processes and intervening cautiously and with regard to the broader ecological context, can also be a *theocentric* ethic. Both ecocentric and theocentric perspectives take seriously the possibility that, as Gustafson puts it, "in the sphere of nature, our interactions frequently require greater conformity to the ordering that is present than they permit a new ordering or even a mastery of the ordering that exists."[53] Humans as finite beings are participants in natural processes that we did not create and cannot fully control. Processes such as natural selection, and the natural ordering more generally, have value that is independent of humans and relative to God, even while humans may find this ordering disconcerting or even, at times, wasteful and improvident. Interdependence in nature is inseparable from conflict, and both are born of the same natural processes. A truly ecological, evolutionary ethic must take all of the available facts of nature into account, not merely those that point to the sort of world we wish we lived in.

NOTES

1. Lynn White, "The Historical Roots of Our Ecologic Crisis," repr. in *This Sacred Earth: Religion, Nature, Environment* (ed. Roger S. Gottlieb; 2nd ed.; New York: Routledge, 2004), 189.

2. Ibid., 191.

3. Lance Nelson, "The Dualism of Nondualism: Advaita Vedanta and the Irrelevance of Nature," in *Purifying the Earthly Body of God* (ed. Lance Nelson; Albany: State University of New York Press, 1998), 62.

4. Judith Scoville, "Fitting Ethics to the Land: H. Richard Niebuhr's Ethic of Responsibility and Ecotheology," *Journal of Religious Ethics* 30.2 (2002): 225.

5. Carolyn Merchant, *The Death of Nature: Women, Ecology, and the Scientific Revolution* (San Francisco: HarperCollins, 1980).

6. White, "Historical Roots," 188.

7. Merchant, *Death of Nature*, 215.

8. Merchant has since backed off slightly from what she now sees as a "declensionist" narrative of *The Death of Nature* and has taken a somewhat different tack in her recent book *Reinventing Eden: The Fate of Nature in Western Culture* (New York: Routledge, 2003).

9. Sallie McFague, *Super, Natural Christians: How We Should Love Nature* (Minneapolis: Fortress, 1997), 20.

10. Rosemary Radford Ruether, *Gaia and God: An Ecofeminist Theology of Earth Healing* (San Francisco: Harper, 1992), 47.

11. Ibid., 57.

12. Ibid., 54.

13. Charles Birch and John B. Cobb Jr., *The Liberation of Life: From the Cell to the Community* (Cambridge: Cambridge University Press, 1981; repr. Denton, TX: Environmental Ethics, 1990), 22.

14. McFague, *Super, Natural Christians*, 158.

15. Ibid., 169.

16. Ibid., 32.

17. Frederic Ferré, "Religious World Modeling and Postmodern Science," in *The Reenchantment of Science: Postmodern Proposals* (ed. David Ray Griffin; Albany: State University of New York Press, 1998), 94.

18. Paul R. Gross and Norman Levitt, *Higher Superstition: The Academic Left and Its Quarrels with Science* (Baltimore: Johns Hopkins University Press, 1994), 152.

19. Ibid., 176.

20. Jürgen Moltmann, *God in Creation* (Minneapolis: Fortress, 1993), 147.

21. Sallie McFague, *Models of God: Theology for an Ecological, Nuclear Age* (Philadelphia: Fortress, 188), 8–11.

22. Birch and Cobb, *Liberation of Life*, 89.

23. David Ray Griffin, "Of Minds and Molecules: Postmodern Medicine in a Psychosomatic Universe," in *The Reenchantment of Science: Postmodern Proposals* (ed. David Ray Griffin; Albany: State University of New York Press, 1998), 143.

24. Birch and Cobb, *Liberation of Life*, 68.

25. Alan Sokal, "Transgressing the Boundaries: Toward a Transformative Hermeneutics of Quantum Gravity," repr. in Alan Sokal and Jean Bricmont, *Fashionable Nonsense: Postmodern Intellectuals' Abuse of Science* (New York: Picador, 1998), 239.

26. Sokal and Bricmont, *Fashionable Nonsense*, 185.

27. David Bohm, "Postmodern Science and a Postmodern World," in *The Reenchantment of Science: Postmodern Proposals* (ed. David Ray Griffin; Albany: State University of New York Press, 1998), 65.

28. David Kinsley, *Ecology and Religion: Ecological Spirituality in Cross-Cultural Perspective* (Upper Saddle River, NJ: Prentice-Hall, 1995), 137.

29. All of the ecotheologians cited here claim that their ecological views are scientifically informed. In fact some explicitly make claims to a Darwinian worldview.

30. Ruether, *Gaia and God*, 213.

31. McFague, *Super, Natural Christians*, 158.

32. Moltmann, *God in Creation*, 102.

33. Michael Northcott, *The Environment and Christian Ethics* (Cambridge: Cambridge University Press, 1998), 194.

34. Charles Birch, "Christian Obligation for the Liberation of Nature," in *Liberating Life: Contemporary Approaches to Ecological Theology* (ed. Charles Birch, William Eakin, and Jay B. McDaniel; Maryknoll, NY: Orbis, 1990), 67.

35. Tom Regan, "Christianity and Animal Rights," in *Liberating Life: Contemporary Approaches to Ecological Theology* (ed. Charles Birch, William Eakin, and Jay B. McDaniel; Maryknoll, NY: Orbis, 1990), 87.

36. Larry Rasmussen, *Earth Community, Earth Ethics* (Maryknoll, NY: Orbis, 1998), 347.

37. McFague, *Super, Natural Christians*, 7.

38. Ibid., 162.

39. Ibid., 170.

40. Sallie McFague, "The Scope of the Body: The Cosmic Christ," in *This Sacred Earth: Religion, Nature, Environment* (ed. Roger S. Gottlieb; 2nd ed.; New York: Routledge, 2004), 293.

41. Birch and Cobb, *Liberation of Life*, 3.

42. Ibid., 135.

43. Ibid., 107.

44. Ibid., 153.

45. Ibid., 155.

46. Birch, "Christian Obligation for the Liberation of Nature," 58.

47. Clare Palmer, *Environmental Ethics and Process Thinking* (Oxford: Oxford University Press, 1998).

48. For examples of such (real-world) cases in wildlife management where conflicts occur between sentient and nonsentient, native and nonnative organisms, and so on, see Holmes Rolston, *Conserving Natural Value* (New York: Columbia University Press, 1992).

49. Birch and Cobb, *Liberation of Life*, 155 (emphasis added).

50. Birch, "Christian Obligation for the Liberation of Nature," 68.

51. Holmes Rolston, "Wildlife and Wildlands," in *After Nature's Revolt* (ed. Dieter Hessel; Minneapolis: Fortress, 1992), 131.

52. See James M. Gustafson's two-volume *Ethics from a Theocentric Perspective* (Chicago: University of Chicago Press, 1984) as well as *A Sense of the Divine: The Natural Environment from a Theocentric Perspective* (Cleveland: Pilgrim, 1994). I certainly do not mean to suggest that humans should consent to natural processes in all realms of life and all situations, such as in cases of human health or disease treatment and prevention, but rather in our dealings with what remains of our natural world and wild animal species. For a fuller discussion of such an ethic, see Lisa H. Sideris, *Environmental Ethics, Ecological Theology, and Natural Selection* (New York: Columbia University Press, 2003).

53. Gustafson, *Ethics from a Theocentric Perspective*, 1.283.

PART III

RELIGIOUS
ENVIRONMENTAL
ACTIVISM

CHAPTER 21

RELIGIOUS ENVIRONMENTALISM IN ACTION

ROGER S. GOTTLIEB

We come here as multi-racial and multi-cultural witnesses to shed light on long-term health consequences associated with toxic waste.... We are here to name the sin of environmental racism and to renew our call for real and lasting environmental justice in order that the burden of toxic waste will be shared by all— and not just some.

> —The Rev. Henry Simmons, board chair, Justice and
> Witness Ministries of the United Church of Christ,
> protesting a plan to incinerate nerve gas in East St. Louis,
> April 20, 2002

By depleting energy sources, causing global warming, fouling the air with pollution, and poisoning the land with radioactive waste, a policy of increased reliance on fossil fuels and nuclear power jeopardizes health and well-being for life on Earth.

> —"Let There Be Light," an open letter to President George
> W. Bush, signed by thirty-nine heads of denominations
> and senior leaders of major American faith groups,
> May 18, 2001

When unethical harvesting of trees is infringing on the health of the land, on sacred mountains, then we have to protect it.

—Earl Tulley, Diné (Navajo) Committee Against Ruining the Environment, quoted in Ernie Atencio, "After a Heavy Harvest and a Death, Navajo Forestry Realigns with Culture," *Western Roundup*, October 31, 1994

Washington, DC, May 3, 2001

AFTER two days of meetings and lobbying, prayer services and press releases, fifty-year-old Episcopalian priest Margaret Bullitt-Jonas, accompanied by twenty-two other ministers, priests, rabbis, and lay people, many in their clerical robes, moved toward the gates of the Department of Energy. Nearly 150 supporters looked on as the slender, dark-haired Bullitt-Jonas knelt in prayer. Capital police demanded that she and her fellows leave, and when they refused, they were arrested.

This action, which drew participants from as far away as Alaska and California, was focused on the energy policy of the Bush administration, and in particular on its stated goal of drilling for oil in the Arctic National Wildlife Refuge. The protest had been organized by Religious Witness for the Earth, a network of religiously oriented environmentalists from diverse faith traditions "dedicated to public witness in defense of God's creation." Signed by Protestants, Catholics, Jews, Buddhists, and Muslims, the call to demonstrate had declared that ANWR should be protected as a sacred place for its native inhabitants, a haven for wildlife, and a "cathedral for the human spirit to glory in God's handiwork." "As a born-again Christian, President Bush must understand that creation is sacred," said Rev. Fred Small, cochair of RWE. "His drill-and-burn energy policy endangers not only the wonders of nature but human existence itself. Despoiling the earth is sacrilege, and exhausting its resources is theft from our own children."

For seventeen of the twenty-two RWE members arrested, it was their first time behind bars, something that ordinarily would be considered shameful. Yet, Bullitt-Jonas tells us, she was inspired to participate because during Easter Holy Week, while pondering the sufferings of Jesus, she felt that she needed to witness against today's "greedy mindset that the earth is ours to devour."[1] "I felt," she writes, "as defiant as a maple seedling that pushes up through asphalt. It is God I love, and God's green earth. . . . We may have nothing else, but we do have this, the power to say, 'This is where I stand. This is what I love. Here is something for which I'm willing to put my body on the line.' "[2]

HEADWATERS FOREST, HUMBOLDT
COUNTY, CALIFORNIA, JANUARY 26, 1997

It was Tu B'Shvat, the Jewish "New Year of the Trees," and 250 celebrants had enjoyed a traditional ritual meal honoring the place of trees in human life and the bounty of God in providing them.[3] Then about ninety people walked over a boundary line into a six-thousand-acre section of old-growth redwoods owned by the Maxaam Corporation, a section that Maxaam planned to cut. In defiance of Maxaam's orders, the celebrants-turned-demonstrators planted redwood seedlings in a denuded stream bank.

Over the preceding eighteen months, three local rabbis had pursued Maxaam CEO Charles Hurwitz, a leading member of the Houston Jewish community. They had asked him to "turn completely," in Hebrew to make *tshuvah,* and cease cutting the ancient trees. His willingness to despoil the area, they argued, violated Jewish ethics. They wrote to him that they were praying that God would "soften your heart and give you clear guidance so that your future actions might reflect the wisdom and generosity of Jewish tradition."[4] They had taken out ads in national papers arguing that the old-growth redwoods of Headwaters provide critical habitat for several endangered species and pointing out that whereas only 150 years ago the redwood forests of Oregon and California covered 2 million acres, now less than 4 percent remain.

Perhaps because many of the protestors wore traditional Jewish prayer shawls, perhaps because they had been careful to inform Maxaam and the police beforehand of their plans, the authorities allowed the civil disobedience to proceed without arrest. "At a place where demonstrators before have been met with billy clubs, nightsticks, and arrests, we are now walking freely," said a local environmental activist who had been struggling to protect the areas for years. "It reminds me of the parting of the Red Sea."[5]

Religious environmentalism is a worldwide movement of political, social, ecological, and cultural action. As expressions of a particular religion, in ecumenical alliances with other traditions, through loose networks of spiritually committed activists, and in coalitions with secular environmental organizations, hundreds of groups have resisted global warming, destructive economic "development," dangerous toxic waste dumps, reckless resource extraction, mindless consumerism, and simple waste. In a wonderful pattern of interfaith cooperation, believers have shown that they are capable of actively working with people whose theologies are quite different from their own. Contrary to the widespread secular liberal belief that religion is inherently antidemocratic, religious environmentalists have shown both a broad openness and a deep civic concern.

The two actions described above are themselves part of ongoing movements and campaigns. Religious Witness for the Earth continues its work to this day, its accomplishments including a 2002 Interfaith Service of Prayer and Witness for Climate Action, held inside the Massachusetts State House and in the capitols of every other New England state. The goal was to call on the New England governors to actually implement their agreed-upon climate change action plan. Later, in well-publicized actions in Northampton and Lynn, Massachusetts, activists challenged fuel-hogging SUVs and confronted representatives of automobile companies. In 2003, a public Witness for Creation at the UN drew three hundred participants from surrounding areas.

RWE's leaders have consistently claimed the mantle—and the tactics—of committed religious social activism. "I wanted to explore," said Fred Small, "how to apply the lessons of Gandhi and Martin Luther King, Jr. to a challenge of comparable moral urgency."[6] In their commitment to direct action, says cofounder and United Church of Christ minister Dr. Andrea Ayvazian, they have "upped the ante" for religious leaders, demanding political action along with a greener theology.[7]

The "Redwood Rabbis" actions in Northern California are part of environmentalist efforts by the larger American Jewish community. In these efforts, two organizations have been central. The Shalom Center, headed by veteran social activist and Jewish Renewal rabbi Arthur Waskow, helped design the Tu B'Shvat seder and has long been engaged in bringing a Jewish presence to key environmental issues. The Center's activities have ranged from promoting "Olive Trees for Peace" in an effort to forge peaceful ties among Jews and Palestinians, to its current campaign against the uncontrolled power of "Big Oil," which, they charge, "incites war, endangers the earth, intensifies the asthma epidemic, corrupts U.S. politics, and shatters indigenous peoples in Africa, Latin America, and Asia."[8]

The Coalition on the Environment and Jewish Life (COEJL) began in the spring of 1992 when, partly at the invitation of Al Gore and Carl Sagan, the leadership of the major organizations in American Jewish life, eminent rabbis, denominational presidents, and Jewish U.S. senators gathered in Washington, DC, to create a specifically Jewish response to the environmental crisis. The following year, COEJL was established as a joint project of the Jewish Council for Public Affairs, the Religious Action Center of Reform Judaism, the Jewish Theological Seminary, and several other organizations. The Coalition is now a national organization of Jewish environmental concern, education, and activism, with chapters in twenty-five states and Canada, national offices in New York City, San Francisco, and Washington, and a yearly budget of nearly $1 million. It draws institutional support from twenty-nine national Jewish organizations, including some of the largest and most powerful. On the national scene, it has prodded Detroit to produce cleaner and more fuel-efficient cars and offered free congregational energy guides to make synagogues greener. Action by its forty regional affiliates range from the Boston group's integral role in publicizing environmental justice issues, to Philadelphia's help in organizing a demonstration against the disastrous environmental policies of the Bush administration, and Vancouver's publicizing of eight environmental actions for the eight nights of Chanukah ("turn down the thermostat," "skip a car trip," "recycle your paper," etc.).[9]

Recently, COEJL has been instrumental in facilitating information exchange and joint projects between environmentalists in the United States and Israel. Its most important contribution, says its representative in Washington, is "the fact that the American Jewish community has become increasingly focused on energy policies, global warming, conservation, and on the environment in Israel. People don't wonder who COEJL is and what the environment has to do with being Jewish any more."[10]

Much of what follows in this chapter focuses on the United States, but a good deal does not, for, contrary to stereotypes of environmentalists as effete, politically correct, white liberals from rich countries, religious environmentalism is truly a *global* phenomenon, involving members of virtually every religious group, race, and culture on the planet.

To get an idea of this global character, consider some of the work of the internationally oriented and UK-based Alliance for Religion and Conservation, which emerged in 1996 after the World Wildlife Fund convened a meeting of five major world faiths to discuss their relation to ecology. Since then, the member faiths of the Alliance have grown to nine, and the organization—often in partnership with local governments, environmental groups, development programs, and even the World Bank—has initiated and supported projects throughout the world. In 2000 the ARC hosted a celebratory meeting, "Sacred Gifts to a Living Planet," in Nepal to honor actions to care for the environment undertaken by religions throughout the world. A brief description of a few of the thirty-six gifts ARC has recognized will indicate how widespread religious environmentalism is.

In 2000, Madagascar fishermen were convinced to stop dynamiting the ocean for fish, a practice with disastrous long-term results to fish populations and undersea coral ecology, when local Islamic authorities ruled that the practice violated the Qur'an's injunctions against wasting God's creation. The fishermen had been blithely ignoring both government pamphlets and strict laws forbidding the use of dynamite. As ARC leader Martin Palmer put it, "By throwing sticks of dynamite into the sea, they could haul in almost guaranteed catches and it took so little time."[11] It was only when their sheiks—who had been brought together by joint efforts of the ARC, the London-based Islamic Foundation for Ecology and Environmental Science, the World Wildlife Fund, and CARE International—applied the Qur'an to dynamite fishing and declared the practice decidedly un-Islamic that things began to change. Since then, dynamite fishing has been dramatically lessened and plans for sustainable fishing have emerged. (A remarkably similar story unfolded among Hindu fishermen and an endangered shark species off the coast of India.)[12]

Appeals to Islamic teaching are also central to Saudi Arabia's commitment to protect its biodiversity, a project directed by its National Commission for Wildlife Conservation and Development. "The reserve's creation and management embodies specific Islamic rulings relating to the sustainable management of natural resources especially through the concept of 'hima,' a traditional method of protecting range land and water resources."[13] Explaining its goals, the commission directly links theology and ecology:

> Islamic teachings maintain that nothing has been created without value and purpose; all creatures are signs of the Creator and glorify Him in unique ways, and all have been given roles by which they contribute to the common good. Hence man in his role as steward on the earth is obliged to conserve them in all their forms.... The Commission acts as the custodian over the integrity of the biodiversity of the Kingdom.... Although man has the right to use these resources, he is not permitted to abuse them. He is required to pass them on to future generations in an unimpaired condition.[14]

Half a world away from the Saudi desert, researchers at the Beijing School of Traditional Chinese Medicine are trying to protect endangered species by looking for alternative ingredients for traditional medicines. Treatments for a variety of illnesses call for components such as tiger penis, bear gall, and rhinoceros horn. Despite international bans on the hunting of many of these animals, the high price they fetch encourages widespread poaching. Arguing that use of endangered species violates Buddhist and Taoist principles of balance in nature, and thus is bad for both the environment and the soul, these world-renowned Taoist physicians are changing long-used prescriptions. The wide-ranging authority of the Beijing School will lessen the use of endangered species by traditional practitioners and perhaps save a few from extinction.

From fishermen eking out a living on the African coast and poachers drawing a bead on endangered rhinos in Africa we go to India, where the Sikh community has committed itself to a three-hundred-year project of energy conservation. Through their network of 28,000 temples, Sikhs provide free meals for tens of millions of people a day. By adopting solar power and fuel-efficient technology, they hope to reduce energy consumption by at least 15 percent. Some of the largest temples have also initiated a series of projects to raise ecological awareness, reduce pollution, and improve damaged ecosystems. Actions include tree planting, promoting solar energy, encouraging recycling, and improving water management.[15]

Forests in Sweden, Japan, and Lebanon protected by Lutherans, Shintoists, and Maronite Christians, Jewish synagogues in England and Buddhist pagodas in Cambodia proclaiming a new green gospel, a rubbish dump converted to a park by Muslims in Cairo and U.S. Methodists confronting Staples about selling paper whose production causes toxic dioxins in the water and air—the list continues. These examples show that religious environmentalism, though not necessarily stemming the tide of environmental destruction, has become a worldwide force for a cleaner, healthier planet and for more moral relations among human beings and between human beings and other species.

In the rest of the chapter, I explore some examples of this exciting movement in greater detail. We will first see how activists in southern Africa and in small Catholic communities in the United States care for the land on which they live, responding to immediate contexts of life and livelihood. I contrast their intensely local actions with the campaigns of international networks aimed at the critical but more remote issue of global climate change. We will also get a sense of how significant political differences can exist within the environmentalism of one particular faith, comparing an emphasis on individual environmental action by a Taiwanese Buddhist organization with the truly revolutionary economic and social perspective of Sri Lanka's Sarvodaya Movement. Next I briefly describe the decisive role of the United Church of Christ in the emergence of environmental justice as a critical concept for religious and secular environmentalism alike, a concept that has literally transformed much of the world environmental community. As a special case of environmental justice concerns, I then examine indigenous environmental activism, whose unique character is rooted in the distinctly ecological character of indigenous spiritual traditions and the intense cultural connections between indigenous peoples and their land. Finally, we will see some examples of the important ways religious environmentalists have made common cause with secular ones.

Activist religious environmentalism goes beyond theology and public declarations. It is directly aimed at changing the world: by making new laws, stopping harmful practices, creating better ways to produce and consume, healing the earth, and nurturing human beings in their relations with the rest of life. *Politically* it seeks to generate a collective force of voters, demonstrators, long-term activists, tree planters, and energy conservers. *Ecologically* it treats the earth with care and respect, hoping to replace our current system with organic agriculture, habitat restoration, the conservation of biodiversity, alternative technology, and renewable energy. *Morally* it pursues justice in the distribution of negative ecological effects.

ZIMBABWE

Chivi District, April 8, 1993

Having taken communion at a tree-planting Eucharist conducted by a member church of the African Association of Earthkeeping Churches (AAEC), the Rev. Solomon Zvanaka addresses the seedling he is about to plant: "You, tree, I plant you. Provide us with clean air to breathe and all the other benefits which Mwari [God] has commanded. We in turn will take care of you, because in Jesus Christ you are one with us. He has created all things to be united in him. I shall not chop down another tree. Through you, tree, I do penance for all the trees I have felled."[16]

Shrine of Mwari, Matopo Hills, Masvingo Province, January 17, 1992

After years of brutal drought that left nearly half a million people receiving food handouts, the spirit mediums of the Association of Zimbabwean Traditional Ecologists (AZTREC) have come to hear instructions from Mwari issue from a mysterious cult cave. An ancient female voice tells them, "The world is spoilt. I shall give you only sparse rains. . . . Persevere with the planting of trees! I shall keep my hand over you."[17]

These two scenes highlight the remarkable coalition of Independent African Christian churches and traditional African religions in a groundbreaking coalition to repair the ravaged landscape in southern Zimbabwe's Masvingo province. The coalition, formally known by the slightly intimidating title of the Zimbabwean Institute of Religious Research and Ecological Conservation (and more easily as ZIRRCON), has played an important role in reversing ecological decline and galvanizing African peasants to act in their own defense. ZIRRCON was the initiative of Marthinus L. Daneel, a Rhodesian-born professor of theology who developed contacts with both Christians and traditionalists while researching the role of religion in Zimbabwe's independence struggle in the 1980s. With some prompting by Daneel, communities facing denuded countryside, eroded hillsides, and deteriorating riverbanks committed themselves to a "war of the trees" in 1988. Mobilized by a common threat, this cooperative effort between unlikely allies grew to a province-wide organization with forty salaried employees, the majority of chiefs and mediums in the area, 150 churches with nearly 2 million members, eighty women's clubs, and thirty youth clubs. It has planted over 8 million trees in thousands of woodlots, raised awareness about and applied religious sanctions against damaging wildlife and water practices, placed ecological issues in a framework of social action and political liberation, and roused peasants from resigned passivity.[18]

The accomplishments of ZIRRCON embody many of the distinct characteristics of religious environmentalism. Perhaps most dramatic, this is a strikingly *ecumenical* effort, one all the more remarkable because it involves strikingly different religions. Far more than a matter of Methodists working with Catholics or Christians with Jews, ZIRRCON is an alliance between Christianity and spirit- and ancestor-centered religions that have wildly different religious beliefs. It is the kind of alliance that for centuries would have been unthinkable, for Christians would simply have rejected the traditionalists as ignorant. Yet, in the context of their environmental vocation, most of the members of the Earthkeeping Christian churches have maintained an attitude of respectful comradeship with the non-Christians. As Daneel said in an early assessment of the partnership, "The mediums should wage war against deforestation in terms of their own beliefs and the churches should do so on Christian Principles. Each movement should have its own religious identity but they should recognize the value of each other's contribution."[19] Bishop Machokoto, after being elected AAEC president, warned his fellow Christians, "We must be fully prepared to recognize the authority of our krallheads and chiefs. For if we show contempt for them, where will we plant our trees?... Let our bishops in their eagerness to fight the war of the trees not antagonize the keepers of the land.... Let us fully support our tribal elders in this struggle of afforestation."[20] Comparable moves were made on the traditional side. For example, it was Daneel's role in the founding of ZIRRCON that enabled him to be the first white man to attend the ceremony of the cult oracle described above. Similarly, though conflicts among different Christian denominations did not disappear, they tended to pale into insignificance in the context of meetings, conferences, shared rituals, and ecological efforts.

This was not the first time either the traditional or the Christian churches had responded to a critical social issue. Their political sensibilities and capacity for decisive action had been proven in the struggle for independence. In that struggle, both groups had made important contributions, including supporting military actions, providing resources for fighters, and lending their social prestige to the struggle. Just as Fred Small appeals to the history of social activism of Gandhi and King, and (as we will see below) some of the "Green sisters" in the United States see themselves as having been formed by the civil rights, antiwar, and feminist movements of the 1970s, so ZIRRCON could view its war of the trees as a continuation of activist religion from the past. Its religious environmentalism is, and has to be, political.

We have seen that religious environmentalism is both rooted in tradition and a creative transformation called forth to meet the demands of the environmental crisis. This creative tension was present in ZIRRCON as well. For a start, traditional religion had always involved beliefs and practices to help conserve land, water, and wildlife. As one observer says of some remaining forests of the Zambezi basin, not too far from ZIRRCON's Masvingo province, despite the economic pressure to use the land for cotton, a few key areas—because they were religiously protected—have been preserved: "If they weren't sacred, the forests would have been long gone."[21] Yet, as valuable as traditional values and practices were, they depended on a steady-state subsistence economy. Increases in population, profit-oriented deforestation, overuse of water for commercial farms, soil erosion, and a decade-long drought created something profoundly new. In response, traditional religious leaders have changed Africa's "age-old religio-ecological values into a modern programme of environmental reform."[22] Mwari morphed from a rain god into a god of ecology, and it turned out that the spirit ancestors would no longer be satisfied with the observance of age-old ecological taboos against taking certain game or felling sacred trees. They now demanded that people heal the land, reforest the earth, and protect the water. Believers could not simply perform religious rituals to convince God to bring rain; actual ecological work (i.e., tree planting) had to be done to help make it happen.

Although the African Independent Churches already had a strong sense of social engagement, they, too, had to change. The face of Jesus had to be seen in the trees and the water, and the power of God had to be understood as an Earth-keeping power that Christians were compelled to manifest in their own lives. Every Christian was urged to recognize his or her responsibility for the health of the land. In a sermon in 1991, Daneel reminded his listeners, "Whenever you celebrate Holy Communion, be mindful that in devastating the earth we ourselves are party to destroying the body of Christ. We are all guilty in this respect."[23] And indeed, it is an essential part of the compelling ritualistic innovation of the tree-planting Eucharist that *all* participants, from the most humble peasants to the most senior bishops, confess their ecological sins of cutting trees without planting, overgrazing the land, or injuring the riverbanks—and *then* proceed to plant a tree.

Concern with the land is a life-and-death matter for these residents of Masvingo. But their environmentalism goes far beyond purely instrumental or conservationist values. They know that caring for the earth is good for people—but that is not the only reason they do it. In the quote that began this section, it is striking that Rev. Zvanaka did not simply ask God to bless the seedling in the hope that his parishioners would have healthier soil and more rain. Rather, he talked directly *to* the seedling, including it as part of a community sanctified by Christ. Environmentalism for these churches is a matter of love for creation as well as enhancing their own material well-being. As God's love was manifest in the creation of the earth, so those who believe in God are to imitate God by showing their love for what God created.

We find in the case of ZIRRCON that once again religion has a particularly important role to play in environmentalism because of its distinct capacity to motivate. When the spirit mediums of AZTREC told their followers that they were obeying the demands of their ancestors, unprecedented ecological activity followed. When the African Independent Churches took on the task of environmental stewardship, people acted, something that usually did not happen in response to the appeals of governmental experts or secular environmental organizations.

The fusion of tradition and innovation, human and ecological concern, ecumenism and political action that we find in ZIRRCON is not limited to one province of Zimbabwe, but is echoed in many other parts of the region. South African churches, many of which played a crucial role in the struggle against apartheid, for example, are engaged in a variety of ecological activities.[24] In the 1990s, the Faith and Earthkeeping Project functioned in almost every South African province, stimulating interest in environmental care in religious circles and working with religious groups to develop community-based environmental conservation projects, including tree planting, water protection, urban greening, and recycling. In the town of Philadelphia, a Dutch Reformed congregation led a successful struggle against building a toxic waste facility. In the poor, rural district of Umzimvubu, Anglican bishop Geoff Davies, after a decade of working to bring ecological concern into the heart of the church, formed the Umzimvubu Sustainable Agriculture and Environmental Education Programme, which supports local communities wishing to start sustainable gardens and offers advice on earth care, land reclamation, and recycling. Davies, popularly known as the "Green Bishop," also drew wide notice for his outspoken criticism of a government plan to build a large toll highway though Umzimvubu, arguing that it would be an ecological and human disaster, flagrantly disregard international conferences on biodiversity held in South Africa, and violate South Africa's own laws.[25]

More recently, an umbrella organization, the Network of Earthkeeping Christian Communities in South Africa, has organized conferences, sent representatives to national and international meetings, and chronicled detailed struggles in local towns and villages. The stated commitment of the Network is ecojustice: in an inclusive vision, love of God's creation takes its place alongside resistance to unjust corporate power, foreign investment, distorted forms of economic development, the oppression of women, and genetically modified crops.[26] Here environmentalism is a key element in determining how economic development affects people's day-to-day existence. In the words of a retired priest and development worker from Umzimvubu, "Sustainable agriculture can liberate us from the chains of dependence and starvation."[27]

Traditional African religions have always been keyed to the health of the land. Any serious Christian who cares about the life and death of his or her neighbors must sooner or later recognize that responding to environmental damage is part of that care. Therefore, it is not surprising that from Masvingo to Umzimvubu—and in countless other places—religious leaders, congregations, and organizations are an essential part of environmental activism in Africa. Whether they can do enough, soon enough, in the face of that continent's enormous poverty, illness, and widespread governmental and ethnic violence, remains to be seen.

FAYETTEVILLE, ARKANSAS,
JULY 15–18, 2004

The sixth annual and tenth anniversary meeting of Sisters of Earth, a loose network of nuns, combined "panels and presentations about sustainability, institutional and congregational greening, eco-spirituality, earth literacy, bioregionalism, and social justice with ritual, celebration, and song."[28] An altar was constructed with material from the different bioregions of the nearly ninety participants, who joined in a newly created ceremony honoring endangered species, focusing specifically on select animals and plants from the United States and Iraq. The directors of the Denver-based grassroots organization EarthLinks shared how they combined devotion to the earth with care for people, seeking to empower the socially powerless by connecting them to earth community.[29] Other speakers criticized genetic engineering and described their resistance to mining companies that literally cut the tops off mountains and leave behind millions of tons of refuse in nearby communities.

Like ZIRRCON, the Sisters of Earth focus their attention on the human connection to the earth, seeking to repair a deeply injured relationship. Yet, as inhabitants of the world's richest country, their immediate context takes a very different shape. Whereas many in Africa find that ecological damage threatens them on the basic level of food and water, most in North America face sterile urban or suburban settings, an ever increasing sprawl in which everything looks increasingly alike. Unless we live in California or Florida, the vast majority of our food comes from hundreds or thousands of miles away, most of it nearly tasteless, laced with additives, or genetically engineered. We are much more likely to confront obesity than starvation, and farming is something conducted by multibillion-dollar agribusinesses using poor migrant workers made sick by pesticide exposure. Our closest connection to the land is at best a small garden and lawn on which we lavish chemical fertilizers and carcinogenic pesticides. Ravaged landscapes are pawned off on the poor and people of color, especially on Native Americans, whose lands bear huge burdens from mining, toxic incineration, and coal-fired power plants. And all this is embedded in a cultural frame in which humans have pretty much unchallenged rights to use, consume, and abuse the rest of the earth at will.

Against this ecological and social background the Sisters of Earth, an informal network of some three hundred Catholic nuns, took shape in the early 1990s. Many of the women were inspired by the ecotheological teachings of Thomas Berry, particularly his stress on human kinship with the rest of life and the way he situated human beings within the cosmic history of the universe and the path of evolution. Some were long-term veterans of progressive politics, having been active in the antiwar, women's, and human rights movements from the 1970s on. Some were recently awakened, partly by the statements of Pope John Paul II and bishops from Appalachia to the Philippines and partly by recognition of the seriousness of the environmental crisis and the immorality of American individualism, consumerism, and anthropocentrism. These diverse origins have led them to a common goal: to, as one observer writes, "reinhabit the earth," to live and teach a form of life in which humans treat the rest of the earth with respect and care, integrating themselves into their own places sustainably and gently.[30]

Sisters of Earth try to infuse these values in their personal lives, the way their communities function, and in contributions to surrounding towns. Their activities include organic gardening, land conservation, reducing consumption, using alternative building materials, solar heating, and hybrid vehicles, and building wildlife sanctuaries. As Carol Coston, cofounder of Santuario Sisterfarm in Texas, says, they want to "find a way to live lightly on the earth."[31] Chris Loughlin, director of the Crystal Spring Earth Literacy Center in Massachusetts, estimates that more than five hundred people help support their community farm and enjoy its organic produce; New Jersey's Genesis Farm serves six times that number and has educated nearly seven hundred people in its extended "earth literacy" courses. EarthLinks of Denver created a "BioBox" program that teaches elementary school students about their bioregion and links children from different regions. They also involve homeless people in gardening projects to help raise both their income and their pride, while connecting them to the healing benefits of gardening.[32] Santuario takes as its focus the preservation and promotion of diversity, in social as well as ecological forms. Quoting Indian ecologist and globalization critic Vandana Shiva, they maintain, "An intolerance of diversity is the biggest threat to peace in our times; conversely, the cultivation of diversity is the most significant contribution to peace—peace with nature and between diverse peoples." To this end, they seek to "bring awareness of the dangerous loss of biodiversity and the exploitation of economically impoverished peoples by multinational corporations that have been usurping the seed lines developed over centuries by small farmers and indigenous peoples around the world."[33] The political implications of this project are reflected in its name, the Rosay Martín Seed Project, which echoes the names of two Peruvian Dominican priests who sided with native peoples against colonialism.

While the sisters applaud the work of groups like the Sierra Club and the Audubon Society, they seek a format for their work in which the spiritual dimension of a kind of universal respect and care is essential. As Mary Romano of EarthLinks puts it, any real change in our environmental practices "must come from a place of love." Echoing Thomas Berry, Chris Loughlin says she hopes Crystal Springs activities—community-supported agriculture, hosting retreats on ecocosmology—will help people develop a new sense of self and create caring relationships with the web of life. To take one small example of this attitude of ecological respect: parts of Genesis Farm are simply off-limits to people. While the farm is committed to growing organic food for human consumption, they also want to leave part of the earth to itself.[34]

The Sisters of Earth also seek to be a force for this changed consciousness in their local communities. Mary Romano, who practices an eclectic spirituality, says that EarthLinks' greatest accomplishment is the way it helps "people establish a personal relation with the natural world, which then enables people to make positive changes in personal life." Miriam MacGillis, who oversees Genesis Farm's 140 acres and nearly quarter-million-dollar annual budget, takes deep joy in the way the farm serves as an "amazing developer of community. Individuals, families, children—they grow up with a cultural life tied to the farm."[35]

Along with their stress on moral change and community organizing, many of these women have a clear sense of the larger social and political implications of what they are doing. They may not couch their concerns in leftist rhetoric, but they have no doubt that society, no less than their organic gardens, requires decentralization, diversity, and interdependence.[36] As politically committed environmentalists, they engage in "disrupting shareholder meetings of corporate polluters, contesting the construction of garbage incinerators, and combating suburban sprawls."[37] Although most activity is local, some have been active with national or international agencies. Jane Blewett of Earth Community Center in Maryland monitors UN debates on sustainable development, having spent years working for an internationally oriented Catholic social justice organization. She regularly conducts workshops on "Justice for People, Justice for Earth: Two Sides of the Same Coin." Other sisters are engaged with Worldwatch, the Environmental Defense Fund, and Greenpeace.[38] Gail Worcelo of Vermont's Green Mountain Monastery, which she cofounded with Thomas Berry, puts it simply: her goal is the health of the total earth community, and thus she resists processes and products that unbalance that health or privilege one part over another. Dominican sister Carol Coston offers a direct criticism of global capitalism: "Its tendency to look only for the bottom line makes it impervious to its effects on local economies. Instead of growing for their families, people are made to grow for export." Further, "when agribusinesses develop genetically modified seeds, they threaten the livelihood of small farmers throughout the world, as well as pose a danger to seed stock biodiversity."[39]

Finally, Sisters of Earth manifest a profound ecumenical respect for other religious paths. Many are open about what they have learned from Buddhist meditation or earth-honoring indigenous peoples. Non-Catholics work at their centers; yoga and meditation are taught at their conferences. "We welcome all people of goodwill," says Miriam MacGillis. Like the different groups in ZIRR-CON, the Sisters of Earth have both maintained and transformed their religious allegiance. They remain Catholic, but Catholicism for them now unfolds in the context of a 14-billion-year-old universe. They believe in the Trinity, but now see Father, Son, and Holy Spirit as permeating all of life, including human beings who have different names—or no names at all—for God.

If this small group of activist women is not likely to lead a revolution, they are keeping alive the hope that human relations with the earth can be repaired. Each of the centers is a small oasis amid the temples of hyperconsumption and the cavalier abuse of the land. Each center offers a place where we can honor something of which many have only the faintest memory. Ultimately, what they teach and the way they live is part of the promise that we can change—here and now—some of what we are doing. However many lives they touch, this is a profound contribution to global environmental action.

For the most part, African Earthkeepers and the Sisters of Earth are responding to their immediate surroundings: planting trees and protecting local wildlife and water, growing food with love and intelligence rather than chemicals, and trying to live a bit lighter on the earth. On the other end of the spectrum of environmental concern are global problems, the causes and effects of which may be separated by thousands of miles. These problems, too, are being addressed by a variety of religious environmentalists.

The most momentous of such problems is global warming. By all serious scientific accounts, this process is already well under way. The many aspects of this calamity—warmer temperatures to be sure, but also droughts, extreme weather events, and increases in insect activity and disease—are affecting people from hurricane-ravaged New Orleans to flood-ravaged Bangladesh. As with most social problems, those in the third world suffer disproportionately, enduring suffering all the more unjust because it is first world industrialization (with the United States alone responsible for 25 percent of greenhouse gas emissions) that is the source of third world misery.[40] Near the Arctic Circle, the Inuit are facing deformed fish, depleted caribou herds, dying forest, starving seals, and emaciated polar bears and losing entire coastal villages as the ice melts and the water level rises.[41] Island nations in the South Pacific are simply disappearing, as thousands of natives have to leave their islands because the waves come ever higher. In previously agriculturally self-sufficient southern African Lesotho, a long-term drought driven by higher temperatures has created the looming threat of famine.[42] And so it goes.

Of the many religious groups that have spoken out about global warming, the World Council of Churches, representing faith groups encompassing some 400 million members, has been particularly clear and decisive. The Council, whose strong and actually quite radical public statements on environmentalism we have encountered already, has (along with a number of associated and subsidiary groups) manifested a significant public presence in this area. The Council's position on global warming, as on environmental issues in general, evolved from the 1970s on.[43] Because it forms the context for its global warming work and exemplifies the hallmark pattern of religious environmentalism's evolution, that evolution is worth recounting.

In the early 1970s, the widely publicized Club of Rome report "Limits to Growth," probably the first instance an internationally respected group asserted that the earth could not support unending industrial development, helped spark environmental concern in the WCC. From then on, its major programs included reference to environmental issues, beginning with the significant but clearly anthropocentric goal of "just, participatory and sustainable societies," and moving by the 1980s to the more ecologically inclusive values of "justice, peace and the integrity of creation." As we have seen in everything from the ecological evolution of Pope John Paul II to the way Genesis Farm restricts part of its land from human contact, one hallmark of religious environmentalism is a deep commitment to acknowledging the value of all life. WCC's transition from "sustainable societies" to "the integrity of creation" signals that movement in the context of this international Christian alliance.[44]

Another crucial aspect of religious environmentalism is the development of an environmental justice perspective. From an initial sense of environmentalism as concerned with how we treat nature, there arises recognition of the connections between environmentalism and more familiar social justice issues. During the 1980s, intense discussions on the "relationship of socioeconomic justice and ecological sustainability" unfolded at WCC congresses and group meetings.[45] And at its historic 1990 world gathering in Seoul, the WCC was able to affirm ten principles linking the economy, justice, ecological health, war, and racism. Seeking to sharpen its position in response to the new world economy, the 1998 General Assembly in Zimbabwe adopted a long-term program to critically assess globalization, paying special attention to its intertwined economic, ecological, and social effects. By 2004, the Council's subgroup on environmental justice could sum up globalization's most damaging practices: multinational corporations moving outlawed operations to developing countries, the shipping of toxic wastes from industrialized nations to the economic south, free trade agreements that restrict the capacity of national governments to adopt environmental legislation, destruction of southern rain forests to provide exotic timber for northern consumers, and pressure on poor nations to engage in ecologically destructive agricultural practices to produce cash crops for export in order to service foreign debt payments.[46]

The WCC's work on global warming embodies moral respect for nature and political criticisms of globalization. Its observers have been present at many of the major international conferences and meetings on climate change, for example, sending representatives to monitor, advocate, and lead religious services at UN climate change negotiating sessions. During the widely publicized 1992 Conference on Environment and Development meetings in Rio, the Council stated in no uncertain terms that climate change was a moral and theological matter, not simply a scientific or economic one. In these international contexts, the Council's representatives constantly emphasize the negative effects of climate change on human beings, assert that the world's poor should not be expected to suffer for the industrialization and unsustainable consumer needs of the rich, and offer a vision of a society that is sustainable for humans and nature alike.[47]

The truly global nature of climate change provides a remarkable, perhaps unparalleled, motivation for ecumenical work. If there were ever an issue that clearly reveals the commonalities of human beings despite differences in religious belief, ideology, or culture, this is the one. Thankfully, many people of faith have recognized this fact, and have forged interfaith coalitions as well as alliances with secular environmentalists.

As one example, consider the Interfaith Global Climate Change Network, itself a joint project of the Eco-Justice Working Group of the National Council of Churches, COEJL, and the National Religious Partnership on the Environment. Together, NCC and COEJL have organized eighteen statewide interfaith climate change campaigns. These groups see themselves squarely in the emerging tradition of religious care for the earth and concern for the connections between humanity and nature. The North Carolina chapter, for example, is a coalition of "various spiritual traditions" committed to "turning human activities in a new direction for the well being of the planet" and for the sacred task of *preserving all eco-systems* that sustain life."[48] The effects of global warming, the coalition warns, will "fall disproportionately upon the most vulnerable of the planet's people: the poor, sick, elderly."[49] The thirty-six signers listed on the group's Web site include rabbis, Buddhist priests, Roman Catholic and Episcopal bishops, and ministers from the Lutheran, Unitarian Universalist, Quaker, Baptist, Methodist, and United Church of Christ denominations. The frequent use of the term "spiritual" in the group's call signals an acceptance of the variety of paths to God; acknowledgment of the sacredness of the earth announces an end to theological anthropocentrism; naming the special vulnerability of the poor opens the way for an account of irrational and unjust social institutions and for common work with secular liberal to leftist organizations. The challenge to existing political and economic arrangements is direct and serious.

The nearly twenty chapters of the coalition have organized meetings with business leaders, local governments, and congressional representatives. They have made visible public statements, educated local congregations, and offered practical ways for religious buildings to become more energy-efficient. Several kindred Interfaith Power and Light organizations offer detailed energy audits of congregational buildings and provide technical help in utilizing renewable energy sources.

Along with their own efforts, religious environmentalists have engaged in many coalitions with secular environmentalists, particularly on the issue of climate change. Perhaps the most interesting of these alliances involves widely publicized joint efforts with—of all groups—scientists! If we remember the centuries of conflict between religion and science—from the Catholic Church's punishment of Galileo to religious resistance to Darwin's theory of evolution—we will be struck by the cultural significance of cooperative efforts between the two. These efforts began in the early 1990s, when thirty-four internationally recognized scientists wrote an "Open Letter to the Religious Community," appealing for a combined effort in defense of the environment, an effort much in need, these *scientists* asserted, of a "religious dimension" and a "sense of the sacred."[50] Several hundred religious leaders from around the world responded, and there followed in June 1991 the "Summit on the Environment," a joint meeting of scientists and religious leaders. The summit issued a "Joint Appeal," initiating perhaps the most high-level cooperation between these long-standing cultural antagonists in history. The appeal acknowledged the scope of the environmental crisis and called for a variety of "diverse traditions and disciplines" to respond to it. Pride of place in the list of environmental concerns was global warming, with its expected consequences of increased drought, depleted agriculture, destruction of the "integrity of ecosystems," and creation of "millions of environmental refugees." Along with noted scientists, signers included the leadership of the WCC, the Rabbinical Council of America, the American Baptist Church, the National Conference of Catholic Bishops, the Greek Orthodox Archdiocese, and the Episcopal and Lutheran Churches.

Thirteen years later, another joint science and religion statement, focused exclusively on global climate change, was sent to the U.S. Senate to urge passage of the Climate Stewardship Act, legislation committing the United States to restrict its greenhouse gas emissions. In the context of broad consensus among scientists worldwide, the letter asserts:

> The United States has both responsibility and opportunity. With 4% of the world's population, we have contributed 25% of the increased greenhouse gas concentration which causes global warming. Moreover, we uniquely possess technological resources, economic power, and political influence to facilitate solutions. However, policies that devalue scientific consensus, withdraw from diplomatic initiative, and seek only voluntary initiatives do not seem to us adequate responses to this crisis.[51]

Signers included the head of the American Association for the Advancement of Science, the founder of the Wood's Hole Oceanographic Institute, a Nobel prize–winning chemist, the president of the National Council of Churches, and the general secretary of the United Methodist Church.

These statements, it should be emphasized, are more than just new theology. By joining forces with scientific leaders, widely considered our society's arbiters of rationality and its best sources of sound public policy, religious leaders are announcing a decisive intention to influence social life in a way that *combines* a spiritual vision with empirical science. Faith here is not a substitute for or alternative to science, but a way of understanding and working with it. Faith expresses the indispensable factor of the human response to the world—both the natural world, which requires protection, and the human world, which, in its unrestrained industrialization, has gone astray and needs new direction. The fact that 4 percent of the world's population produces 25 percent of the greenhouse gases does not by itself tell us that something must change; only a commitment to basic human moral equality will do that. If we do not believe that we should love our neighbors as ourselves, or that all people are made in the image of God, or that the rich nations do not have a presumptive right to inflict ecological disasters on poor ones, then the bleak truths of global warming will mean little. When religious leaders speak out in this context—offering their vision, trying to affect public policy—they are in effect saying, "These facts have powerful ethical implications; here is how we should respond to them." Their voices move arcane issues of industrial policy and economics, energy sources and conservation, into the realm of ethical life, personal and collective responsibility, and even (gasp) sin, a movement with potentially quite powerful political implications. For it is typically only when long-established ways of life, especially those that benefit the socially powerful, receive a *moral* challenge that they can be *politically* changed. Freeing the slaves, granting equal rights to women, and now making fundamental changes in our relation to nature are possible only if we change our moral assessment of slavery, male domination, and unrestrained production, consumption, and pollution.

The language of the religion-science statements was necessarily measured, somewhat cautious, and noninflammatory. More flamboyant and exuberant, as well as controversial and just plain fun, was the Evangelical Environmental Network's "What Would Jesus Drive?" campaign. Endorsed by hundreds of ministers and lay leaders of Evangelical Christianity, promoted by a tour through Bible Belt centers in Texas, Arkansas, Tennessee, and Virginia that ended at the country's largest Christian rock festival, "WWJD?" promoted "ways to love your neighbor as we strive together to reduce fuel consumption and pollution from the cars, trucks, and SUVs we drive." The campaign started in February 2002 with a Detroit press conference, meetings with auto industry executives, and support from non-Evangelical religious leaders. Over the ensuing months it received massive press coverage, with thousands of newspaper, radio, and TV stories in the United States and throughout the world. Its guiding document made the religious implications of gas guzzlers crystal clear:

> Obeying Jesus in our transportation choices is one of the great Christian obligations and opportunities of the twenty-first century. Pollution from vehicles has a major impact on human health and the rest of God's creation. It contributes significantly to the threat of global warming.... Making transportation choices that threaten millions of human beings violates Jesus' basic commandments: "Love your neighbor as yourself" (Mark 12:30–31); and "Do to others as you would have them do to you" (Luke 6:31).[52]

The campaign was significant for a number of reasons. For a start, Evangelical Christians are generally socially conservative, far from the usual collection of Volvo liberals, aging hippies, and young "crunchy" types identified as environmentalists. Also, the campaign was not limited to a mild-mannered and widely acceptable celebration of the beauties of nature or bland generalities about "stewardship" and "creation." This was (by God!) an effort of "biblically orthodox" folk, an in-your-face challenge to religious conservatives who believe that "religion in politics" means being against abortion rights, pornography, and gay marriage and for the nuclear family, tax breaks for religious groups, and prayer in schools. "WWJD?" redefined Christian morality to include pressuring government and business leaders to increase fuel efficiency and develop mass transit and, as un-American as it sounded, encouraged their fellow citizens to walk, bike, or take the bus instead of driving. It carried a deadly serious message in a slightly playful way and bore the unmistakable stamp of religion entering the public arena to demand a change in business—or at least driving—as usual.

I have made much of the political significance of religious environmentalism, stressing how it leads to serious criticisms of the existing social order. Yet it would be a mistake to think that every religious environmentalist, or every activist campaign by a religious group, carries this stamp. Within the ranks of religious environmentalists there are mild-mannered reformers as well as wild-eyed radicals, advocates of small, personal, local change as well as those who would initiate an ecological revolution and remake the world.

To see how these variations can coexist within the same religious tradition, it will be instructive to compare two sizable, nationally important, and internationally influential social movements from the world Buddhist community. Both are well-supported, influential, grassroots organizations started by far-thinking and charismatic leaders. Both are sustained by the integrity of their guiding principles and the moral commitment of their members. Yet their respective places on the political spectrum between individual and institutional change could hardly be greater.

Taiwan

A central player in the development of the new Southeast Asian economies, Taiwan is also massively polluted, facing dying rivers, contaminated soil, and dangerously poor air quality.[53] In the 1980s, an environmental movement emerged, and among its ranks were a number of Buddhist and Christian organizations. The largest and most influential of the former was the Buddhist Compassion Religious Tzu-Chi Foundation, which had been founded as a neighborhood philanthropic association in the mid-1960s by a Buddhist nun, Cheng Yen. Beginning with a group of thirty housewives who would put aside 13 cents a week for donations, Tzu-Chi has grown to a major force in Taiwan and a presence in several foreign countries. It has established hospitals and a medical/nursing school, distributed over $20 million in charity funds in Taiwan and abroad, and is currently engaged in a wide variety of educational, cultural, and health care projects. A truly international force, it has provided flood relief in Thailand and Mexico, free medical care in Indonesia and California, helped handicapped children in Malaysia, and organized a beach cleaning in Singapore (after which volunteers watched a video of Cheng Yen speaking on "Compassion for Mother Nature").[54]

Tzu-Chi's guiding principles are taken directly from the moral teachings of Mahayana Buddhism.[55] Members commit themselves "to support one another through love and wisdom, and to walk hand in hand on the Path of the Bo-dhisattvas," which means, quite simply, to seek to end the suffering of every living being in the universe. They seek "Purity in our minds, Peace in the society, and a disaster-free world," and they hope to achieve these by "Kindness, Compassion, Joy, and Giving through helping the poor and educating the rich." Oriented toward enlightenment rather than devotion to God, Buddhist environmentalism is based as much in personal virtues such as nonattachment and wisdom as in the imperative to care for nature. Yet the mental outlook these personal virtues promote—for example, detachment from compulsive cravings—provides the basis for a radical critique of any culture that seeks the endless multiplication of desires and requires correspondingly high levels of production and consumption.

With this general perspective as a foundation, Cheng Yen committed herself to applying Buddhist insights to the social world. Such an application, she taught, calls for deep personal change if it is to succeed. "Environmental protection must start from the mind...if everybody can get rid of greed, anger, delusion, and pride, then all people can help each other and work together to open up a piece of clean land."[56] A clean environment is an essential element of physical health, itself a corollary of the mental purification taught by Buddhist meditation.

Cheng Yen's explicit calls for environmental protection began in 1990 and have since focused on recycling and avoiding the use of polluting products. Because of her track record of integrity and generosity, Tzu-Chi was able to mobilize wide support in response to her call. "Between July 1990 and November 1996, 6,000 to 7,000 people per month were involved with the [recycling] program....About 275,000 tons of paper, aluminum cans and metal cans were collected per month," saving more than 30,000 trees. In a single day in 1992, 176 tons of waste paper was collected in six hours.[57] The communal center of Tzu-Chi is run by principles like those of the Sisters of Earth: paper is recycled, organic cleaners are used, car travel is minimized. Nuns have "collected waste paper from the trash, wood chips from wood shops, wooden molds from building sites, and wooden boxes. This trash, which would otherwise be dumped, is recycled by the Tzu Chi environmental protection volunteers."[58]

Cheng Yen's vision for Tzu-Chi focused on compassion and concern, support for those in need, and an associated joy of selflessness which is the hallmark of the serious Buddhist. Although Tzu-Chi has created some institutions (hospitals, schools), it does not focus on a political transformation of society. Its offers no pointed critique of globalization, does not protest World Bank development schemes, and leaves issues of democratic control of major social institutions to others. It helps to clean up parks, not to punish polluters, to promote the inner joy that comes from generosity rather than directly confronting the conditions that make generosity so desperately needed. For this it has been criticized by a Taiwanese Buddhist scholar for too much stress on inner transformation rather than social change, for allowing the government and the industries to pollute rather than confronting them.[59] Despite these limitations, its contributions are undeniable.

Sri Lanka

Like Tzu-Chi, the Sarvodaya Movement is also rooted in Mahayana Buddhist principles of compassion, generosity, and personal contentment. Yet it resides at the other end of the political spectrum, seeking nothing less than a full-scale, nonviolent social revolution that will fundamentally reshape modernization both in its country and throughout the developing world. The Sanskrit word *sarvodaya* was used by Gandhi to mean "the benefit of all." A. T. Ariyaratne, the Sri Lankan science teacher whose vision brought Sarvodaya into existence, uses it in a self-consciously Buddhist sense: "Everyone wakes up." In 1957, Ariyaratne sparked the movement by arranging for student volunteers from the college he taught at to work—building roads, improving sanitation, teaching basic literacy—in a few of the poorest and lowest caste of Sri Lanka's 24,000 villages. Out of this simple beginning there grew with remarkable speed a national organization coordinating volunteer efforts in thousands of villages, creating preschools, offering agricultural education, forming groups for women and teenagers, aiding countless development projects, and touching the lives of nearly 4 million people.[60]

Sarvodaya was always distinguished by the idealism of its workers, its focus on the poor, and its inclusion of women and members of minority Tamil and Muslim communities. Even more important, however, were the distinct goals that shaped its activities. For Ariyaratne personally and Sarvodaya as an organization wanted a transformation of Sri Lanka into a society governed by broadly interpreted Buddhist ideals, brought about by Gandhian methods of nonviolence, spiritual discipline, and inclusion. Central to this process, they taught, was a subjective awakening: to a sense of self-worth, compassion for and cooperation with others, and active engagement in community life. This awakening would create a society very different from the acquisitive, high-technology nation sought by the Sri Lankan government, and taken by the dominant institutions of global capitalism to be the hallmark of a successfully "modernized" country. In a truly radical political and spiritual stance, Ariyaratne flatly rejects this image of success, opting instead for a humanly and ecologically sustainable society in which there is neither Western-style affluence nor crushing third world poverty. "In production-centered society, Ariyaratne says clearly, "the total perspective of human personality and a sustainable relationship between man and nature is lost sight of."[61] As a commentator puts it, "Sarvodaya's main message is that human suffering cannot be alleviated merely by material means.... All its projects are meant to serve the specific needs of a local community that has been reawakened ... to the ancient virtues of interdependent sharing and caring, joint suffering, and compassionate interaction."[62]

As Tzu-Chi began in charity work, Sarvodaya initially focused not on environmental issues but in support of a comprehensive, morally oriented reshaping of Sri Lankan modernization. It envisaged quasi-independent villages controlled by local citizens through an engaged democracy, in which all basic human needs would be met, many by subsistence labor rather than the marketplace. The result would be a kind of Buddhist socialism, for the goal of economic life would not be continual industrial expansion and ever growing consumer "needs," but balanced support for all facets of a moral and humane life. As abstract as this sounds in principle, Sarvodaya's success has always depended on its being clearly grounded in the details of village life. It seeks the "liberation of the goodness that is in every person," but sees that process as unfolding in concrete and highly practical actions such as preventing soil erosion, teaching literacy, building schools and roads, purifying water, and conserving biodiversity. For Sarvodaya, the problems of economic development, in fact, parallel those of spiritual development. "The root problem of poverty," argued Ariyaratne, is "personal and collective powerlessness." Yet awakening is not an isolated process of spiritual practice, but something that arises out of "social, economic, and political interaction...interdependent with the awakening of one's local community."[63]

Environmentalism is a natural consequence of this program. Thus, it is not surprising that Sarvodaya constantly makes references to environmental protection and respect for nature. Its goals for national awakening include protection of the environment, biodiversity, the use of appropriate technology "without destruction of nature or culture," and avoiding dependence on "exploitative international economic relationships."[64] The maldevelopment pursued by the central government and international agencies such as the World Bank were not just against Buddhist principles, they damaged human beings. "We believe," wrote Ariyaratne, "that poverty, powerlessness, and related conditions are directly linked with affluence imbalances, and injustices in the exercise of political and economic power and other advantages enjoyed by the few over the many. What is necessary is not a palliative, but a strategy for a total, nonviolent revolutionary transformation." Waging peace in the face of a protracted and bloody, ethnically based civil war, Ariyaratne called for ecological sustainability along with economic and political justice as part of a comprehensive peace plan.[65]

During the 1960s and 1970s, much of Sarvodaya's expansion was supported by a consortium of European donor agencies, which were impressed by its large number of volunteers and effectiveness at creating village-wide organizations. It also received government cooperation and support to help build centers and staff its operations. Yet with this growth and dependency inevitably came conflict. Ariyaratne's stress on the integration of religious values with economics and his vision of decentralized political power and civically active peasantry were ill adapted to mainstream models of modernization. The donor consortium, for instance, aimed to separate economic development from other kinds of growth, valuing projects that generated income rather than community-supporting subsistence labor or improvements in women's social position. Seeking a society with neither wealth nor poverty, Sarvodaya rejected the exorbitant, technologically sophisticated projects favored by the international agencies and its own government. Buddhist virtues of modest consumption would not fit Sri Lanka for participation in a high-tech global economy; microcredit schemes would not lead to large industrial projects.

From Sarvodaya's point of view, such projects typically devastated the local ecology and did little for the people who were most closely affected by them. Ariyaratne lamented, "By the side of gargantuan dams are parched fields that poor farmers watch disconsolately and with mounting discontent. Under the electricity lines which carry power from the dam to the cities and factories live people who have no permanent structures to call homes and hence are not eligible for that electricity."[66] Guided by traditional Buddhist ideals of universal compassion, and applying those ideals through engagement with the critical social problems of his society and nation, Ariyaratne was able to chart a course past destructive models of what third world communities needed. In doing, so he has committed himself, and with him a remarkably large and widespread organization, to a fundamental transformation of social life. Although not defined by its environmentalism, concern for ecology and biodiversity is integral to this transformation. Sarvodaya's religious cast gives it its distinctive character, and its enormous range of positive contributions indicates just how important religious environmentalism can be.

EAST ST. LOUIS, ILLINOIS, APRIL 20, 2002

Facing a decision by the U.S. Army to have Onyx Environmental Services incinerate tons of neutralized nerve gas, residents of one of the nation's most environmentally contaminated neighborhoods protested. Among their supporters were members of the United Church of Christ's Justice and Witness Ministries, who led a public demonstration in opposition. Drawing on the UCC's history of involvement in environmental issues, Rev. Bernice Powell Jackson, then executive minister of the Ministries (and now president of the National Council of Churches!), asked some pointed questions:

> Did East St. Louis' high asthma rate among its children factor into Onyx's decision? Or was that fact even considered at all? Were all segments of the community—especially those most likely to be affected—involved in the decision making process? Can the burden of disposing potentially toxic wastes be shared equally among all communities and not borne by the most vulnerable members of our society? Should East St. Louis continue to bear an unfair burden for our nation's waste?[67]

This last pointed question is a hallmark of the environmental justice movement, a movement spearheaded initially not by Greenpeace, the Sierra Club, or the World Wildlife Federation, but by the United Church of Christ. Environmental justice and the related idea of environmental racism center on the simple but crucial fact that environmental burdens are distributed unequally: people of color and the poor are much *more* likely to face polluted air and water in their communities and much *less* likely to have attention paid to their plight by either government institutions or environmental organizations. African American, Native American, and Latino communities have served as the dumping grounds for industrial waste, pollution from production and incineration, and the toxic by-products of mining. The integration of an environmental justice perspective has been a crucial part of the development of all aspects of religious environmentalism, from ecotheology and institutional commitment to political activism. It has decisively enlarged a conservationist ethic that had focused almost exclusively on the fate of nature and more traditional political agendas that had been limited to people. For secular environmental organizations that place environmental justice alongside nature preservation, it means a potentially much larger constituency. Simultaneously, socially marginalized groups of African Americans in East St. Louis, Latinos in New Mexico, or Native Americans in Wyoming can now feel that environmentalism concerns their lives as well as pandas and rain forests. In general, recognition of the class and racial nature of the environmental crisis was among the most important steps in helping environmentalism move into the mainstream of political life as a potentially unifying focus of political action, rather than remain the province of comparatively privileged groups.

From what is perhaps the first environmental struggle in which race played a key role, religion has been an essential part of the environmental justice movement. The struggle was sparked by North Carolina's 1982 decision to dispose of its toxic PCBs (a suspected cause of cancer and reproductive, immune, and endocrine problems) in Warren County, the area with the highest percentage of African Americans in the state. Residents accused the state government of picking their county cause of its racial makeup. Warren County Concerned Citizens, a biracial group based in a local Baptist church with the active leadership of its minister, began a lengthy process of protest, resistance, and civil disobedience. More than five hundred residents and supporters were arrested, including activist members of the United Church of Christ and a U.S. congressman. The campaign received national attention and helped spark an increase in political militancy and political representation for Warren County's black community.[68]

Having taken a committed role in the Warren Country struggle, the UCC's Commission for Racial Justice (formed in the early 1960s to work for "justice and reconciliation" in both the church and the broader society) began to study patterns of environmental contamination in U.S residential areas. Five years later it issued its landmark study, *Toxic Wastes in the United States*, which conclusively documented that race was the most important variable in determining the location of hazardous waste facilities, even taking precedence over socioeconomic status. What this meant for people of color was frightening: 60 percent of "black and Hispanic Americans and about half of Asian/Pacific Islanders and Native Americans lived in communities with uncontrolled toxic waste sites"; three of the five largest commercial hazardous waste landfills, accounting for almost half of capacity in the country, were in predominantly black or Hispanic communities; the more toxic facilities a community had, the more likely it was to have a high percentage of racial minorities.[69]

Along with its wealth of technical details, *Toxic Wastes* helped begin the critical process of thinking about the relation between race and pollution. It discussed how economically and socially marginalized racial minorities frequently lacked the social power, financial resources, and government connections possessed by white communities. To poor people, offers of jobs and tax revenues from toxic facilities could be attractive, and legal resources for resistance were often lacking. "Poverty, unemployment, and problems related to poor housing, education and health" meant that attention usually focused on immediate problems of survival, rather than on longer-term issues of environmental protection.[70] Such communities, in short, were vulnerable—and a consequence of that vulnerability was poisoned air, water, and earth.

Building on the national publicity of *Toxic Wastes*, the UCC convened a historic meeting, the first National People of Color Environmental Leadership Summit, in 1991. Called by the UCC's Benjamin Chavis, who had overseen the preparation of the 1987 report, the Summit gathered six hundred people from all over the United States, Canada, and the Pacific. Delegates heard Cherokee chief Wilma Mankiller describe resistance to Sequoyah Fuels' uranium conversion facility, Dolores Huerta of the United Farm Workers talk of the effects of pesticides on farm workers, and Pat Bryant from Louisiana's Gulf Coast Tenants

Organization talk about a "billion pounds of poisons" dumped into the "cancer alley" between New Orleans and Baton Rouge.[71]

On the basis of their experience, the delegates affirmed seventeen "Principles of Environmental Justice." These are worth quoting at some length, because in a remarkably clear way they show the integration of the spiritual and the political, concern for nature and for human beings, challenges to corporations and governments that has been the most important hallmark of religious environmentalism:

1. Environmental justice affirms the sacredness of Mother Earth, ecological unity and the interdependence of all species, and the right to be free from ecological destruction....

3. Environmental justice mandates the right to ethical, balanced and responsible uses of land and renewable resource in the interest of a sustainable planet for humans and other living things.

4. Environmental justice calls for universal protection from nuclear testing and the extraction, production and disposal of toxic-hazardous wastes and poisons....

7. Environmental justice demands the right to participate as equal partners at every level of decision-making including needs assessment, planning, implementation, enforcement and devaluation....

10. Environmental justice considers governmental acts of environmental injustice a violation of international law....

14. Environmental justice opposes the destructive operations of multinational corporations.[72]

Both *Toxic Wastes* and the Summit—organized, overseen, and financially supported by the United Church of Christ—have had a remarkable impact. The Summit itself had outside observers from Greenpeace, the Sierra Club, the National Resources Defense Council, and the Environmental Defense Fund. These groups had been alerted by the earlier report and also by a letter directed to the ten major U.S. environmental groups and signed by hundreds of activists in 1990 who charged the groups with failing to recognize environmental racism, a lack of diversity in their organizations, and often making policy decisions without including those affected in the process.[73] In the next few years, virtually all of these groups made significant policy shifts. Feature articles on environmental racism and the environmental justice movement appeared in their publications. They began to acknowledge and support campaigns in places and with groups they had previously ignored, vigorously seeking participation from people of color.[74]

Having given the problem a name and begun the process of understanding it, the environmental justice work of the UCC prompted further research. For example, a study chaired by environmental scholar Robert Bullard (who earlier had connected toxic waste siting and race in Houston)[75] revealed that penalties for violating hazardous waste laws in areas having a high percentage of whites were 500 percent higher than those applied in areas with large minority populations. The fines averaged $335,566 for Euro-American districts, $55,318 in minority ones.[76] Countless other studies have been done since then, and "environmental justice" has become part of the standard lexicon of the environmental movement everywhere it is studied.

Whether or not the specific *terms* are used, the *concepts* of environmental racism and environmental justice have spread far beyond the borders of the United States. Global environmental activists can see environmental injustice in the way Shell Oil and the Nigerian government collude in the oil production that has devastated the Ogoni peninsula. They can ask by what right does Philadelphia dump 15,000 tons of its toxic ash on Kasai Island off the mainland capital of Canabry, Guinea. And why the United States is the only industrialized county that has refused to ratify the Basel Convention, which forbids rich countries from exporting toxic waste to poor ones.[77] Such questions continue the initial connections made in Warren County and *Toxic Wastes in the U.S.*

The change registered on the governmental level as well. In 1994, less than three years after the Summit, President Bill Clinton issued an executive order directing federal agencies to "make achieving environmental justice part of [their] mission by identifying and addressing disproportionately high and adverse human health or environmental effects of [their] programs, policies, and activities on minority populations and low-income population."[78] Although the George W. Bush administrations have worked to undo Clinton's order, and although there are questions about how much was accomplished even under Clinton, it is nevertheless true that the concept of environmental justice has entered the mainstream of even comparatively conservative governmental environmental policy.[79] It is no accident that in 1998 EPA head Carol Browner issued an unprecedented order overruling Louisiana's approval of a PVC plant in Convent, a largely African American region that already had several toxic facilities. Sixteen years after Warren County, a real victory in the global war for environmental justice was won. Religion—in the form of the United Church of Christ's efforts, insights, and energy—could take a significant amount of credit for that victory.

This victory did not stem from a new perspective of world-famous theologians or the somewhat impersonal rhetoric of an Earth Day celebration. It was achieved because of direct connections among people who knew each other. And the churches' role depended on their moral rootedness in the everyday lives of their communities. Countless comparable actions, less well-known but equally the result of the simple moral ties between local religion and the daily life of community throughout the world, could be added. In India, the Sankat Mochan Foundation, led by Hindu priest and civil engineer Veer Bhadra Mishra, has received international recognition for pioneering work in organizing to restore ecological health to the Ganges River.[80] In southern Brazil, local church activists joined in an antidam movement, first to help protect peasants' land and later in support of a more inclusive concern with the region's ecology.[81] As the most important institution of civil society, poised uneasily between the formal structures of government and the private life of families, religion is at times the most powerful resource in any struggle against entrenched injustice.

Indeed, several commentators have argued that for environmentalism to succeed, it must be rooted in community life, moving beyond centralized organizations, national laws or policies, and single-issue campaigns keyed to a particular wilderness area, pollutant, or endangered species. Labeled by some "civic environmentalism," this model relies on informal networks of people concerned with the enduring existence and sustainable development of a particular locale.[82] It involves creative planning for the future as much as stopping some practice that is damaging, and the health of the human community as much as the health of the land. It focuses precisely on what is close at hand, in the hope that *this particular place* can be restored and sustained. Civic environmentalism can be found in a neighborhood coalition in Oakland, California, locally based conservation in rural Colorado, urban agriculture in Boston's Dudley Square, and in Africa, when natives are integrated into the ecotourism of national parks instead of being expelled so that the "wilderness" will be purely "natural" (except for the white tourists).[83] Religious environmental activists, connected to their neighbors through myriad congregational activities, church suppers, and midnight masses, are particularly suited to be active participants in civic environmentalism. As this chapter indicates, they have done so throughout the world already.

DURANGO, COLORADO

Before she could walk, Lori Goodman was taught that all parts of the world are connected, and that as a matter of course, you are to show respect for all your elders, human and nonhuman alike. A religious environmentalist virtually all her life, the connection between spirituality and caring for the earth was not something she had to realize or develop. Our true worth, she learned, depends on the quality of our relationships to people, animals, and the land—and not on how much "we can hoard and keep." As she grew up, however, she came to see the technical details clearly: "If you damage one part, all the other parts get damaged as well. For example, toxic waste dumps will come back to your water."[84]

It is not surprising that Goodman is a Native American, a Navajo, for proper relations with the earth are central to most native religious traditions. As Gail Small, whose work for the Northern Cheyenne parallels Goodman's with the Navajo, puts it, "Environment, culture, religion, and life are very much interrelated in the tribal way of life. Indeed they are often one and the same. Water, for example, is the lifeblood of the people. . . . Indeed, there is a profound spiritual dimension to our natural environment and without it, the war [to protect their lands] would not be worth fighting."[85]

Goodman's activism has centered on Diné (a Navajo word meaning, roughly, "the people") Citizens Against Ruining our Environment (Diné CARE). For nearly twenty years, Diné CARE has sought to defend the Navajo and their lands against a variety of threats.[86] In 1988, it successfully organized in the town of Dilkon to prevent the siting of a toxic waste incinerator and dump, overcoming pressure from the tribal governments to acquiesce and fears of their own powerlessness. As cofounder of the Indigenous Environmental Network, CARE has helped scores of native environmental activists exchange information, resources, and support. In the early 1990s, they kept an asbestos dump from their sacred mountains and resisted a tribal timber industry that was literally clear-cutting the reservation's forests and destroying the character of some of their tradition's most important sacred sites.[87] In the course of the struggle, which pitted Diné CARE against a Navajo-operated sawmill, one of its leading activists died mysteriously. Since then, the assault on Navajo forests has declined significantly.

Perhaps most important, from 1998 to the present Diné CARE has worked to bring justice to victims of uranium mining by forcing modification of the Radiation Exposure Compensation Act. Uranium had been mined on Navajo land since the late nineteenth century, but during World War II and the cold war the intensity of mining increased dramatically. Yet the Navajo people were never told of the dangers of uranium mining, nor how to lessen its effects on miners, their families, and their communities. In Navajo communities near the "tailings" left from old mines, the cancer rate can be as much as seventeen times the national average, and contaminated abandoned mines fill with rainwater and get used by livestock. Leading a coalition of radiation victims from the ranks of uranium miners as well as the "downwinders" who had been exposed to radioactive fallout from weapons tests in Utah, Diné CARE helped make government compensation for these victims more accessible. It has also tried to completely prohibit any future uranium mining on Navajo lands.

The struggle for sustainable forests, like the management of other resources on indigenous lands, typically pits the long-term interests of native culture and community against timber or mining industries that have a commitment only to short-term profit. Because short-term profit can be sizable, there are often conflicts within tribes as well as between tribes and white-owned corporations, as certain members of the community—often those with political power—opt to avail themselves of a percentage of the money flowing in. In Diné CARE's case, for example, pushing to curtail timber sales and prevent waste dump siting set them at odds with tribal leaders. It was only their ability to generate publicity and rally large numbers of ordinary Indians that enabled them to succeed.

The situation faced by Lori Goodman and Diné CARE is replicated throughout North American native communities, and indeed much of the world. On one side are native groups whose culture and history tie them to a place, an ecology, a way of life. As Goodman says, "Our land is our sacred books. We know the places we walk on."[88] *Where* they are is as essential to them as the Bible or the Qur'an is to Jews, Christians, and Muslims. On the other side is a world of nation-states, global corporations, and culturally alien communities, all seeking to extract as much from the land as possible. Timber in Minnesota, rivers in northern Quebec, coal in Montana and Wyoming, rain forests in Brazil—these are often the last frontiers for low-cost mining, cutting, damming for "cheap" power, or cattle raising. Further, as the political power of previously colonized countries throughout Africa, Asia, and the Middle East grew, the comparative powerlessness of native groups made them attractive sites for "development"—or for destructive use. Native peoples, like African Americans in urban settings from East St. Louis to Chester, Pennsylvania, are almost always marginalized groups, less liable to put up a fuss when they are ravaged by pollution. Suffering from poverty, lack of education, emotional depression, and cultural dislocation, they can be easy to divide.[89]

Finally, like Sri Lankan peasants whose rivers are dammed to make electricity for someone else, native peoples are for the most part suffering pollution while someone else benefits. Gail Small has worked for more than a decade to resist the effects of coal and gas production complexes that power Los Angeles and Seattle but not her own reservation. The Western Shoshone Defense Project, focused on tribal lands in Nevada and Southern California, has resisted nuclear testing and nuclear wastes, threats that are part of someone else's foreign policy and someone else's nuclear energy.[90]

Despite the David and Goliath feeling of many native struggles, some impressive victories have been achieved. Through aggressive court action, for example, Native Action got the Environmental Protection Agency to classify the Cheyenne region a Class-One airshed, which meant that nearby strip-mining was subject to vastly stricter EPA air quality standards. They also "launched court proceedings that resulted in a nationwide moratorium on all federal coal leases.[91]

Pollution, of course, is bad for everyone: we all suffer from higher cancer rates, more birth defects, the exorbitant costs of global climate change, and the loneliness that comes as biodiversity diminishes. Yet environmental destruction of native lands is also a kind of cultural genocide. When these lands are contaminated, rivers dammed, and traditional game rendered toxic through mercury or PCBs, indigenous groups can simply die as a people. Given the role of culture in sustaining people's sense of identity and personal well-being, it is often a kind of quasi-physical genocide as well, as the loss of culture all too often leads to epidemics of depression, suicide, family instability, and drug and alcohol abuse.

For generations indigenous groups have been guided by perspectives that stress ecological balance, preservation, and reciprocity rather than the characteristically modern attitude of "improvement," "development," and taking as much as we can as fast as we can. As a number of writers have suggested, this is often not so much a matter of finding the sacred in nature, as if the Western concept of God were now identified with the earth. Rather, it is a sense of respect and care—a moral, psychological, and spiritual relationship that can be characterized as "social" as much as "holy."[92] A marvelously clear (and clearly racist) statement of the distinct environmental consequence of this worldview can be found in an 1874 newspaper editorial condemning Indians for resisting a gold rush into the Dakotas: "What shall be done with these Indian dogs in our manger? They will not dig the gold, nor let others dig it.... They are too lazy and too much like animals to cultivate the fertile soil, mine the coal, develop the salt mines, bore the petroleum wells or wash the gold."[93]

As one historian puts it, "Virtually all Americans at the time saw the 'Indian problem' as the perpetuation of Indian cultural patterns and land use incompatible with those of the larger society."[94] Yet, what is from the dominant white point of view a willful and unreasoned refusal to "develop" is from the native point of view an ongoing commitment to their *particular* place. Moving "somewhere else" after having ruined this piece of land is simply not an option—or at least, not an option that is compatible with the continued existence of the group as it understands itself. As the Central Land Council of Australian Aborigines makes clear, "For us, land isn't simply a resource to be exploited. It provides us with food and material for life, but it also provides our identity and it must be looked after both physically and spiritually. If we abuse our land, or allow someone else to abuse it, we too suffer."[95]

Thus, for most native traditions, "religious environmentalism" is redundant. Their traditions simply are environmentally oriented. Unlike the dominant themes of Western religions, which stress heavenly salvation and purely interpersonal morality, or the Hindu and Buddhist pursuit of liberation from the sufferings of embodied existence, native traditions believe that the well-being of people and nature are inextricably linked:

> Tribal members observed elaborate systems of proscription—taboos—governing the essential activities of hunting, planting and harvesting, and insuring that individuals would not disturb the web of agreements by which animals and plants consented to offer themselves up to meet human needs. Through traditions of myth, ritual, community identity and moral action, then, tribes were often not easily inclined to see themselves in positions to sell land or extinguish title to other human beings, but rather as dependent upon the greater-than-human power embodied in the land.[96]

The set of distinct attitudes and practices is one reason "Native American environmentalism" is not limited to defensive actions in response to pollution or clear-cutting. Rather, the intelligence of the native worldview is at times shown in their management of their own lands. To take but one example: the Menominee tribe has been in control of its forest resources since 1854, when the Wolf River Treaty created their reservation. In appearance, its 220,000-acre forest in northeast Wisconsin looks wild and untouched. Yet in truth, it has been highly managed—in a sustainable way—by the tribe. Over 2 *billion* board feet of lumber have been removed, yet the total mass of wood in the forest is greater than it was a century and a half ago. Towering white pines fill the forests, along with ruffed grouse, eagles, ospreys, woodcocks, bobcats, coyotes, otter, marten, fox, white-tailed deer, and bears. With many of its members still living close to the land, the forest provides meat and furs, maple syrup and ginseng. Continual monitoring assures the Menominee that their forest will continue for generations, serving human needs while maintaining its own ecosystemic integrity. As one elder said, "Everything we have comes from Mother Earth—from the air we breathe to the food we eat—and we need to honor her for that. In treating the forest well, we honor Mother Earth."[97]

Of course, one should not romanticize native traditions nor overgeneralize about them. There are, for instance, eyewitness accounts of Indians killing large numbers of buffalo and taking only a small piece of the carcass. Some evidence suggests that as many as fifteen large mammal species were wiped out by native hunting between 12,000 B.C.E. and 1000 C.E.[98] It may be that native environmentalism had to be learned, that the knowledge was not some magical, spontaneous genetic endowment.[99] As radical environmentalist Dave Foreman asks, "Is the land ethic of the Hopi a result of a new covenant with the land following the Anasazi ecological collapse seven hundred years ago? Would the hunt ethics of tribes in America (and elsewhere) have been a reaction to Pleistocene overkill?"[100]

Indeed, this may well be the case for indigenous peoples throughout the world: perhaps they learned to honor the earth after having some experiences of the consequences of *not* honoring it. The point, nonetheless, is that for the most part, they learned the lessons of their histories and over the course of generations developed a deeply religious sense of their interdependence with the earth, a sense that was expressed not only in myth and ritual but in limitations on hunting, fishing, and farming, in literal management of the ecosystem.[101] Given the difference between their technological powers and our own, we can only hope that contemporary societies will learn such lessons without making mistakes of comparable scope—where we have not done so already.

The insatiable demand of the dominant society for land and production, of course, does not make the prospects of native environmentalism particularly bright. Key victories have been won, but throughout the world the onslaught continues. Thankfully, the same sensibility that leads indigenous activists to the struggle sustains them in it. When I asked Lori Goodman what she does about despair over her losses, she seemed genuinely surprised. "We don't feel despair," she said. "No matter what happens. We know we are not going to beat them, but at best we can delay them, and perhaps end up with something sustainable. In the meantime our ceremonies, our prayers, our celebrations—they keep us going."[102]

If there is hope for native peoples throughout the world, it will come because they have aligned themselves with nonnative environmentalists, in particular the environmental justice movement of the larger societies in which they live. Throughout the world, cooperation among different types of people is often the key to success in environmental struggles. As we have seen in the example of the religion-science cooperation on global warming, religious environmentalists are actively engaged in working alliances with and becoming members of secular environmental organizations and campaigns. Often, there is no clear division between the secular and the religious in the environmental community (even if, as many of the Sisters of Earth feel, some need an organizational home base with a strong spiritual dimension). In fact, although in some countries (e.g., Islamic ones) religious motives will be sufficient to mobilize political change, in those in which religion and government are separated it is in the broadest possible coalitions that religious environmentalism can make the largest possible impact. This fact has not been lost either on religious environmentalists or on secular ones, and their increasing cooperation, as we'll now see in two telling examples, is one of the most hopeful signs in the world environmental movement.

In January 2002, a remarkable ad appeared on TV stations in Georgia, Arizona, North Dakota, Indiana, Missouri, and Delaware. With a background of uplifting new age music and a series of breathtaking images of mountains, rivers, and tundra, the ad told viewers, in language familiar to virtually every secular environmental organization, that America's energy needs could be better met by conservation and renewable sources than by drilling in the Arctic National Wildlife Refuge. And there was more: in language that would have fit perfectly with the Religious Witness's civil disobedience at the Department of Energy, the ad also proclaimed that Americans have a deep obligation not to ruin the land that is in our trust, but to "keep our promise to care for creation."

The combination of secular and religious vocabulary was no accident, because the ad was a joint project of the Sierra Club and the National Council of Churches. Sierra's executive director Carl Pope, feeling a real need to reach out to communities of faith, had initiated contact with the NCC's secretary general, former Pennsylvania congressman Bob Edgar. After taking, Edgar says, "about thirty seconds to think about it" before agreeing, the NCC provided advice on how to phrase the issues and lent its name and influence. The Sierra Club put up the money and directed production. Interestingly, within the Sierra Club there was some serious discussion about whether to do the joint ad and, if so, how: the word "God" was taken out at the last minute, but the word "creation," though opposed by some, remained. Afterward, a large volume of positive responses came to both groups.[103] And the follow-up on both sides has been impressive. The

Sierra Club now spends upwards of $100,000 a year in outreach to faith communities and to offer support to religious environmentalists. The NCC and its members partner with Sierra in several contexts, from climate change work in New Mexico to clean water campaigns in Arizona and environmental justice work in Louisiana.

The Sierra-NCC ad was one moment in a long campaign to protect ANWR and redirect America's energy policy. Focused on India, the extensive struggle to defeat the Narmada complex has had a distinctly international character. The religious element in this struggle has been one piece of the puzzle, and its role indicates something of the importance, as well as the difficulty and the complexity, of the real-world politics in which religious environmentalists are involved.

In the late 1980s, the Indian government proposed its largest ever public works project: a series of 30 large, 135 medium, and 3,000 small dams to harness the waters of the Narmada and its tributaries, ostensibly for flood control, irrigation, and electric power. Alarmed over the possible human and ecological effects of the project, long-time Gandhian nonviolent social activist Baba Amte, whose extensive political and humanitarian work included successfully resisting large dam projects in Maharastra in the early 1980s, organized a meeting of dozens of India's leading environmentalists. They estimated that the projects would flood more than 200,000 acres of pristine forest and fertile farmland, threaten the existence of endangered species, and displace as many as 300,000 local villagers, most of them tribal people or untouchables. Out of that meeting was born Narmada Bachao Andolan (Save the Narmada), an ongoing organization that has opposed the dam complex by a wide variety of means, from legal challenges and public protests to international pressure and civil disobedience.[104]

Opposition to the Narmada dams was based in a fascinating mix of religion, science, and simple human decency. Like the Ganges, the Narmada has a sacred character for the tribal peoples who live near it. For them, the dams are a kind of desecration. Further, the benefits of this sacrilege, according to critics, were vastly overrated and would in any case be selectively distributed to the corporations paid for building the dams and the comparatively small percentage of city dwellers who would receive the electricity. Interestingly, opposition to large dams in India was justified as much by science as by religion, for large dam projects, so beloved by development agencies and central governments in the third world, have for the most part proved to be environmental and human disasters. They have destroyed ecosystems, created tens of millions of "refugees from development," and produced comparatively little in the way of benefits. Riverbanks are eroded for tens or even hundreds of miles downstream; fertile sediment stays behind the dam, worsening fishing and farming; and submerged villages release toxic chemicals into the water. Even in terms of irrigation, small-scale dams tend to work much more efficiently.

When the negative environmental and human effects and the astronomical costs of construction are factored in, the electricity produced is neither cheap nor clean. Ironically, the World Bank, long a strong supporter of large dams (including the Narmada project), in 2002 undertook its own study of their effectiveness and concluded that by and large they had had unjustified ecological and human effects.[105] (Strangely, this has not stopped them, along with numerous other private and public development organizations, from funding the Narmada project.) In his challenge to the Narmada dams, Baba Amte put this matter clearly nearly fifteen years before the World Bank got the point: "I will not let my beloved state of Gujarat fulfill a death wish by adopting an antediluvian technology. The science of large dams now seems to belong to the age of superstition; the coming century belongs to the technology of mini and micro dams and watershed development ensembles."[106]

Narmada Bachao Andolan has been struggling for seventeen years, trying to halt the dams, limit their height, or at least get recompense for the refugees they create. And they have not done so alone. The movement against the Narmada dams has been truly global in scope. International environmental NGOs, such as Oxfam, the Environmental Defense Fund, and the International Rivers Network, publicized the case, offered support, and publicly protested at World Bank meetings. Indian environmentalists and civic activists have joined with tribals in demonstrations and protests, challenging state and national governments.[107]

Like other international struggles—against the James Bay hydroelectric plant in northern Quebec, or nuclear power, or global warming—the Narmada campaign incorporates secular analysis and religious reverence, a sophisticated critique of failed technology and a near instinctive sense that some rivers should be left intact. The worldwide scope of these campaigns indicates, against all dire warnings to the contrary, that there is no inherent incompatibility between science, technology, the political defense of human rights, environmentalism—and religion. Religious environmentalists—activists defending God's creation or Mother Earth, the entire globe or their own villages—have become an essential part of an international movement for a sustainable future.

NOTES

1. Jan Nunley and Jerry Hames, "Episcopalians Join Protest against Drilling in Arctic Refuge," Episcopalian News Service, May 14, 2001; Massachusetts Conference Edition, United Church News, "Four Conference Clergy Members Arrested at Protest," June 2001; "22 Arrested in Peaceful Civil Disobedience at Dept. of Energy: Religious Leaders Pray for Arctic Refuge," Arctic Truth, May 3, 2001. Information and links to news stories at Religious Witness for the Earth, http://www.religiouswitness.org/.

2. Margaret Bullitt-Jonas, "Conversion to Eco-Justice: Reflections on the Inner Journey," based on remarks given at the Costas Consultation in Global Mission 2002–2003, "Earth-Keeping as a Dimension of Christian Mission," February 28–March 1, 2003, unpublished ms.

3. From Shin Shalom, twentieth-century Jewish poet: "On Tu B'Shvat . . . an Angel descends, ledger in hand, and enters each bud, each twig, each tree, and all our garden flowers. . . . When the ledger will be full, of trees and blossom and shrubs, when the desert is turned into a meadow and all our land is a watered garden, the Messiah will appear." In Stein, *A Garden of Choice Fruit*, p. 65.

4. Letter to Charles Hurwitz from Naomi Steinberg for Children of the Earth, reproduced at Green Yes, http://greenyes.grrn.org/1997/0248.html.

5. Seth Zuckerman, "Redwood Rabbis," *Sierra* (November/December 1998); Margaret Holub, "Redwoods and Torah," *Tikkun* (May/June 1999).

6. Religious Witness for the Earth, http://www.religiouswitness.org/.

7. Andrea Ayvazian, interview with author, October 14, 2004.

8. Shalom Center, http://www.shalomctr.org/index.cfm/action/read/section/aboutus/article/article711.html.

9. COEJL, www.coejl.org.

10. Hadar Susskind, interview with author, May 31, 2005. COEJL's Israel program is the Jewish Global Environmental Network.

11. Palmer, *Faith in Conservation*, p. 3.

12.

Dwarka, India—A few weeks ago, the crew of an Indian fishing boat in the Arabian Sea thought they had the biggest catch of their lives. A 40-foot-long unsuspecting whale shark had entered their nets on a still night. But instead of killing the creature, known as the gentle ocean giant, the captain called the boat owner who promptly told him to let it go. "I may have lost a lot of money. But I'm happy that I could play a role in saving the protected fish," Kamlesh Chamadia, the boat owner said. Two years ago, Chamadia, like hundreds of other fishermen along the Saurashtra coast of India's western Gujarat state, would have had little hesitation in killing whale sharks. But a lively campaign by a wildlife group and a popular religious leader has helped reduce the killing. . . . The whale shark protection campaign in Gujarat has also got a religious flavor to it after Murari Bapu, a popular Hindu preacher, agreed to be its brand ambassador.

Bapu, who holds his audience spell-bound with his narration of the stories from the Hindu epic Ramayana, likens the migration of the whale shark to pregnant daughters coming to their parental home for the delivery. "Would you ever think of any harm to your daughters, let alone killing them. Whale sharks are your daughters and you should take good care of them," Bapu told a gathering of hundreds of fishermen in Dwarka, a coastal pilgrimage city said to be founded by Lord Krishna. Activists have also staged a street drama depicting a pregnant daughter pleading with her fisherman father not to kill a whale shark trapped in his net. The play, which has been performed in towns across the Saurashtra coast, strikes an emotional chord with the fishermen and four towns have adopted the shark as their mascot.

Thomas Kutty Abraham, "Whale Shark Finds New Friends in Indian Fishermen," Reuters, December 10, 2004. Thanks to Cynthia Read for bringing this story to my attention.

13. Alliance for Religion and Conservation, www.arcworld.org/projects .asp?projectID=173.

14. National Commission for Conservation, http://www.saudinf.com/main/a63.htm.

15. Alliance for Religion and Conservation and the Khalsa environmental project, www.khalsaenvironmentproject.org/project_environment.htm.

16. Martinus L. Daneel, *African Earthkeepers: Wholistic Interfaith Mission* (Maryknoll, NY: Orbis Books, 2001), p. 185. African Independent Churches "are all-African churches founded *by* Africans *for* Africans, and have emerged during the twentieth century as thoroughly inculturated movements throughout Africa, south of the Sahara, representing more than half of African Christianity in Zimbabwe and South Africa." This quote is from Martinus L. Daneel, "African Initiated Churches as Vehicles of Earth-care in Africa," in *The Oxford Handbook on Religion and Ecology*, ed. Roger S. Gottlieb (New York: Oxford University Press, forthcoming).

17. Daneel, *African Earthkeepers*, p. 15.

18. M. L. Daneel, "African Earthkeeping Churches," in Taylor, *Encyclopedia of Religion and Nature.* To get an idea of what this number of trees means, the Green Belt Movement of Kenya, whose leader, Wangari Maathai, won the 2004 Nobel Peace Prize, planted approximately 30 million trees over a much longer period of time.

19. Daneel, *African Earthkeepers*, p. 55. "Instead of withdrawing from the traditionalist practitioners of ancestor veneration to demonstrate its rejection of 'heathenism,' the prophetic earthkeeping church underscored at least ecological solidarity with its traditionalist counterparts in the green struggle" (Daneel, "African Initiated Churches").

20. Daneel, *African Earthkeepers*, p. 155.

21. Bruce Byers, "Mhondoro: Spirit Lions and Sacred Forests," *Camas* (Fall 2002),[0] reprinted in Gottlieb, *This Sacred Earth*, 2nd ed., p. 659.

22. Daneel, *African Earthkeepers*, p. 104.

23. Ibid., p. 151.

24. Ernst Conradi, Charity Majiza, Jim Cochrane, Weliel T. Sigabi, Victor Molobi, and David Field, "Seeking Eco-Justice in the South African Context," in *Earth Habitat: Eco-Injustice and the Church's Response*, ed. Dieter Hessel and Larry Rasmussen (Minneapolis: Fortress, 2001). The essay lists other interesting projects, including the Catholic-based "Planters of the Home," which focuses on urban agriculture and greening, and the Methodist-based Khanya Program, which joins sustainable agriculture, appropriate housing, and micro-industries (Khanya program, http://gbgm-umc.org/umcor/stories/khanya.stm).

25. See comments by Geoff Davies at NECCSA, http://www.neccsa.org.za/Issues-Toll%20road.htm. Also see Save the Coast Campaign, http://www.wildlifesociety.org.za/SWC%20press%20Bishop.htm.

26. "The Earth Belongs to God: Some African Church Perspectives on the World Summit on Sustainable Development (WSSD) 2002 and Beyond," statement adopted at the African Regional Consultation on Environment and Sustainability, Machakos, Kenya, May 6–10, 2002, at Network of Earthkeeping Christian Communities in South Africa, http://www.neccsa.org.za/Documents-Earth%20belongs%20to%20God.htm. Ernst Conradie brought the Network to my attention.

27. Quoted at http://www.uspg.org.uk/news2004/newsaut2.htm. For a detailed history of activities in the province, I was fortunate to consult Andrew Warmback, "The Diocese of Umzimbuvu," chapter 4 of *The Household of God: The Environment, Poverty and the Church in South Africa*, unpublished doctoral dissertation.

28. Tovis Page, "Reflections on the 6th International Sisters of Earth Conference July 15–18, 2004," unpublished ms.

29. EarthLinks, www.EarthLinks-Colorado.org.

30. These remarkable women were brought to my attention by the immensely valuable research of Sarah Macfarland Taylor. See Sarah Taylor, "Reinhabiting Religion: Green Sisters, Ecological Renewal, and the Biogeography of Religious Landscape," in Gottlieb, *This Sacred Earth*, 2nd ed.

31. Carol Coston, interview with author, September 23, 2004.

32. See S. Taylor, "Reinhabiting Religion," and Web sites of EarthLinks, Genesis Farm (www.genesisfarm.org/), Crystal Spring Earth Learning Center (http://home .comcast.net/~cryspr/), Santuario Farm (http://www.sisterfarm.org/), and Sisters of Earth (http://www.sistersofearth.org/).

33. Santuario Sisterfarm, http://www.sisterfarm.org/.

34. S. Taylor, "Reinhabiting Religion," p. 618.

35. MacGillis interview.

36. Rosemary Radford Ruether, "Sisters for [*sic*, should be "of"] Earth," Faith in Place, http://www.faithinplace.org/sisters-for-earth.php.

37. S. Taylor, "Reinhabiting Religion," p. 613.

38. Ruether, "Sisters."

39. Coston interview.

40. Given the pace of recent industrialization in India and even more in China, in a decade or so this may be somewhat less true.

41. Ross Gelbspan, "Slow Death by Global Warming," *Amnesty International* (Fall 2004): 6. See also *Time* magazine's September 2004 issue on global warming in Alaska. For a concentrated picture of global events, see Max Seabaugh, "Feeling the Burn," *Mother Jones* (May–June 2005): 46–47.

42. Andrew Simms, "Farewell Tuvalu," *The Guardian* (Manchester, UK), October 29, 2001; Michael Grunwald, "Bizarre Weather Ravages Africans' Crops: Some See Link to Worldwide Warming Trend," *Washington Post*, January 7, 2003, http:// www.massclimateaction.org/African_crops_WashPost010703.htm.

43. David G. Hallman, "Climate Change and Ecumenical Work for Sustainable Community," in Hessel and Rasmussen, *Earth Habitat*.

44. For a compressed account of this history, see Hessel, "The Church's Eco-Justice Journey."

45. Hallman, "Climate Change," p. 126.

46. World Council of Churches, http://www.wcc-coe.org/wcc/what/jpc/economy .html.

47. World Council of Churches, http://www.wcc-coe.org/wcc/what/jpc/cop6-e.html.

48. Web of Creation, http://www.webofcreation.org/ncc/climate.html.

49. Web of Creation, www.webofcreation.org.

50. See an account of this development and the statement itself in Mary Evelyn Tucker, *Worldly Wonder: Religions Enter Their Ecological Phase* (Chicago: Open Court, 2003), pp. 116–123.

51. National Religious Partnership for the Environment, http://www.nrpe.org/.

52. What Would Jesus Drive?, http://www.whatwouldjesusdrive.org/tour/intro.php.

53. My account of Tzu-Chi is indebted to Wan-Li Ho, "Environmental Protection as Religious Action: The Case of Taiwanese Buddhist Women," in *Ecofeminism and Globalization: Exploring Culture, Context, and Religion*, ed. Heather Eaton and Lois Ann Lorentzen (Lanham, MD: Rowman and Littlefield, 2003). On Taiwanese environmentalism in general, see Hsin-Huang Michael Hsiao, "Environmental Movements in Taiwan," in *Asia's Environmental Movements: Comparative Perspectives*, ed. Tok-shiu F. Lee and Alvin Y. So (Armonk, NY: M. E. Sharpe, 1999).

54. Tzu-Chi, July 2, 2004, http://www.tzuchi.org/global/news/articles/20040602.html.

55. Tzu-Chi, http://www.tzuchi.org/global/about/index.html.

56. Cheng Yen, "Let's Do It Together: Environmental Protection for Mind, Health, and the Great Earth." Tzu-Chi, http://www.taipei.tzuchi.org.tw/tzquart/96winter/qw96-2.htm.

57. Tzu-Chi, http://www.tzuchi.org/global/about/missions/culture/index.html#tagstay.

58. Chang Shun-yen, "Environmental Protection at the Abode of Still Thoughts," *Tzu-Chi Quarterly* (Summer 1997), Tzu-Chi, http://taipei.tzuchi.org.tw/tzquart/97summer/97summer.htm.

59. Ho, "Environmental Protection," p. 132.

60. Sarvodaya, http://sarvodaya.org/Introduction/sarvodayaover.html.

61. Quoted in George D. Bond, *Buddhism at Work: Community Development, Social Empowerment, and the Sarvodaya Movement* (Bloomfield, CT: Kumarian Press, 2004), p. 16.

62. Detlef Kantowsky, *Sarvodaya: The Other Development* (New Delhi: Vikas Publishing House, 1980), p. 166.

63. Sarvodaya, http://sarvodaya.org/Introduction/sarvodayaphilos.html.

64. Sarvodaya, http://sarvodaya.org/Library/Overview96/Paurushodaya.htm.

65. Quoted in Bond, *Buddhism at Work*, pp. 50, 33.

66. Quoted in ibid., p. 50.

67. "Justice and Witness Ministries board members witness in opposition to a U.S. Army plan to incinerate tons of neutralized nerve gas near East St. Louis, Illinois." United Church of Christ, http://www.ncccusa.org/.

68. Dollie Burwell, "Reminiscences from Warren County, North Carolina," in *Proceedings of the First National People of Color Environmental Leadership Summit*, ed. Charles Lee (New York: United Church of Christ Commission for Racial Justice, 1992). Earlier studies had documented the existence of racial or class imbalances in this area, starting with an article in *Science* in 1970 and including a General Accounting Office study in 1983. See Robert J. Brulle, *Agency, Democracy, and Nature: The U.S. Environmental Movement from a Critical Theory Perspective* (Cambridge, MA: MIT Press, 2000), pp. 210–213. For a survey of the history of the attempt to connect environmental and other political issues, see Robert Gottlieb, *Forcing the Spring: The Transformation of the American Environmental Movement* (Washington, DC: Island Press, 1993), pp. 235–269.

69. Commission for Racial Justice, *Toxic Wastes and Race in the United States* (New York: United Church of Christ, 1987), pp. xiii–xiv.

70. Ibid., p. xii.

71. Lee, *Proceedings*, p. 85.

72. Ibid., pp. xiii–xiv; the Principles are also in Gottlieb, *This Sacred Earth*, 1st and 2nd eds.

73. Letters initiated by the Southwest Organizing Project and the Gulf Tenants Association. See Dana Alston, "Moving beyond the Barriers," in Lee, *Proceedings*.

74. The Sierra Club, for instance, adopted an environmental justice policy in 1993 and created a specific environmental justice program. For an overview of its campaigns since then and interviews with some of its staffers, see Jennifer Hattam, "Why Race Matters in the Fight for a Healthy Planet," *Sierra* (May–June 2004). On the other hand, relative to other environmental concerns, the environmental justice movement is woefully underfunded. See Daniel Faber and Deborah McCarthy, *Green of Another Color: Building Effective Partnerships between Foundations and the Environmental Justice Movement*, a research report initiated by Nonprofit Sector Research Fund of the Aspen Institute, April 2001, available at Northeastern University, http://www.casdn.neu.edu/~socant/misc/Another%20Color%20Final%20Report.pdf.

75. Robert Bullard, "Solid Waste Sites and the Black Houston Community," *Sociological Inquiry* 53 (Spring 1983).

76. Robert Bullard, *Environmental Justice Project* (Washington, DC: Lawyers' Committee for Civil Rights Under the Law, 1993). For a brief but effective account of the history of environmental justice, see Target Earth, http://www.targetearth.org/about/human_rights.html. For a description of the emergence and importance of the movement, see Gottlieb, *Forcing the Spring*, 1993; 2nd ed., 2005.

77. *Rachel's Environmental & Health Weekly* 595 (April 23, 1998), http://www.rachel.org/bulletin/index.cfm?St=4. The United States has been joined in this dubious distinction by Afghanistan and Haiti. See Basel Convention, http://www.basel.int/ratif/frsetmain.php.

78. U.S. Forest Service, http://www.fs.fed.us/land/envjust.html.

79. Conclusions critical of implementation of the policy, including the Bush administration's stated policy of redefining "environmental justice" to apply to "everyone," are contained in a report from the inspector general of the Environmental Protection Agency: "EPA Needs to Consistently Implement the Intent of the Executive Order on Environmental Justice," Office of the Inspector General, U.S. Environmental Protection Agency, March 1, 2004. See "Reverse Environmental Justice by Bush EPA," Greenwatch, http://www.bushgreenwatch.org/mt_archives/000070.php. A detailed analysis, and criticism, of the policy implementation is also given by the U.S. Civil Rights Commission report *Not in My Backyard: Executive Order 12,898 and Title VI as Tools for Achieving Environmental Justice*, October 2003, U.S. Civil Rights Commission, http://www.usccr.gov/pubs/envjust/main.htm.

80. Mishra was identified as one of seven "Heroes of the Planet" by *Time* magazine for his work in protecting the fresh waterways of the earth. See Sankat Mochan, http://members.tripod.com/sankatmochan/index.htm.

81. Franklin Rothman and Pamela Oliver, "From Local to Global: The Anti-Dam Movement in Southern Brazil, 1979–1992," in *Globalization and Resistance: Transnational Dimensions of Social Movements*, ed. Jackie Smith and Hank Johnston (Lanham, MD: Rowman and Littlefield, 2002), pp. 119–123.

82. William A. Shutkin, *The Land That Could Be: Environmentalism and Democracy in the Twenty-First Century* (Cambridge, MA: MIT Press, 2000). See also Gottlieb, *Forcing*

the Spring; Dewitt John, *Civic Environmentalism: Alternatives to Regulation in States and Communities* (Washington, DC: Congressional Quarterly Press, 1994).

83. Raymond Bonner, *At the Hand of Man: Peril and Hope for Africa's Wildlife* (New York: Knopf, 1993).

84. Lori Goodman, interview with author, September 28, 2004. See also Winona LaDuke, *All Our Relations: Native Struggles for Land and Life* (Boston: South End Press, 1999). This book and *The Winona LaDuke Reader* (Stillwater, MN: Voyageur Press, 2002) were major sources for this section. See also the large number of links on the Web site of the Indigenous Environmental Network, www.ienearth.org/.

85. Gail Small, "The Search for Environmental Justice in Indian Country," *Amicus Journal* (March 1994).

86. Diné CARE, http://dinecare.indigenousnative.org/.

87. Ernie Atencio, "After a Heavy Harvest and a Death, Navajo Forestry Realigns with Culture," *Western Roundup, October 31, 1994.*

88. Goodman interview.

89. An aptly titled and wide-ranging anthology making reference to more than twenty geographical contexts is Mario Blaser, H. A. Feit, and G. McRae, eds., *In the Way of Development: Indigenous Peoples, Life Projects and Globalization* (London: Zed Press, 2004).

90. "Since the 1950s, the U.S. has tested hundreds of nuclear weapons on Western Shoshone homelands and disposed of thousands of metric tons of radioactive waste in unlined trenches at the Nevada Test Site. Currently Congress has passed legislation to dump all of the nuclear industry's high-level nuclear waste at Yucca Mountain." Western Shoshone Defense Project, http://www.wsdp.org/index.htm.

91. Winona LaDuke, "Native Environmentalism," *Earth Island Journal* 8, no 3 (Summer 1993).

92. For example, Szerszynksi, *Nature, Technology and the Sacred.*

93. Donald D. Jackson, *Custer's Gold: The United States Cavalry Expedition of 1874* (New Haven: Yale University Press, 1966), pp. 8–9.

94. Matthew Glass, "Law, Religion, and Native American Lands," in B. Taylor, *Encyclopedia of Religion and Nature.* I found the newspaper quote in Glass's excellent article.

95. Central Land Council of Aborigines of Australia, *Our Land, Our Life* (Alice Springs, Northern Territory, Australia: Central and Northern Land Councils, 1991), quoted in Fabienne Bayet, "Overturning the Doctrine: Indigenous People and the Wilderness—Being Aboriginal in the Environmental Movement," in *The Great New Wilderness Debate*, ed. J. Baird Callicott and Michael P. Nelson (Athens: University of Georgia Press, 1998), p. 319. Compare this quote from Paula Gunn Allen: "We are the land. To the best of my understanding, that is the fundamental idea that permeates American Indian life; the land (Mother) and the people (mothers) are the same.... The earth is the source and being of the people and we are equally the being of the earth. The land is not really a place separate from ourselves.... Rather for the American Indians ... the earth *is* being, as all creatures are also being: aware, palpable, intelligent, alive" (*The Sacred Hoop: Recovering the Feminine in the American Indian Traditions* [Boston: Beacon Press, 1989], pp. 60, 119). See also John A. Grim, "Indigenous Traditions: Religion and Ecology," in Gottlieb, *Oxford Handbook*; Kenneth M. Morrison, "The Cosmos as Intersubjective: Native American Other-than-human Person," in *Indigenous Religions: A*

Companion, ed. Graham Harvey (London: Cassell, 2000); J. Baird Callicott, "American Indian Land Wisdom? Sorting Out the Issues," in *In Defense of the Land Ethic: Essays in Environmental Philosophy* (Albany: State University of New York Press, 1989).

96. "Native Traditions and Land Use," in B. Taylor, *Encyclopedia of Religion and Nature*.

97. Paula Rogers Huff and Marshall Pecore, "Case Study: Menominee Tribal Enterprises," Menominee, http://www.menominee.edu/sdi/csstdy.htm#A.

98. On the buffalo massacre, see Momaday, "A First American Views His Land." On megafauna extinctions, see Ted Steinberg, *Down to Earth: Nature's Role in American History* (New York: Oxford University Press, 2002), pp. 126–127.

99. Commenting on the sources of Native American knowledge, John A. Grim observes: "These forms of interactive knowledge may not be scientific as that term describes Western modes of empirical, falsifiable, experimental investigation, but indigenous knowledge has its own modes of empirical observation, acquisition through lived experience, and testing in the context of one's community." Grim, "Indigenous Traditions."

100. Dave Foreman, "Wilderness Areas for Real," in Callicott and Nelson, *The New Wilderness Debate*, p. 402.

101. For a highly detailed account of traditional native management practices in a specific region, see M. K. Anderson, *Tending the Wild: Native American Knowledge and the Management of California's Natural Resources* (Berkeley: University of California Press, 2005).

102. Goodman interview.

103. Melanie Griffin, Sierra Club, interview with author, October 4, 2004; Bob Edgar, National Council of Churches, interview with author, December 10, 2004.

104. There are many accounts of this struggle. Two are Friends of the Narmada, http://www.narmada.org/, and Arundhati Roy, *Power Politics* (Boston: South End Press, 2001).

105. International Rivers Network, http://www.irn.org/index.asp?id=/basics/damqa.html; see report on large dams.

106. Baba Amte biography, at Maharogi Sewa Samiti, http://mss.niya.org/people/baba_amte.php.

107. For a thoughtful account of the different dimensions of the resistance to Narmada, see William E. Fisher, "Sacred Rivers, Sacred Dams: Competing Visions of Social Justice and Sustainable Development along the Narmada," in Gottlieb, *This Sacred Earth*, 2nd ed. There are many other internationally oriented efforts, including those that focus on the Brazilian rain forest and stopping the James Bay dam project in Canada.

CHAPTER 22

....

RELIGION AND ENVIRONMENTAL STRUGGLES IN LATIN AMERICA

....

LOIS ANN LORENTZEN
SALVADOR LEAVITT-ALCANTARA

MILLIONS of indigenous people and peasants crowd into urban slums throughout Latin America. Displaced from their land, they struggle to provide for their families in cities that offer them air pollution, contaminated water, minimal shelter, and inadequate sanitation services. Mining, logging, ranching, the clear-cutting of forests, soil erosion, and the polluting of lakes and rivers force others into desperate battles for survival in the countryside. Latin America faces environmental crises that directly affect the health and well-being of its people, especially the poor.

This essay tracks the involvement of religious groups in the myriad environmental struggles found in Latin America today. We assume that religion, through myth, rituals, and narratives, provides a framework by which one can understand a group's relationship to nonhuman nature and to actions on behalf of the environment. We realize that analyzing the beliefs and actions of religious actors is complicated terrain; each tradition or group will be more complex than evidenced by the brief treatment given in this essay. Religious traditions, identities, and institutions are continually contested and reshaped to fit historical, social, economic, and cultural conditions. With this caveat, however, we contend that taking religious traditions and actors into account deepens understanding of environmental struggles and movements within Latin America. This essay explores religion and environmental struggles in Latin America by charting religious beliefs and practices of indigenous religions, the Roman Catholic church, liberation theologians, ecofeminist movements, Protestant faith traditions (emphasizing evangelical and Pentecostal Protestantism), and diaspora religions of Latin America and the Caribbean. In each case we analyze religious symbols, theologies, myths, narratives, and rituals as they relate to the nonhuman world. We track links between theory and practice as we discuss the involvement of religious institutions and individual actors in environmental struggles. The essay concludes by discussing common themes found among the religious traditions, debates within the scholarly literature, and issues for further study.

INDIGENOUS RELIGIONS

One could easily make the claim that religion plays a more central role in the environmental struggles faced by the indigenous peoples of Latin America than it does for other groups. Some contend that it is impossible to separate indigenous struggles on behalf of the environment from religion, since religious beliefs play a major role in current indigenous political activism.[1] We hesitate to make too many generalizations about indigenous religions, given the literally thousands of different groups present in Latin America and the heterogeneity of beliefs and practices. Yet, we believe that common themes emerge when analyzing the indigenous relation to nature. Brandt Peterson writes:

> The tremendous diversity of Latin American indigenous peoples is reflected in the heterogeneity of their religious beliefs and relations to nature. Yet Indians throughout the America share a basic experience of colonization and social, political, and economic marginalization in which assimilationist efforts to eradicate indigenous belief systems have persisted from missionary colonists through post-independence education policies, as have the dispossession and destruction of Indian lands by outsiders. For many indigenous peoples religion as an expression of a unique identity and a philosophy of connections to particular territories and place is central to their struggles to secure and protect their rights as distinct peoples.[2]

Pre-Columbian cosmovisions, whether Aztec, Inca, or Maya, spoke of an earth in which humans were inextricably interconnected with sacred landscapes, animals, and spirits. The cosmic order for the Aztecs could not be taken for granted; human blood obtained through ritual sacrifice was offered to keep the sun moving, for example, and human actions were seen as inseparable from the forces of nature. Deities and animals in turn were believed to assist humans, as evidenced in an Aztecan myth chronicling the discovery of corn: "Again, they said, 'Gods what will they eat? Let food be looked for.' Then the ant went and got a kernel of corn out of Food Mountain. . . . Then the gods chew them and put them on our lips. That's how we grew strong."[3] Mayan creation myths found in the *Popol Vuh* show the connections among humans, the earth, and deities, clearly depicting a spirituality centered on the natural world. Humans and animals celebrate the dawn of the world: "And here is the dawning and showing of the sun, moon, and stars. . . . And then, when the sun came up, the animals, small and great were happy. The eagle, the white vulture, small birds, great birds spread their wings, and the penitents and sacrificers knelt down. They were overjoyed, together with the penitents and sacrificers of the Tams, the Ilocs."[4] The Spanish also encountered Inca and other Andean peoples who believed that they were the offspring of the sun, moon, stars, thunder, and lightning. Latin America's pre-Columbian population inhabited a world both natural and sacred.

Contact with Europeans devastated most of Latin America's indigenous groups as the new arrivals introduced diseases and spread epidemics. Native peoples were killed through warfare, exploited economically, enslaved, and converted to Christianity. The nature-based religious systems of indigenous peoples were viewed by the new arrivals with suspicion and hostility. Father Hernando Ruiz de Alarcón wrote a typical response to indigenous practices in 1629: "In this New Spain, as in all the other heathen lands, they . . . still today hold the sun in great veneration, doing so as if it were God . . . and thus they attribute a rational soul to the sun and the moon and animals, speaking to them for the purpose of witchcraft."[5] This hostility toward native beliefs meant that indigenous peoples throughout the Americas received punishment for their religious practices and were forced to go underground or mix their local beliefs with Christianity. Given this history of conquest and loss of people, land, and practices, it is amazing that indigenous peoples and religions still survive.

For most contemporary indigenous religions, the relationship between nature and religion remains central in myths, narratives, and rituals. While recognizing the great diversity of beliefs and practices and the aftermath of colonization, most scholars still claim that indigenous worldviews encourage concern for nature and, by extension, practices that are not environmentally exploitative. Leslie E. Sponsel writes of the Amazon jungle, for example, "Indigenous environmental impact is usually negligible to moderate."[6]

For most indigenous groups the sacred permeates nature. Sharp divisions between nature and humans or between the wild and civilized are rarely made. Rather, "indigenous peoples have evolved complex relationships based on systems of ecokinship with the elements of the world that surround them."[7] The creation myth of the Huarorani of Ecuador, for example, features a giant tree linking the earth and sky. The earth is part of this tree of life; humans and forest animals are connected through their place on the tree of life. This sense of ecokinship yields norms valuing reciprocity or balance between humans and the nonhuman world. Juan Carlos Galeano reports that for Amazon indigenous groups "many of the tales which illustrate such mythological importance of reciprocity and balance with nature are constructed with the fabric of direct experiences of fishermen, loggers, rubber tappers, hunters, intruders, and other forest dwellers."[8] Some Andean communities believe that mountains will kill people through landslides, floods, and illness if they do not nourish mountains that feed and shelter them. Mining, road building, and the clear-cutting of forests potentially throw sacred mountains into imbalance.[9] The U'wa of Colombia believe that the universe will end if humans, nature, and deities fall out of balance. Contemporary Nahua of Central America may ask permission before cutting trees or plowing the ground in order to plant corn, saying, "Because the tree is the brother of *Cemanahuac* (that which surrounds us), one must ask permission before using its wood. If this is not done, the substance that lives within the tree and in the forest can do harm to the peasant or his family by causing disease and even death. If the gifts of the forest are overexploited, there will come a time when the forest will cease to produce, because all living beings become tired."[10]

Rituals are often the means by which indigenous groups express the balance and reciprocity that characterize their relation with nature. Religious ceremonies, rituals, and practices honor the human-nature bond, recognizing interdependence, a desire to maintain balance, and deep connections with place. Myths and narratives relate humans to their natural surroundings. The Maya, for example, consider corn to be religiously important; they call themselves people of the corn. A local mountain or grove may be sacred, and animals frequently appear in indigenous myths as moral voices. Narratives may express geographically specific ways of relating to nature that are framed in religious terms.

Most indigenous religious traditions reflect geographical boundedness, yet also incorporate a range of religious symbols and practices. Traditional elements and Christian symbolism are mixed frequently, yielding a religious syncretism in service of ecological values. Mayan priests in Guatemala may also be practicing Catholics, and throughout the Americas indigenous groups have incorporated local myths into the worship of Christian saints. The Virgin Mary is often represented with symbols for local indigenous goddesses; Guadalupe, for example, incorporates aspects of the Aztecan goddess Tonantzin, as will be discussed later in this essay. Incan symbols of the sun and moon, representing Coya, a daughter of the moon, appear in depictions of the Andean Virgin Mary. A fluidity of forms and mixing of indigenous and Christian practices is common.

The indigenous of Latin America have maintained their ethnoreligious identity and close connections to the land over centuries of oppression. Contemporary ecoindigenous religion begins with the experience of colonization and economic, social, and cultural marginalization. Increased modernization and relentless development grew during the last decades of the twentieth century, threatening indigenous lands and lifeways. Indigenous resistance movements have grown throughout the region, and religious practices enable native peoples to resist ongoing attempts at further colonization and erasure as external actors press for resource extraction on their lands and as governments initiate assimilation projects. The patenting of traditional resources by Western companies also concerns many indigenous groups; they view such appropriation as an exploitation or "biopiracy" of indigenous knowledge. Indigenous protests and movements to protect lands from oil-drilling, deforestation, and mining, to protest "bioprospecting," and to resist the introduction of genetically modified crops have emerged throughout Latin America over the past several decades. Thousands marched from the Amazon jungle to Quito, Ecuador, in 1992 to demand protected territory for local indigenous groups. Maya in Guatemala protest oil-drilling projects in Lake Petén Irzá, the Huarorani resist oil-drilling in their forests, the Tukanoan organize to halt deforestation created by the planting of coca in their lands, and the Kogi of Colombia struggle to maintain their lifeways and their mountain lands. The U'wa utilize civil disobedience and lawsuits to resist oil-drilling in their mountain ranges; they have threatened mass suicide if drilling proceeds.

Increasing threats to the land and lifeways of many of Latin American's indigenous peoples have led to the growth of panindigenous activism. The Zapatista uprising in Chiapas, Mexico, for example, united traditional indigenous peoples, Roman Catholics, and evangelical Maya to protest the imperatives of the global economy as expressed in the North America Free Trade Agreement. The production of maize, sacred to Mayan communities in southern Mexico, decreased dramatically following NAFTA, as the United States flooded Mexico with cheap corn. In addition, the Mexican government eroded communal ownership patterns. Faced with similar threats throughout the Americas, native groups have attempted to promote solidarity and activism among heterogeneous groups with divergent religious practices. The International Working Group on Indigenous Affairs and Survival International were formed in the 1960s and 1970s as indigenous rights and activist organizations. In 1992, 650 representatives met at the World Conference of Indigenous Peoples on Territory, Environment, and Development held at Kari Oca, Brazil, during the week before the Earth Summit (the U.N. Conference on Environment and Development). The group issued the Kari Oca Declaration and the Indigenous Peoples' Earth Charter. These documents affirmed the connections among land, spirituality, and self-determination, promoted a pan-Indian religious perspective and ethic, and linked indigenous religion and ethics with the defense and protection of natural resources.

Increasingly, environmentalists from more affluent nations have joined Latin American indigenous land struggles to bring international attention to their efforts to protest deforestation, resist mining, and protect intellectual property. These alliances have often proven successful. The danger exists, however, that nonnative outsiders, with a superficial understanding of indigenous religions and lifeways, may objectify native religions and indigenous in their search for the "pure" environmentalist. John Grim writes that "this romantic exploitation of indigenous religions typically accentuates a perceived native ecological wisdom as having been genetically transmitted."[11]

Indigenous peoples throughout Latin America fight against mining, logging, ranching, and the damming of rivers to protect their land, ways of life, and religious sensibilities. Cleary and Steigenga write, "Indigenous mobilization cannot be understood without a careful consideration of religious factors. While specific political openings and social and economic processes facilitated the indigenous resurgence, religious institutions, beliefs and practices provided many of the resources, motivations, identities, and networks that nurtured the movement."[12] Land, life, the sacred, and the environment remain inseparable for the indigenous peoples of Latin America.

ROMAN CATHOLICISM

Roman Catholicism remains the dominant religion in Latin America. Although *campesinos* ("peasant farmers") may practice a "folk Catholicism" that is tied to nature, the Catholic church in Latin America itself does not, as Anna Peterson writes, "have a long tradition of explicit theological and moral reflection about the natural world."[13] Early Christian and medieval theologians such as Bonaventure, Francis of Assisi, and Hildegard of Bingen expressed appreciation for nature in their theologies and presumed that a harmonious order among humans and the natural world corresponded to God's design. Their views, however, did not reflect dominant theologies at the time of the conquest. The post-Reformation Catholicism that reached the Americas presumed that the domination of nature and other peoples by Christians reflected God's will. Extraction of resources, destruction of land, and colonizing of "savages" thus posed few theological problems.

A notable exception to triumphalist theologies of domination was that of Spanish Dominican Bartolomé de las Casas, who both defended the rights of indigenous peoples and extolled the natural world they inhabited. De las Casas vehemently protested against the destruction of indigenous peoples and their lands and engaged in numerous debates concerning the Spanish Christian right to colonize, enslave, and evangelize them. Eduardo Mendieta writes that "in De las Casas we find an explicitly articulated theology and missiology that for the first time combines reverence for the dignity of human beings along with their'natural environment,'without which human beings would not be able to live and flourish."[14] Contemporary liberation theologians, such as Gustavo Gutierrez, have championed De las Casas as an apostle for the poor, the indigenous, and the land.[15]

The legend of the apparition of the Virgin of Guadalupe in 1531 symbolizes for many a softened theological posture toward native peoples and their lands. On 8 December 1531 the Virgin of Guadalupe appeared to a Christianized native, Juan Diego, on the hill of Tepeyac near Mexico City. According to the legend, she spoke to Juan Diego in Aztec and asked him to tell the bishop of Mexico, Juan de Zumárraga, to build a chapel in her honor at Tepeyac. Juan Diego failed on two successive visits and met the apparition for the third time, telling her that the bishop had demanded signs from her. Guadalupe then told Juan Diego to gather roses for the bishop from Tepeyac. Upon opening his cloak in front of the bishop, the roses fell out, revealing a life-size image of Guadalupe imprinted on his rough cactus-fiber cloak. The bishop recognized the miracle and placed the image in a cathedral built at Tepeyac.

Significantly, the Guadalupan tradition was introduced thirty-five years after conquest. Native peoples at times resisted domination overtly, but were more likely to subvert their conqueror's intentions covertly through merging local beliefs and lifeways with Christian images and practices. Cults of Mary brought by the Spaniards merged with pre-Columbian earth deities such as Tonantzin, the "revered mother" of Tepeyac, to create Guadalupe, for example. The Roman Catholic church used Guadalupe as a means of evangelization of native peoples, whereas the indigenous worshiped Guadalupe as a way to continue pre-Christian practices that linked her with sacred space and power coming from the earth. As the tradition evolved in the seventeenth century, Guadalupe became associated with Mexican patriotism, nationalism, and the interests of creoles (Mexican-born Spaniards), rather than with the struggles of indigenous peoples. It was not until the twentieth century that the Virgin of Guadalupe again became explicitly linked to the rights of native peoples, the poor, and the land. The image of the Virgin of Guadalupe, named the "Patron Saint of the Americas" by Pope John Paul II, has been a complicated symbol in Latin America's history, standing for conquest, earth goddesses, nature, the modern nation-state, and indigenous rights.

LIBERATION THEOLOGY

Bartolomé de las Casas and the Virgin of Guadalupe are exceptions, however, to dominant Roman Catholic theologies and practices that persisted in Latin America for nearly five hundred years. The Roman Catholic church tended to ally itself with local power elites, governments, and the wealthy.[16] The emergence of liberation theology marked a sea change in the role of the Roman Catholic church in Latin America. Liberation theology grew in Latin America during the late 1960s as a response by activist priests and concerned laypeople to increased poverty, the failed promises of modernization, and the brutality of military dictatorships. It grew quickly over the next decades, spread from Latin America to other less affluent nations, and with the publication of *A Theology of Liberation* by Gustavo Gutierrez in 1971 was introduced to an even wider audience.[17] Liberation theology did not address ecological concerns in its early years, focusing instead on the social, economic, and political dimensions of the oppression of the poor. Increasingly, however, liberation theologians recognized that the destruction of the earth and the oppression of the poor were linked; the poor's liberation was seen as impossible without defense of the environment. Liberation theologians now frequently promote ecological understanding as a paradigm for interpreting social realities.

Liberation theology claims that God sides with the most oppressed. The Latin American Bishops Conference meetings in Medellín, Colombia, in 1968 and in Puebla, Mexico, in 1979 underscored this "preferential option for the poor" as being at the heart of Christian theology and the gospel mandate. Brazilian liberation theologian Leonardo Boff writes, "Liberation Theology is born from the efforts to listen to the cry of the oppressed... there is an immediate relationship between God, oppression, liberation: God is in the poor who cry out."[18] The poor, the true subjects of liberation, must be understood within a context of socioeconomic, political, and cultural oppression according to liberation theologians. This understanding of the central subject of liberation theology, the poor, yields hermeneutic tools for theology. The poor and oppressed are hermeneutically privileged, and all social analysis must begin with their experience. This hermeneutical privilege holds true for environmental issues as well as for theology. Eduardo Bonnin writes that the option for the poor is the "hermeneutic key that helps us realize the importance of an authentic solution to the ecological problem for the construction of the Kingdom of God in Latin America."[19] Just as the poor of the land are central to theological discourse, they must also be central to ecological discourse.

Privileging the poor led to particular forms of social analysis for Latin America's radical Catholics. Liberation theology grounded itself in a socioeconomic analysis of the plight of the poor. In its early stages it relied heavily on a Marxist social analysis, claiming that this analysis best described the material conditions of Latin America's poor. Consequently, the eschatology espoused by liberation theology was fully "this worldly." The reign of God occurs on earth as a process that makes historical, social, economic, and political progress central. A community in the Amazon River basin, for example, brings the kingdom of God to earth as it fights to preserve lakes and rivers from the destructive practices of the market as embodied by commercial fishermen. Heidi Hadsell writes that community members view their struggle as Christian because "this Christianity is marked by a this-worldly eschatology requiring Christian involvement in historical projects of political and economic justice by and for the excluded. Toward this end it takes as a central ethical task the responsibility to facilitate political and economic education and analysis, as well as transformative action, both by the poor themselves, who are viewed as primary historical agents, and by others, who act in solidarity with the poor and the excluded."[20]

Social analysis, although a critical starting point, is insufficient without praxis (struggle with and for the oppressed), given liberation theology's eschatology. Thus, orthopraxis (correct action) is viewed as equal to or more important than orthodoxy (right thought or doctrine). A theological vision emphasizing the poor and an eschatology centered in this world lead to political action, community involvement, and a religious and political accompaniment of a community in its struggles. Clodovis Boff articulated liberation theology's three methodological phases as the social analytical, the hermeneutical, and the practical-pastoral.[21] Believers are to reflect on their social situation, often using a Marxist analysis, interpret scripture in light of social analysis, and then act in the world. Salvadoran environmental activists, for example, who were engaged in protesting the destruction of one of the country's last remaining forest preserves and promoting appropriate technologies realized, according to Lois Lorentzen, that "Christians are to actively engage in the work of the kingdom of God, an earthly labor which involves changing unjust social structures."[22]

Liberation theologians emphasize social sin and structural injustice over individual wrongdoing. They claim that environmental exploitation stems from structural injustices that affect both the poor and the nonhuman world; ecological problems cannot be resolved until structures of exploitation and domination are transformed. Leonardo Boff writes that "ecological injustice is transformed into social injustice by producing social oppression, exhaustion of resources, contamination of the atmosphere and the deteriorating quality of life."[23] Social sin is not just the poverty and exploitation of people, but also the contamination of their resources. Ricardo Navarro, director of the Centro Salvadoreño de Tecnología Apropiada and founding member of the Salvadoran Ecological Union, claims that polluting rivers through the excessive use of pesticides is a social sin equal to denying food to people through unjust economic and social structures; both cause death.[24] The violation of nature is a religious offense.

The emphasis on structural injustice and economic, political, and social institutions led liberation theologians to propose social ecology as the philosophical movement within environmentalism that best expressed a liberation perspective. The United Nations'first international conference on the environment held in Stockholm in 1972 had a great influence on theologians such as Carlos Herz and

Eduardo Contreras of Peru and Eduardo Guaynas of Uruguay. Participants from less affluent countries called poverty an environmental problem and claimed that the poor suffer most from socioenvironmental deterioration. Following the conference, Guaynas wrote, "We define social ecology as the study of humans, individually and socially, interacting with the environment. The land shapes human cultural and social manifestations. The human interacts intensely with the environment. Neither can be studied in isolation."[25] The political, social, economic dimensions of environmental degradation and crisis are thus primary; both social ecologists and liberation theologians agree that no divide exists between social and environmental issues.

Sharp criticisms of more affluent countries emerge from an analysis based in social ecology and liberation theology. Relations between rich and poor countries are characterized as neocolonial and exploitive. Tony Brun writes, "As opposed to the North, where the environmental crisis is felt in a context of material well-being, in the South it is closely related to poverty. In Latin America, the dramatic situation of its natural ecosystems is related to the profound social problems."[26] Liberation theologians uniformly denounce the neoliberal model of development and global capitalism for their "antiecological character."[27]

As ecotheology evolved in Latin America, prominent theologians such as Ivone Gebara and Leonardo Boff departed from a strict social ecology perspective. Boff was one of the first to connect liberation theology with environmental concerns, writing an article as early as 1976 on the environment and Franciscan spirituality. He embraced social ecology in his early writings, employing a Marxist analysis to illuminate the plight of the poor and to argue that dominant groups exploit both the poor and the land. Boff took a more ecocentric or holistic approach in 1993 with the publication of *Ecology and Liberation* and claimed that ecology is fundamentally theological. Both the trinity and ecology demonstrate to Boff that "everything that co-exists, pre-exists. And everything that co-exists and pre-exists subsists by means of an infinite web of all-inclusive relations. Nothing exists outside relationships. . . . All being constitutes a link in the vast cosmic chain. As Christians, we may say that it comes from God and returns to God."[28] He used the Amazon of Brazil as a concrete case to articulate an "ecology-based cosmology, rooted in evolutionary process in which sin is defined as breaking connectedness," with the 1995 publication of *Cry of the Earth, Cry of the Poor*.[29] Boff's articulation of an ecospirituality that looks more to ecocentric and biocentric models in environmental thought still remains firmly rooted in historical analysis however. He writes, "This violence was planted in Latin America with the sixteenth-century standard of labour and a relationship with nature that implied ecocide, the devastation of our ecosystems."[30]

Historically grounded liberation theologies, whether adopting social ecology or bio- and ecocentric models, embrace praxis as the true test of faith commitment.

Christian base communities became the ideal loci for articulation and praxis of an informed ecotheology. Initially formed in areas underserved by priests, Christian base communities provided space for reflection for those generally excluded from theological discourse. Participants in Christian base communities reflected upon Christian scriptures in light of their life situations and started organizing grass-roots projects to meet local needs. Many groups and individuals moved on to political activism. Increasingly base communities addressed environmental issues such as air pollution, water contamination, sanitation services, soil erosion, mining, the use of chemical pesticides, logging, and other ecological issues that directly affected their communities'health and well-being. Heidi Hadsell writes that in Brazil issues include "the land-tenure patterns . . . which result in a relative few owners of large pieces of land and millions of rural peasants who have no land to farm at all, and many millions of urban poor who endure substandard housing and sanitation, poor or nonexistent health and education provisions, and who often can find no employment. Another ethical theme is that of the plight of Brazil's indigenous populations which are steadily dwindling through violence and disease as their lands are encroached upon by loggers, ranchers and farmers, and as what were once remote areas come under the embrace of significant capitalist expansion."[31]

Christian base communities in Brazil's Amazon River basin have supported and organized rubber tappers and other poor landholders in struggles against ranchers. Rubber tappers have participated for decades in nonviolent efforts to halt deforestation and the destruction of their way of life. At times these peaceful protests are met with violence, most notably in the case of the murder of Chico Mendes. The Pastoral Land Commission of the Roman Catholic church of Brazil was formed to work on environmental and other issues with landless peasants. Since its formation it has worked to protect fishing habitat, to gain land for peasants, and to protect the Amazon jungle for rubber tappers and indigenous peoples. The well-known Movement of the Landless, although technically a secular movement, is also supported by the Roman Catholic church and Christian base communities. Throughout Latin America, churches have formed ecological committees in order "to promote conversion to ecological community. This conversion denounces environmental damage as a serious affront against the creator, promotes technologies that respect the land and local material cultures, practices agriculture that is sustainable and for local consumption, and actively resists mainstream neoliberal development policies."[32] Churches, base communities, and Catholic organizations participate with popular social and environmental movements throughout Latin America to fight for clean water and air, to halt deforestation, and to confront the numerous environmental issues affecting the region's rural and urban poor. Religious belief and practice are seen as inseparable from environmental struggle.

RELIGIOUS ECOFEMINISM

Religious ecofeminism belongs to what Costa Rican theologian Elsa Tamez and Brazilian Ivone Gebara term the third stage of feminist theology in Latin America. Women theologians in the first phase (1970–80) according to Tamez and Gebara, tended to see themselves as liberation theologians and enthusiastically participated in the growing Christian base community movement. An explicitly feminist consciousness grew during the second phase (1980–90), and the current third phase (1990 onward) is marked by "challenges to the patriarchal anthropology and cosmovision in liberation theology itself and by the construction of a Latin American ecofeminism."[33] Most Latin American ecofeminists came from Christian base communities (and may still be very active within them) and were influenced by liberation theology. Many still consider themselves liberation theologians, or more appropriately ecofeminist liberation theologians. Gebara, a Brazilian Sister of Our Lady and professor for decades at the Theological Institute of Recife, Brazil, is the most widely known spokesperson for ecofeminist theology from a Latin American perspective. Gebara gained international attention in 1995 when the Vatican, under the auspices of the Congregation of the Doctrine and Faith, silenced her for two years. Gebara had claimed that liberation theology needed to be tolerant of women's choice for abortion given the hardship of raising children in the context of desperate poverty. The congregation instructed Gebara not to speak, teach, or write for two years and sent her to France for theological reeducation. She returned to Brazil in 1997 and again became active in writing ecofeminist theology and environmental activism.

Ecofeminist theologians such as Gebara, while influenced by liberation methodology, want to move beyond what they see as androcentric tendencies in many theologies of liberation. Gebara defines ecofeminism as the "thought and social movement that refers basically to the ideological connection between the exploitation of nature and the exploitation of women within a hierarchical/patriarchal system. From this philosophical and theological viewpoint, ecofeminism may be considered as a knowledge that hopes to heal the ecosystem and women."[34] Latin American ecofeminist theologians locate women's experience at the center of liberation discourse in order to uncover the ways in which women's oppression and daily reality measure all paradigms, including those of liberation theology.

Latin American ecofeminists contend that not only are women and nature linked ideologically and conceptually, but also that environmental destruction affects women differently than men. Women are more likely to provide family sustenance and thus depend on a healthy environment. They must provide clean water for their families; in the countryside they need trees for fuel, food, and fodder. They bear the brunt of childcare and care of the sick and elderly; thus polluted waters that give family members cholera or diarrhea (the largest cause of child death in poor countries) affect them directly.

Ecofeminist theologians share with liberation theology the idea of hermeneutic privilege. They contend, however, that the poor women of Latin America are the oppressed within the oppressed. Mary Judith Ress writes that the ecofeminist collective Con-spirando, for example, is "deeply marked by the concrete, lived experience of women. Influenced by liberation theology's option for the poor, feminist theologians stress the feminization of poverty:'the poor have a face and it is the face of a woman and her children,'has become the starting point for much of our theological work."[35] *La vida cotidiana* ("the daily life") of poor women is the microcosm that questions the macrocosm according to Gebara and thus becomes central to her epistemology. She writes that "women, children, and African and indigenous populations are the first victims and the first to be excluded from the goods of the earth."[36] For Gebara, an ecofeminist epistemology based in the daily life of Latin America's poorest women yields ambiguous and complex understandings rather than fixed ideologies. It is a "contextual epistemology, which seeks to take the lived context of every human group as its primary and most basic reference point."[37] The methodology developed by Latin America's ecofeminist theologians puts women's corporality (sexuality, sex, body, etc.) at its center and explores the relationship between the daily life of women and systemic forms of oppression, thus connecting women's exploitation with environmental and economic exploitation.

The central aspect of daily life is interdependence or interrelationship, according to Gebara. Interdependence expresses both the way people know and the way they should act in the world and thus is core for Gebara's ecofeminist epistemology and ethics. This insistence on *la cotidianidad* allows Gebara to criticize linear paradigms of history, dualistic approaches to matter/spirit, reason/body, while helping her build an ethics of solidarity with others and with creation. Her holistic approach to ecology based on *la cotidianidad* stems from the interdependence that emerges from the "simple fact of sharing life."[38] Ress writes, "Ecofeminism's greatest insight is the . . . notion that everything is connected and therefore everything is sacred. Ecofeminists make the connection that oppression of women by a system controlled by ruling-class males and the devastation of the planet are not only two forms of violence that reinforce and feed upon each other, but that they come from a terribly misguided sense of the need to control, to dominate the Other, that which is different."[39] Interdependence and interrelatedness do not stop with other humans but encompass nature, yielding an holistic epistemology. Ecofeminists share liberation theology's suspicion of mainstream economic and development models. Gebara criticizes the "religion of the market," which reduces suffering and questions of existence to technical problems.[40] She writes, "All this makes clear who is the most responsible for the catastrophic destruction of the ecosystem. The poor do not destroy natural springs or watersheds; these have long since been taken away from them. The poor don't use powerful electric saws to cut hundred-year-old trees, because they don't own the chainsaws."[41]

The Con-spirando women's collective started in 1991 to create a network of women concerned with the themes of ecofeminism; it remains the most visible movement reflecting ecofeminist theology. The collective publishes a quarterly journal, *Con-spirando: revista latinoamericana de ecofeminismo, espiritualidad y teología* ("Con-spirando: A Latin American Magazine of Ecofeminism, Spirituality, and Theology"), offers a summer school and workshops, and sponsors a year cycle of rituals. A diversity of class, race, age, and culture characterizes collective members, although most come from the Christian tradition. Many members considered themselves liberation theologians at some point in their development and were active in Christian base communities (some remain involved), although all criticize the patriarchal underpinnings of Christian theologies, including liberation theology. Con-spirando self-consciously shapes rituals that mix Christian and indigenous practices. Cofounder Mary Judith Ress writes of ecofeminist rituals, "Sensitive to the indigenous roots of Latin America, we have present in the circle's center the four elements and always salute the four directions. Many rituals concentrate on reconnecting with our ancestors and with the broader community of life."[42]

The collective works on environmental issues with grassroots groups and women's centers in slums, Christian base communities, and universities. A network of collectives similar to Con-spirando exists throughout Latin America, connecting active ecofeminist movements in Peru, Brazil, Bolivia, Uruguay, Venezuela, Mexico, Honduras, El Salvador, Guatemala, and Chile. The ecofeminist groups address a range of environmental issues. In El Salvador and Honduras, popular ecological resistance by women's groups stemmed directly from the need to provide their families and communities with basic sustenance.[43] Gebara writes that working with poor women in slums shapes the issues addressed by a "social ecofeminism." Garbage that overflows the streets, inadequate health care, and the struggle to find potable water—all daily survival crises faced by poor slum women as they provide for families—are central issues addressed by ecofeminist activism and theology.

PROTESTANTS IN LATIN AMERICA

Missionaries from mainline Protestant denominations did not arrive in Latin America until the end of the nineteenth century. Many governments welcomed them, believing that they would assist with the liberal project of modernization. Indeed, Protestant missionaries built schools and clinics and encouraged development projects. Their efforts at evangelization, however, remained largely unsuccessful until late in the twentieth century when a new group of Protestant missionaries arrived on the scene. Fresh waves of evangelical Protestants had arrived in Latin America in the 1950s, laying the groundwork for the rapid expansion of evangelical Protestantism, especially Pentecostalism, in the late twentieth century. The rapid expansion of Pentecostalism in Latin America has been well documented, and most agree that its current growth is homegrown rather than a "northern invasion" as was assumed in the 1980s. Scholars try to account for the rapid growth of Pentecostalism by pointing to economic marginalization, globalization, rapid social change, displacement, and failed modernization.[44]

Recent scholarship on evangelical Protestantism cautions against generalizations, given the variety of forms it takes. What evangelical Protestant groups hold in common, however, are two major tenets: the Bible is the authoritative guide for faith and life, and individual salvation comes through recognition of Jesus as savior. Most evangelical Christian groups do not use nature-based symbols or rituals in worship; religious practices generally do not reflect connections to the natural world. Protestants in general rarely speak of sacred spaces or connections to animals, especially in their spirit forms.

Latin America's Protestants are not uniformly politically conservative, as is often assumed. In Venezuela, for example, evangelical Protestants tended to support Hugo Chavez in elections.[45] Emmanuel Baptist Church in San Salvador, El Salvador, consistently promotes an activist, leftist brand of evangelical Christianity. The Progressive Evangelical Movement of Bahia, Brazil, unites progressive evangelical activists in their struggles against racism and construction of an ethnoreligious identity. The activism is grounded in the two tenets of evangelical Protestantism; for Brazil's black progressive evangelicals "Jesus was a leftist," and members "stress the scriptural basis for their progressive politics."[46] These examples, while demonstrating the diversity among evangelical groups, may be exceptions. Stephen Selka writes in the Brazilian case, "As progressive evangelicals'constant efforts to distinguish themselves politically from Pentecostals suggests, most evangelicals in Brazil are conservative."[47] Active Protestant involvement in environmental struggles, for example, is not characteristic.

Evangelical Protestantism has grown rapidly among indigenous communities, especially in Central America. Indigenous evangelical Protestants tend to abandon traditional nature-based religious practices as they adopt Protestant doctrines and theologies concerning nonhuman nature. Virginia Garrard-Burnett writes, "While Mayan Protestants officially subscribe to the biblical teaching that God gave humankind dominion over the earth (Genesis 1:28), Mayan Protestants are likely to interpret this'dominion'as a benign guardianship."[48] Garrard-Burnett claims that in spite of their dominion ecotheology, indigenous Christians must still work in the natural world and therefore are likely to protect it.

Evangelical Protestant groups, however, are generally not actively engaged in environmental struggles. Exceptions exist, of course, including Floresta, an evangelical Protestant nonprofit environmental organization. Floresta started in the United States but now boasts locally run projects in Mexico, the Dominican Republic, and Haiti. A local church or evangelical organization invites Floresta to a region and then runs environmental and agricultural projects locally. Floresta's twin aims are reforestation and sustainable agriculture; to date they claim to have planted 2.8 million trees. The Evangelical University of El Salvador is a member of the Salvador Ecological Union, an activist group of nineteen organizations in El Salvador. Mayan evangelicals joined the Zapatista movement to protest the impact of North America Free Trade Agreement on native lands.

Mainstream Protestant groups are more likely to articulate an environmental ethos than their evangelical counterparts however. The most notable example is Roy H. May Jr. of Costa Rica's Instituto Biblico Latinamericano (Latin American Biblical Institute). The institute began as an evangelical Protestant seminary. When key faculty became influenced by liberation theology and started teaching a Protestant version of liberation theology, the institute was asked to leave the local evangelical association. The institute continues to train Protestants (and some Catholics). Institute faculty member May wrote a Protestant ecotheology from a liberation perspective. His ecotheology is Bible based: "For the poor, the land is a major concern, and they want to know what the church and the Bible have to say about it."[49] May looks at Christian scripture and finds a biblical tradition that depicts land as both site of struggle and God's presence.

Land in Latin America becomes an important hermeneutic tool by which May understands current socioeconomic reality: "The biblical tradition of land as inherited soil that signifies God's presence, and therefore hope and future, is what makes that tradition meaningful to Latin American peasants. Seen from their context of marginalization and struggle, they discover in the Bible a historic dimension of the land. At the same time, the biblical story of struggle for land becomes their struggle for land even today."[50] May also links land explicitly to conversion and redefines salvation from an ecological perspective: "Without land, peasants and Amerindians have nothing. Land means future, and for them, salvation. The identification of the land with salvation can be explicit. As a Mexican peasant said following a conflict with a large landholder, 'Jesus saves me when they give me land.'"[51] May takes the two central tenets of Protestant Christianity, biblical authority and salvation, and reinterprets them utilizing his land-based hermeneutic. When Latin American Protestants are environmentally active, they generally employ a biblically based hermeneutic to argue for ecological stewardship.

DIASPORA RELIGIONS

Millions of African slaves were brought forcibly to the Americas, carrying their religions with them. African religions merged with indigenous and Christian practices in their new countries, yielding religious syncretic practices. These "diaspora religions" include Candomblé and Umbanda in Brazil, Voodoo in Haiti, Rastafarianism in Jamaica, and Santería in Cuba. Candomblé grew in the early nineteenth century in Brazil as a blend of Yoruba religion and Christianity. Umbanda emerged in Brazil in the early twentieth century as a mix of African religions, indigenous practices, Roman Catholicism, and spiritism. It shares features with Candomblé, such as the importance of *axé* and *orixás*, although its most important spirits come from Brazil itself. Haitian Voodoo mixes west and central African religious practices with Roman Catholicism. Santería, which grew in Cuba, maintains to this day, religious symbols and divination systems stemming directly from west Africa. The Rastafari movement grew in the early twentieth century among the poor of Jamaica. Slaves from central and west Africa added Christian symbolism to earlier practices as a means of maintaining traditional African religions, demonstrating a remarkable "cultural flexibility and adaptability that allowed Africans and their descendents to forge a novel African-derived belief system in the Americas."[52]

Given the diverse beliefs and practices that arose in contexts as far flung as Brazil, Cuba, and Haiti, we should be slow to make too many generalizations. Yet, as Terry Rey writes:

> In spite of the wide diversity of Caribbean religious cultures, taken as a whole the region's peoples generally share a deep sensitivity to nature as an expression of and gift from God. . . . Caribbean believers have always viewed God, spirits (or the Holy Spirit, the *laws* or the *orishas*) and the dead as manifest in nature. Understanding, communing with, and living in harmony with the sacred is thus only possible because of nature and the eternal living force, or *ashé* that inhabits it.[53]

Diaspora religions frequently use natural symbols in their ritual practices and images. For Candomblé, *orixás* (the gods and goddesses who interact with humans) are frequently associated with a natural element such as water or vegetation that gives them their power or energy, *axé*. The force *axé* flows through all things, animate or inanimate, and thus links all beings in the cosmos. Each *orixa* is identified with a Christian saint and offers followers help with daily problems such as health or money.

Nature is revered as a place inhabited by spirits and ancestors. Practitioners of Voodoo believe that ancestors and spirits inhabit the natural world, which is thus revered. Practitioners must take care to preserve harmony with the spiritual and ancestral world; natural elements such as trees or the herbs used for healing are believed to maintain balance with nature. Santería also features worship of the spirits and ancestors found in the natural world. Voodoo and Santería both believe that one communicates with spirits or ancestors through sacrifices, divination, or possession.

The natural world is seen as good and sacred, inhabited as it is by spirits and ancestors. Rastafarianism claims that Jah (God) created nature, thus affirming its goodness. It holds the hope of a return to Zion, the promised land, which is envisioned as a place of nature, forests, and freedom. Babylon, on the other hand represents, according to Rey, "the city, the West, the colonized world, the U.S. and Britain, and other locations of suffering."[54] Trees and the forest are central in Haitian Voodoo religious imagination and mythology, revered as the residence of spirits and ancestors.

A norm of balance or harmony with nature and its deities characterizes Latin America's diaspora religions. For Santería, humans must pursue harmonious relations with the *orishas* ("deities") who reside in nature, especially in the "wildest" parts of the natural world. All of creation is full of the life force, *ashé*; humans must take care to perform rituals and dances to honor and maintain harmony with the *ashé* of the ocean, the mountain, the forest. Manuel Vasquez claims that Umbanda "has at its core a relational ethos that tightly links humans, spirits, and the natural world."[55]

Although diaspora religions acknowledge ties to nature, practitioners other than Rastafarians are not known for a high level of environmental activism. Umbanda's environmental ethic, for example, is mixed; it tends toward anthropocentricism as its practitioners manipulate elements of nature for personal gain, even as it reveres aspects of the natural world. On the other hand, the Rastafari movement possesses a direct lived environmental ethic. The ideal Rasta lifestyle promotes an organic vegetarian diet, traditional farming, and a life lived as close to nature as possible. Rastafarians have been involved in environmental movements, including Earth First!, Green Parties, Friends of the Earth, the Wild Greens, and other green activist and anarchist groups.

COMMON THEMES

Survival

Environmental struggles in Latin America emerge first and foremost out of a desperate need for survival. May writes: "Quite simply, ownership and control of land determines who lives and who dies."[56] Environmental problems directly affect human survival in ways not always so visible in more affluent nations. Contaminated water causes gastrointestinal illnesses and cholera; deforestation means that the poor lack fuel and food and that mudslides and flooding are more likely; soil erosion leads to poor crop yields and inadequate food supplies; and air pollution contributes to respiratory problems. Environmental issues in Latin America revolve around the struggle to provide basic human needs in the context of environmental destruction. This is not to say that religious motivations are not important; rather that religions articulate their vision of the natural world in the context of a damaged environment. Bron Taylor writes: "To acknowledge that basic human needs provide the most decisive impetus to ecological resistance (especially in less affluent countries) is not incompatible with recognizing that moral and religious idea motivations are deeply intertwined with the material motivations or that popular ecological resistance cannot be accounted for if moral and religious variables are overlooked . . . human motivations are embedded both in material interests and in ideal factors."[57] The U'wa claim that disharmony with nature signals the end of the world. Although not always expressed this graphically, most religious groups link environmental deterioration with spiritual disease.

The Legacy of Conquest

All groups, with the possible exception of evangelical Protestants, highlight the ongoing effects of colonization. Boff writes: "This violence was planted in Latin America with the sixteenth-century standard of labour and a relationship with nature that implied ecocide, the devastation of our ecosystems."[58] Contemporary views of the human relationship with nonhuman nature can only be understood in the context of colonization, as Gebara notes: "We can state without fear of contradiction that the domination of women and nature accelerated under colonization, as part of its political ideology."[59] The nature of environmental struggles and religious responses to the environment clearly reflect a postcolonial posture. Most groups criticize the neoliberal model of development as a new form of colonization by more affluent nations. Boff writes: "The avid search for unlimited material development has led to inequality between capital and work, creating exploitation of workers and the accompanying deterioration in the quality of life."[60] Nearly all religious traditions surveyed in this essay agree that the neoliberal model pursued by Latin American governments yields overexploitation of both humans and nature. They utilize religious symbolism, theologies, and practices to provide alternative ways of thinking about development and the land, claiming that the land itself provides a lens with which to understand sociohistorical reality.

Religious Syncretism and Sacred/Secular Blurring

Most groups mix elements from various religious traditions. Christian symbols take on indigenous characteristics, native religions adopt Christian saints, African practices borrow from Christianity, and so on. In some cases this happened over the course of centuries, in others, such as the mixing practiced by the ecofeminist collective Con-spirando, it happens quite self-consciously in the present as they incorporate indigenous myths with Christianity (the religious background of most members). Gebara calls this an "ethics of biodiversity" in which religions, regions, and cultures are connected to give a larger vision. Not only are religious traditions mixed together, a blurring occurs between the sacred and the secular as well. Ingemar Hedstrom writes: "It isn't possible to construct a separation between the environmental and human problem on one side and the theological on the other."[61] Many members of Latin America's secular environmental movements claim that their inspiration came from liberation theology, for example. In the offices of Centro Salvadoreño de Tecnología Apropiada, a secular environmental organization in El Salvador, one finds pictures of Archbishop Romero, who was assassinated in 1980 for his defense of the poor.

DEBATES

Epistemology and Hermeneutics

Liberation theology begins with the poor as its underlying hermeneutical principle. Liberation ecotheologies stem from a methodology in which the poor become the center of ecological discourse. The underlying epistemology and hermeneutics then lead to particular modes of environmental activism, including land reform, participation of Christian base communities, political activism, and other environmental issues that relate directly to social systemic problems. Ecofeminist theologies begin with women, the oppressed within the oppressed, as its guiding hermeneutic principle. Women's corporality and daily life experiences are the center of its methodology; from this vantage point it criticizes the androcentrism found in paradigms and systems, including those of liberation theology.

Ecoindigenous religions often begin with nature itself or with the indigenous as colonized and oppressed subjects. Religious syncretism is also characteristic, as most groups combine indigenous and Christian values. Indigenous religions are relatively more eco- or biocentric than either liberation or ecofeminist theologies. Evangelical Protestantism on the other hand, utilizes a hermeneutic based in

Christian scripture and argues for biblical authority and the centrality of individual salvation. Its ecotheology is thus anthropocentric and theocentric.

The various religious traditions clearly disagree over which hermeneutic principles and epistemologies offer the most environmentally friendly theologies, myths, rituals, religious symbols, and practices. The differing methodologies may also result in greater weight being placed on particular environmental issues and struggles.

The Link between Environmental Ethics and Action

It should not be taken for granted that the high regard for nature expressed in religious thought and practice necessarily translates into positive environmental actions or sustainable ways of life. Brandt Peterson writes:

> For the most part, anthropology, ecology, and other disciplines remain ambivalent about the links between spirituality or religion and ecological sustainability in indigenous communities. While many indigenous traditions express respect for non-human life or "the environment," the extent to which these expressions predict ecologically wise and sustainable practices is uncertain. Understanding the natural world as sacred does not necessarily call for an ethic of environmental protection or stewardship. . . . Specifically religious responses may not address ecological problems in some cases, and the "ecological balance" that many see expressed in indigenous religious traditions may be the result rather than the cause of particular practices that are ecologically sustainable and sensible.[62]

A sustainable lifestyle may simply be due to lack of access to destructive technologies, rather than to particular religious beliefs. If scholars look primarily to religious beliefs to explain indigenous relations with the natural world, for example, we overlook a complex history, in which indigenous people have been excluded from development due to racist policies rather than indigenous choice. Furthermore, localized religious practices do not necessarily provide resources to deal with environmental issues (such as global warming) that reach beyond a particular sacred grove, mountain, or river. And, the poor everywhere are forced into antiecological practices in order to survive even when their religious practices evidence a reverence for nature. Rey writes that in Haiti peasants have been forced to cut trees for charcoal, deforesting most of the land, even though Voodoo is "deeply sensitive to nature."[63]

Ecofeminist Principles

Scholars and activists who consider themselves feminists and environmentalists still may disagree with the principles of ecofeminism as outlined in this essay. The debate centers on the core conceptual claim that women are identified with nature and that both nature and women are devalued in the process. It is not the case that all indigenous groups who value nature also highly value women. It also may be difficult to maintain the notion of poor women's greater care for nature vis-à-vis men. Women may indeed suffer disproportionately from environmental damage. However, theorizing certain groups as being closer to nature runs the risk of essentializing women (and nature) and replicating the dualism that ecofeminism hopes to overcome. In fact, "claims made about indigenous and Third World women may actually serve to reassert patriarchal beliefs about women."[64]

RECOMMENDATIONS FOR FURTHER STUDY

The Religious Attitudes and Environmental Action Connection

As noted above, too often scholars in religious studies and/or theology emphasize worldviews, myth, narrative, and ritual and do not take the further step of linking religious teachings to concrete behaviors, including participation in environmental struggles. Conversely, social-movement theorists rarely consider religion when they look at environmental movements. Rey writes that "this rooting in and respect for nature of Caribbean religious cultures has not, however, ever inspired broad environmental activism anywhere in the Caribbean."[65] Rigorous case studies should be conducted to study the connection between religious teachings and practice and concrete environmental action.

These case studies could include comparisons of anthropocentric and bio- or ecocentric worldviews. Many environmentalists claim that bio- or ecocentric worldviews are the most pure and thus most likely to lead to the strongest environmental actions. Latin America's liberation theology, however, has an anthropocentric approach based in the survival of the poor that has led to activist struggles on behalf of the environment, including some of the most sustained environmental struggles of the last three decades.

Understudied Religions

Little research has been conducted on the environmental attitudes and actions of Latin American evangelical Protestants and Pentecostals. The same could be said for diaspora religions. Although the use of natural elements has been studied in diaspora religions, the role of these traditions in Latin America's environmental struggles remains understudied.

The poor of Latin America face severe environmental crises. In the face of deforestation, desertification, and a desperate struggle for survival, they look to religious traditions for guidance. Taking religious traditions and actors into account deepens our understanding of environmental movements within Latin America and the Caribbean. Although religion may play a role in environmental struggles, the poverty of much of the region may undermine efforts to consider ecosystem health a primary concern.

NOTES

1. Edward L. Cleary and Timothy J. Steigenga, "Resurgent Voices, Indians, Politics, and Religion in Latin America," in *Resurgent Voices in Latin America: Indigenous Peoples, Political Mobilization, and Religious Change* (ed. Edward L. Cleary and Timothy J. Steigenga; New Brunswick: Rutgers University Press, 2004), 1–24.

2. Brandt Peterson, "Indigenous Activism and Environmentalism in Latin America," in *ERN* 838.

3. John Bierhorst, ed., *History and Mythology of the Aztecs: The Codex Chimalpopoca* (Tucson: University of Arizona Press, 1992), 146–47.

4. Dennis Tedlock, trans., *Popol Vuh: The Definitive Edition of the Mayan Book of the Dawn of Life and the Glories of Gods and Kings* (New York: Simon & Shuster, 1985), 181.

5. Hernando Ruiz de Alarcón, *Treatise on the Heathen Superstitions That Today Live among the Indians Native to This New Spain* (trans. and ed. J. Richard Andrews and Ross Hassig; Norman: University of Oklahoma Press, 1984), 70.

6. Leslie E. Sponsel, "Amazonia," in *ERN* 38.

7. Henrietta Fourmile, "Indigenous Peoples, the Conservation of Traditional Ecological Knowledge, and Global Governance," in *Global Ethics and Environment* (ed. Nicholas Low; London: Routledge, 1999), 219.

8. Juan Carlos Galeano, "Amazonian Folktales," in *ERN* 41.

9. Lisa Maria Madera, "Andean Traditions," in *ERN* 62.

10. Javier Galicia Silva, "Religion, Ritual, and Agriculture among the Present-Day Nahua of Mesoamerica," in *Indigenous Traditions and Ecology: The Interbeing of Cosmology and Community* (ed. John A. Grim; Cambridge: Center for the Study of World Religions, Harvard Divinity School, 2001), 319.

11. John A. Grim, "Introduction," in *Indigenous Traditions and Ecology: The Interbeing of Cosmology and Community* (ed. John A. Grim; Cambridge: Center for the Study of World Religions, Harvard Divinity School, 2001), xxxvi.

12. Cleary and Steigenga, "Resurgent Voices," 17.

13. Anna Peterson, "Roman Catholicism in Latin America," in *ERN* 1408.

14. Eduardo Mendieta, "Casa, Bartolomé de las (1495–1566)," in *ERN* 271.

15. Gustavo Gutierrez, *Las casas: In Search of the Poor of Jesus Christ* (Maryknoll, NY: Orbis, 1993).

16. See the history of the Roman Catholic church's relationship with Latin American governments in Jose Miguez Bonino, *Toward A Christian Political Ethic* (Minneapolis: Fortress, 1983).

17. Gustavo Gutierrez, *A Theology of Liberation* (Maryknoll, NY: Orbis, 1973).

18. Leonardo Boff interviewed by Mev Puelo, *The Struggle Is One: Voices and Visions of Liberation* (Albany: State University of New York Press, 1994).

19. Eduardo Bonnin, "La problemática del ambiente en el documento de Puebla," in *Volveran las golondrinas? La reintegracion de la creación desde una perspectiva latinomericana* (ed. Ingemar Hedstrom; San Jose, Costa Rica: Departamento Ecuménica de Investigaciones, 1990), 241.

20. Heidi Hadsell, "Profits, Parrots, Peons: Ethical Perplexities in the Amazon," in *Ecological Resistance Movements: The Global Emergence of Radical and Popular Environmental Movements* (ed. Bron Raymond Taylor; Albany: State University of New York Press, 1995), 77.

21. Iain S. Maclean, "Christianity-Liberation Theology," in *ERN* 358.

22. Lois Ann Lorentzen, "Bread and Soil of Our Dreams: Women, the Environment, and Sustainable Development-Case Studies from Central America," in *Ecological Resistance Movements: The Global Emergence of Radical and Popular Environmental Movements* (ed. Bron Raymond Taylor; Albany: State University of New York Press, 1995), 64.

23. Leonardo Boff, "Social Ecology: Poverty and Misery," in *Ecotheology: Voices from the North and South* (ed. David G. Hallman; Maryknoll, NY: Orbis, 1994), 244.

24. Ricardo Navarro A., "Urge desarrollar una teologia ecologica y la iglesia se vuelve ecologista," in *El pensamiento ecologista* (ed. Ricardo Navarro, G. Pons, and G. Amaya; San Salvador: El Centro Salvadoreño de Tecnología Apropiada, 1990), 64.

25. Quoted by Ingemar Hedstrom, *Volveran las golondrinas? La reintegracion de la creación desde una perspectiva latinomericana* (San Jose, Costa Rica: Departamento Ecuménica de Investigaciones, 1990), 44.

26. Tony Brun, "Social Ecology: A Timely Paradigm for Reflection and Praxis for Life in Latin America," in *Ecotheology: Voices from the North and South* (ed. David G. Hallman; Maryknoll, NY: Orbis, 1994), 82.

27. Boff, "Social Ecology," 244.

28. Leonardo Boff, *Ecology and Liberation: A New Paradigm* (trans. John Cumming; Maryknoll, NY: Orbis, 1995 [orig. 1993]), 7.

29. Iain S. Maclean and Lois Ann Lorentzen, "Boff, Leonardo (1938–)," in *ERN* 208.

30. Boff, "Social Ecology," 239–40.

31. Heidi Hadsell, "Brazil and Contemporary Christianity," in *ERN* 213.

32. Lois Ann Lorentzen, "Radical Catholicism, Popular Resistance, and Material Culture in El Salvador," in *Technology and Cultural Values: On the Edge of the Third Millennium* (ed. Peter D. Hershock, Marietta Stepaniants, and Roger T. Ames; Honolulu: University of Hawai'i Press, 2003), 259.

33. Lois Ann Lorentzen, "Gebara, Ivone (1944–)," in *ERN* 689.

34. Ivone Gebara, *Intuiciones ecofeministas: ensayo para repensar el conocimiento y la religion* (trans. Graciela Pujo; Trotta, 2000), 18.

35. Mary Judith Ress, "The Con-spirando Women's Collective: Globalization from Below?" in *Ecofeminism and Globalization: Exploring Culture, Context, and Religion* (ed. Heather Eaton and Lois Ann Lorentzen; Lanham, MD: Rowman & Littlefield, 2003), 152.

36. Gebara, *Intuiciones ecofeministas*, 25.

37. Ivone Gebara, *Longing for Running Water: Ecofeminism and Liberation* (trans. David Molineaux; Minneapolis: Fortress, 1999), 61.

38. Ivone Gebara, "Women Doing Theology in Latin America," in *With Passion and Compassion: Third World Women Doing Theology* (ed. Virginia Fabella and Mercy Amba Oduyoye; Maryknoll, NY: Orbis, 1988), 126.

39. Ress, "Con-spirando Women's Collective," 156.

40. Gebara, *Intuiciones ecofeministas*, 101.

41. Ibid., 5.

42. Mary Judith Ress, "Con-spirando Women's Collective (Santiago, Chile)," in *ERN* 419.

43. Lorentzen, "Bread and Soil of Our Dreams."

44. Lois Ann Lorentzen, "El milagro está en casa: Gender and Private/Public Empowerment in a Migrant Pentecostal Church," *Latin American Perspectives* 32.1 (Jan. 2005): 58.

45. David A. Smilde, "Contradiction without Paradox: Evangelical Political Culture in the 1998 Venezuelan Elections," *Latin American Politics and Society* 46.1 (2004).

46. Stephen L. Selka, "Ethnoreligious Religious Identity Politics in Bahia, Brazil," *Latin American Perspectives* 32.1 (Jan. 2005): 80, 77.

47. Ibid., 86.

48. Virginia Garrard-Burnett, "Mayan Protestantism," in *ERN* 1086.

49. Roy H. May Jr., *Poor of the Land* (Maryknoll, NY: Orbis, 1991), xii.

50. Ibid., 50.

51. Ibid., 106.

52. Robert Voeks, "Candomblé of Brazil," in *ERN* 264.

53. Terry Rey, "Caribbean Cultures," in *ERN* 269.

54. Terry Rey, "Rastafari," in *ERN* 1345.

55. Manuel Vasquez, "Umbanda," in *ERN* 1675.

56. May, *Poor of the Land*, 5.

57. Bron Taylor, "Popular Ecological Resistance and Radical Environmentalism," in *Ecological Resistance Movements: The Global Emergence of Radical and Popular Environmental Movements* (ed. Bron Raymond Taylor; Albany: State University of New York Press, 1995), 336.

58. Boff, "Social Ecology," 239–40.

59. Ivone Gebara, "Ecofeminism: An Ethics of Life," in *Ecofeminism and Globalization: Exploring Culture, Context, and Religion* (ed. Heather Eaton and Lois Ann Lorentzen; Lanham, MD: Rowman & Littlefield, 2003), 169.

60. Boff, "Social Ecology," 240.

61. Hedstrom, *Volveran las golondrinas?* 43.

62. Peterson, "Indigenous Activism and Environmentalism," 835.

63. Terry Rey, "Trees in Haitian Vodou," in *ERN* 1659.

64. Lois Ann Lorentzen, "Indigenous Feet: Ecofeminism, Globalization, and the Case of Chiapas," in *Ecofeminism and Globalization: Exploring Culture, Context, and Religion* (ed. Heather Eaton and Lois Ann Lorentzen; Lanham, MD: Rowman & Littlefield, 2003), 67.

65. Terry Rey, "Caribbean Cultures," in *ERN* 268.

CHAPTER 23

AFRICAN INITIATED CHURCHES AS VEHICLES OF EARTH-CARE IN AFRICA

MARTHINUS L. DANEEL

Earthkeeper's Call

In the beginning
Earth was formless and void
and the Spirit of God moved over the waters.
God said: "Let there be light!"
And there was light.

After *chimurenga*
The earth was scorched and barren
and the Spirit of God urged prophets:
"Cry, the empty gullies, the dying plains—
clothe naked land of the forebears!"
And hope returned
Healing hands, young leaves of trees.

Heeding the call
they came:
black multitudes

churches of the poor
billowing garments . . .
red, white, blue, resplendent green
bearing holy staves, cardboard crowns.
Cursed descendants of Ham
rejects of white mission
lift the fallen banner of Spirit
kingdom's cornerstone
where souls of people, tree souls meet.

Prophets shouted:
Repent! Confess!
I bare earth with axe and fire
rape forests without return
sledge-rip gullied meadows
turn earth's water to trickling mire.
Confess and baptise . . .
the wizards, the land!
Oust demons of neglect.
From Jordan emerge
with bonded hands, new earth community
Touch childless womb, touch hapless soil
—seedlings of love.

Proclaim new heaven,
new earth in black Jerusalem
"Come Mwari!
Come Son! Come Spirit!"
Bare feet touch sacred soil
where rhythmic bodies sweat and sway.
where weary traveler
finds cool in shade
fountains spring
clear water of life.[1]

In the post-*chimurenga* years of independent Zimbabwe, the African Initiated Churches (AICs) indeed heeded the prophetic call to earth-keeping. They joined forces with practitioners of traditional religion—the chiefs, headmen, spirit mediums, and ex-combatants of the country's liberation struggle—and formed their own wing of the green army. Thus, under the auspices of the Zimbabwean Institute of Religious Research and Ecological Conservation (ZIRRCON),[2] two religiously distinct movements—the Association of Zimbabwean Traditionalist Ecologists (AZTREC)[3] and the Association of African Earthkeeping Churches (AAEC)[4]—joined forces to wage a new *chimurenga*, a struggle for the liberation of creation, particularly the rehabilitation of the degraded environment of Zimbabwe's overcrowded communal lands, under the banner "war of the trees."

Who then are the AICs? They are all-African churches founded *by* Africans *for* Africans and have emerged during the twentieth century as thoroughly inculturated movements throughout Africa south of the Sahara. They represent a deep-seated reaction to colonial oppression and alien, Western forms of religion, as well as a creative, contextualized response to their own independent interpretations of scriptures. In Zimbabwe and much of southern Africa today the AICs total between 50 percent and 60 percent of African Christianity. Their overall growth rate appears to be outstripping that of the so-called mainline mission churches. Based on their own missionary initiatives, and despite obvious limitations, manifest in periodic schismatic fragmentation and a leadership invariably bereft of theological training, they count as a Christian force to be reckoned with.

Typologically, the AICs can be classified in two main categories: the Ethiopian-type, or nonprophetic, churches whose patterns of worship and doctrines reflect those of the Protestant mission churches from which they have evolved; and the Spirit-type, or prophetic, churches (the vast majority of all AICs)—mainly Zionist or Apostolic movements—with indigenized Pentecostal traits, focused on Holy Spirit manifestations, glossolalia, and prophetic healing. Some of the latter churches develop leadership patterns with messianic or iconic features.[5] A cross-section of all these different types of AICs joined the earth-keeping movement through AAEC membership. This included two messianic movements, the Zion Christian Church of Bishop Mutendi (counterpart of the multimillion membership Zion Christian Church of Lekganyane in South Africa) and the African Apostolic Church of Johane Maranke, popularly called the *va-Postori*, the Apostles—the two largest AIC movements in Zimbabwe. In terms of the number of affiliated AAEC member churches and influence in the green campaign, the prophetic Zionist churches took the lead.

What then did this multifaith green movement achieve? During its meteoric growth over a fifteen-year period (1988–2002), ZIRRCON expanded from a relatively small research unit to an ecoadministration comprising more than fifty salaried staff members (organized into an executive board, women's desk, development desk, youth organization, theological training unit, research department, nursery keepers, etc.). The traditionalist stakeholders in AZTREC consisted of the majority of senior chiefs, headmen, kraalheads, spirit mediums, clan elders, and traditionalist villagers throughout the Masvingo Province and its bordering territories. Their Christian counterparts in the AAEC comprised some 180 churches, mainly AICs, representing in total an estimated three million members. As a united force this green army developed sixteen main nurseries near growth points and towns in Masvingo Province and a host of satellite nurseries through the efforts of thirty newly formed youth clubs at schools and eighty women's clubs. The latter clubs also engaged in income-generating projects, such as orchards, vegetable gardens, sunflower-oil production, cattle breading, bakeries, soap making, and clothes manufacturing—all in a holistic combination of eco-economic endeavor. In this period an estimated nine million to ten million seedlings—exotic and indigenous—were planted in several thousand woodlots over a wide range of communal lands and farms, where the majority of participant rural earth-keepers reside. Gully formation and soil erosion was arrested, moreover, through stone-filling and the planting of trees and *vertiver* grass in the affected areas. In addition, the preparatory work for the development of a major conservancy in Masving-East was done. This project comprised some twenty farms and involved a cross-section of African, colored, Indian, and white farmers who were collaborating on a reconciliatory basis. Unfortunately, however, the farm invasions unleashed by Mugabe inhibited the implementation of this project to the extent of it being placed on hold indefinitely.

Unfortunately, too, the movement started to stagnate and disintegrate soon after my withdrawal as founder-director and fundraiser. Several factors contributed to the malaise. Unrealistic economic expectations of local stakeholders in AZTREC and AAEC, stakeholders who could barely survive on the basis of their own subsistence farming in a fast-deteriorating and abused economy, placed the resources available to ZIRRCON under pressure. A local leadership which failed to meet the administrative requirements of financial efficiency, transparency, and accountability jeopardized the solid relations which ZIRRCON had initially established with donors abroad. Consequently funding started to dry up. Misplaced donor intervention in some respects contributed toward expectations and disillusionment in rural areas, with destabilization of the earth-keeping movement as a result. Ultimately, however, the breakdown of social and ethical norms in a society bankrupted by corruption and a weakened judiciary rendered restraint, legal redress, and recovery impossible.

There are a number of reasons why the ZIRRCON "experiment" continues to be relevant for the development of an African and global ecotheology. At its peak this movement was rated by internationally prominent ecologists, such as John Grim, Mary-Evelyn Tucker, and Larry Rasmussen, who came out to Zimbabwe to observe its activities, as one of the best—in terms of religious commitment, innovation, and scope of inculturated ecological activities—that they have seen in the developing world. At the time it was certainly one of the most prominent and numerically strongest earth-care endeavors launched by rural African communities in all of southern Africa.[6] Even if the movement is in structural decline, a tradition of religiously inspired earth-care has been established, which is bound to continue influencing and inspiring rural communities, both Christian and Traditionalist, in their environmental attitudes and activities. But more important from a theological viewpoint than the ecological achievements of the green *chimurenga* is the way in which the participant AICs reinvented and reimaged their churches as *vehicles of earth-care* in their attempts to achieve the goals of protecting and nurturing the creation of God, to which they had committed themselves. The main ecclesiological features of an "African earth-keeping church" as it emerged in the war of the trees and characteristic elements of an enacted ecotheology integral to it will therefore be focal in this essay. One can only hope that the brief and admittedly incomplete narrative that follows will convey something of the challenge and inspiration unto a living ministry of earth-stewardship, to the Christian church in Africa, the world church, and all forms of African religion sensitive to the needs of a broken earth.

Why then consider the green contribution of AICs if they are considered by some Western-oriented mission churches as heretical or sectarian splinter groups of dubious or overly syncretic Christian nature? After all, these churches are not noted for written contributions in the fields of African or environmental theology. The reasons, however, for such consideration are briefly the following. First, the escalation of environmental destruction the world over has taken on such proportions in our time that any serious engagement in earth-care should be worthy of note, even if only for its challenge and inspirational value in the global village. Second, academically engaged African theologians have often referred to the AICs as containing the building blocks for the development of an inculturated African theology.[7] Few of them, however, have studied the AICs in sufficient depth to plumb and record convincingly the latter's unwritten, praxis-oriented theologies. Any attempt, therefore, by African theologians, to produce an authentic African ecotheology,[8] will do well to explore the innovative and richly diversified earth-keeping legacies of AICs, especially those in Zimbabwe, where pioneering tree-planting activities had given rise to new perspectives on a Savior-God, his kingdom and church in a beleaguered creation. Third, the earth-keeping AICs of Zimbabwe have hardly had any exposure to ecotheological literature and can therefore be said to have developed earth-care concerns in direct response to their own environment and to what they themselves considered as biblical injunctions relatively free from Western influence.[9] Fourth, the environmental ministry developed by the AAEC related directly to African peasant perceptions and experience of ecological deterioration, including such problems as deforestation, water pollution, drought, and depleted wildlife resources. Consequently one obtains an idea of the spontaneous development of a grassroots theology born of existential need related to local conditions rather than one based on abstract reflection and burgeoning ideological conditioning. Fifth, we need to keep the reflective and practical dimensions of an earth-keeping mission in a meaningful dialectic with each other. Invariably, however, our theological deliberations tend to be eschewed by rationalistic and academic considerations at the expense of insights drawn from praxis, or we refrain from ecological action once we have delivered the "ecologically correct" message. We therefore need to trace more deliberately the movement of God's earth-keeping Spirit in today's world as it is already manifest in Christian communities, if we are to revision and understand his and the church's mission in the context of a ravaged earth. Finally, the AICs have been characterized by observers as "protest movements" rather than as missionary institutions in their own right.[10] Nevertheless, these churches have a rich tradition of missionizing activity in Africa.[11] This factor causes the Independents themselves to integrate their earth-keeping ministry with what they understand as mission—in itself an insight which broadens the missionary task of the church beyond the basic concern for human well-being and salvation, as it extends the salvific good news of Christ to the entire earth community.

The Church as Environmental Healer/Liberator

Ever since the inception in Zimbabwe of the Spirit-type churches, early in the twentieth century, the pervasive image of the Christian church in these circles was that of a "healing institution."[12] Members popularly referred to their churches as "hospitals" (*hospitara*), an understandable qualification if one considers that the comprehensive faith-healing practices of prophets encompassed all spheres of life, including harmonious relations in family life, the finding and keeping of a job or spouse, opening of business ventures, pregnancies, childbirth and childcare, the achievement of social standing or political power in society, and so forth. Healing as a kind of guided accompaniment through life elevated the church as a *protective institution* in relation to all destructive forces threatening human well-being, such as witchcraft, anger, jealousy, spirit-forces causing illness or social conflicts, drought in frail subsistence economics, and the processes of individual alienation in the harsh contexts of urban industries. Inevitably prophetic healing became the major attraction and recruitment device in these churches,[13] a means of articulating in the indigenous idiom and acting out the good news in a colonially oppressed society. Due to the wide connotations attached to healing, the term *healing* became virtually synonymous with the biblical concepts of redemption and salvation. To be healed and to achieve a form of well-being in the church despite a situation of poverty and limited opportunity was as good as or a sign of being saved by God in the here and now. Salvation was concretized in this existence in a form of "realized eschatology," where the features and caregiving presence of Christ became manifest in the healing hands and persuasive commitment of prophetic healers.

Due to the engagement of prophetic leaders in the sociopolitical fields, healing also became closely associated with the African struggle for liberation. The very act of forming and maintaining churches independent of white missionary endeavor was interpreted as a move toward liberation, an authentic exodus of God's people from the slavery in Egypt, in this instance, from the paternalistic tutelage of white church leaders. Healing an invaded society—torn by divisive loyalties, alien forms of worship and religious control, and invasive processes of acculturation—meant the creation of safe havens based on African initiatives, African control systems, and inculturated forms of worship based on self-styled interpretations of vernacular scriptures. Incorporated therefore in the preoccupation of prophets with the healing of individual afflictions was the ministry of religiocultural liberation. Once AICs became involved in the countrywide struggle for political independence by way of participation in the politicized *pungwe* night vigils,[14] guerrilla-instated kangaroo courts, and the feeding, hiding, and healing of guerrilla fighters in rural villages, the image of these churches as resisting colonial rule, providing the seal of ritual approval for the military struggle, and generally being the religious vehicles for the wide-ranging cause of African political liberation became more pronounced. Healing the country meant liberating it from its colonial yoke!

Given this background to the development of specific features of the AICs in relation to historically determined African needs, the wholesome response of these churches to the earth-keeper's call came as no surprise. The need for environmental repair was in evidence throughout the country in the aftermath of war. Political control over the "lost lands" did not alter the fact that the communal lands, due to population pressure and poor land husbandry, were ecologically becoming "more lost" then before. AIC subsistence farmers knew only too well that "salvation here and now" in the agricultural world meant nothing if the tired and overused earth—the sustainer of their lives—was not reclaimed and restored. Nothing short of another liberation struggle, of the same magnitude as *chimurenga*—but this time on behalf of the abused earth—was required if a meaningful future was to be obtained for Zimbabwe's people living on the land. Key to their determination to start mobilizing their churches in the green struggle was the formation of the AAEC, joining hands with their traditionalist counterparts who had already, a few years previously, declared the war of the trees. Such commitment forged an even deeper understanding in the ranks of the AICs of the church's healing and liberative ministry. This time the focus and intensity of AIC activity shifted noticeably to the healing of a suffering creation. A new partnership in the divine-human encounter, with an overriding emphasis on Christian stewardship of nature prompted by God's indwelling Spirit, was taking shape.

Tree-planting sermons provide illuminating clues about the way in which the key role-players themselves interpreted the church's earth-healing ministry. AIC leaders used such occasions as "teaching sessions" to instruct and mobilize their followers. In January 1991, for example, Bishop Wapendama, leader of an

Apostolic Church and influential member of the AAEC executive committee at the time, preached as follows at a tree-planting ceremony:

> Mwari [God] saw the devastation of the land. So he called his envoys to shoulder the task of deliverance. Come, you messengers of Mwari [ZIRRCON/AAEC representatives], come and deliver us. Together with you, we, the Apostles, are now the *deliverers of the stricken land.* Let us go forth and clothe [i.e., heal] *Mwari's spoilt land.* This is not a task which will enrich you. No! The deliverers were sent by Mwari on a divine mission. He said: "You go to Africa, for the land is ravished!" Peace to you, people of Mwari!
>
> Deliverance, Mwari says lies in the trees, but in the first place the people have to obey. Mwari therefore sends his deliverers to continue here on earth with his own work, with all the work that Jesus Christ started here. Jesus said: "I leave you, my followers, to complete my work." *And that task is the one of healing! We are the followers of Jesus and have to continue his healing ministry.* You are the believers who will see his miracles in this afflicted world. *So let us all fight, clothing the earth with trees!* Let us follow the example of the deliverers who were sent by Mwari. God gave this task to a man of his choice. Because this man responded, the task is proceeding as you can see for yourselves today.
>
> It is *our* task to strengthen this *mission* with our numbers of people. You know how numerous we are. Sometimes we count ten thousand people at our church gatherings. If we work with enthusiasm we shall clothe the entire land with trees and drive off affliction [evil]. . . . Just look at the dried out and lifeless land around you. *I believe that we can change it.* Because of our repairing the damage, because of our doing penance for our guilt in land destruction, God will heed our wish and give us plentiful rain.[15]

Bishop Wapendama's exposition was representative of viewpoints frequently expressed at the time in the tree-planting services of fellow AIC earth-keepers. God takes the initiative to restore the ravaged earth, but the responsibility to deliver the stricken earth from its malady lies with the Christian body of believers, the church. The deliverance of the earth takes the form of *kufukidza nyika* ("clothing the land") with trees. This mission was clearly seen as an extension of Christ's healing ministry, which his disciples must fulfill in this existence. That it is a communal obligation was highlighted by Wapendama's reference to the large church meetings (usually at paschal celebrations) which, in his view, was to serve as a platform, a liberating force for the deliverance of nature.

Wapendama's confidence that the army of believers could repair the damage as an act of penance, whereby God's wrath at human guilt of earth destruction would make way for renewed benevolence—to be revealed, for instance, in God's breaking of the drought—also reflected a positive attitude commonly found in the ranks of the earth-keeping churches. This attitude derived from both biblical notions of God responding to the actions of his people and ingrained traditional beliefs about reciprocity in the divine-human encounter (e.g., gifts of the right-minded complementing rain requests at the oracular shrines of the traditional deity, Mwari, at

Matonjeni and eliciting a positive divine response in the form of ample rains). The assumption of a new divine-human partnership was all too clear in the bishop's understanding of nature's deliverance. True to prophetic perceptions of salvation as human well-being secured through healing in this existence, the earth itself was to be salvifically restored under directives from Mwari—a new order which would not be unilaterally ushered in by God, but which also depended on being "worked out," even "brought about," by human endeavor.

As fellow tree-planter I used preaching opportunities during tree-planting eucharists to emphasize the link between the church's ecological healing ministry and Christ's presence in creation. The New Testament passage Colossians 1:15–18 often featured as the basis for such sermons. In 1991, for instance, I preached at the First Ethiopian Church headquarters in the Bikita district:

> In Colossians 1:15–18 the body of Christ is explained in a special way. He is the image of the unseen Mwari [God], the firstborn of all creation. All things are created *in* Him and *for* Him, the seen and the unseen. Because of this *all things hang together in Christ* [v. 17]. Through Christ's death and resurrection, which we commemorate today in our sacrament of bread and wine, all power in heaven and on earth has been given to Him [Matt. 28:18]. From all this we conclude that Christ is not only Lord of creation, but that His body *is* all of creation. All created things are part of His body! The implication for us as stewards of creation is that if we fell trees indiscriminately we are "killing" [abusing] the body of Christ.... We have learnt in the past that the church of believers is Christ's body. That truth we celebrated in our sacrament of bread and wine in the past. That remains, for it is good and not wrong. It is just that [in our tree-planting eucharist] we are reminded these days of something we have neglected. We are healing and restoring that part of Christ's body [Him as earth] which we have unwittingly abused.... We are *all guilty* in this respect, for, both the whites and blacks are exploiters of the earth. That is the message I leave with you today: *Clothe the barren earth! Heal the earth!* Doing so is fully part of our lives as Christians.[16]

This christological focus within the church's earth-healing ministry became widely accepted and propagated in AAEC circles. There were those AIC leaders who used the same idiom of qualifying earth-care as the church's task of "tending" to Christ's body. Said Rev. Tawoneichi during a tree-planting ceremony:

> Earthkeeping is part of Christ's body ... because we as humans are part of His body and the trees are essential for us to breathe, to live.... The random felling of trees hurts the body of Christ. Therefore the church should heal the wounded body of Christ. Tree-felling is only justified when there is a sound purpose aligned to God's will. Otherwise God is angered and will punish us. One of the signs of such punishment is the continuing drought. No trees, no rain! Mwari is disturbed.

Like Wapendama's observations, Tawoneichi's thoughts illustrate an understanding of the close correlation between ecological stewardship and God's response. Environmental abuse is sinful. It causes Christ to suffer and God to judge, drought being God's disapproval in the traditional sense. By its very nature the church in this context is or should be protector and healer of an overexploited earth, the "wounded" body of Christ.

Other AIC leaders preferred less controversial descriptions of Christ's body. Bishop Farawo, leader of the Zion Apostolic Church of Zimbabwe and for many years an influential figure among AAEC nursery keepers, for instance, defined his views as follows:

> The earth we abuse is not Christ's body, for it is the creation of Mwari. Yet creation is like a person, the image of the body of Christ. Look at the trees. They breathe like humans. So if we fell them we hurt the Spirit of God, because his Spirit is in the trees. . . . Earthkeeping is like an expression of Christ's body. Tree-planting during the eucharist is *not really part of Christ's body*, but it pleases Christ because we are clothing his earth.[17]

Whatever the views of earth-keepers about the exact relationship between their green activities and the living body of Christ, the centrality of the Christ figure in African guise at tree-planting events was always in evidence. Not only was he the wounded healer who set the tone for a sacrificial ministry of earth-care, but he also, through his perceived suffering in creation, provided the incentive for a compassionate ministry of reconciliation between earth-keepers and a depleted earth community. Bishop Chimhangwa of the African Zion Apostolic Church preached on occasion: "All of us are guilty of the crime of deforestation. . . . So, today we plant trees as an act of reconciliation between us and all creation, in Jesus Christ. We thank Him for His atonement, whereby this act of reconciliation is made possible." Here ecological healing obtains salvific dimensions. The intuition behind these words suggests a widened perception of salvation: the realized presence of God's kingdom encompassing all creatures in this existence, rather than a narrow focus on a future salvation of human souls.

The impact of an earth-healing ministry on AIC perceptions of the nature of the church was quite noticeable. A new type of healing colony was in the make, one which did not override the existing prophetic preoccupation with human afflictions, but extended such mission to incorporate, on a broad front, therapeutic responsibilities for an ailing environment. In a sense the new and trendsetting model of an earth-keeping church headquarters, another kind of "holy city" revolving around the activities of a founder "prophet-healer," emerged at the ZIRRCON headquarters in Masvingo town. Without any pretence of controlling or dominating an extensive ecological strike force of some 180 churches, the ZIRRCON-AZTREC-AAEC administrative center nevertheless figured as a central earth-keeping nexus.

It started with a difference. My house in town was the first headquarters-cum-nature healing center. But the original earth-keepers were traditionalist spirit mediums, chiefs, and tribal elders—not AIC Christians!—who crowded my house during weekends to discuss the strategies for a green *chimurenga*.[18] The white "prophet," in this instance, was nicknamed Muchakata (the "wild cork tree"; protected in traditional religion as a holy ancestral tree used for ritual purposes) to signal the connection with the guardian ancestors of the land in the greening of their territories. The concept of a healing center, moreover, was already implicit in the development of our first nursery in my backyard, so that we could clothe the barren earth with young seedlings. This basic idea was expanded, once the AIC earth-keepers had joined us, in the building of a new earth-keepers' center, away from my house, in Masvingo town, with three major nurseries—named Muchakata (in honor of my contribution), Muuyu (Baobab nursery), and Muonde (Wild Figtree nursery)—on the outskirts of town, as the extended "healing colonies" of earth community.

The new headquarters housed the traditionalist (AZTREC) and Christian (AAEC) green armies, both of which operated in a united struggle while retaining their respective religious identities. Insistence on the part of the AIC earth-keepers that I be called "Bishop Moses"—after the Old Testament leader of the Israelite exodus from Egypt—during church ceremonies of earth-care underscored their underlying determination not to compromise religious identity in the midst of sustained ecumenical, traditionalist, and Christian interaction. Thus, in typical Pauline fashion I became a traditionalist to the traditionalists, that is, Muchakata, and a "prophet" to the AIC prophets, that is, Bishop Moses.

Taking their cue from the ZIRRCON headquarters in town—where church conferences were conducted, ecological policies were formulated, and infrastructures were devised for nursery development, woodlot monitoring, wildlife management, and so on—several prophetic leaders out in the rural areas started to extend the healing colonies at their own church headquarters as "environmental hospitals." The patient in such instances became the denuded land, and the "dispensary" (i.e., the faith-healing "medicinal" arsenal of holy cords, holy water, staffs, paper, and other symbols of divine healing power) became the nursery where the correct "medicine" for the patient, in the form of a wide assortment of indigenous, exotic, and fruit trees, was cultivated. The entire church community—both at headquarters and in outlying congregations, residents, and visiting patients—now became the healing agent under the guidance of the church's principal earth-healer. Consistent aftercare of budding woodlots provided proof of the church's dedication, the woodlot itself becoming the focus of testimony sermons and a source of inspiration for an expanding ministry, as the testimonies of healed human patients in the past contributed both to a reaffirmation of belief in God's healing powers and to the church's recruitment of new members in its expansionist drive.

Consider, for instance the escalation of earth-keeping activities at Bishop Wapendama's Sign of the Apostles Church. The bishop had developed a model nursery at his headquarters, where the nucleus of his church's leadership was engrossed in pot-filling, seed germination, and watering and nurturing the seedlings. Drawing on the resources not only of the AAEC but also of his followers, the bishop modernized his nursery equipment, which included the installation, at his own expense, of a costly water pump. Having established woodlots of red mahogany and other indigenous trees in his outlying congregations, the bishop's annual itinerary now included numerous ecologically motivated visitations to ecclesiastic outposts to monitor expanding afforestation projects. Far from interfering with the pastoral care of his flock and preaching appointments, the bishop's earth-healing ministry appeared to stimulate the spiritual lives of his followers and to trigger new recruitment of members as the church was seen to strive valiantly and concertedly in the war of the trees. There were similar indications of intensified growth in the Zionist churches of bishops Machokoto and Marinda, former AAEC president and general secretary, as a result of their high profiles as earth-keepers in rural society, where subsistence farmers were increasingly appreciative of environmental reform.

The person whose private and church life was probably the most drastically affected by AAEC developments was Bishop Farawo of the Zimbabwe Zion Apostolic Church. Having moved from his church headquarters in the Bikita district to Chivi district's local government headquarters to take control of the AAEC's largest nursery, the bishop and his family had turned the nursery complex with its dwellings and toolshed into a veritable little "Zion city of the trees." Bishop Farawo ministered to the members of his church in the district and at his new headquarters, but afforestation became focal to his entire ministry. He collaborated with forestry representatives and land extension officers on a regular basis. He mobilized school communities to help collect seeds for the nursery and establish woodlots at their schools. Apart from becoming expert at germinating a wide variety of indigenous tree seeds, the bishop also supervised numerous tree-planting ventures throughout the district during the rainy season. He supplied seedlings not only to AAEC members but also to other churches, associations, clubs, and even traditionalist elders associated with AZTREC. Thus a distinctly Zionist Christian ministry of afforestation served a religiously pluralistic society, in the process contributing not only to a growing network of ecumenical ties among churches but also to cooperation and goodwill between Christians and African traditionalists.[19]

"Greening" Sacraments

Baptism

Committed AAEC-affiliated prophets over the years tended to blend their ecological insights with their moral guardianship over their churches. This was reflected in the "greening" of their sacraments. In the baptismal context they started to insist that converted novices not only confess their moral sins in relation to fellow human beings but also their *ecological sins*: felling tress without planting any in return, overgrazing, riverbank cultivation, and neglect of contour ridges, with soil erosion as a result—in other words, taking the good earth for granted and exploiting it without honoring or nurturing it.

At "Jordan" (any river or dam chosen for the baptismal ceremony) it made sense to the newly converted to confess ecological guilt, there where the treeless plains, erosion gullies, abused riverbanks, and clouds of windswept dust were clearly in evidence. Crossing Jordan in baptism subsequent to ecological confessions meant more than just individual incorporation into the body of Christ and the prospect of personal salvation. It also required the new convert's commitment to active participation in restoring creation as part of God's plan and as a sign of genuine conversion in recognition of the gift of God's free grace.

To many independents, baptism is also a healing ceremony because of a persistent magical worldview which attributes divine powers to the baptismal water. Baptisands therefore drink the life-giving water of Jordan, filled by the Holy Spirit, for individual cleansing and curative purposes. It follows that the ceremony offers a unique opportunity for interpreting the Spirit as healer of both the people and the land. In this respect baptism became yet another feature of an extended ministry of healing, a changing ecclesiology. In that case, the drinking of Jordan water symbolized a shift from the baptisand's personal benefit from the Holy Spirit's healing and salvific powers to a ritual affirmation of solidarity with all creation, a new commitment through individual conversion, to earth-healing.[20]

Tree-Planting Eucharist

In terms of its centrality in earth-care and its attraction as a means of drawing large numbers of tree-planters, the tree-planting eucharist was the most significant ecoliturgical innovation introduced by the AAEC. The participation of a variety of local churches in each sacramental ceremony and the sharing of ritual roles on an interchurch basis strengthened an environmentally focused ecumenism. Through the integration of sacrament and tree-planting, environmental stewardship was effectively drawn into the very heartbeat of church life and biblical spirituality. To a large extent this eucharist also presented the participant earth-keepers with a staying platform for the enactment of a relevant AIC theology of the environment.

The event itself was always carefully prepared by the host church and supportive ZIRRCON staff members. The latter assisted with food provisions and the digging of holes for the trees to be planted in a new woodlot. Seedlings also had to be transported to the site concerned. While the communion table—bearing neat tablecloths, bread, wine, and some seedlings—was being prepared, groups of dancers danced around the seedlings stacked nearly. Dance and song brought praise to Mwari, the great earth-keeper, encouraged the green fighters to be vigilant in the struggle, and even entreated the young trees to grow well. The service itself contained several earth-keeping sermons by AAEC bishops and ZIRRCON staff members. It invariably also included speeches by local tribal elders, visiting government officials, and representatives of the departments of Forestry, Education, Parks and Wildlife, as well as the Natural Resources Board. School children used to sing in choirs and recited their own poems on trees, village life, and wildlife. For a eucharist, the ceremony itself was a remarkably open-ended event which provided opportunity for outsiders to contribute and, importantly, for AIC preachers to witness to their faith in Christ, whose redemptive death on the cross they were about to celebrate. This latter aspect rendered the entire eucharist a missionary outreach event, the proclamation of Christ, the earth-keeper's good news, in an existentially relevant idiom among rural subsistence farmers.

The sacrament itself started with the confession of ecological sins. All communicants, church leaders included, would line up behind a band of prophesying prophets to confess their guilt and receive prophetic admonition as they slowly filed past the prophets before picking up a seedling and moving to the communion table to partake in the sacrament. Doing so with seedling in hand symbolized the earth-keeper's embrace of earth community, drawing its "members" so to speak into the union and *communitas* of Christ's body, proclaiming and reasserting in Christian fashion the cosmic holism which the practitioners of African traditional religion had intuited all along.

The official tree-planting liturgy of the AAEC, read in full or in abbreviated form on these occasions, started with Mwari declaring to his tree-planters the value of trees. Then followed the trees and the earth-keepers, speaking in the first person:

> They will provide you with shade
> to protect you from the heat of the sun.
> They will give you fruit, for you to lead healthy lives.
> These trees will clothe the barren earth,
> protecting it against soil erosion,
> preventing it from turning into a desert,
> keeping the moisture in the soil.
> Look at the stagnant water
> where all the trees were felled
> Without trees the water holes mourn;
> without trees the gullies form
> for the tree roots which hold the soil . . .
> are gone!

These friends of ours
give us shade.
They draw the rain clouds,
breathe the moisture of rain.

I the tree...I am your friend
I know you want wood
for fire
to cook your food,
to warm yourself against cold.
Use my branches...
What I do not need
you can have.

I, the human being,
your closest friend
have committed a serious offence
as an *ngozi*, the vengeful spirit,
I destroyed you, our friends.
So the seedlings brought here today
are the bodies [*mitumbu*] of reparation
a sacrifice to appease
the vengeful spirit.
We plant these seedlings today
as an admission of guilt
laying the *ngozi* to rest,
strengthening our bonds with you,
our tree friends of the heart.

Let us make an oath today
that we will care for God's creation
so that he will grant us rain.
An oath, not in jest...
but with all our heart
admitting our guilt,
appeasing the aggrieved spirit,
offering our trees in all earnestness
to clothe the barren land.

Indeed, there were forests,
abundance of rain.
But in our ignorance and greed
we left the land naked.
Like a person in shame
our country is shy
of its nakedness.

Our planting of trees today
is a sign of harmony

> between us and creation
> We are reconciled with creation
> through the body and blood of Jesus
> which brings peace,
> he who came to save
> all creation (Col. 1:19–20).

Once the ceremony at the communion table was completed and all the communicants had congregated next to the newly dug holes in "the Lord's acre" (the new woodlot), the lead bishop walked through the woodlot sprinkling holy water on the ground and on some of the seedlings and saying:

> This is the water of purification and fertility.
> We sprinkle it on this new acre of trees.
> It is a prayer to God, a symbol of rain
> so that the trees will grow,
> so that the land will heal
> as the *ngozi* we have caused withdraws.

The bishop then led the green army into the Lord's acre to do battle against the earth's nakedness. The seedlings were addressed one after another as they were placed in the soil:

> You, tree, my brother ... my sister
> today I plant you in this soil.
> I shall give water for your growth.
> Have good roots
> to keep the soil from eroding.
> Have many leaves and branches
> so that we can
> breathe fresh air
> sit in your shade
> and find firewood.[21]

During the actual tree-planting, the traditionalist counterparts representing AZTREC, who up to that point had attended the proceedings only as respectful observers, would join and assist their fellow fighters as participants in the planting process. This was the reciprocal response to AAEC Christians who attended and observed introductory ancestral rituals without drinking sacrificial beer during AZTREC's tree-planting ceremonies, prior to helping their traditionalist friends plant trees in a woodlot dedicated to the guardian ancestors of the land. Retention of religious identity, therefore, was not felt to frustrate religious tolerance or to obstruct the joining of ranks in the united battle of liberation against earth destruction. Toward sunset the proceedings culminated in a blending of earth-healing and human-healing. Then afflicted earth-keepers knelt in rows in front of healer prophets, to the rhythm of muted drumming, song, and dance, for the curative laying-on of hands and sprinkling with holy water. That no distinction was made in this, the final healing ceremony, between Christians and traditionalists undergoing therapeutic treatment once again underscored the multifaith unity in purpose and action of the stewards of God's creation, and through them the extension of God's limitless mercy to the entire earth community.

The liturgy at first glance may appear to be relatively simple and straightforward. Yet, it is a thoroughly inculturated expression of commitment. The close identification between earth-keepers, water, soil, and trees—to the extent of relevant intercommunication—reflects African religious holism. Here the intuitions of the African past are taken to a level where mutual dependence is eloquently verbalized. In the ritually declared friendship, subsequent to the tree-planters' confession of guilt in the mindless destruction of nature, mutual responsibility is reaffirmed: the new trees to provide shade, the rooted bonding of soil against erosion and of unpolluted air to sustain healthy life for humans, and the earth-keepers to water and protect their budding friends in the secure space of the Lord's acre. The liturgy assumes responsible aftercare (normally the Achilles' heel of most African tree-planting endeavors at the grassroots)—in itself a strong incentive to the woodlot keepers not to let the green army and its monitoring agents down.

It is no coincidence that water, the age-old symbol of purification, of healing and life itself, should feature so prominently in the ritual preparation of the acre to be "healed" and in the faith-healing restoration of strength in the afflicted bodies of earth-keepers. Sprinkling of holy water in the AICs protects infants from the onslaught of evil spirits, cleanses the sick from contamination, prepares barren women for childbearing, and liberates unwilling hosts from plaguing spirits. As such, holy water, prayed over by the prophet-healer, is the most potent symbol of God's reign, a divine rule capable of healing and restoring all of life. Thus the sprinkling of holy water on afflicted earth-keepers renews the impetus of the green struggle as much as it revitalized the guerrilla fighters in the front during the military struggle for the "lost lands." Likewise, in the Lord's acre it signals the casting out of the soil's denuded barrenness, of the vengeful spirit that causes drought, and it prepares the soil for full recovery under the cover of new trees.

Impersonating the vengeful *ngozi* spirit in terms of earth destruction is as potent a way of accepting full responsibility for deforestation as the confession of ecological wizardry. The *ngozi* is an aggrieved spirit of a murdered person or someone who has been the victim of a grave injustice prior to death.[22] In customary law and traditional religion the *ngozi*, which wreaks havoc in the offender's family through illness and death, has a legitimate claim to full compensation in the form of up to ten sacrificial beasts called *mutumbu* (lit., "corpse" or "body," since they pay for the corpse of the deceased). In some cases the offender's relatives also provide the *ngozi* with a young wife, who must sweep and tend the small hut specifically erected for her disgruntled "spirit-husband."

Presenting the trees to be planted as *mutumbu* compensation for the *ngozi* spirit provoked by wanton tree-felling is an illustration of thoroughly contextualized appeasement between humans and environment. The ritual, moreover, expresses compassion for the badly abused friends: trees, soil, water, and, by implication, all of life in nature.

The *ngozi* concept has several subtle connotations in the liturgy. It reflects the ruthlessness of the supposed human stewards of the earth who attack nature with a vengeance, like that of the *ngozi*, hence the seedlings are legitimate sacrificial substitutes for the stricken tree trunks or "corpses." Then, in sprinkling the water over God's acre, the words "it [the water] is a prayer to God, a symbol of rain" implicitly suggested that God himself/herself turns *ngozi* against the ecological offenders by retaliating with severe drought. This interpretation tallies with the still persistent and widespread traditional belief that the creator-God punishes transgressions against nature and the guardian ancestors of the land, who are responsible for ecological equilibrium, by withholding rain. In the admission of guilt, in the ritual plea for termination of divine discipline, and in the renewal of human resolve to heed the environment as ordained by deity and ancestors, absolution is found. God responds by sending life-giving rain. Transformed as they are in the Christian liturgy, some of these traditional notions are still in evidence.

In the tree-planting eucharist the close identification of Christ's body—"in whom all things hang together" (Col. 1:17)—with the abused and barren soil makes sense. Christ's incarnation where the broken earth "lies shy in its nakedness" places him in direct relationship with the ancestral guardians of the land. These *varidzi venyika* traditionally belonged to the soil; they *are* the soil! Their ecological directives issued from the soil, as expressed in the saying *ivhu yataura* (lit., "the soil has spoken"). In a sense Christ in this context is both guardian and the soil itself. New conceptions of Christ's lordship and his salvation for all creation can develop from this essentially African expression of his pervading presence in the cosmos. In African peasant society at any rate, Christ's reign as *muridzi* ("guardian") of the land is an essential part of the good news, for he is the one who is believed to consciously strike a balance between exploitive agricultural progress and altruistic, sacramental restoration of the land.

An Ecumenically Rooted Mission

The foundational text for the first ecumenical movement among Zimbabwean AICs, called Fambidzano and founded in 1972,[23] was John 17:21–23. This text refers to Christ's high priestly prayer to his Father in which he requests unity among his disciples so that the world can *see* and *believe* that he had been sent to the world. These were taken as the classical biblical texts forging a link between ecumenism and mission, unity among Christians and the church's outreach into the world. Fambidzano's participating AICs accepted that without a solid basis of loving care in unity among Christ's followers, in this instance among African churches, their witness to the world would be unconvincing and suspect.

As many of the AICs who joined the AAEC had formerly belonged to Fambidzano, the integral relationship between ecumenism and mission was familiar and readily adopted in the earth-keeping movement. The difference between the two movements in terms of ecumenical vision was essentially that of scope. Whereas Fambidzano's specialization in theological education for AICs did not require a broad ecumenical base for its teaching to be effective, the determination of the earth-keepers to unleash a green *chimurenga* countrywide stimulated a vision in the AAEC for a massive Christian movement operating from a united platform. The latter vision included the formation of an African Earthkeeper's Union with an earth-keeping affiliation and environmental praxis stretching continentwide. Whereas the mission of Fambidzano was the improvement of AIC leadership through theological education, a goal which specifically included the improved effectiveness of the church's missionary witness in the African world, the AAEC mission, by contrast, was the extension of Christ's salvific message to all creation through a ministry of comprehensive earth-care. It is significant to note at this point that the mission of earth-keeping in the AAEC was at all times seen as an extended mandate complementing and not substituting the original one of Christ sending out his disciples to convert people and baptize them in the name of the triune God (Matt. 28:19). In the circles of the earth-keepers the conviction seemed to grow, however, that the conversion and life of the Christian convert was somehow flawed it if did not include the sharing of ecoresponsibilities with fellow Christians and/or humans of all walks of life.

From the outset the first AAEC president, Bishop Machokoto, propagated the need for strong ecumenical foundations in our common ecological struggle. "What I asked of God," he said, "is a true sense of unity amongst us. We have to work together to avoid all forms of confusing conflict. Our unity must rest on convincing work. . . . The basis of our work, according to God's Word, is love, a love which reveals itself in work. . . . We, the [AAEC] churches will have to make sacrifices for the cause to which we have pledged ourselves. Therein lies our unity." My own expositions of a *divine mandate* for our work, with reference to Isaiah 43, as well as its christological basis, implied wide ecumenical interaction.[24]

This call for united action provided a key to interpret the ecumenism developed by the AAEC. Representing a predominantly peasant society which confronts the hazards and threats to subsistence on a deforested, overgrazed, and overpopulated land, the AIC bishops and their churches did not join forces to realize some abstract ecumenical ideal or for the sake of church unity as an end in itself. It was rather an ecumenism shaped by churches sharing a newly identified and common commitment—that of healing the earth. In a sense, therefore, the realization of a common quest, the action in the field, gave expression to the love in unity to which Bishop Machokoto referred. One can say that the escalating battle of the trees, the development of nurseries with many thousands of seedlings, and the preaching and ritual celebration of tree-planting and tree-watering in the newly established woodlots ameliorated interchurch conflicts. The earth-keeping church was purged from within of isolation and self-centered ambition as it bonded with other churches in an all-absorbing ministry of environmental stewardship.

Not that all the differences and conflicts of the past were suddenly resolved. The prophetic churches tended to accuse the nonprophetic movements that they were not fully Christian because they did not heed the work of the Holy Spirit, and the Ethiopian-type churches in turn accused the Zionists and Apostles that their prophecies were products of traditional ancestral or alien spirits, not of the Holy Spirit. It was rather that these old conflicts paled into insignificance as the green revolution unfolded, at the annual conferences and executive meetings and in the joint labor and ritual celebration of earth-healing ceremonies. Thus a new comradeship transcending traditional ecclesiastic divisions had started to evolve between creator-God, earth-keeping humanity, and the trees, plants, and wildlife.

Apart from being drawn into closer union with other AICs, the earth-keeping church was also challenged to participate in spontaneous ecumenism by interacting in a common cause with traditionalists, that is, with AZTREC. Once the AIC bishops felt secure about retaining their own brand of Christian identity in the green struggle, they were eager to promote wider religious unity, or at least meaningful cooperation with their AZTREC counterparts, the chiefs, spirit-mediums, and other traditional authorities. Said Bishop Machokoto in his key address at the inception of the new ecumenical movement:

> We must be fully prepared to recognise the authority of our kraalheads and chiefs. For if we show contempt for them, where will we plant our trees? A Christian attitude is required towards the rulers of the land. Let our Bishops in their eagerness to fight the war of the trees not antagonize the keepers of the land. If you are a church member, yet try to place yourself above the laws of the land, you are not a true convert. Let us fully support our [tribal] elders in this struggle of afforestation, so that the ZIRRCON-AAEC objectives may be realised in practice.[25]

This plea certainly did not imply a sweeping compromise of the prophetic AICs' general confrontational approach to the ancestral religion represented by their traditionalist cofighters in the green struggle: the chiefs and spirit-mediums. Yet the call for submission to the laws of the land and for cooperation with the chiefs reveals a growing tolerance and a preparedness to move beyond the stereotyped prophetic restraints of in-group dynamics in the interest of a stricken environment. Instead of withdrawing from the traditionalist practitioners of ancestor veneration to demonstrate its rejection of so-called heathenism, the prophetic earth-keeping church underscored at least ecological solidarity with its traditionalist counterparts in the green struggle. Such religioecumenical openness made it possible for ZIRRCON to call the first combined annual conference of AZTREC and the AAEC in May 1993, comprising contingents of some fifty chiefs, headmen, and spirit-mediums on the one hand and fifty AIC Bishops and prophets on the other. An interesting feature of the conference was interfaith dialogue and the Christian witness by some of the AIC representatives during plenary sessions. The mediums responded to this by conveying their ancestors' wishes in a state of trance.

Such interfaith encounter included both heated debates on highly divergent customary and Christian approaches to environmental projects and agreement on common ground for future cooperation. On the issue of establishing game sanctuaries in communal lands, for instance, the traditional spirit-mediums pointed out that the guardian ancestors of the land could be provoked into mystical retaliation if the holy groves (*marambatemwa*) containing their graves were game-fenced. The AIC prophets for their part supported game-fencing, provided that they could have their own game sanctuaries, which they intended to call the "Lord's acre" or the "Lord's dwelling place" in contradistinction to the traditional holy groves. A common purpose between AZTREC and AAEC representatives to escalate the struggle in unison was described by both parties after the conference as a major earth-healing and religioecumenical breakthrough, the first they had ever witnessed in Zimbabwe.

Another significant breakthrough was the sending of both traditionalist and Christian delegates toward the end of the tree-planting season to report to the high-god Mwari at the Matopo shrines about the achievements of the green *chimurenga*. Zionist delegates, as in the traditional tree-planting ceremonies, were not obliged to participate actively in the rituals, but they were placed in a position to observe firsthand how their fellow AZTREC tree-planters reported their work at the cave-shrines and to listen directly to the oracular deity's spoken response. This experience to them was a far cry from the bitter conflicts which had raged in earlier years between the cult priests and Zionist prophets over rainmaking.[26] The Zionist earth-keepers actually established ties of friendship with the Matonjeni priests and spent long hours discussing religious matters with them. On occasion quite a number of AAEC bishops and AZTREC chiefs and mediums together accompanied me to Matonjeni to plant trees around the homestead and the grave of our friend, the—at that point—recently deceased senior Mwari priest Jonas Chokoto. At the time it was noticeable that the Zionist contingent visiting the Matopo shrines was on the whole perceiving of such visits as a form of privileged pilgrimage, an opportunity to be in touch with the roots of the religion of their forebears.

In this kind of interfaith climate, ecclesiological perceptions were bound to change, both in terms of an increased sense of belonging to and playing a meaningful role in the Christian household of God and in terms of respecting the divergent faiths of all humankind so as to secure the widest possible front against environmental degradation.

Missiologists have on the whole tended to exclude earth-keeping from the study of missions: its theory and praxis. Jongeneel's encyclopedic survey of mission studies over the past two centuries reveals a lack of literature on the relationship between missions and earth-care.[27] Kirk suggests that a number of missiologists today are skeptical of including environmental care in the missiological discipline, as it would overextend the scope of mission.[28] Kirk himself, however, devotes an entire chapter in his book *What Is Mission?* to this subject. He argues that environmental conservation relates to mission in the same way as does peace building and activities focused on the acquirement of economic and political justice.[29] Darr concurs; then he adds a historical argument for such inclusion, namely, that "Protestant Missions [in southern Africa] have included earth care in the scope of mission from the very beginning."[30]

To the earth-keeping AICs of Zimbabwe the battle of the trees emerged as an extended mission mandate of the church, not so much on account of careful theological reflection but as a spontaneous response to existential environmental realities, a response guided somehow by the life-giving Spirit of God. As propounded by Bishop Wapendama, the healing ministry of Christ remained focal to the church's mission but, in the AAEC context, it included more explicitly than before the deliverance and salvation of Mwari's stricken land. This form of salvation became manifest to the extent that the church fulfils its role as the keeper of creation, a mission in which its entire membership was harnessed to active agents instead of just a few specialized missionaries.

Despite the challenge to churches and believers to engage in the earth-keeping mission, a pervading sense of being mandated and sustained by God was noticeable at most AAEC tree-planting ceremonies. For all his vision and enthusiasm, Wapendama repeatedly reminded his audience that they were facing a formidable task, one which could be accomplished only in full recognition of dependence on God. "I beseech you," he said, 'to place yourselves in the hands of Mwari.' He alone can give us the strength to endure in this struggle."

In the Ethiopian-type churches one found similar emphases on Mwari's initiative, Mwari's care for creation, and Mwari's earth-care commission to his church. In concluding a tree-planting ceremony at Rev. Zvobgo's African Reformed Church at Shonganiso Mission, Rev. Mandondo claimed:

> Today we have done God's work. You will see, in a short space of time the trees will grow tall. And we shall say: God surprises us. God exists. God does what pleases him. Today we have done his bidding. Today we have learnt that if we want to be God's children, we must do his work. We are the inheritors, existing

by virtue of the inheritance—in this instance the fruit of trees. Today we did what God sent us and commanded us to do. It is not so much a matter of success or failure, but in the first place complying with God's will, giving *Him* joy through our obedience.

Here clearly the church's earth-keeping mission derived from God's commission and the obedient response of God's people to his divine promptings. It was verbalized by a church leader who showed profound awareness of divine presence in all of nature:

> Up in the mountains I can see Mwari. In the rocks and the trees I see Mwari. There his strength and his works are revealed.... Whose strength do those massive trees [at Mount Selinda] reveal? Mwari's, of course.... His work is clearly seen in the things he has created. Follow the rivers and observe the running waters. Whose work do you think it is? Mwari's!

Mention was already made of liberationist connotations in the mission of earth-care. Historically, the church's ecological mission stood fully in the tradition of *chimurenga*. That the earth-keepers drew inspiration from the struggle prior to Independence in 1980 was reflected in the fighters' use of *chimurenga* tree-names to indicate a preferred trait in action or the nature of individual contribution: *Muchakata* ("wild cork tree") to invoke ancestral protection; *Murwiti* ("black ebony") to show strength of character and resolution in the struggle; *Muchecheni* ("wait-a-bit thorn tree") to show resolute recruitment of green fighters by literally "hooking" them into the struggle; *Mushuku* ("the wild loquat"; most popular and common of all wild fruit) to feed the entire populace; and *Mukute* ("the water-berry tree") to show the softness or coolness of reconciliation between humans and earth; and so forth.

Asked about the connection between the church's green mission and *chimurenga*, AIC leaders invariably made statements such as the following: "We are engaged in the *chimurenga* of clothing the land"; "we fight this second *chimurenga* to liberate the land"; "we are showing our strength in this liberation struggle which covers and protects the soil." Bishop Machokoto said: "There is absolutely no doubt about the connection...this war [of the trees] is the most important one following the first *chimurenga*. We are all committed to this struggle to restore the vanquished land through afforestation.... Trees draw the rain clouds. Trees provide medicine for the sick. Tree leaves cleanse the air for us to breathe properly. Nothing surpasses the value of trees! Trees are our life. We say, 'a ward with dense forests knows no death.' Even President Mugabe and the government know that the earth cannot be the earth, and we cannot be people, without trees."

From these responses it was evident that the church's liberationist task was extended without hesitation from the sociopolitical to the environmental sphere. Without obscuring the futuristic dimension of people yet to be saved in the coming kingdom—as AAEC churches tended to preach—the earth-keeping mission focused on liberation, as a form of salvation, in this existence by materially improving the quality of life of all creation, protecting nature and thereby the lives of people at the behest of God himself/herself.

NEW ETHICAL CODES AND DISCIPLINE

"The Church is the keeper of creation," said Bishop Nhongo of the Zion Christian Church. "All churches now know that they must empower their prophets to expose the wizards [*varoyi*] who kill the land. These people who willfully defile the church through their destruction of the earth should be barred from participation in the holy communion. If I was the One who owned heaven I would have barred them from entry. The destroyers of the earth should be warned that the 'blood' they cause to flow will be on their own heads...."

The Bishop's son, evangelist Samuel Nhongo, concurred: "Jesus said to Simon Peter: 'I give you the keys to lock and unlock!' It is in this light that I see the earth-destroyers whom we expel from the church.... The tree-fellers who persist in their evil ways should be locked out of the church.... This war [of the trees] must be fought on all fronts and with severity. The church's new ecological laws should be universally known and respected. Otherwise we will be merely chasing the wind."[31]

To voice protest against environmental abuse in the drastic terms of expulsion from the church and even from heaven raises questions. To propose the purging of the church of the wizards of the land sounds like the idiom of the witch hunt and the imposed legalism of medieval patriarchs. Until, of course, one sees the devastation of deforestation in the immediate vicinity of Bishop Nhongo's village; until one feels the helplessness of villagers observing the mindless onslaught on the remaining trees by those who have no hope of earning the necessary school fees for their children unless they sell firewood; until one listens to the sighs of tired village women who have to walk miles each day to gather firewood and fetch water. Then the radical and irritated attitudes of earth-keepers in relation to those who refuse to heed the environment—and, by implication, their communities—start to make sense.[32]

Given a situation of population pressure on the land, depleted soils, and treeless plains, with diminishing returns in subsistence farming as a result, the determination of rural earth-keepers to retain at least some form of control over resources was understandable. In this context the proclamation of Christ as *muridzi venyika* ("guardian of the land") and the refashioning of his church as "keeper of creation" made sense. Here, too, a realistic ethic of the environment, including strict measures of discipline, took shape. Bishop Nhongo's preoccupation with earth destruction as a form of wizardry, *uroyi*, reflected the new idiom. Sin no longer concerned only the antisocial acts committed directly against fellow human beings or God, but also and particularly offenses against the environment. These included the indiscriminate felling of trees, refusal to participate in curative conservationist activities such as ZIRRCON's gully-reclamation programs, and all agricultural activities causing soil erosion, for instance, riverbank cultivation, neglect of contour ridges, the use of sledges, the pollution of water resources, and so forth. The dire implications of such transgressions against the entire earth community—people, animals, vegetation, and all animate or inanimate objects belonging to creation—were considered in such serious light that willful perpetrators were indeed branded as traitors to the green cause, as destroyers, or wizards, of the land. The discerning ministry of green prophets who urged the confession of environmental sins prior to church services and earth-keeping rituals enabled the church to identify the enemy outside and within its own ranks. Identification of the wrongdoers illuminated publicly the church's concern for ecojustice and concretized its ethical code and control system. This development was reminiscent of the *chimurenga* struggle, during which the counterrevolutionaries and collaborators with the Smith regime were branded as wizards.[33]

In my discussions with AAEC key figures, the consensus generally was that the earth-keeping church should draft and enforce its own nature-protective laws. Disciplinary action, many claimed, had to assume the form of prophetic elicitation of illegal tree-felling confessions, whereafter the church council should force wanton tree-fellers to plant and care for new trees as a form of punishment and recompense for the damage done. Bishop Chimhangwa insisted that all church laws on earth-keeping should be written up in church books at once, thus reinforcing the gospel message on the earth's salvation. He considered many people still to be ignorant of the "gospel of the trees" (*evangheri yemiti*). Consequently "the threat of the destructive axe must be repelled!" He also claimed that many of the AAEC earth-keepers themselves were still secretly making use of the "destructive axe." Mrs. Chimhangwa, the bishop's wife, felt so strongly about unchecked deforestation through illegal use of the "destructive axe" that she suggested that the church should have trespassers thrown into jail until the urgency of environmental care was fully understood.

The assumption underlying Mrs. Chimhangwa's terse remarks was that church and state should cooperate in bringing the earth destroyers to book. In this connection Bishop Machokoto was emphatic: "We need both church and civil laws together to protect nature. . . . We shall ask ZIRRCON to tell the government about our churches' insistence on more effective legislation. Then it should be published for our people to be taught." A variation on this theme was that AAEC bishops and AZTREC chiefs, the latter as representatives of local government, should cooperate in meting out punishment to willful tree-fellers.

In practice such collaboration was already taking place prior to ZIRRCON's decline. Spirit mediums, some of whom were keepers of holy ancestral groves, and AIC prophets coordinated their efforts to detect and report on illegal tree-felling at local church councils and chief's courts. Not only did such action strengthen multifaith ecumenical ties, but it also led to the levying of heavy fines at chief's courts, in itself a deterrent for environmental abuse.

THE CHURCH AS VEHICLE OF THEOLOGICAL ORIENTATION

Engaged in earth-care programs based on an expanded missionary mandate, the earth-keeping church of necessity had to monitor and assess its growing ministry.[34] This was a spontaneous, reflective process relating to and emanating directly from praxis—there where tree-planting and tree-nurturing took place in a newly ritualized context. Instead of appointing theological committees or experts to theologize by way of written texts, as often happens in the West, the earth-keeping church empowered its key figures to relate praxis to scriptural truth at the point of engagement in the field. Often the result was a kind of reenactment and improvisation of biblical history, in which African church leadership and Christian communities identified with biblical figures and/or events.

Elsewhere I have discussed some of the implications of an emergent environmental theology for the understanding of the nature and activities of the triune God of the Bible—in terms of an immanent Godhead who is present in the crops, the rain, the trees, wildlife, and the lives of people; Christ the keeper and guardian of creation, who heals and saves a degraded planet unto a new heaven and a new earth; and a life-giving Holy Spirit who guides the church in its comprehensive mission.[35] This trinitarian perspective was related to the publications of prominent African theologians in an attempt to stimulate discourse in the development of an authentic African ecotheology. Here, however, I merely sketch a few characteristic Old Testament themes as they featured in earth-keeping discussions and tree-planting sermons.

The creation story and Adam's ecological responsibility in the garden of Eden were virtually always given prominence in green sermons. Textual interpretation and contextual application varied considerably, yet common traits were noticeable. Sauro Masoro's sermons were classic examples of the straightforward manner in which humanity's earliest perception of ecological stewardship as portrayed in

Genesis 2 was directly linked to AAEC preoccupation with trees. Once, preaching in the shade of a *muchakata* tree, the Zionist minister proclaimed that the people present and the tree under which they were sitting were one. Without the tree the people would not be able to breathe, a fact which he considered to be underscored by God's first creating the garden of Eden as a necessary condition for living before he could create a human being to inhabit it. God's act of creation, therefore, implied total interdependence between humans and vegetation, something which was not sufficiently recognized in the past. Genesis 2:9 was given immediate relevance when he insisted that "the *muchakata* fruit we eat from this tree you see here, is medicine which heals us. The *matamba* fruit we eat is medicine which heals us. Even if we eat a mango, guava, or orange, it is still healing medicine to us."

In this AIC "garden of Eden" theology, Adam and Eve did not figure as the crown of creation, or as rulers over nature, but as the equals of animals and birds and fully identified with their life. In conversation, Zionist Bishop Machokoto even went so far as to say that "human beings were created for the purpose of caring for all of creation," thus interpreting the meaning of human existence basically in terms of ecological stewardship, service to creation. Seen thus, there is a joyous relationship between humans and plant life, which Machokoto described in terms of communication and mutual respect. The trees being addressed as brothers and sisters by tree-planters during tree-planting ceremonies had acquired a sense of dignity and value, because they were no longer objects of mindless destruction by humans.

In his sermon at Shonganiso, Rev. Mandondo of the African Reformed Church embroidered on the same theme along slightly different lines. God featured as the first "tree-planter":

> He made the trees his children. We human beings, in our turn, are the inheritors of this garden, this kingdom of God consisting of trees and animals. Inheriting this kingdom means that we are responsible for the continuation of the work God started. We say that as Christians we are the inheritors belonging to God. If we are serious about this claim, it means that we, too, are children of God and as such have to proceed with the task of planting trees and taking care of living things. Genuine inheritors are stewards of the land.

Here, once again, plant life and human beings featured as equals. In fact, the kingdom of God was portrayed as starting with the garden of Eden in which trees and (by implication) animals were all "God's children." Then came humans, who happened also to be God's children. They inherited rather than reigned over God's kingdom, and their very inheritance qualified them as stewards who gave, restored, and protected rather than took or invaded the life of God's creatures on this earth. In this earth-keeper's view the imagery of God's kingdom was not spiritualized in the inner world of individual believers, but was concretely observable in all of nature. Far from the triumphalist attitude of the technological age, which purports to conquer and reign over nature, the appropriate position of God's earth-keepers was one of humble respect for the fellow "citizens" of God's kingdom, the trees and the animals.

Another feature of the earth-keeper's garden of Eden theology which was gaining prominence was that Adam's original sin obtained contextualized African ecological connotations. Bishop Farawo suggested, for instance, that because God planted the first trees of creation, he was and remained particularly jealous and protective of all his trees. In a near pantheistic, immanentist perception of the presence of God's Spirit in trees, Farawo literally considered God to be hurt and anguished whenever trees are felled. It is this love of God for his creation which Adam disregarded and offended when he first sinned against God:

> When there was harmony between God and Adam, God was happy to observe the wellbeing of his animals and trees. But when Adam sinned everything was spoilt. *Mwari's disappointment caused him to withdraw and become remote.* Even today we are still far away from Mwari because we sin against him much more grievously than Adam did. Take, for example, Masvingo province, especially Chivi district. The land is barren because all God's trees were felled. So God is absent.

In Farawo's views one was struck by the parallel with traditional African creation myths, which invariably featured the theme of the creator-God's anger and withdrawal from creation because of some mishap or human misbehavior. To the present-day earth-keeper the cardinal sin against God was disrespect for his presence in nature and mindless provocation of his protective jealousy of his forests—contemporary symbolic extensions of the garden of Eden—as evidenced by the deforested regions of the Chivi district, where Farawo himself led the church's war of the trees. The bishop did not judge and reject Adam's sin, moreover, but identified with it, giving it particular poignancy in a rural context where environmental destruction reflected God's anguish and withdrawal. Absolution and deliverance, it seemed, related directly to restoring Mwari's creation, thus restoring harmony and closeness between creator and human beings.

Adam's sin also became the focal point in defining the earth-keeping church's task of restoring creation. Zionist Bishop Nhongo illustrated this point as follows:

> In the garden of Eden God only forbade Adam to eat form the tree of life. But the snake got into that tree and tempted Adam to transgress Mwari's law. As a result of Adam's sin the garden of Eden was destroyed. Likewise, nature was destroyed by greedy human beings over the entire earth. ZIRRCON and the church's objective is to restore the earth to what it was originally in Adam's time. The entire land is to be clad in green vegetation once again and wildlife will abound.

Somewhat naïve as this view may seem in continental or global perspective, it provides a theologically understandable point of departure in the world of dispossessed peasants, where planting and nurturing a single woodlot represents a tangible, meaningful signpost of a better future. Once again the common guilt arising from Adam's transgression and subsequently compounded by the sins of all humanity was asserted—hence the somewhat utopian ideal of restoring creation to its original unspoilt state.

Next to Adam the most frequently mentioned Old Testament figures were Noah and Moses, both of them liberators and "men of God" in their own right. What virtually amounted to a new Africanized mythology was developed around Noah's ecologically salvific work. His ark, in the earth-keeper's view, contained *all* animals and seedlings of *all* plants. ZIRRCON or the AAEC churches were likened to Noah's ark in that they became the protectors of Mwari's creation in the deluge of environmental destruction. The uniqueness of Moses's call to liberate Israel was seen to derive from the fact that God addressed him from a burning bush and gave him a wooden staff as symbolic affirmation of his task. Great importance was attached to God's choice of a tree as a symbol of divine presence and power during Israel's exodus.

In some interpretations God's presence in a tree signified equal status between humans and the rest of creation, the implication being that both were equally dependent on divine liberation from enslavement. Said Rev. Masoro:

> The Israelites complained to Moses that he alone was conversing with Mwari. They, too, wanted to communicate directly with God. So God said: "Let them wash and prepare themselves before we converse." But God did not speak out in the open plains. Whenever he spoke he was hidden in a *denhere* [clump of trees]. And the people had to lie prostrate in his presence. This shows that the tree and the human being are one [in status and need of deliverance].

From the ecological significance of the Old Testament figures Adam, Noah, and Moses, the attention often switched directly to Christ, the true source of all life and guardian of all creation. Union with Christ (John 15), poignantly expressed in the imagery of the vine and its shoots or the *muchakata* tree with its branches laden with fruit (suggestive of the original harmony between God and humans in the garden of Eden), provided life and, to the earth-keeper, empowerment to heal creation. Said Rev. Masoro in concluding his tree-planting oration: "We cannot bear fruit if we are not in Christ, the true vine. If we do not go and ask for trees to plant we shall not have the trees which heal and clean us." Pragmatic as this motive for tree-planting seemed, it nevertheless flowed from new life in Christ. Obedience to Christ inevitably implied Christian responsibility for all creation, especially tree stewardship.

Instead of this story ending with a summary of the features of our earth-keeping church, it stutters to a halt, as did the war of the trees when the drums fell silent. Anguish over a beloved country which seemingly cannot afford or sustain the good news of a revived environment, celebrated by its own people, draws analysis to a premature close. But the story persists . . . lives in our hearts . . . draws us back to those late afternoon eucharistic ceremonies, lest we falter in the telling, in the global village:

> Through the procession of communicants going into the Lord's Acre to plant their seedlings Christ the wounded earthkeeper addresses the eroded soil. It is *He* who, in the tree-planter's dialogue with nature, implores the seedlings to grow strong roots to prevent further soil erosion. It is *His* healing hands which plant the seedlings in the soil, and with them the promise of protection and a new cycle of life.[36]

Finally, as the tired tree-planters return to their meeting place where the sacrament was administered, the healing cycle comes full circle. Women with sick children, elderly people with ailments, young people with problems—all the *maporesanyika* ("healing the land") fighters who seek help—flock to where the prophetic healers are getting ready to tend to the needy. Once again the drums of Zion, of the AAEC, can be heard beating rhythmically; rattles are shaken and the women start swaying in song and dance. The healers shake and speak in tongues to confirm the presence of the Holy Spirit. The spirit of Christ the healer moves in our midst. The late afternoon sun rays beaming through the flimsy leaves of seedlings in the new woodlot and in the expectant eyes of afflicted people tell their own story: Christ's healing of the land and his healing of people join hands, are one!

I stand among the healers, praying and laying on hands, sprinkling people with holy water, blessing newly filled bottles of water, healing and being healed. I have never claimed any healing vocation or powers, but I participate never-theless, having learned about diagnosis of illness in terms of African worldviews and about the value of the AIC symbols of Christ's healing and protective power. I am no longer surprised at hearing patients answer affirmatively when I ask questions about vengeful spirits, spoiled relations, or *uroyi* ("wizardry"). Feeling the straining body in a state of possession or the heat of witchcraft medicine in a victim's system no longer disturbs me. Every time I am thus engaged the nagging questions of my rational mind recede, and I find quiet and peace in the doing. Indeed, as one of the villagers observed: "We have come to be healed, together with the land."[37]

In front of us is a sea of black heads, faces with hopeful eyes, eyes full of pain, fear, rejection, or joy. Hands, black and white, are laid on those heads in prayer, conveying a message of reconciliation and wholeness. They are the hands of icons, illuminating the presence, empathy, love of Jesus. They are wounded hands, suffering hands, still dusty from planting seedlings in the soil. In the dying sun they are the reassuring hands of the Lord himself, holding a promise of new life and resurrection—where all evil ceases.

NOTES

1. Marthinus L. Daneel, *African Earthkeepers*, vol. 2: *Environmental Mission and Liberation in Christian Perspective* (Pretoria: University of South Africa Press, 1999), 1–2. The story of Zimbabwe's African Earthkeepers had been told and retold in various publications. For this paper I made use mainly of the materials presented in the first two chapters of my *African Earthkeepers*, vol. 2. The text has been abridged in places, summarized where necessary, and in a number of instances reproduced with only minor editorial changes. Despite the addition of a few new insights there is no pretence here of adding an entirely new contribution to my earth-keeping narrations of the past. Nevertheless, I take pride in retelling this story, if not so much for its academic value then at least, on behalf of the Zimbabwean earth-keepers, for its inspirational challenge in the contexts of world church and global village.

2. ZIRRCON was the institutionalized and extended version of my field research team. Founded in 1984, this body, which basically at the time was engaged in studying the role of religion in *chimurenga* (the Zimbabwean liberation struggle), took responsibility for the initiation and development of two, religiously distinct, earth-keeping movements.

3. AZTREC began in 1988 as the Association of Zimbabwean Spirit Mediums and was later renamed the Association of Zimbabwean Traditionalist Ecologists. Composed of chiefs, spirit mediums, headmen, tribal elders, and ex-combatants of *chimurenga*, its stronghold was the Masvingo Province.

4. AAEC was officially formed in 1991 at the request of AIC members who had participated in tree-planting activities since the movement's inception. For a full description of the organization and constitutional development of AZTREC and AAEC, see M. L. Daneel, *African Earthkeepers*, vol. 1: *Interfaith Mission in Earth-care* (Pretoria: University of South Africa Press, 1998), chap. 3.

5. Marthinus L. Daneel, *Quest for Belonging: Introduction to a Study of African Independent Churches* (2nd ed.; Gweru: Mambo, 1991), 38–42.

6. Rosemary R. Ruether, *Women Healing Earth: Third World Women on Ecology, Feminism, and Religion* (Maryknoll, NY: Orbis, 1996), 117–18. Ruether claimed that "the most important mobilization of the churches for environmental protection in South Africa has taken place in Zimbabwe. . . . The AAEC," she said, "is a fascinating example of the symbiosis of Christian and African cosmology applied to practical restoration of the environment."

7. Fashole-Luke saw the AICs as providing the "raw material" for African theology, and Carr found that African theology came to life in the music, song, prayers, sacraments, liturgy, church structures, and community life of the Independent Churches. See E. W. Fashole-Luke, "The Quest for African Christian Theologies," 144, 148; and Burgess Carr, "The Relation of Union to Mission," 162—both in *Mission Trends No. 3: Third World Theologies* (ed. Gerald H. Anderson and Thomas F. Stranshy; New York: Paulist Press/ Grand Rapids: Eerdmans, 1976).

8. See, for instance, Samson K. Gitau, *The Environmental Crisis: A Challenge for African Christianity* (Nairobi: Acton, 2000).

9. As founder of the Zimbabwean Earthkeeper's movement and director of ZIRRCON, I have admittedly influenced the nature and interpretation of its religioecological programs. Yet, my contribution was more that of stimulating motivation for

environmental reform and providing financial empowerment through fundraising than attempting to provide a theological blueprint for all activities. Instead, I encouraged local initiative and contextualization. Consequently my own proposals, when accepted, tended to be absorbed and to be creatively inculturated by the key figures of the AICs involved, whereby there was no question of the imposition from above of stereotype ecological models.

10. For a discussion of the viewpoints of observers such as Sundkler, Balandier, Andersson, and Mair, who basically present the AICs as sociopolitical protest movements, see Daneel, *Quest for Belonging*, 69, 73–75. David Barrett, in turn, emphasizes religious protest as the root cause of Independentism, due to the failure of Christian missions to demonstrate consistently the biblical concept of love; see *Schism and Renewal in Africa: An Analysis of Six-thousand Contemporary Religious Movements* (London: Oxford University Press, 1968), 156.

11. I have indicated that the characterization of AICs predominantly "protest movements" is flawed and that AIC missionaries on the whole are inspired in their evangelistic outreach by the gospel. Hence the central hypothesis in all my work on AIC growth that church expansion—despite recognizable features of reaction to aggravating sociopolitical and religious conditions—took place largely as a result of Africanized missionary strategies and praxis; see M. L. Daneel, *Old and New in Southern Shona Independent Churches*, vol. 2: *Church Growth: Causative Factors and Recruitment Techniques* (The Hague: Mouton, 1974), chaps. 2 and 5; idem, "Missionary Outreach in African Independent Churches," *Missionalia* 8.3 (1980); idem, ed., *African Christian Outreach*, vol. 1: *African Initiated Churches* (Pretoria: South African Missiological Society, 2001); and idem, "AICs in Southern Africa: Protest Movements or Missionary Churches," in *Christianity Reborn: The Global Expansion of Evangelisation in the Twentieth Century* (ed. Donald Lewis; Grand Rapids: Eerdmans, 2004).

12. This section depends on Daneel, *African Earthkeepers*, 2.37–48.

13. Daneel, *Old and New in Southern Shona Independent Church*, 2: chap. 3.

14. Mafuranhunzi Gumbo, *Guerrilla Snuff* (Harare: Baobab, 1995).

15. Daneel, *African Earthkeepers*, 2.39.

16. Ibid., 42.

17. Ibid., 43 (emphasis added).

18. For a full description of AZTREC's formation, see ibid., 1.81–87.

19. Ibid., 2.46–47.

20. Ibid., 67.

21. Ibid., 75–77.

22. Gelfand 1959: 153; and Daneel 1971: 133–40.

23. Marthinus L. Daneel, *Fambidzano: Ecumenical Movement of Zimbabwean Independent Churches* (Gweru: Mambo, 1989), 30.

24. Daneel, *African Earthkeepers*, 1.101–2.

25. Ibid., 109.

26. For a discussion of the conflict between Bishop Mutendi of the Zion Christian Church and the Mwari cult officials, see Marthinus L. Daneel, *The God of the Matopo Hills: An Essay on the Mwari Cult in Rhodesia* (The Hague: Mouton, 1970), 64–71.

27. Jan A. M. Jongeneel, *Philosophy, Science, and Theology of Mission in the Nineteenth and Twentieth Centuries: A Missiological Encyclopedia*, vol. 1 (New York: Lang,

1995); in Richard S. Darr, "Protestant Missions and Earthkeeping in Southern Africa, 1817–2000" (PhD thesis, Boston University, School of Theology, 2005), 285.

28. Andrew J. Kirk, *What Is Mission? Theological Explorations* (Minneapolis: Fortress, 2000), 164–83; and J. J. Kritzinger, "Mission and the Liberation of Creation: A Critical Dialogue with M. L. Daneel," *Missionalia* 20 (Aug. 1992); referred to by Darr, "Protestant Missions and Earthkeeping," 285.

29. Kirk, *What Is Mission?* 166.

30. Darr, "Protestant Missions and Earthkeeping," 285.

31. Interviews with Bishop Nhongo and Samuel Nhongo, quoted in M. L. Daneel, "Earthkeeping Churches at the African Grassroots," in *Christianity and Ecology* (ed. Dieter T. Hessel and Rosemary Radford Ruether; Cambridge: Center for the Study of World Religions, Harvard Divinity School, 2000), 531.

32. Ibid., 532.

33. Ibid., 537.

34. This section is a reproduction, with only minor alterations, of Daneel, *African Earthkeepers*, 2.59–65.

35. Ibid., chaps. 3–5.

36. Ibid., 202.

37. Ibid., 203.

CHAPTER 24

THE SCIENTIST
AND THE SHEPHERD

The Emergence of Evangelical Environmentalism

CALVIN B. DEWITT

Amidst sweeping corporate globalization and relentless pressing of Earth for resources, a shepherd cries, "Care for God's Creation." Another voice, muffled by powers-that-be, decries global warming and species extinctions. The first is a cry of the Shepherd, the second a cry of the Scientist.

As I write these words on this wonderful morning on the great marsh, autumn is all ablaze, the cranes are bugling above layered ancient peats, geese are calling overhead as they wing their way toward the lake. Handel's *Messiah* is playing in the background. A few minutes earlier I had walked outside to scan these magnificent wetlands with wondering eyes, walked back into the house, and wrote the words on top of this page: "The Scientist and the Shepherd."

I am humming Handel's tune as I write: "*For we like sheep have gone astray... have gone astraaaaay... have gone astraaaaay.... We have turn-ed, we have turn-ed, every-one to his own way.*" As I write, the *Messiah* sinks in: "For we like sheep have gone astray." "The Lord has laid on him the iniquities of us all." It is an inspirational beginning for my writing today, yet sobering. Classical marsh; classical music—playing remarkably well in the mind's eye and ear on autumn's glorious morning. My prayer for forgiveness and blessing ascends from the great and vibrant wetlands whose peats reach back to times well before Christ's sojourn on earth.

Why is evangelical environmentalism emerging in our day, we may ask? It has to do with sheep, the scientist, and the shepherd. But this question does not have a handy answer. Instead, the answer will progressively unfold as we trace the development of this movement. The full answer can come only at chapter's end.

Two Great Needs

In building the movement called evangelical environmentalism,[1] there were two great needs, and both have been increasingly met over its twenty-five years of development: First, was the need to build a creation theology, and more specifically a creation-care theology. This had to be rooted in a thorough and scholarly biblical theology. Most importantly, it also had to be coherent with an integrated scientific understanding of the biosphere and the world. This was achieved by an academy of evangelical scientists, ethicists, and theologians who grew in numbers and publications from 1980 to the present and, becoming aware of itself as such, officially became the Academy of Evangelical Scientists and Ethicists in 2005.

Second, was the need to find ways to put this evangelical creation-care theology into practice. Already, many organizations were engaging the world in practical ways: World Vision and MAP International, for example. Needed beyond these practical evangelical agencies, however, was that of networking evangelical practitioners around matters of creation-care. Various work to meet this need was attempted in the late 1980s and early 1990s, with the World Evangelical Fellowship creating the first evangelical environmental network—the International Evangelical Environmental Network (IEEN)—in 1992. This, in turn, led to the formation of the Evangelical Environmental Network (EEN) in 1993.

The formation and development of evangelical environmentalism has a complex history that has been thoroughly studied and reported by David Larsen in his scholarly two-volume dissertation "God's Gardeners." In addition to what Larsen presents in his thorough and comprehensive treatise is the important observation that the emergence of this movement, as with many other movements, was complex and complicated. The work of the gospel, by definition, is the work of self-giving and service. And what this movement continually found was that it was difficult both to be in control and simultaneously serve in Christ-like humility.

This problem of achieving power coupled with need to be humble servants, of course, is why Jesus is such a great model for the evangelical environmental movement. Jesus did not control by organizational hierarchy or by corporate power. Christ's control came through the hearts of people who received his message of passion, compassion, love, and care, and all of these reside at the core of creation care. Temptations to grasp for power were overcome by the passion for caring—caring for each other and for the whole creation. The sometime attempts at human "grasp" were overcome by God's "gift" (to use Walter Brueggemann's terms),[2] even as the problem of power-seeking versus service-giving will likely surface periodically. The Lord's beatitude, "The meek shall inherit the earth," raised an important question for this movement: How does one make progress when meekness is the quality and behavior we seek? Evangelical environmentalism is attempting the difficult work of doing just that—to achieve power and influence, and yet do so through meekness. This essay is an early one in what hopefully will prove to be a continuing story of the power of meekness, love, and care.

A Sweeping Secularization

The twentieth century experienced a widespread secularization of American society. Many American colleges and universities severed or weakened their religious roots; many separated from the seminaries they once held in partnership; many churches diminished their community services as human needs were increasingly met by insurance policies, retirement plans, and government assistance. Toward the conclusion of the century many mainline denominations declined in membership and also in social and political influence, as graying congregations no longer held the youth that would have succeeded them. Biblical studies in colleges and universities during this period not only were diminished, extinguished, or signed over to seminaries and Bible colleges, but the Bible lost much of its credence in an increasingly secularized and urban America.

Yet there remained a significant population of Americans that continued to take the Bible seriously, maintaining that its study and exposition was vital and necessary for right living on earth and in society. These people and their worshiping congregations were not very much in touch with each other during most of the century. They found themselves pretty much at the fringe of American society, with little influence. Among them were many who in their relative isolation preferred "not joining hands with the world." As the secular society, however, gained greater prominence—even as the Bible was diminished as a rule of faith and practice—these people, facing a common challenge to their biblical worldviews, formed organizations to address common needs and interests. While "agreeing to disagree" on many fundamentals of the Christian faith, they worked together, particularly in relief and development and in Christian higher education.

Faced with a common challenge of secularization and the common need to address major issues, they came together in various ways. Among the organizations they formed were the Association of Evangelical Relief and Development Organizations and the Council of Christian Colleges and Universities. As these people became more visible within the wider society they came to be called evangelicals.

THE MEANING OF *EVANGELICAL* AND *EVANGELICALISM*

In 1974 evangelical scholar Carl F. H. Henry wrote that "evangelical Christians are thus marked by their devotion to the sure Word of the Bible; they are committed to the inspired scriptures as the divine rule of faith and practice. They affirm the fundamental doctrines of the Gospel, including the incarnation and virgin birth of Christ, His sinless life, substitutionary atonement, and bodily resurrection as the ground of God's forgiveness of sinners, justification by faith alone, and the spiritual regeneration of all who trust in the redemptive work of Jesus Christ."[3]

Although evangelicals have not generally identified themselves as environmentalists or with environmentalism, their growing work on "caring for creation" has been given the designation *evangelical environmentalism* by Larsen and other observers of their emergence as a political and environmental force in American society. While they first were largely identified with personal and family ethical issues, their increasing identification with environmental issues makes it appropriate and important to address evangelical environmentalism as a present and future force in shaping environmental policy and the environmental movement.

The focus of evangelicals on the Bible, including extensive biblical teachings on caring for the earth and its creatures, is important in our assessment. As they hold firmly to the belief that the Bible and its sixty-six canonical books are vital for faith and practice and the authoritative source for defining how rightly to live on earth, evangelicals also believe that the good news of the Bible should not be selfishly kept. Instead, the good news must be proclaimed, with this belief reflecting the Greek derivation of "evangelical," *euangelion*, from *eu* ("good") and *angelos* ("a messenger or bearer of news"). This news is *good* and it is *good* for every creature. The reason for this news being good is because it brings restoration and reconciliation of all things, countering and undoing human-wrought degradation.

The personal harbinger of this good news is the lord of creation who brings "joy to the world" and "makes his blessings flow far as the curse is found."[4] This is the biblical teaching in the New Testament canon as it asserts that the reach of this good news is as great and wide as is human-wrought degradation. Evangelicals "repeat the sounding joy" because the restorative reach of the second Adam (Jesus Christ) is as great and wide as the damaging reach of the first Adam (see 1 Cor. 15:45).

At the heart of this good news as proclaimed in evangelicalism is salvation. Salvation is a saving from degradation—a saving offered to those who are committed to follow in the footsteps of Jesus, the savior and reconciler of all things (*ta panta*). Believers of this good news bring joyful service to humanity, to every creature, and to all creation. Telling this good news is seen as a worthy service that works to fulfill the eager expectation of the whole creation for the coming of God's children. Evangelicals then, are people who see themselves as bringers of good news in the footsteps of Jesus, are serious about honestly describing the way that things really are, and are visionary toward the way things ought to be. They are followers of Jesus who seek religiously to bring food and the means of its production to the hungry, compassionate care and the means of healing to the sick, and thoughtful restoration to a degraded creation. In doing this they join with other people who are committed to making things right in society and the creation.[5]

Evangelical Relation to Human Authority

While largely identified with specific congregations, churches, and denominations, evangelicalism can be found in individuals, groups, and congregations in nearly every denomination across America, including mainline Protestants and Roman Catholics.

A distinctive feature of evangelicalism is that it widely distrusts human authority and ecclesiastical hierarchy. This distrust is reflected in congregational and institutional polity, with many congregations operating as independent entities and others but loosely organized into associations or joined together in denominations with limited hierarchy. Many in America are associated with the National Association of Evangelicals, but not all.

In evangelicalism there usually is no human "word from above" spoken by bishops or prelates. Instead there is *the* word—the Bible. The way this takes shape in practice is serious and continuing Bible study, individually and in fellowship with others. The purpose of Bible study extends beyond edification to discovering biblical teachings and applying these to personal lives, society, and the rest of creation.

Fear and mistrust of earthly authority sometimes have separated evangelicals from authoritative sources of knowledge: knowledge of biblical teachings on environmental stewardship and environmental degradation, knowledge of ecology and other natural sciences, and sometimes even knowledge of biblical material in their early Hebrew and Greek texts. As their knowledge and understanding of these subjects increases, however, through the growth and interconnections of their educational and relief and development institutions, evangelicals have become a powerful influence.

Evangelical Sources for Knowledge and Praxis

Evangelical growth in knowledge and understanding is published and promoted by publications such as *Christianity Today* and by their numerous and growing evangelical colleges, universities, and seminaries that convey knowledge through professional and popular writing, educate pastors and teachers, and prepare social, scientific, medical, legal, business, and environmental professionals. Among evangelical colleges and universities, many have gained depth and influence in the study of creation and Christian environmental stewardship through partnership with the Au Sable Institute of Environmental Studies, whose participating evangelical colleges and universities had grown to sixty in 2005.

Putting evangelical faith and belief into practice is expedited worldwide through evangelical organizations and agencies dedicated to medical services, disaster relief, development work, food production, environmental restoration, and word and deed ministries. The Association of Evangelical Relief and Development Organizations is a principal networking organization that develops knowledge and understanding of human and environmental needs around the world. In turn, the Association of Evangelical Relief and Development Organizations is linked with the EEN, which networks across institutions and agencies to mobilize and serve evangelical leaders, churches, and denominations in environmental stewardship and caring for creation (see list of EEN participants below). Among the accomplishments of the EEN was a successful effort in 1996 that prevented the U.S. Congress from weakening the Endangered Species Act and produced *An Evangelical Declaration on the Care of Creation* that came to be highly regarded and later was supported with an authoritative book-length commentary.[6]

Evangelicalism's commitment to taking the Bible seriously has important implications for its contributions to understanding God, abundant human life, and caring for creation. Since evangelicals measure their faith and practice against biblical standards and believe in continuous adjustment, correction, and conversion in response to their falling short of biblical requirements, their growing knowledge base and networks in education and responsible practice substantially increase their strength for addressing care for creation, environmental degradation, and ecological sustainability.

Two-Books Theology and
Sola Scriptura

Contrary to what one might first expect, evangelical adherence to the centrality of the Bible, often identified with the phrase *sola Scriptura*, does not close the window on learning from society and creation. On the contrary, it is the window through which the world and God's creation are seen. This window is a biblical window—in the book of God's word—that opens the book of God's works—God's creation. More specifically, the biblical window is Romans 1:20, teaching that everyone is left without excuse from knowing God's divinity and everlasting power through the testimony of God's creation. This adherence both to the book of God's word and the book of God's works is sometimes referred to as two-books theology—a theology that reveals God both through the Bible and creation. The Confession of Faith of 1651, for example, presents "The Means by Which we Know God" as follows:

> We know him by two means:
>   First, by the creation, preservation, and government of the universe, since
> that universe is before our eyes like a beautiful book in which all creatures,

great and small, are as letters [Lettres] to make us ponder the invisible things
of God: his eternal power and his divinity, as the apostle Paul says in
Romans 1:20.

All these things are enough to convict men and to leave them without
excuse.

Second, he makes himself known to us more openly by his holy and divine
Word, as much as we need in this life, for his glory and for the salvation of
his own.[7]

The consequences of learning both from God's word and God's works are
profound. It means, for example, that locating a village on the slope of a volcano
does not prevent volcanic eruptions. Settling of human beings in river floodplains
will not prevent floods. Neither will God render powerless in our bodies the
poisons we design to kill other creatures, like pesticides. At base in evangelicalism,
knowledge of God's word, accompanied by knowledge of God's works, leads to
interfacing with the rest of creation wisely.[8] Both books must be read, and they
must be read coherently. Failure to do so will disorient people physically, spiritually, and morally and will ultimately bring disaster; and it will diminish people's
and the rest of creation's capacity to praise and worship God.

Evangelical Religion: A Passion
to Live Rightly before God on Earth

In setting forth the groundwork for understanding evangelical environmentalism
as a significant religious, ethical, and political voice, it is important to reflect on
the meaning of the word *religion* in ways that are consistent with a biblical
understanding of the world and life. The definition of Wayne C. Booth comes the
closest, I believe:

> Religion is the passion, or the desire, both to live right—not just to live but to live
> right—and to spread right living, both desires conceived as responses to some
> sort of cosmic demand—that is, to a demand made to us by the way things are,
> by the way the world is, by the nature of Nature (as some would say) or by God
> himself (as explicitly religious people put it).[9]

For evangelicals this definition can be modified to read:

> Religion is the passion, or the desire, both to live right—not just to live but to live
> right—and to spread right living, both desires conceived as responses to a cosmic

demand made to us by the way things are created and ordained by God and as responses to God's word and its expectation that we will, as image-bearers, image God's love for the world in our caring for each other and the rest of creation.

This means, in more general terms, that understanding of both science and ethics is required for a proper understanding of living rightly in the world (praxis). Three things, then, need to be tied together: science, ethics, and praxis. These three need to be ligated to form a science-ethics-praxis triad, with each corner accompanied by a corresponding question, illustrated as follows:

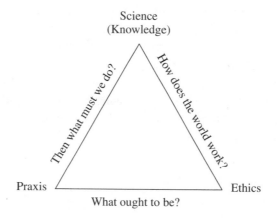

Science, ethics, and praxis are tied together in this diagram by three ligaments. When any ligament is broken or torn, one or all corners are freed or distanced from the constraints and contributions of the others. Such brokenness may produce a science unconstrained by ethics that, for example, may degrade human relationships with the rest of creation. It may produce an ethics that is uniformed about the material world that, for example, might advocate removing dead wood from a forest, thereby denying the truth that death and decay of trees is necessary for forest soil-building. It may produce a praxis cut free from science and ethics that may, for example, allow residential and commercial developments on barrier islands, river floodplains, geological fault lines, or soil types defined by soil science as "liquid when wet." All three—science, ethics, and praxis—need to be connected. Where the ligaments have been broken they need to be religated. Religation of science, ethics, and praxis, in this view, is necessary for right living on earth, and it would seem that a principal responsibility of religion and religious people is to keep these three bound together into a wholesome and fulfilling relationship.

Religating of science, ethics, and praxis has been a particularly compelling task for evangelical Christianity and is vital to the effectiveness of evangelical environmentalism. This religation—restoring and strengthening the connections among science, ethics, and praxis in caring for creation—is the principal reason for the emergence of an academic evangelical environmentalism.

PRE-1980 ROOTS OF ACADEMIC
EVANGELICAL ENVIRONMENTALISM

In 1967 medieval historian Lynn White Jr. wrote what has become the most frequently reprinted article in the journal *Science*: "The Historical Roots of Our Ecologic Crisis." In it he laid the blame for the environmental crisis largely on the Judeo-Christian religion, with this centered primarily on the dominion passage of Genesis 1:28, where he interpreted the word *dominion* as meaning "domination." White's claim, and his failure to put this passage into its broader biblical stewardship context, generated an immediate and dramatic response, including articles and papers by evangelicals who advanced the biblical concept of stewardship as a better interpretation of this and other related biblical passages. Among these was evangelical Francis Schaeffer and his book *Pollution and the Death of Man*.[10] Its publication in 1970 coincided with the national environmental movement that swept across America, much of which stemmed from the response of the American public and the Congress to Rachel Carson's *Silent Spring*, published in 1962.[11]

Along with Schaeffer, evangelicals joined the environmental movement in America that created widespread environmental awareness and extensive national legislative work that in the 1970s produced the Clean Air Act, the Water Quality Improvement Act, the Water Pollution and Control Act Amendments, the Resource Recovery Act, the Resource Conservation and Recovery Act, the Toxic Substances Control Act, the Occupational Safety and Health Act, the Federal Environmental Pesticide Control Act, the Endangered Species Act, the Safe Drinking Water Act, the Federal Land Policy and Management Act, and the Surface Mining Control and Reclamation Act. Early among these was the National Environmental Policy Act, signed into law on 1 January 1970, the same year that saw the first Earth Day. Its stated purposes were "to declare a national policy which will encourage productive and enjoyable harmony between man and his environment; to promote efforts which will prevent or eliminate damage to the environment and biosphere and stimulate the health and welfare of man; to enrich the understanding of the ecological systems and natural resources important to the Nation; and to establish a Council on Environmental Quality." Evangelicals joined in, and the evangelical magazine *Christianity Today* published a stream of editorials and articles on the Christian role in caring for the environment.

As the decade progressed, American culture turned to other issues, and so did the evangelicals. Larsen writes, "Like the rest of the country, evangelicals were swept up into the outpouring of environmental concern that attended Earth Day 1970, though they often professed to be more concerned about 'moral pollution' and tended to proffer uniquely evangelical solutions, chiefly conversion.... Moreover, conservative evangelicals tended to disregard environmental threats either out of a preoccupation with the Second Coming or a belief that these threats were exaggerated."[12]

Development and Emergence of Academic Evangelical Environmentalism

The year 1980 marked a number of events that make it a milestone for academic evangelical environmentalism. Principal among these was the publication of a multiauthored book, *Earthkeeping*, as a product of a full year of interdisciplinary study by a team of evangelical scholars.[13] The authors of *Earthkeeping* had been brought together in 1977 as the first group of fellows of the newly created Calvin Center for Christian Scholarship to work in a full year of discussion and study on the topic "Christian Stewardship and Natural Resources." The center was established by Calvin College "to promote rigorous, creative, and articulately Christian scholarship which is addressed to the solution of important theoretical and practical issues" and that would be "focussed on areas of life and thought in which it is reasonable to expect that a distinctively Christian position can be worked out."[14] The Calvin Center governing board selected the topic as one that would meet the objectives of the new center. This work, published in *Earthkeeping* and its revised edition of 1991, put into place much of the academic groundwork for evangelical environmentalism.

A second event extended the participants in the academy of scholars that grew out of the Calvin Center group, this being the first Au Sable Forum held in 1980. Three members of the Calvin Center fellows were joined by twenty-three other scholars, among them Wesley Granberg-Michaelson (who would organize the Au Sable Forum that produced *Tending the Garden* in 1989) and Orin Gelderloos (who would become chair of the Au Sable board of trustees). This forum began what would become a long series of academic meetings, the Au Sable Fora, most of which produced books that contributed to the development of a creation-care theology. In the conclusion of the book produced from the first forum, Gelderloos wrote prophetically, "The growth of membership in the conservative churches and associated religious colleges may be a beginning which will lead to a significant influence of these institutions on environmental and societal affairs if the revival does not become a narcissistic movement."[15]

A third event established the program of environmental stewardship by the newly established Au Sable Institute, whose participating colleges grew in number to eighteen from 1980 to 1985 and in successive five-year milestones to twenty-seven, thirty-eight, forty-five, and sixty participating colleges in 2005, each college with a scientist member of their faculty as an official representative. These scientists then met annually for professional development in environmental stewardship and to guide and direct the institute's academic program. During the same period, from 1980 to 2005, the institute faculty and fellows grew from twelve to seventy professors, sixty-eight of whom are scientists and two are ethicists, and thirty-five of which teach at the institute in any given year. Accompanying this growth, many of these scientists, joined by students returning to their colleges from Au Sable, formed environmental-studies programs at their home campuses, all with significant creation-care content.

In 2005 the academic evangelical environmentalism that had been growing steadily from its beginnings in 1980 formally organized into the Academy of Evangelical Scientists and Ethicists, made up of scientists and ethicists who have earned the highest degrees in their fields.[16] A statement that these academics hold in common is the following:

> Together we believe we have the privilege and responsibility to care for God's Creation, and to do so recognize in our academic work and our lives that we are following Jesus Christ by whom all things were created, and held together, and are reconciled to God in full accord with Colossians 1:15–20. Moreover, believing on Jesus and recognizing his love and work as a servant, we seek also to be servants of him, the triune God, and God's whole Creation.
>
> Our service in the academy, as evangelicals, ethicists, and responsible stewards includes a faith commitment to protect and care for the totality of God's Creation. There is a richness, depth, and beauty in Creation that goes well beyond any of our abilities to express. There also is a richness, depth, and beauty in Christ's work—from the foundations of Creation—also inexpressible. Yet it is from that richness that we have come to be God's stewards with love, devotion, and dedication.
>
> This richness in Creation and in Christ's redeeming work is something to which we as evangelical scientists and ethicists can speak and about which we can teach and provide serving leadership. Beyond our ability to convey this richness, we also believe we must exercise that ability to the glory of God.[17]

DEVELOPMENT AND EMERGENCE OF THE EVANGELICAL ENVIRONMENTAL NETWORK

The EEN had its origin in an International Consultation held on 26–31 August 1992 and cosponsored by Au Sable Institute as one of the Au Sable Fora. The

World Evangelical Fellowship's Unit on Ethics and Society had formed a Committee on Evangelical Christianity and the Environment that, in partnership with Au Sable, invited sixty people from eight countries and five continents to meet on the topic "Evangelicals and the Environment."[18] This proved to be a seminal forum, with thirteen contributions coming from it that were published in the *Evangelical Review of Theology* in 1993[19] and with its decision to create the IEEN to "disseminate information" and "share insights and experiences among the worldwide evangelical family in the care for the creation." The EEN is a fellowship of believers who...

- Declare the lordship of Christ over all creation. He is the firstborn over all creation. For by him all things were created. All things were created by him and for him. He is before all things, and in him all things hold together (Col. 1:15b, 16a, c, 17). These believers...
- Deepen their walk with the Lord and the life of their churches through joy-filled worship, Bible study on the topics of creation's care, and prayer that God's will "be done on earth as it is in heaven" (Matt. 6:10).
- Show the compassion of Christ for people who suffer from creation's destruction (Prov. 14:31).
- Demolish strongholds of sin that tarnish the glory and integrity of God's good creation (2 Cor. 10:4–5).
- Build our Lord's kingdom by active service to restore and renew the works of his hands (Matt. 6:33; Eph. 2:10).
- Share the gospel with those who do not know that Jesus Christ is the ultimate hope for creation groaning under our sin and the only hope for our own souls (Rom. 8:19–21; Col. 1:20, 27).

The IEEN, formed well before the launch of the U.S. activity, led to formation of the EEN in America. It also resulted in the U.K. EEN and the *Creation Care* newsletter,[20] that later becoming the title of the EEN periodical published in the United States. In preparing the way, the World Evangelical Fellowship in its formation of the IEEN, concluded in its summarizing report that the Christian community "must dare to proclaim the full truth about the environmental crisis in the face of powerful persons, pressures and institutions which profit from concealing the truth."[21]

It also declared that "such recognition of hard truths is a first step toward the freedom for which creation waits." Of particular significance to the formation of both the IEEN and EEN was the World Evangelical Fellowship's conclusion that "Christians need to form and join environmental organizations that apply explicitly Christian principles to environmental problems."[22]

With its strong perspective on global environmental justice, the 1992 World Evangelical Fellowship/Au Sable Forum concluded that "the Christian community must be willing to identify and condemn social and institutionalized evil, especially when it becomes embedded in systems. It should propose solutions which both seek to reform and (if necessary) replace creation-harming institutions and practices."[23]

Addressing the appearance and programs of churches, the forum also concluded that "churches should seek to develop as creation-awareness centers in order to exemplify principles of stewardship for their members and communities, and to express both delight in and care for creation in their worship and celebration." And, "they should particularly aim to produce curricula and programs which encourage knowledge and care of creation." "In particular, Christian colleges and seminaries should provide teaching in this area. The church's goal should be the growth of earthkeepers, both in the habits of everyday life, and in the provision of leadership for the care of creation."[24]

The derivative EEN held its first meeting at Au Sable in the summer of 1993, soon after the Rio de Janeiro Earth Summit. Subsequent to the Au Sable meeting, Loren Wilkinson prepared a draft entitled *Evangelical Declaration on the Care of Creation* that became basis for a daylong meeting of several American evangelical leaders, among them leading evangelical scholar Kenneth Kantzer and InterVarsity Christian Fellowship president Steve Hayner, that produced the final document.[25] The resulting declaration was published the following year in *Christianity Today*, a statement signed by hundreds of evangelical leaders as well as a statement that became the common confession of the many partners that make up the EEN (see list of EEN participants below).

The EEN works from the *Evangelical Declaration on the Care of Creation* to define and summarize the evangelical framework out of which creation care emanates and also demonstrates the depth and passion of the evangelical environmentalism that is emerging in America. "Because we await the time when even the groaning creation will be restored to wholeness, we commit ourselves to work vigorously to protect and heal that creation for the honor and glory of the Creator," the declaration states with confident religious conviction.[26]

Being extremely careful to avoid pantheism or idolization of creation—both of which evangelicals see as highly problematic—they exercise "creation care" under God, who both transcends and is immanent in creation: "Our creating God is prior to and other than creation, yet intimately involved with it, upholding each thing in its freedom, and all things in relationships of intricate complexity. God is transcendent, while lovingly sustaining each creature; and immanent, while wholly other than creation and not to be confused with it." They also boldly confess that degradation of creation by human society and their pursuit of gain is sinful: "The earthly result of human sin has been a perverted stewardship, a patchwork of garden and wasteland in which the waste is increasing." They extend their concern for the creation to the poor and downgraded people of the earth, whom they see as fully part of God's creation. Justice to creation and to people is intertwined, and "one consequence of our misuse of the earth is an unjust denial of God's created bounty to other human beings, both now and in the future."

What is the purpose of the EEN? It is "God's purpose," and "God's purpose in Christ is to heal and bring to wholeness not only persons but the entire created order." Citing the great Christ hymn of Colossians (Col. 1:15–20), "for God was pleased to have all his fullness dwell in him, and through him to reconcile to himself all things, whether things on earth or things in heaven, by making peace through his blood shed on the cross," those affirming the declaration follow their namesake in reconciliation and restoration of creation, including its poor and downtrodden.

A highly important development from the declaration was the work of R. J. "Sam" Berry of the University College–London to develop a commentary on the declaration by scientists, ethicists, and theologians from the United Kingdom and elsewhere around the world. His book *The Care of Creation* puts the declaration to the test of highly competent scholars of great substance. Among the many affirmations of the declaration was that of prominent German theologian Jürgen Moltmann: "The *Declaration* gives a fresh perspective on the ecological problems of the modern world, by presenting them through the experience of the healing God. In Christ Jesus we experience not only the forgiveness of our sins but also the healing of our wounds. It is only logical to 'extend Christ's healing to the suffering creation.'" Moltmann added that we must "overcome the hedonism of the consumer society, lest we become blind and numb and careless." In his affirmation of the declaration he observed that it "rightly calls for a rebirth of our relationship to God, to each other and to the community of all God's creatures."[27]

For evangelical environmental missions entrepreneur Peter Harris, the declaration compelled him to face "an uncomfortable paradox." His conclusion is that he is "more and more convinced that the urgent task of changing the way we live as evangelical Christians has to begin with believing differently, and not simply obeying new rules. It is not that we need to adopt an updated legalistic code for the contemporary Christian life: thou shalt recycle thy toothpaste tube, thou shalt not covet thy neighbour's bicycle, and thou shalt make sure everyone else notices that thy yoghurt is organic." Harris writes that he is "puzzled over the causes for current (not historic, incidentally) evangelical indifference to creation as we have encountered it with A Rocha [his organization]," and he concludes that "it has become uncomfortably clear that its roots lie in biblical unbelief. It is not that evangelicals shrink from paying a price in lost comfort for a change of lifestyle. There are many wonderful examples of how the Christian church worldwide lives very sacrificially in the face of human need, and in many places it is a compelling stimulus for social change and redemption." Instead, Harris finds that "the problem is that we do not extend that commitment and concern to the wider creation, nor are we persuaded that God cares about it."[28]

Affirming the declaration, Harris concludes that it gives us "a foundation for a better biblical understanding, and a guide to more biblical living." It challenges "the careless or wilful adoption of human-centered thinking into the heart of the church" as the source of "Christian indifference to creation." He adds that "in its extreme manifestation, salvation becomes a kind of fashion accessory in the main project of our individual self-fulfillment: Gucci for the body, the media for the mind, and Jesus for the soul. Grateful for the crumbs falling from the postmodern table, the church has been tempted to settle happily into the role of provider of 'spiritual' need. In denial of the force of the biblical argument, 'spiritual' becomes equated with 'nonmaterial.'" Observing that "the foundational biblical affirmation about God is that he is the Creator, not only of the human race, but of all life and everything there is," Harris finds that "this places us in a completely different relational context; no longer are we the starting point of our existence, but we are the creation of God, together with not only everyone else on the planet, but all of the universe."[29]

Berry, the editor of this important commentary on the *Evangelical Declaration on the Care of Creation*, concludes with two final points: "As Christians we can and should make common cause with others in the care of creation, not least because the Genesis commands to have dominion and to tend the garden were given to all human beings at our birth as men and women in God's image." Responding to critics who say that "stewardship is an unhelpful concept," he notes that "stewarding is only part of our role; our God-given purpose is worship of the Creator in the company of all created beings (not merely as human beings sometimes meeting in church). God's covenant is with us as creatures in his image; and not only that, but also 'with every living creature' (Gen. 9:10). We share our stewardship with our fellow human beings (whether or not they acknowledge the Creator); we add to their work a recognition and acknowledgment that our work is part of our commitment and worship of the one who made us in his own image, who is Creator of all, and who has reconciled 'all things' to himself through Christ's death on the cross (Col. 1:20)."[30]

"This places the *Declaration* as indisputably 'evangelical,'" writes Berry, "because it proclaims truly 'good news.' This is backed by John Stott's definition of 'evangelical' as incorporating 'the revealing initiative of God the Father, the redeeming work of God the Son, and the transforming ministry of God the Holy Sprit.'" "The secular world sees little more than decay when it views creation; Christians see the same failures, but for them there is also the confident hope that 'the universe itself is to be freed from the shackles of mortality...to enter upon the glorious liberty of the children of God' (Rom. 8:21 REB). Our part is to accept and rejoice in God's work, and to be light for all the world as obedient stewards."[31]

Highly respected evangelical theologian John Stott in his foreword to this commentary on the declaration affirms this by observing that "we human beings find our humanness not only in relation to the Earth...but in relation to God, whom we are to worship; not only to the creation, but especially in relation to the creator. Only then, whatever we do, in word or deed, shall we be able to do it to the glory of God (1 Cor. 10:31)."[32]

EEN Partner Organizations

Focus of Organizations	Organizations
creation care and evangelical action evangelical relief, health, and environment	• Evangelicals for Social Action • Restoring Eden • Target Earth • Association of Evangelical Relief and Development Organizations • Christian Reformed World Relief Committee • Floresta • Habitat for Humanity International • MAP International • Marah International • World Relief • World Vision
denominational agencies and churches	• Baptist General Convention of Texas • Los Angeles Metropolitan Churches
Bible and missions	• International Bible Society • InterVarsity Christian Fellowship • Mission Society for United Methodists • Mission Training International
youth camps and youth missions	• Christian Camping International • Youth with a Mission
science and higher education	• American Scientific Affiliation • Au Sable Institute of Environmental Studies • Council for Christian Colleges and Universities • Zhaniser Institute

THE SCIENTIST AND THE SHEPHERD

We began by asking, Why is evangelical environmentalism emerging in our day? We have seen the answer unfolding during the course of this chapter, and now we can firm it up. Evangelical environmentalism is elicited from the growing realization that human beings, including evangelicals, are coming to be numbered among those who, corporately and individually, have become destroyers of creation.

Earth is crying out—divested by us of its divine wonder. As the shepherd and scientist point to a larger economy whose workings run the whole world and all the life of the biosphere, we know it is an economy under assault. Its fabric of interrelated species is being torn, most of its fisheries have collapsed, microbialization of the oceans is rampant. Moreover, the great stores of carbon and heavy metals that the peatlands and coal swamps sequestered ages ago as troubling excesses from an earlier atmosphere are being opened. They are being forcibly removed from the great stores beneath earth's surface and compelled by us to fuel the fires of progress and release that troubling excess of carbon and heavy metals back into the skies. Most regretfully not only have people reconceived creation as resources, but these are people all of whom are made in the image of God—image-bearers given the gift and capacity to image God's love for the world. Human image-bearers—increasingly also reconceived as resources—are transformed by a lesser economy into consumers of the earth and its greater economy.

But there is the shepherd—announced in the book of God's word—who is beckoning the sheep who have gone astray. As the shepherd seeks to lead, the scientist helps read the lettres, helps the image-bearers once again to behold and wonder, helping us to ponder the works of God, to understand the degradation and destruction, to understand creation care. And a few come forth, out of the great pool of human resources once again to sing the forgotten hymn, "Beautiful Savior, King of Creation."

In concert with religious people around the globe, evangelicals are at the verge of refreshing their awe and wonder for creation, refusing to be consumers of the world, seeking first the kingdom of God, and committing themselves to imaging God's love for creation. It is out of passion for the creator and the passion of the shepherd that evangelical environmentalism is emerging in our day.

As I write this dark autumn evening on the great marsh, its blazing color diminished by the softer light of the moon, a buck snorts angrily as he bolts from our glacial drumlin into the marsh; an owl hoots high in the oak, even as the giant puffballs in the yard glow an eerie white. The day has died in the west; the marsh has become a frightening mire. The *Messiah* is now playing, "Surely, surely he has borne our griefs. . . . *Surely* he has borne our griefs and carried our sorrows." Is all well with the world? We all wait for the morning, expecting the sun to rise again.

NOTES

1. The term *evangelical environmentalism* was coined by David K. Larsen in "God's Gardeners: American Protestant Evangelicals Confront Environmentalism, 1967–2000" (PhD diss., University of Chicago, 2001). It is not a term I really welcome, largely because evangelicalism works to see the creation whole; there is no "us" versus the environment. Rather, human beings are part and parcel of the creation even as they are made in the image of God, and the creative system is not separable into us and everything else.

2. Walter Brueggemann, *The Land: Place as Gift, Promise, and Challenge in Biblical Faith* (2nd ed.; Overtures to Biblical Theology; Fortress, Minneapolis, 2002).

3. Carl F. H. Henry, "Evangelical," in *The New International Dictionary of the Christian Church* (ed. J. D. Douglas; Grand Rapids: Zondervan, 1974), 358–59.

4. These phrases are from the popular carol "Joy to the World" or "Joy to the Earth," whose words were penned by Isaac Watts in 1719.

5. An example of this joining to work toward creation's integrity is the Noah Alliance (www.noahalliance.org), a cooperative venture formed by the Academy of Evangelical Scientists and Ethicists and scientists and rabbis from the Coalition on the Environment and Jewish Life (COEJL) to prevent weakening of the U.S. Endangered Species Act.

6. See R. J. Berry, ed., *The Care of Creation: Focusing Concern and Action* (Leicester, England: Inter-Varsity, 2000), for a copy of the "Evangelical Declaration on the Care of Creation." This entire book is devoted to an evangelical analysis of the declaration.

7. Belgic Confession, art. II, from the *Psalter Hymnal* (Grand Rapids: CRC, 1988), 818.

8. It is because of this development of wisdom that comes from a coherent reading within and across both of the two books, that my book subtitled "A Biblical Response to Environmental Issues" is entitled *Earth-Wise* (Grand Rapids: CRC, 1994). This book was written for church discussion groups and may be consulted for a summary of creation care that is developed from both biblical and scientific understanding.

9. This is Booth's restatement of Ernest Hocking: "If, to agree on a name we were to characterize the deepest impulse in us as a 'will to live,' religion also could be called a will to live, but with an accent on solicitude—an ambition to do one's living well. Or, more adequately, religion is a passion for righteousness, and for the spread of righteousness, conceived as a cosmic demand." See W. C. Booth, "Systematic Wonder: The Rhetoric of Secular Religions," *Journal of the American Academy of Religion* 53 (1984): 677–702. I am indebted to Peter Bakken for bringing Booth's work to my attention.

10. Francis Schaeffer and U. Middelman, *Pollution and the Death of Man* (Wheaton, IL: Crossway, 1970).

11. Rachel Carson, *Silent Spring* (Boston: Houghton Mifflin, 1962).

12. Larsen, "God's Gardeners," xi.

13. Peter A. DeVos, C. B. DeWitt, Vernon Ehlers, Eugene Dykema, Dirk Perenboom, Aileen VanBeilen, and Loren Wilkinson, *Earthkeeping: Christian Stewardship of Natural Resources* (Grand Rapids: Eerdmans, 1980).

14. Orin G. Gelderloos, "Leadership in Environmental Ehtics," in *The Environmental Crisis: The Ethical Dilemma* (ed. Edwin R. Squires; Mancelona, Michigan: Au Sable Trails Institute of Environmental Studies, 359.

15. DeVos, et al., *Earthkeeping*, xii.

16. The Merriam-Webster definitions of an academy include "a society of learned persons organized to advance art, science, or literature" and "a body of established opinion widely accepted as authoritative in a particular field." From 1980 to 2005, then, the academy had become the Academy.

17. Organizational and Incorporation Papers, Academy of Evangelical Scientists and Ethicists, 2508 Lalor Road, Oregon, Wisconsin.

18. Chris Sugden, "Guest Editorial: Evangelicals and the Environment in Process," *Evangelical Review of Theology* 17.2 (1993): 119–21.

19. J. Mark Thomas, ed., "Evangelicals and the Environment: Theological Foundations for Christian Environmental Stewardship," special issue of *Evangelical Review of Theology* 17.2 (1993): 241–86.

20. In the first issue of the *Creation Care* newsletter, its publisher and writer, Bob Carling, wrote that this newsletter "aims to act as a communication vehicle for all who are concerned with environmental issues and wish to see a specifically Christian viewpoint developed. It is primarily for those involved in environmental concerns at a professional level and seeks to work with rather than duplicate the efforts of other more grassroots environmental organizations. Written from a biblical perspective, it includes news and views, summaries of reports in the media, occasional papers on specific issues (e.g., transcripts of key speeches by leading Christian environmentalists), notices of forthcoming meetings, reports of meetings, selected book reviews and some full-length articles on environmental issues."

21. "Summarizing Committee Report of the World Evangelical Fellowship Theological Commission and Au Sable Institute Forum," *Evangelical Review of Theology* 17.2 (1993): 132.

22. Ibid.

23. Ibid.

24. Ibid.

25. Loren Wilkinson, "The Making of the Declaration," in *The Care of Creation* (ed. R. J. Berry; Leicester, England: Inter-Varsity, 2000), 50–59.

26. "An Evangelical Declaration on the Care of Creation," in *The Care of Creation* (ed. R. J. Berry; Leicester, England: Inter-Varsity, 2000), 18.

27. Peter Harris, "A New Look at Old Passages," in *The Care of Creation* (ed. R. J. Berry; Leicester, England: Inter-Varsity, 2000), 133.

28. Ibid.

29. Ibid.

30. R. J. Berry, "Conclusions," in *The Care of Creation* (ed. R. J. Berry; Leicester, England: Inter-Varsity, 1999), 28.

31. Ibid., 183.

32. John Stott, "Foreword," in *The Care of Creation* (ed. R. J. Berry; Leicester, England: Inter-Varsity, 1999), 28.

CHAPTER 25

RELIGION AND ENVIRONMENTALISM IN AMERICA AND BEYOND

BRON TAYLOR

THE ENVIRONMENTAL PROTECTION AGENCY: TROJAN HORSE FOR PAGAN ENVIRONMENTALISM?

A few years into the twentieth-first century, I had a discussion with Ike Brannon, a well-placed economist in the administration of President George W. Bush. He was in a position that has much to do with federal environmental policies. He is also a Christian and a political and fiscal conservative, but of a libertarian bent. Socially and religiously moderate, some of his views would be considered liberal. He is also a man with whom environmentalists can find much common ground, for both can readily agree that much governmental spending leads to environmental injury and that such spending should be arrested. He rides a bike or uses mass transit, does not own an automobile, and lives a lower-impact lifestyle than many if not most who would embrace the environmentalist label. He also takes pride in working well with some environmental groups, such as Environmental Defense.

Despite these characteristics, Brannon would not accept the environmentalist moniker, because he disagrees with what he considers its central tenet, that economic growth jeopardizes the environment. He believes, instead, that small government helps to create the growth and prosperity that is a prerequisite to being able to afford environmental protection. He is a man who cannot be simplistically characterized as pronature or antinature.

Brannon once indicated to me that he and his Republican colleagues are often frustrated when dealing with the Environmental Protection Agency. He and his peers often say, he told me, "The EPA is a religion, not a job, to its people." He and others of his colleagues felt that such religious devotion to the earth made it impossible to negotiate in cost/benefit terms.

Whatever their accuracy, these perceptions provide a window into the religious dimension of environmental conflict in America. It is not difficult to find many more examples. Some conservative Christians, for example, consider environmentalism to be a Trojan horse that threatens Western civilization with a revitalized paganism.[1] Even among religiously and ideologically complex characters like my economist friend, such views can be found.

In this essay I take up such assertions, and I argue that religious perceptions and practices *have* decisively shaped American environmentalism and to such an extent that much environmentalism can be considered a nature religion. I then characterize three major types of what I call "green religion" that have emerged in American culture, reflecting on their impacts, both domestically and internationally, while speculating on their long-term influences upon religion and environmental politics, both in America and beyond. The examination of the history and present manifestations of green religion and nature religion suggests that such phenomena, which have been very important in North American religious history, will continue to shape environmentalism both domestically and internationally; indeed, such phenomena will do so in ways that are, at present, barely on the radar of historians and social scientists.

RELIGION AND NATURE RELIGION

My analysis must begin by clarifying the terms *religion*, *green religion*, and *nature religion*.

With regard to *religion* in general, I will follow David Chidester, who once defined it as "that dimension of human experience engaged with sacred norms" that are related to experiences of extraordinary, transformative, or healing power.[2] While this may be a circular definition (religion is whatever people consider sacred and the sacred demarcates the religious realm), it does reflect the ways in which people actually speak about religion, as Chidester points out. It has the added advantage of allowing us to sidestep thorny issues such as whether one must have otherworldly divine beings to have religion, for such a definition requires the observer to take sides in implicitly religious disputes about what constitutes an authentic religion. Deploying a vague and admittedly circular definition allows our analytic, religion-focused lens to be employed whenever people use rhetoric of the sacred (or its opposite) to describe and promote what they most deeply feel and experience.

I will use the terms *nature religion* and the plural *nature religions*, meanwhile, simply as umbrella terms for religious perceptions and practices that, despite substantial diversity, are characterized by a reverence for nature and consider nature to be sacred in some way. Another sense I will have in mind when I think of religion (in general) harkens back to its Latin root, *religre*, which originally meant to tie fast or to bind. With this in mind, nature religion involves the feeling some people have of being bound, connected, or belonging to nature.

By *green religion* I refer to religious sensibilities that consider environmental concern a religious duty, regardless of whether nature itself evokes reverence or is considered sacred.[3]

European American Attitudes and Practices toward Nature from European Contact to the Twentieth Century

Scholarship exploring the religious dimensions to attitudes and practices toward nature in North America since the arrival of significant numbers of Europeans generally provides the following narrative.

European Americans were deeply conditioned by the attitudes typical of the continent from which they had come. Consequently, their perceptions and feelings regarding nature were often characterized by fear and hostility, or at least by a deep ambivalence toward the wild landscapes that differed greatly from the domesticated agricultural and pastoral ones they left behind. Fear, hostility, and ambivalence were also among the central European attitudes toward the continent's aboriginal peoples. These attitudes were shaped decisively by Christianity, and especially Puritanism, whose devotees sometimes viewed American Indians as not only physically but spiritually dangerous, even in league with Satan. Such beliefs played an important role in the violent subjugation of American Indians and the land they inhabited.

Thus Christianity in general, and Puritanism in particular, provided a cosmology and theology that reinforced the general impetus among European settlers to consider land not as something sacred and worthy of reverence, but as a resource to be exploited for both material and spiritual ends. For such Christians, both the material and spiritual ends had something to do with glorifying and satisfying a deity who resided beyond the earth and, thus, should not be too closely identified with it. Nevertheless, people could learn about this deity through nature. Moreover, people could grow spiritually through the challenges and dangers posed by nature. The natural world was therefore not only a material but a spiritual gift. Moreover, Puritans such as Cotton Mather and especially Jonathan Edwards, who published their most important works in the first half of the seventeenth century, promoted a Platonic doctrine of correspondence, where nature on earth was seen to correspond to divine realities. This helped fertilize American ground for the appreciation of nature that was soon to grow wildly, including in transcendentalism and the many other forms of nature religion that would follow.[4] According to such interpretations, nature in early European America was invested with complicated religious meanings.[5]

Perhaps the most prevalent scholarly narrative suggests that American attitudes did not shift toward nature appreciation until this had first occurred in Europe, first with the Enlightenment and then with the romantic reaction to it. Roderick Nash, for example, spotlights figures like English philosopher Edmund Burke, whose *Philosophical Enquiry into the Sublime and Beautiful* (1757), combined with Immanuel Kant's *Observations on the Feeling of the Beautiful and Sublime* (1764), linked the esthetic and the sublime.[6] Meanwhile, Jean-Jacques Rousseau, who did as much if not more than any other figure to precipitate the Romantic Movement, located a sacred nature at its core. Nash puts it this way: "Rousseau argued in *Émile* (1762) that modern man should incorporate primitive qualities into his presently distorted civilized life. And his *Julie ou la nouvelle Héloïse* (1761) heaped such praise on the sublimity of wilderness scenes in the Alps that it stimulated a generation of artists and writers to adopt the Romantic mode."[7]

Soon European literary figures would visit America, where wildlands could be experienced directly, helping to reshape the perceptual possibilities for the intelligentsia of America's cities during a period when wild landscapes were retreating rapidly. By the late eighteenth century, accelerating during the nineteenth, nature was increasingly perceived as sublime in American arts and letters. Deism made its own contribution to nature religion in America, for with such religion the God who established the laws of nature was revealed exclusively through them.

Many figures and movements could be featured in this history in which an appreciation for the sacred dimensions of nature gained momentum. Some of these include the deistic third president of the United States, Thomas Jefferson, whose *Notes on the State of Virginia* (1785) linked the "sacred fire" of liberty to the connection of the people to the land;[8] Christian thinkers, including a number of prominent figures from the Religious Society of Friends (popularly known as the Quakers), who found nature sacred and articulated what could be called a kinship ethic with nonhuman creatures;[9] artists, beginning with Thomas Cole and the Hudson River School of artists (from the early nineteenth century), who depicted wildlands as mysterious, sacred places; poets, beginning perhaps with William Cullen Bryant's "A Forest Hymn" (1825), which expressed the view that peace and harmony, and the creator's hand, could be found in the very forests that most European Americans had previously found perilous, and others, including Walt Whitman, who wrote famously in *Leaves of Grass* (1855), "This is what you shall do: love the earth and sun and animals," thus articulating an early, religious kinship ethic toward all creatures; and novelists, such as James Fenimore Cooper, whose Leatherstocking tales (published in five novels between 1823 and 1841) not only expressed a reverence for nature but also appreciation for the Native American lifeways, which Cooper understood to be deeply dependent on and embedded in nature.[10] This perspective in turn would become a feature typical of much environmentalist thinking, from the early 1830s, when George Catlin, "an early student and painter of the American Indian," first promoted the idea of setting aside large national parks that, in his view, should include both wild natural beauty as well as Indians and wild animals.[11]

Better known are developments that gained momentum in the second half of the nineteenth century, some of which were precipitated by the rise of the transcendentalists, led by Ralph Waldo Emerson in *Nature* (1836), who argued that all natural objects can awaken reverence, "when the mind is open to their influence."[12] Indeed, at times, Emerson sounded an animistic tone, speaking of the spiritual truths conveyed by nonhuman beings. Other times he struck a pantheistic note, speaking effusively about the beauty and sublime character of nature. Although some interpreters say he was a Platonist, viewing nature more as the pathway to spiritual truth than a spiritual end, he nevertheless contributed decisively to the dramatic rise in nature appreciation in the latter decades of the nineteenth century.

Emerson's influence includes his impact on Henry David Thoreau and John Muir, both of whom were, by most accounts, far more interested in nature for its own sake than was Emerson. Indeed, in *Walden* (first published in 1854), Thoreau seems to provide an early example of what I call nonsupernaturalistic nature religion (see below). He also represented an archetype, even if he only partly embodied it himself, of the personal, spiritual return to nature. By the end of the century and into the next, John Burroughs would more consistently represent such impulse, fusing a scientific naturalism and a religious pantheism in what was perhaps the best early example of the back-to-the-land movement that would become an important characteristic of the American counterculture in the second half of the twentieth century.[13] Burroughs, Thoreau, and Muir were all naturalists who were more scientifically inclined than Emerson. They fueled a more naturalistic form of nature spirituality than did Emerson and most other transcendentalists; indeed, they pioneered an approach more amenable to the veneration of nature through environmental activism and bioregional experimentation than those of a more transcendentalist bent (or those today who we might consider New Age in their spiritual orientation).

This race through time brings us to the cusp of the age of conservation, which is most often traced to Thoreau, Muir, and the utilitarian forester and founder of the U.S. Forest Service, Gifford Pinchot, who, for his own part, was influenced by the social-gospel movement that flourished at the turn of the century.[14] A central reason for sketching this history is to underscore that by the beginning of the twentieth century most of the main features of religious environmentalism had been presented and had begun to spread. A key rationale for this presentation is to emphasize that the American experience is a result of complex reciprocal influences between European and American experiences, perceptions, and worldviews and to suggest that the land itself, so wild and different than the domesticated landscapes in Europe, had its own influence on the minds and hearts of those engaged with it.[15]

Indeed, by the time that historian Fredrick Jackson Turner proclaimed and lamented the end of the American frontier (circa 1890 with the end of significant, armed, Indian resistance), many Americans were regretting the steady decline of wild landscapes.[16] Turner himself asserted that this loss threatened the virility and spiritual health of the nation. Such anxieties helped set the stage for the further development and increasing diversity of nature religions, many of which during the twentieth century also came to promote a spiritual or religious environmentalism.

NATURE, RELIGION, AND THE SEEDS OF ENVIRONMENTALISM IN THE TWENTIETH CENTURY UNTIL EARTH DAY

The first two-thirds of the twentieth century were characterized by developments that built on those sketched above. I will briefly mention three examples.

Among the most noteworthy was the emergence of nature-based youth movements such as the Boy Scouts and Indian Guides. Ernest Thompson Seton set this trend in motion when during the first decade of the new century he established the "Woodcraft Indians" movement and, subsequently, helped shape the Boy Scouts (founded in 1910) and Indian Guides (1925). Through these movements Seton spread his view that American Indians provided ideal models for spirituality and ethics. Influenced by Darwinian thought, Seton also wrote widely about the emotional and altruistic lives of animals, although his critics claimed he did so in an unscientific and anthropomorphic way. Whatever the truth in these criticisms, Seton fostered a more positive evaluation of the continent's first inhabitants, promoted a nature-based spirituality that understood nature as a sacred place where God can be encountered and virtue developed, and thereby helped shape the century's environmentalist discourse.

Better known is John Muir, who was born in Scotland, immigrated to rural Wisconsin as a youngster, wandered widely as a young adult, eventually finding his own Shangri-la in the wildlands of California's Sierra Nevada mountains. He subsequently founded the Sierra Club, which articulated the preservationist ethic that would undergird the wilderness movement in general and, at least in theory, guide the management of the National Park Service. Using both theistic and nontheistic religious language (he often sounded more pantheistic and animistic than theistic), Muir described his own spiritual experiences in nature, called on Americans to travel to wild places to experience the earth's sacred voices for themselves, and battled the more politically powerful Gifford Pinchot on a number of issues related to wildlands management. The most famous conflict was over a dam that Pinchot sought to build at Hetch Hetchy Valley in Yosemite National Park.

The nation's first forester who became its initial U.S. Forest Service director, Pinchot had a more orthodox, anthropocentric, and utilitarian religious ethic than did Muir. He succeeded in establishing the "multiple use" doctrine as the management philosophy for governing most federal lands. Although Muir lost most of the specific battles he waged, he contributed to the more preservationist ethic undergirding the National Parks and designated wilderness areas in the United States and helped to make the wilderness movement an important social force, if not a new religious movement.

Some scholars, such as Stephen Fox and Michael P. Cohen, portray Muir as more of a pagan than a theist. Fox builds a compelling case that Muir's use of theistic tropes was motivated more out of a desire for political effectiveness (convincing a predominantly Christian nation to protect his own sacred places) than out of a theistic cosmology. Whatever one's judgment about whether Muir remained theistic in some way or became more of a pantheistic or animistic pagan, the tension between Muir and Pinchot can be seen as archetypal all the way into the twentieth-first century. Conflicts among those who are more traditionally theistic (if not always orthodox), but believe there is a religious duty to conserve nature (at least for people's sake), and those who believe that nature is directly sacred in some way have contributed to at least some of the internal tensions among environmentally concerned people in America. So the battles between Muir and Pinchot, which were especially heated in the first dozen years of the new century, are noteworthy for presaging much of what was to follow later in the century.[17]

The third development was to be found in two increasingly well-known scientist writers, Aldo Leopold and Rachel Carson. Their nature-related spiritualities and influence on later environmentalists are too little appreciated.

Leopold, who died in 1948, shortly before the publication of *A Sand County Almanac*, is now considered the greatest ecologist/ethicist of the twentieth century, in large measure for articulating a holistic and biocentric "land ethic," for example, in these famous passages:

> The land is one organism.... If the land mechanism as a whole is good, then every part is good, whether we understand it or not. If the biota, in the course of eons, has built something we like but do not understand, then who but a fool would discard seemingly useless parts? To keep every cog and wheel is the first precaution of intelligent tinkering.[18]

> All ethics so far evolved rest upon a single premise: that the individual is a member of a community of interdependent parts.... [And] the land ethic simply enlarges the boundaries of the community to include soils, waters, plants, and animals, or collectively: the land.[19]

> A thing is right when it tends to preserve the integrity, stability, and beauty of the biotic community.[20]

Less well known is the influence on his thought by Russian mystic Pyotr Demianovich Ouspensky, who was in turn influenced by theosophy as well as other Russian mystical thinkers, including Georges Ivanovitch Gurdjieff. Leopold's holism, which anticipated James Lovelock's Gaia Hypothesis and considered life to be integrally connected and interdependent, can be traced to these influences.[21] Also not well known is that Leopold confided to his daughter that his own religion came from nature or that and his family considered him and themselves also as sharing a pantheistic spirituality in which there was no personal God orchestrating things, but nevertheless, a sense of the land as sacred and having value apart from human needs.[22]

Carson also had a deep, nature-based spirituality. This is less known than her role in precipitating the modern environmental movement through *Silent Spring* (1962), which assimilated the available science to express alarm about the threats posed by pesticides and herbicides.[23] Of course, when writing *Silent Spring*, Carson was concerned about her scientific credibility, and so her spirituality and ethics were only subtly expressed. But careful reading of her work, especially her earlier books focused on the oceans and talks presented to women's organizations, make her own nature religion clearer, as well as the deeper reasons for her environmental activism.[24] Her spirituality did not involve extraworldly divine beings. But her language sometimes expressed a subtle animistic perception,[25] as well as a spirituality that appreciated the miracle and mystery of life on its own terms. She even dedicated *Silent Spring* to Albert Schweitzer, whose reverence for life ethics had become famous, thereby making clear her own affinity with such ethics.

Anthropologist and naturalist Loren Eiseley, in his wide-ranging books and essays, expressed a similar reverence for life and its mysteries. He came to this reverence in no small part through scientific inquiry, for although he was a believer in evolution, he averred that science was unable to fully explain the beauty, value, and mystery of life. He wrote, for example,

> I am an evolutionist . . . [but] in the world there is nothing to explain the world. Nothing to explain the necessity of life, nothing to explain the hunger of the elements to become life, nothing to explain why the stolid realm of rock and soil and mineral should diversify itself into beauty, terror, and uncertainty.[26]

> No utilitarian philosophy explains a snow crystal, no doctrine of use or disuse. Water has merely leapt out of vapor and thin nothingness in the night sky to array itself in form. There is no logical reason for the existence of a snowflake any more than there is for evolution. It is an apparition from that mysterious shadow world beyond nature, that final world which contains—if anything contains—the explanation of men and green leaves.[27]

Eiseley concluded his life without any pretension that he understood what the explanation for life was, yet he never wavered regarding his intuition that it was a miracle and worthy of reverent care.[28]

NATURE, RELIGION, AND ENVIRONMENTALISM FROM THE 1960S AND EARTH DAY (1970) FORWARD

Between the end of the Second World War and the publication of *Silent Spring*, Americans focused more on building and recovering from the trauma of war, and environmental issues were little discussed. The 1960s brought many upheavals, from those surrounding the civil-rights movement to Cold War ideology. The Cold War precipitated both a reliance on weapons of mass death for security and the bloody war in Vietnam. These events not only led to widespread distrust of the government and its anticommunist ideology; they also forced many into a serious reappraisal of Christianity, the country's dominant religion.

Those looking for new forms of religious meaning and political engagement found a greater array of religious alternatives than had ever been the case

previously. This was in part due to immigration from Asia and also to the proliferation of new religious movements that appealed directly to those disenchanted with the dominant, organized, institutional religions.

Less often noticed, for it was not always understood as religious, was the
growth of a spiritually holistic environmentalism that challenged the anthropocentric ethics of the culture's mainstreams. The movements and figures previously
discussed made available such religious and ethical alternatives, and when the
broader social and environmental conditions were ripe in the 1960s, they grew
more rapidly and widespread than ever before.

The next three sections provide a tripartite typology of religious environmentalism and activism that came more clearly into view from that time and until
the early twentieth-first century. The subsequent section examines the increasing
global impact of such green religion.[29]

Green Religions as Environmentally
Concerned World Religions

A great ferment over religion and nature emerged in America during and since the
1960s that was rooted in scholarship focusing on the relationships between religions, cultures, and the earth's living systems.[30] Among the most interesting work
was that of anthropologists, including Roy Rappaport, Marvin Harris, and Ger-
rardo Reichel-Dolmatoff, who focused on the role that ritual and religion can play
in regulating ecosystems in a way that supports traditional livelihoods and pre-
vents environmental deterioration.[31] But these theorists had little impact on the
popular green religion.

A number of other scholars, however, engendered a strong reaction among
both scholars and some laypeople by arguing that some of the world's major
religious traditions are especially responsible for environmental deterioration
because they foster environmentally destructive attitudes and behaviors.[32] The
most commonly identified culprit was Christianity (especially the most powerful
Protestant forms), and sometimes scholars articulated prescriptions to fix such
religions or offered alternatives to them.[33]

Such critiques and suggestions drew two reactions. In the first, religious
thinkers either defended their traditions, arguing that, properly understood, they
are environmentally friendly. Others argued that *some* streams within their traditions provide green alternatives. In the second approach, culpability was forth-
rightly acknowledged, and ideas for the greening or environmentally friendly
reform of the tradition were advanced.

In America, the bulk of such reflection occurred within Christianity. This was
for a number of reasons, including the following: more Americans identify with

Christianity, which had been the tradition most often blamed, so Christians were feeling especially defensive; and the tradition's emphasis on sin and repentance provided an internal, theological basis for reevaluating life and making changes in the light of faith.[34]

Although there was a greater quantity and range of debate within and about Christianity and the environment than regarding other world religions, similar questioning about and soul-searching within other traditions began in earnest. A scholarly field now best known as religion and ecology sprang up, growing especially during the early 1990s, which sought to identify obstacles that the worlds' religions might pose to environmental sustainability, uncover the resources within such traditions for promoting sound environmental behavior, or discover the ways in which these traditions would need to change in order to become more green. By the early twentieth-first century, significant scholarly work had explored the world's major religious traditions to understand their environmental attitudes and impacts, as well as to lay a foundation for steering them in greener directions.[35]

It was the case, however, that reforming religious thinking and action along environmental lines was not easy and did not, by all available evidence, extend widely. Indeed, it has always been difficult to assess the extent to which environmental attitudes and behaviors cohered and under what set of circumstances they were more or less likely to do so.[36] It is clear, however, that some intellectuals and laypeople have been pushing their traditions to understand the quest for environmental sustainability as a central religious imperative.[37]

Green Religions as "Nature-as-Sacred Religions"

In analyzing three general types of green religion since Earth Day, I need to be clear that that the boundaries among them are fluid. This is not a problem if we remember with Jonathan Z. Smith that "map is not territory"; typologies do not perfectly mirror the world but are useful if they help to orient us to the terrain.[38]

Whatever social scientific curiosities may animate scholars involved in the emerging religion and ecology field, and regardless of whether the scholars involved feel loyalty or affinity toward the traditions they focus upon, most of them seem to consider nature to be sacred in some way. This conviction appears to be tethered to ethical concern about the environmental decline. Such religion is the second general type of green religion that has emerged with some force in American culture. Indeed it is what I would call a nature religion.

This definition is apt, I believe, for the perception that nature is sacred and worthy of reverent care is central to the identities of a number of groups whose participants consider themselves to be engaged in what they also sometimes call nature religion. Such religions include paganism, many if not most indigenous religions, and some new religious movements and branches of New Age spirituality. The survey of earlier environmental thinkers and groups revealed that such spirituality is not new in North America, but that such trends have been strengthening since Earth Day.[39]

The revival or reinvention (depending on one's historiography) of paganism is an especially important contemporary manifestation of the sensibility that the earth is sacred.[40] In rhetoric and sometimes in practice, pagans support environmental causes. Indeed, those who consider themselves to be pagan have been deeply involved in radical environmentalism, including participation in Earth First! (from 1980) and the Earth Liberation Front (from the early 1990s).[41] Such groups have alternatively deployed demonstrations, civil disobedience, sabotage, and even arson, as well as more mainstream tactics, from boycotts to electoral politics, in their efforts to save the living beings they consider to be intrinsically valuable and a Mother Earth they consider sacred. Indeed, those who are self-consciously pagan in their religious identity are the ones most likely to be at the forefront of ecological resistance movements. This is because when the earth is itself considered sacred, and not only indirectly so because it was created by a divine being, then the earth itself, and not some creator being, becomes the locus of religious and ethical devotion.

Like some of the earliest environmental advocates in America, politically radical pagans generally express affinity with what they perceive to be the nature-beneficent spiritualities and lifeways of indigenous peoples. Some try to draw on such spiritualities in their own religious lives, even though this is sometimes controversial. They have sometimes, as well, expressed solidarity through political action supporting Native American groups who are struggling for cultural survival and to protect places sacred to them.[42]

Indeed, there has been a stunning revival of indigenous cultures and spiritualities purporting to consider nature sacred and promoting ethics of kinship toward all creatures. This development has nowhere been stronger than in North America. Evidence ranges from the engagements of indigenous peoples in lobbying and other forms of activism seeking to protect places considered sacred and essential to their traditional livelihoods;[43] to scholarly analyses of "traditional ecological knowledge," which argue for the value of such knowledge in the management of ecosystems;[44] to the hard work that goes into preventing the extinction of indigenous languages, upon which ceremonies and connections with the natural and spirit worlds are thought to depend;[45] to other forms of ethnographic and archeological work, which can help indigenous people to protect and reconstruct their traditions.

The impact of what is now several generations of reappraisal of the value of indigenous culture has led to an increasing affirmation in Western popular culture of the nature religions of indigenous peoples.[46] I will mention just two examples that reached large audiences.

The motion picture *Pocahontas* (1995), which celebrated such spirituality, provides our first example. Directors Mike Gabriel and Eric Goldberg reported afterward how they "tried to tap into [Pocahontas's] spirituality and the spirituality of the Native Americans, especially in the way they relate to nature."[47] The animistic nature spirituality and environmental kinship ethic depicted in this Disney-produced movie was reincarnated at its Animal Kingdom theme park in Florida, which opened in 1998. There a performance entitled "Pocahontas and Her Forest Friends" reprised the story and the quest for kinship and harmony among all creatures, beseeching audiences, "Will you be a protector of the forest?"[48]

The opening ceremonies of the 2002 Winter Olympics in Salt Lake City Utah provide a second example, when an assumed American Indian reverence for Mother Earth was celebrated on ice. During the opening ceremony, representatives from five of the region's indigenous nations offered in their own languages both welcoming messages and spiritual blessings; these were then translated into the official Olympic languages. To tunes played on indigenous flutes, skaters in Indian-inspired costumes performed a dance, choreographed in synch with native drumming depicting the "heartbeat of Mother Earth" herself. The narrator underscored the value of this indigenous Mother Earth spirituality, and many of the Native American participants were thrilled to be able to present their cultures in a positive way to a television audience estimated at three billion.[49]

In North America it is not only pagan, radical environmentalist, and American Indian groups, however, who consider nature to be intrinsically sacred. Some of those involved in New Age subcultures and new religious movements share such perceptions, even though these groups have acquired a reputation for otherworldliness and political apathy.

Some in these movements, for example, view consciousness change as a prerequisite to both positive social change and environmental well-being. They also consider efforts to protect and restore nature important ways to educate and foster the needed transformation of attitudes. Some even think that healthy ecosystems themselves contribute critically important energies to the envisioned and needed consciousness transformations, such that without environmental protection and restoration movements, and the resulting healthier natural energies, the envisioned consciousness change has little chance to occur.[50]

When considering together the greening-of-religions phenomena and the growth of nature religions, recognizing that these developments involve a diversity of religionists, activists, and scholars, all of whom believe that the earth is in peril by human behavior yet worthy of reverent care, it is unsurprising that initiatives are underway to bridge the gap between what is (ecological decline) and what ought to be (environmentally sustainable lifeways). The Earth Charter provides an example that might prove globally significant response.

The Earth Charter was first proposed by Maurice Strong, a Canadian who served as general secretary for the 1992 World Summit on Environment and

Development, which was sponsored by the United Nations and held in Rio de Janerio. During the late 1990s the Earth Charter went through an extensive drafting and review process, within a number of nongovernmental, religious, "civil-society" organizations that are engaged with the United Nations.[51] The drafting process was designed to gain maximum support from the international community. A draft presented a decade later to the United Nations during the World Summit on Sustainable Development, a follow-up to the Rio conference, held in Johannesburg, South Africa, spoke of "respect and care for the community of life in all of its diversity" and proclaimed that protecting and restoring the ecological integrity of the earth was a "sacred trust," inseparable from the quest for justice and peace. The document concluded: "Let ours be a time remembered for the awakening of a new reverence for life."[52]

The initiative mimics the strategy that guided the U.N. Declaration on Human Rights. Charter proponents hoped that U.N. ratification would be gained and then leveraged to promote improved environmental practices all around the world.

Early Earth Charter proponents clearly considered nature to be sacred. Strong's speeches were laced with Gaia-theory-inspired earthen spirituality, and the charter's most famous proponent, Mikhail Gorbechev, professed, "I believe in the cosmos... *nature is my god.* To me, *nature is sacred.* Trees are my temples and forests are my cathedrals."[53]

Many would be surprised to hear the former leader of the Soviet Union expressing his religious fidelity to the earth or to learn that he is now the president of Green Cross International, devoting his life to turning international institutions green. It may be equally surprising to hear scientists, some of whom have no belief in supernatural divinities, draw on metaphors of the sacred to express their awe at the wonders of the universe and reverence for life. A statement issued in the early 1990s by a group of prominent scientists, including Stephen Jay Gould, Stephen Schneider, and Carl Sagan, expressed such sentiment: "As Scientists, many of us have had profound personal experiences of awe and reverence before the universe. We understand that what is regarded as sacred is more likely to be treated with care and respect. Our planetary home should be so regarded. Efforts to safeguard and cherish the environment should be infused with a vision of the sacred."[54] David Takacs, when interviewing scientists who had devoted their lives to the conservation of biological diversity, found such sentiments to be common, concluding that "some biologists have found their own brand of religion, and it is based on biodiversity. [They] attach the label spiritual to deep, driving feelings they can't understand, but that give their lives meaning, impel their professional activities, and make them ardent conservationists."[55] Indeed, increasing numbers of books written by scientists illustrate how, for some, science evokes a religious reverence for life.[56]

Green Religions as Postsupernaturalistic "Spiritualities of Connection" to Nature

I have already provided examples where individuals involved in green religious production use rhetoric of the sacred to express awe and reverence toward the "miracle" of life and reviewed early and recent examples where science has been an important source of such perception. This introduced the possibility of nature religion without supernaturalism.

In this section I suggest that nonsupernaturalistic nature religion will likely become an important feature in the religious life of America and beyond and that such religion will increasingly become a wellspring for environmental action based on kinship ethics and a reverence for life. A good way to explore this hypothesis is to address two sets of questions:

- How strong are the trends leading to the three types of green religious production under analysis? More specifically, are the social and environmental conditions that gave rise to them likely to wane or increase and intensify?
- What are the drivers of nonsupernaturalistic nature religion? Are these likely to become more or less important influences on nature-related religion?

With regard to the first set of questions, I think there are two critical factors catapulting forward all three forms of nature religion: environmental degradation and evolutionary science. These variables are intertwined in a complex mix with many other variables but they are, for a number of reasons, becoming increasingly important.

Environmental degradation is increasingly obvious and alarming, for example. This recognition is increasingly grafted onto existing religions (green religion type 1) and mixed in with revitalized and new forms of nature-as-sacred religions (green religion/nature religion type 2). This development represents a significant innovation in the history of religion. Increasingly and for the first time, apocalyptic expectation arises not from the fear of angry divinities or incomprehensible natural disasters, but from environmental science.

With regard to the second set of the preceding questions, while environmental science is reshaping green religions of the type 1 and 2, it is, ironically, built upon a Darwinian foundation. This erodes the supernaturalistic beliefs that usually accompany such religious forms. The point here that is most relevant to speculations about the future of religion (in general) and green and nature religions (in particular) is that evolutionary science is the central driver producing or a least shaping green religions of type 1 and 2—but this science *makes less plausible* the supernaturalistic metaphysical foundations of such religion.

It is a story as old as Galileo and as new as the Hubble Telescope: as human optics improve, the more difficulty people have believing in divinities residing beyond the earth. Science provides a cosmogony that, while leaving many mysteries unexplained, has explanatory power apart from beliefs in intelligent design or other forms of divine creativity. Evolutionary science, ironically, challenges and tends to make incredible supernatural religions, just as some such religion began to turn green through an appreciation of what are, at their roots, evolutionary insights.

Moreover and meanwhile, it seems irrefutable that ecological degradation and evolutionary science will increase in their cultural influence, for these are easier to apprehend than the divinities upon which supernaturalism depends. Indeed, within hardly more than a century, notwithstanding polling data in America revealing that less than half of Americans believe that natural selection provides the best account of the diversity of life and human origins, it appears that evolutionary understandings have gained a solid foothold, if not widespread acceptance, among the world's intelligentsia.[57] Moreover, despite resistance from conservative religionists, evolution is increasingly taught globally and appears to be well along the way toward acceptance among both well-educated and less-educated sectors, at least when taking a long view and realizing that gestalt usually takes time, partly because this requires that resistance must be overturned or at least eroded.

Meanwhile, within only the past several decades, large numbers of people began to recognize that environmental degradation is severe and threatens both the quality of life and life itself. Such recognition appears to have grown even more rapidly than beliefs in evolution, probably because such dynamics are more easily observed by ordinary individuals (often through global media) than are evolutionary processes. Sometimes this degradation is depicted in other moral or esthetic terms, at other times, explicitly or implicitly as desecration, it is viewed with dismay as threatening the beautiful if not miraculous and awe-inspiring diversity of life on earth.

Given the comfort that supernaturalism offers to humans facing an apparent mortality, it would be a fool's errand to suggest its total eclipse. It seems a reasonable hypothesis, however, in the light of recent decades of intellectual, cultural, and religious developments, to expect *at least when thinking in very long time frames* that supernatural religions, including those forms that fit green religion types 1 and 2, will gradually decline.

It may be that the third, nonsupernaturalistic type, now only nascent and growing within small enclaves of devotees around the world, will inherit much of the religious future. With such religion, people feel awe and reverence toward the earth's living systems and even feel themselves connected and belonging to these systems, but without believing in deities they cannot perceive. Such an affectively grounded spirituality of connection might not resemble today's more common supernaturalistic religions. They might, nevertheless, require religious terminology to capture the feelings. It might also require ritual forms to physically venerate the living systems for which the word *sacred* is used as a way to express their ultimate value.[58]

CIVIC EARTH RELIGION
AND ENVIRONMENTAL ACTION

Perhaps even more significant than the emergence of postsupernaturalistic nature religions, or of the more prevalent types 1 and 2 green religions, is how diverse forms of nature-related spirituality are spreading globally, even into international institutions, which may become important carriers of such religion.

I already mentioned how an Olympics ceremony expressed respect for indigenous Mother Earth spiritualities. In my view, the World Summit on Sustainable Development in 2002 signaled an even more important possibility, namely, of the evolution of a global, civic earth religion. Borrowing from but taking the civil-religion thesis in a global and environmental direction, in a civic earth religion, people around the world, despite diverse and sometimes mutually exclusive religious worldviews and national interests, would unite to express and act upon a religious fidelity to the biosphere.[59]

Political theorist Daniel Deudney has argued that such "Terripolitan Earth Religion" (his term for what I am calling civic earth religion) could displace provincialism and nationalism with loyalty to and reverence for the biosphere.[60] Environmental sustainability requires the construction of new international institutions and legal frameworks, Deudney believes, and he contends that given its scientific credibility, "Gaian Earth religiosity" is particularly "well suited to serve as the civic religion for a [desperately needed] federal-republican Earth constitution."[61]

I saw a sign that such a possibility may be more than a utopian dream at the World Summit on Sustainable Development. The Earth Charter, for example, was mentioned positively by a number of world leaders, including the Republic of South Africa's President Thabo Mbecki when he opened the summit. It was also celebrated on a number of occasions during events attended for this purpose by other prominent figures. Moreover, it was mentioned positively by several heads of state during their speeches and drew positive mention in early drafts of the conference's concluding "political declaration," and although mention of it explicitly was deleted from the final declaration, a number of its phrases survived the editing process.

There is even more tantalizing evidence. The opening ceremony was framed by an evolutionary cosmogony, wherein during the official welcoming ceremony, Africa was labeled the "Cradle of Humanity" after a nearby archeological site known locally as Sterkfontein, where the now extinct prehuman *Australopithecus*, dating back four million years, had been found in long-forgotten caves. The caves were designated a World Heritage Site by the United Nations in 2000, and the site's "Cradle of Humanity" name was woven into many of the summit's speeches and ceremonies. It was used especially to express a kinship ethic, which in turn was deduced from the evolutionary, scientific consensus that all of humanity had emerged from Africa. Even more significantly, perhaps, was the pageantlike performance that followed the opening speeches, which expressed a nonsectarian reverence for Mother Earth and all of her creatures.

In this musical theater, a child was found wondering what happened to the forests and to the animals. In response, in prose and song, a cosmogony compatible with both evolution and Gaia spirituality was articulated. The earth was conceived of as a beneficent person, while at the same time the emergence of complex life on earth was depicted in a way compatible with an evolutionary unfolding. The story also provided an explanation for moral and environmental decline. In the words of one character: "Life began with the earth, then came the plants, then the animals, and finally the human beings. We are all children of Mother Earth, which is why we have to take care of her and be her custodians. She is the hand that feeds us and the heart that heals us. But I'm sad to say, we are failing. Greed and foolishness have even dipped into the fabric of humankind. We are failing to love and care for mother earth."

After this speech, dystopian and apocalyptic music and imagery followed, depicting a bleak future. Then a child asked plaintively, "Is there anything we can do?" This question was answered hopefully and positively, citing as evidence the U.N. summit itself: "The nations are gathering, and the life and health of Mother Earth depends on their decisions." The mood shifted from dystopian to utopian during the finale of an opening ceremony extravaganza as a gigantic, iconic earth descended into the assemblage of earth's creatures below her. The dancing denizens of the earth then celebrated (and symbolized) a utopian hope for a planet in ecological and social harmony, with human beings uniting to repay Mother Earth for her many gifts.

Not only was the Cradle of Humanity mentioned in this ceremony and in various speeches at the summit, but pilgrimages were made to it. Famous primatologist Jane Goodall, for example, accompanied the U.N. Secretary General Kofi Annan and Thabo Mbeki. After viewing the *Australopithecus* fossils, Mbeki gave a speech stressing how this and other such sites connect us to our evolutionary history to each other: "When we say that this is your home, it is not merely that we want you to relax and enjoy our hospitality, but it is because in reality, this is in many ways a homecoming—a return to our common ancestors."

He continued explaining that the site "traces the evolution of the significant part of our Earth as well as the interdependence of peoples, plants and animals, thus, in many ways [it is] teaching all of us how we can co-exist and ensure enduring prosperity for all species."[62] Here Mbecki affirmed, as he had earlier in the opening ceremony, an evolutionary cosmogony, while articulating an ethic of evolutionary kinship that had many affinities with the forms of green religion found in American history.

Political leaders such as Mbeki and Annan, as representatives of the earth's diverse religious people, have to be especially careful not to offend people of diverse faiths or exacerbate religious tensions, so they scrupulously avoid sectarian religiosity. Through their support of the establishment of this World Heritage site, their pilgrimage to it, and their endorsement of the scientific narrative associated with it, however, they may well be quietly promoting a new form of civic earth religion and helping to establish sacred places that cohere with this new religion through ritual pilgrimage.[63]

It may be that such a religion, in which an evolutionary story becomes intertwined with reverence for life and combined with practices designed to protect and restore nature, will play a major role in the religious future of humanity.

Whatever the future may hold, it is clear that ever since the arrival of Europeans in North America, the relationships among the continent's diverse peoples, environments, and religions have been complex, contested, and sometimes violent. In a world characterized by an increasing global, intercultural encounter, in which reciprocal influence is the common, people alternatively expressed fear and ambivalence, as well as awe and reverence, at the natural worlds they surveyed, fought over, and inhabited. Thus, the religious history of America has had a great deal to do with nature, and nature a great deal to do with religious history in America. And this relationship has and will continue to have a great deal to do with the rest of the biosphere and the fate of those all who depend upon it.

NOTES

1. See Samantha Smith, *Goddess Earth: Exposing the Pagan Agenda of the Environmental Movement* (Lafayette, LA: Huntington, 1994); and Alston Chase, *Playing God in Yellowstone: The Destruction of America's First National Park* (San Diego: HBJ, 1986), especially chap. 16: "The New Pantheists."

2. David Chidester, *Patterns of Action: Religion and Ethics in a Comparative Perspective* (Belmont, CA: Wadsworth, 1987), 4.

3. See Bron Taylor, "Earth and Nature-Based Spirituality (Part I): From Deep Ecology to Radical Environmentalism," *Religion* 31.2 (2001): 175–93, especially 175–77 for further discussion of definitional conundrums related to religion and nature. This essay

and the second part of the study, "Earth and Nature-Based Spirituality (Part II): From Deep Ecology to Scientific Paganism," *Religion* 31.3 (2001): 225–45, which focuses on the period from 1960 to the present, complement the material discussed in this article.

4. See Catherine L. Albanese, *Nature Religion in America: From the Algonkian Indians to the New Age* (Chicago: University of Chicago Press, 1990), 42–45.

5. In addition to Albanese's *Nature Religion in America*, other commonly cited scholarly sources include Perry Miller, *Errand into the Wilderness* (Cambridge: Belknap, 1984 [orig. 1956]); Carolyn Merchant, *Ecological Revolutions: Nature, Gender, and Science in New England* (Chapel Hill: University of North Carolina Press, 1989); Peter N. Carroll, *Puritanism and the Wilderness: The Intellectual Significance of the New England Frontier, 1620–1700* (New York: Columbia University Press, 1969); and Roderick Frazier Nash, *Wilderness and the American Mind* (4th ed.; New Haven: Yale University Press, 2001 [orig. 1967]). More recently John Gatta in *Making Nature Sacred: Literature, Religion, and Environment in America from the Puritans to the Present* (Oxford: Oxford University Press, 2004), added interesting interpretations of early Christian attitudes toward nature, including a novel reading of Puritan evangelist Jonathan Edwards, whom Gatta claims anticipated Aldo Leopold's biocentric land ethic earlier than John Muir. Albanese recently revisited her work on American nature religion in *Reconsidering Nature Religion* (Harrisburg, PA: Trinity, 2002).

6. Edmund Burke, *Philosophical Enquiry into the Sublime and Beautiful* (New York: Penguin, 1999 [orig. 1757]); and Immanuel Kant, *Observations on the Feeling of the Beautiful and Sublime* (Berkeley: University of California Press, 2004 [orig. 1764]).

7. Nash, *Wilderness and the American Mind*, 49.

8. See Gatta, *Making Nature Sacred*, 27.

9. See ibid., 35–70, for a provocative discussion of the nature-related spirituality of Calvinist and Quaker thinkers in the late eighteenth and early nineteenth centuries.

10. Walt Whitman, *Leaves of Grass* (150th anniversary ed.; Oxford: Oxford University Press, 2005). Gatta's *Making Nature Sacred* provides excellent discussions of these and many other influential artistic figures: Cole, Bryant, and Cooper (71–88); Whitman (110–16); and Herman Melville (116–25). For more discussion of these figures and movements see also Nash, *Wilderness and the American Mind*, 67–83.

11. See Nash, *Wilderness and the American Mind*, 100–101.

12. Ralph Waldo Emerson, *Nature*, in *The Essential Writings of Ralph Waldo Emerson* (ed. Brooks Atkinson; New York: Modern Library, 2000).

13. For an excellent analysis with Burroughs at the center, see Rebecca Kneale Gould, *At Home in Nature: Modern Homesteading and Spiritual Practice in America* (Berkeley: University of California Press, 2005).

14. Keith Naylor, "Gifford Pinchot," in Bron Taylor, ed., *Encyclopedia of Religion and Nature* (New York & London: Continuum International, 2005), 1280–81. [Henceforth *ERN*.]

15. This latter point is subtly made by Gatta in *Making Nature Sacred*, during his analysis of a number of American writers who were moved by the American land and its creatures.

16. Frederick Jackson Turner, "The Significance of the Frontier in American History," in Turner's *The Frontier in American History* (Melbourne, FL: Krieger, 1962).

17. Stephen Fox, *The American Conservation Movement: John Muir and His Legacy* (Madison: University of Wisconsin Press, 1981); Michael P. Cohen, *The Pathless Way: John*

Muir and American Wilderness (Madison: University of Wisconsin Press, 1984); and Steven J. Holmes, *The Young John Muir: An Environmental Biography* (Madison: University of Wisconsin Press, 1999).

18. Aldo Leopold, *The Sand County Almanac with Essays from Round River* (Oxford: Oxford University Press 1949), 190. This classic work is available in many editions.

19. Ibid., 239.

20. Ibid., 262.

21. David Pecotic, "Pyotr Demianovich Ouspensky," in *ERN* 1225–27; idem, "Georges Ivanovitch Gurdjieff," in *ERN* 730–32; and Lovelock's fascinating "Gaian Pilgrimage," in *ERN* 683–85. For his original theory, see James Lovelock, *Gaia: A New Look at Life on Earth* (rev. ed.; Oxford: Oxford University Press, 1995 [orig. 1979]).

22. For the most revealing discussion, see Curt Meine, *Aldo Leopold: His Life and Work* (Madison: University of Wisconsin Press, 1988), 506–7.

23. Rachel Carson, *Silent Spring* (New York: Houghton Mifflin, 1962).

24. For example, see Rachel Carson, *The Sea around Us* (New York: Oxford University Press, 1950). But an even more forthcoming expression of her spirituality can be found in "The Real World around Us," an address she presented in 1954 to nearly one thousand women journalists and finally published in *Lost Woods: The Discovered Writing of Rachel Carson* (ed. Linda Lear; Boston: Beacon, 1998), 148–63, especially 159. Here one finds her expansive love of nature and the mysteries of the cosmos, as well as the underpinnings for her engagement as an environmental activist, for she concludes with a call to action. Also of special interest is her early claim about the superior moral intuition of women, which would be echoed later by at least some of those who would call themselves "ecofeminists.

25. A 1942 memo from Carson to a person in the marketing department of the publisher of her first book, *Under the Sea Wind* (New York: Dutton, 1991 [orig. 1941]), provides a revealing window into Carson's biocentric motive and, arguably, reveals a kind of animistic imagination. For this memo, see *Lost Woods*, 54–62.

26. This passage is from Loren Eiseley's memoir, *All the Strange Hours* (Lincoln: University of Nebraska Press, 2000), 242.

27. Loren Eiseley, "The Flow of the River," in Eiseley's *The Immense Journey: An Imaginative Naturalist Explores the Mysteries of Man and Nature* (New York: Vintage, 1946), 15–27, at 27.

28. See also Eiseley's *Immense Journey*; *The Unexpected Universe* (New York: Harcourt, 1972), which includes "The Star Thrower" (62–92), his best known essay and probably the best place to start when reading his work; and *The Star Thrower* (New York: Harcourt Brace, 1978), which also reprints this essay (169–85) and many others, including "Science and the Sense of the Holy" (186–201).

29. The following analysis is adapted with permission from Bron Taylor, "A Green Future for Religion?" *Futures Journal* 36.9 (2004): 991–1008, which provides additional details.

30. Bron Taylor, "Introduction," in *ERN* vii–xxi.

31. Roy A. Rappaport, *Ecology, Meaning, and Religion* (Richmond, CA: North Atlantic, 1979); idem, *Ritual and Religion in the Making of Humanity* (Cambridge: Cambridge University Press, 1999); Marvin Harris, "The Cultural Ecology of India's Sacred Cattle," *Current Anthropology* 7 (1966): 51–66; Gerardo Reichel-Dolmatoff, "Cosmology as Ecological Analysis: A View from the Rainforest" *Man* 2.3 (1976): 307–18; and idem, *The*

Forest Within: The Worldview of the Tukano Amazonian Indians (Totnes, U.K.: Themis-Green, 1996).

32. Leopold, *Sand County Almanac*; Lynn White, "The Historic Roots of Our Ecologic Crisis," *Science* 155 (1967): 1203–7; Yi-Fu Tuan, "Discrepancies between Environmental Attitude and Behaviour: Examples from Europe and China," *The Canadian Geographer* 12 (1968): 176–91; Paul Shepard, *Nature and Madness* (San Francisco: Sierra Club, 1982); and Merchant, *Ecological Revolutions.*

33. Vine Deloria Jr., *God Is Red: A Native View of Religion* (Golden, CO; Fulcrum, 1994 [orig. 1972]); Paul Shepard, *The Tender Carnivore and the Sacred Game* (New York: Scribner, 1973); René Dubos, *The Wooing of Earth* (New York: Scribner, 1980); J. Baird Callicott and Thomas Overholt, *Clothed-in-Fur and Other Tales: An Introduction to an Ojibwa World View* (Washington, DC: University Press of America, 1982); J. Baird Callicott and Roger T. Ames, eds., *Nature in Asian Traditions of Thought: Essays in Environmental Philosophy* (Albany: State University of New York Press, 1989); and J. Baird Callicott, *Earth's Insights: A Survey of Ecological Ethics from the Mediterranean Basin to the Australian Outback* (Berkeley: University of California Press, 1994).

34. For an excellent starting point on the ferment regarding Christianity, see Elsbeth Whitney, "Thesis of Lynn White," in *ERN* 1735–36; the Christianity section in *ERN* 316–75; and the cross-references found in these entries.

35. The efforts of scholars with the Religion and Ecology group of the American Academy of Religion and Forum on Religion and Ecology (www.religionandecology.org) are especially noteworthy in this regard. For a brief history of the role of religion-related scholarship in fostering environmental concern, see Bron Taylor, "Religious Studies and Environmental Concern," in *ERN* 1373–79, and the cross-references found there.

36. See James Proctor and Evan Berry, "Social Science on Religion and Nature," in *ERN* 1571–77, for an important introduction to the current state and difficulties that inhere to social scientific inquiry into the links between environment-related beliefs and practices.

37. *ERN* has many examples of such efforts, and with regard to all of the world's major religious traditions and many small groups and individuals. A good starting point is with the entries beginning with the tradition one is interested in, following with the cross-references found in them.

38. Jonathan Z. Smith, *Map Is Not Territory: Studies in the History of Religions* (University of Chicago Press, 1993 [orig. 1978]). The phrase was originally coined by Alfred Korzybski.

39. See Bron Taylor, "Resacralizing Earth: Pagan Environmentalism and the Restoration of Turtle Island," in *American Sacred Space* (ed. David Chidester and Edward T. Linenthal; Bloomington: Indiana University Press, 1995), 97–151.

40. There has been an explosion of scholarly and popular books devoted to pagan spirituality. The best include Graham Harvey, *Contemporary Paganism: Listening People, Speaking Earth* (New York: New York University Press, 1997); Ronald Hutton, *The Triumph of the Moon: A History of Modern Pagan Witchcraft* (Oxford: Oxford University Press, 2000); Sarah Pike, *New Age and Neopagan Religions in America* (New York: Columbia University Press, 2004); Jone Salomonsen, *Enchanted Feminism: Ritual, Gender, and Divinity among the Reclaiming Witches of San Francisco* (London: Routledge, 2001); and Michael York, *The Emerging Network: A Sociology of the New Age and Neo-Pagan Movements* (Lanham, MD: Rowman & Littlefield), 1995.

41. For primers and references, see Bron Taylor, "Radical Environmentalism" in *ERN* 1326–35; and idem, "Radical Environmentalism," in *ERN* 518–24. See also the writings on radical environmentalism featured at www.religionandnature.com/bron. For a global overview, see Bron Taylor, ed. *Ecological Resistance Movements: The Global Emergence of Radical and Popular Environmentalism* (Albany: State University of New York Press, 1995).

42. Amanda Porterfield, "American Indian Spirituality as a Countercultural Movement," in *Religion in Native North America* (ed. Christopher Vecsey; Moscow: University of Idaho Press, 1990), 152–62; Andy Smith, "For All Those Who Were Indian in a Former Life," in *Ecofeminism and the Sacred* (ed. Carol Adams; New York: Continuum, 1993), 172–80; Wendy Rose, "The Great Pretenders: Further Reflections on White-shamanism," in *The State of Native America: Genocide, Colonization, and Resistance* (ed. Annette M. Jaimes; Boston: South End, 1992), 403–21; Bron Taylor, "Earthen Spirituality or Cultural Genocide: Radical Environmentalism's Appropriation of Native American Spirituality," *Religion* 17.2 (1997): 183–215; and Bruce Ziff and Pratima V. Rao, eds., *Borrowed Power: Essays on Cultural Appropriation* (Camden: Rutgers University Press, 1997).

43. Jace Weaver, ed., *Defending Mother Earth: Native American Perspectives on Environmental Justice* (Maryknoll, NY: Orbis, 1996); JeDon Emenhiser, "G-O Road," in *ERN* 701–2; Brandt Peterson, "Indigenous Activism and Environmentalism in Latin America," in *ERN* 833–38; and Tom Goldtooth, "Indigenous Environmental Network," in *ERN* 838–45.

44. Fikret Berkes, *Sacred Ecology: Traditional Ecological Knowledge and Resource Management* (Philadelphia: Taylor & Francis, 1999); Darrell A. Posey and William Balée, eds., *Resource Management in Amazonia: Indigenous and Folk Strategies* (New York: New York Botanical Gardens, 1989); Nancy M. Williams and Graham Baines, *Traditional Ecological Knowledge: Wisdom for Sustainable Development* (Canberra: Centre for Resource and Environmental Studies, Australian National University, 1993); Merideth Dudley and William Balée, "Ethnobotany," in *ERN* 617–22; and Fikret Berkes "Traditional Ecological Knowledge," in *ERN* 1646–49.

45. Richard Grounds, "Native American Languages," in *ERN* 1160–62.

46. For a good representative overview of such a view, see David Suzuki and Peter Knudtson, *Wisdom of the Elders: Honoring Sacred Native Visions of Nature* (New York: Bantam, 1992). Some scholars have challenged the "ecological Indian" stereotype. For example, Sam Gill argued that the notion of Mother Earth in Native American cosmology is a relatively recent invention created largely by Westerners and creative Indians responding to them, rather than a long-term aspect of Native American cultures; see *Mother Earth: An American Story* (Chicago: University of Chicago Press, 1987). Shepard Krech has challenged directly the stereotype of the ecological Indian who leaves the land untouched and has always lived in perfect harmony with it in *The Ecological Indian: Myth and History* (New York: Norton, 1999). For good introductions to such claims and rejoinders, see Matthew Glass, "Mother Earth," in *ERN* 1102–5; and Shepard Krech "American Indians as First Ecologists," in *ERN* 42–45.

47. See www.movieweb.com/movie/pocahontas/pocprod1.txt. The quotation is from an anonymously written movie review available in May 2003.

48. The quotation is from park literature describing the show.

49. See, e.g., the article by Kenny Frost of the Southern Ute Drum, published by *Canku Ota* (*Many Paths*) 56 (9 March 2002), an "online newsletter celebrating Native

America" found at www.turtletrack.org. The article conveys the pride that many American Indians felt at the performance. It was reviewed June 2003 at www .turtletrack.org/Issues02/C003092002/CO_03092002_Olympics.htm.

50. James Redfield's series of books on the Celestine Prophecy provide such a perspective. For an introduction to the New Age and nature religion, see Michael York, "New Age," in *ERN* 1193–97; and Bron Taylor, "Celestine Prophesy," in *ERN* 278–80. Redfield's most important and representative books are *The Celestine Prophecy* (New York: Warner, 1993) and *The Tenth Insight* (New York: Warner, 1996). The influence of Redfield's novels, and an increasing number of other books in the New Age genre that express environmental themes, helps to explain both the greening of the New Age movement, as well as why New Age thinking often permeates contemporary environmentalism. On hybridity in contemporary green religion, see also Bron Taylor, "Diggers, Wolfs, Ents, Elves, and Expanding Universes: Bricolage, Religion, and Violence from Earth First! and the Earth Liberation Front to the Antiglobalization Resistance," in *The Cultic Milieu: Oppositional Subcultures in an Age of Globalization* (ed. J. Kaplan and H. Lööw; Lanham, MD: Altamira/Rowman & Littlefield, 2002), 26–74.

51. See Stephen Rockefeller, "Earth Charter," *ERN* 516–18.

52. See www.earthcharter.org for the entire text, from which these quotations are drawn.

53. Mikhail Gorbechev, "Nature Is My God," *Resurgence: An International Forum for Ecological and Spiritual Thinking* 184 (Sept.–Oct. 1997): 14–15, at 15 (emphasis added).

54. Suzuki and Knudtson, *Wisdom of the Elders*, 227, cf. 167.

55. David Takacs, *The Idea of Biodiversity: Philosophies of Paradise* (Baltimore: John Hopkins University Press, 1996), 254–70, at 270.

56. See, e.g., Ursula Goodenough, *The Sacred Depths of Nature* (New York: Oxford University Press, 1998); Edward Osborne Wilson, *Biophilia: The Human Bond with Other Species* (Cambridge: Harvard University Press, 1984); Stephen R. Kellert and Edward O. Wilson, eds., *The Biophilia Hypothesis* (Washington, DC: Island Press, 1993); Stephen R. Kellert and Timothy J. Farnham, eds., *The Good in Nature and Humanity: Connecting Science, Religion, and Spirituality with the Natural World* (Covelo, CA: Island Press, 2002); and Loyal Rue, *Everybody's Story: Wising Up to the Epic of Evolution* (Albany: State University of New York Press, 2000).

57. The Gallup Organizations asked people about their beliefs on evolution and creation in 1982, 1991, 1993, and 1997. The wording of the questions was identical and the responses nearly identical during this period. Looking at the 1997 data for the general public, 44% of adults held a creationist view, 39% held a theistic evolution view, and 10% held a naturalistic evolution view, while among the scientists, 5% held a creationist view and 40% held a theistic evolution view. See also, Cornelia Dean, "Evolution Takes a Back Seat in U.S. Classes," *New York Times* (1 Feb. 2005).

58. For example, see Bron Taylor, "Conservation Biology," in *ERN* 415–18; and William Jordan III, "Restoration Ecology and Ritual," in *ERN* 1379–81.

59. The civil-religion thesis was most famously developed by Robert Bellah in *The Broken Covenant: American Civil Religion in Time of Trial* (New York: Seabury, 1975).

60. For Daniel Deudney's key theorizing in this regard, see "In Search of Gaian Politics: Earth Religion's Challenge to Modern Western Civilization," in *Ecological Resistance Movements: The Global Emergence of Radical and Popular Environmentalism* (ed. Bron Taylor; Albany: State University of New York Press, 1995), 282–99; and "Ground

Identity: Nature, Place, and Space in Nationalism," in *The Return of Culture and Identity in IR Theory* (ed. Yosef Lapid and Friedrich Kratochwil; Boulder, CO: Rienner, 1996), 129–45.

61. Daniel Deudney, "Global Village Sovereignty: Intergenerational Sovereign Publics, Federal-Republican Earth Constitutions, and Planetary Identities," in *The Greening of Sovereignty in World Politics* (ed. Karen Litfin; Boston: MIT Press, 1998), 299–323, at 318.

62. Mbeki's statements are drawn from his 1 Sept. 2002 speech, which was available in June 2003 at www.dfa.gov.za/docs/wssd029g.htm.

63. For a general discussion of how tourist pilgrimages can contribute to civil religion, see John Sears, *Sacred Places: American Tourist Attractions in the Nineteenth Century* (New York: Oxford University Press, 1989). See also Ronald Engel's analyses of World Heritage Sites and Biosphere Reserves as new humanly constructed sacred places in "Biosphere Reserves and World Heritage Sites," in *ERN* 192–94. This entry and a number of its references fit well with speculations about the possible emergence of a civic earth religion. So does *Man Belongs to the Earth* (Paris: UNESCO, 1988), which begins by quoting a speech attributed (erroneously but famously) to Chief Seattle (of the Suquamish Indians): "The Earth does not belong to Man; man belongs to the Earth" (10, 12). For this interesting historical issue and its importance in environmental discourse, see Michael McKenzie, "Chief Seattle (Sealth)," in *ERN* 1511–12.

BIBLIOGRAPHY

This list, which for reasons of space includes only a fraction of the relevant materials, is a joint product of the editor and authors. There are dozens of relevant websites, including those of virtually all major religious denominations. The five major sites listed here provide links to hundreds of others.

WEBSITES

Alliance of Religions and Conservation: http://www.arcworld.org/
Forum on Religion and Ecology: http://environment.harvard.edu/religion/
Indigenous Environmental Network: http://www.ienearth.org/
National Religious Partnership for the Environment: http://www.nrpe.org/
Religion and Nature: http://www.religionandnature.com/

BOOKS AND ARTICLES

Abbey, Edward. *Desert Solitaire: A Season in the Wilderness.* New York: Ballantine, 1968.
Abdel Haleem, Harfiyah, ed. *Islam and the Environment.* London: Ta-Ha, 1998.
Abrecht, Paul, ed. *Faith, Science, and the Future.* Geneva: World Council of Churches, 1978.
Abrey, Rose, ed. *Judaism and Ecology.* London: Cassell, 1992.
Afrasiabi, Kaveh L. "The Environmental Movement in Iran: Perspectives from Above and Below." *Middle East Journal* 57.3 (2003): 432–48.
Aharoni, Yohanan. *The Land of the Bible: A Historical Geography.* 2nd ed. Translated and edited by Anson Rainey. London: Burns & Oates, 1979.
Ahmad, Akhtaruddin. *Islam and the Environmental Crisis.* London: Ta-Ha, 1997.
Albanese, Catherine L. *Nature Religion in American: From the Algonkian Indians to the New Age.* Chicago: University of Chicago Press, 1990.
Allan, Sarah. *The Way of Water and the Sprouts of Virtue.* Albany: State University of New York Press, 1997.
Ames, Roger T. "Taoism and the Nature of Nature." *Environmental Ethics* 8 (1986): 317–50.
Ammar, Nawal H. "An Islamic Response to the Manifest Ecological Crisis: Issues of Justice." Pages 131–46 in *Visions of a New Earth: Religious Perspectives on Population,*

Consumption, and Ecology. Edited by Harold Coward and Daniel C. Maguire. Albany: State University of New York Press, 2000.

Anderson, E. N. *Ecologies of the Heart: Emotion, Belief, and the Environment.* New York: Oxford University Press, 1996.

Anisimov, Arkadii F. "The Shaman's Tent of the Evenks and the Origin of the Shamanistic Rite." In *Siberian Shamanism.* Edited by Henry Michael. Toronto, 1963.

Apffel-Marglin, Frederique, with the Andean Project on Peasant Technologies. *The Spirit of Regeneration: Andean Culture Confronting Western Notions of Development.* New York, 1998.

Atharva Veda. Translated by Devi Chand. New Delhi: Munsiram Manoharlal, 1982.

Attfield, Robin. *Environmental Ethics.* Cambridge: Polity Press, 2003.

Austin, Mary Hunter. *The Land of Little Rain.* New York: Dover, 1996.

Austin, Richard Cartwright. *Beauty of the Lord: Awakening the Senses.* Atlanta: John Knox, 1988.

Ba Kader, Abou Bakr Ahmed, Abdul Latif Tawfik El Shirazy Al Sabagh, Mohamed Al Sayyed Al Glenid, and Mawil Y. Izzi Deen. *Islamic Principles for the Conservation of the Natural Environment.* Gland, Switzerland: International Union for Conservation of Nature and Natural Resources, 1983. 2nd edition online http://www.islamset.com/env/contenv.html.

Bakken, Peter W., Joan Gibb Engel, and J. Ronald Engel, eds. *Ecology, Justice, and Christian Faith: A Guide to the Literature.* Westport, CT: Greenwood, 1995.

Barbour, Ian, ed. *Earth Might Be Fair.* Englewood Cliffs, NJ: Prentice-Hall, 1972.

———, ed. *Western Man and Environmental Ethics.* Reading, MA: Addison-Wesley, 1973.

Barnhill, David Landis, ed. *At Home on the Earth: Becoming Native to Our Place.* Berkeley: University of California Press, 1999.

Barnhill, David Landis, and Roger Gottlieb, eds. *Deep Ecology and World Religions: New Essays on Sacred Ground.* Albany: State University of New York Press, 2001.

Bartram, William. *Travels of William Bartram.* Edited by Mark Van Doren. New York: Dover, 1928.

Basil (Osborne) of Sergievo. "Beauty in the Divine and Nature." *Sourozh* 70 (1997).

Basso, Keith H. *Wisdom Sits in Places: Landscape and Language among the Western Apache.* Albuquerque: University of New Mexico Press, 1996.

Batchelor, Martine, and Kerry Brown, eds. *Buddhism and Ecology.* London: Cassell, 1992.

Bell, Stephen. *Rebel, Priest, and Prophet: A Biography of Dr. Edward McGlynn.* New York: Schalkenbach Foundation, 1968.

Berkes, Fikret. *Sacred Ecology: Traditional Ecological Knowledge and Resource Management.* Philadelphia: Taylor & Francis, 1999.

Bernstein, Ellen, ed. *Ecology and the Jewish Spirit: Where Nature and the Sacred Meet.* Woodstock, VT: Jewish Lights, 1998.

Berry, Thomas. *The Dream of the Earth.* San Francisco: Sierra Club, 1988.

———. *The Great Work: Our Way into the Future.* New York: Bell Tower, 1999.

Berry, Wendell. *Collected Poems, 1957–1982.* San Francisco: North Point, 1984.

———. *Recollected Essays, 1965–1980.* San Francisco: North Point, 1981.

Berthrong, John H. *Transformation of the Confucian Way.* Boulder, CO: Westview, 1998.

Berthrong, John H., and Evelyn Nagai Berthrong. *Confucianism: A Short Introduction.* Oxford: OneWorld, 2000.

Beston, Henry. *The Outermost House: A Year of Life on the Great Beach of Cape Cod.* New York: Holt, 1992.

Bhagavadgita. Gorakhpur, India: Gita, 1996.

Birch, Charles, William Eakin, and Jay McDaniel, eds. *Liberating Life: Contemporary Approaches to Ecological Theology.* Maryknoll, NY: Orbis, 1990.

Birch, Charles, and Lukas Vischer. *Living with the Animals: The Community of God's Creatures.* Geneva: World Council of Churches, 1997.

Black, Alison Harley. *Man and Nature in the Philosophical Thought of Wang Fu-chih.* Seattle: University of Washington Press, 1989.

Blaser, Mario, H. A. Feit, and G. McRae, eds. *In the Way of Development: Indigenous Peoples, Life Projects, and Globalization.* London: Zed, 2004.

Boersma, Jan J. *The Torah and the Stoics on Humankind and Nature: A Contribution to the Debate on Sustainability and Quality.* Leiden: Brill, 2001.

Boff, Leonardo. *Cry of the Earth, Cry of the Poor.* Translated by Philip Berryman. Maryknoll, NY: Orbis, 1997.

———. *Ecology and Liberation: A New Paradigm.* Translated by John Cumming. Maryknoll, NY: Orbis, 1995.

Bond, George D. *Buddhism at Work: Community Development, Social Empowerment, and the Sarvodaya Movement.* Bloomfield, CT: Kumarian, 2004.

Boston Research Center for the 21st Century. *Buddhist Perspectives on the Earth Charter.* Cambridge: Boston Research Center, 1997.

Brock, Sebastian. "World and Sacrament in the Writings of the Syrian Fathers." *Sobornost* 6.10 (1974).

Brown, Lester. "Challenges of the New Century." In *State of the World 2000.* New York: Norton, 2000.

Bruun, Ole. *Fengshui in China: Geomantic Divination between State Orthodoxy and Popular Religion.* Honolulu: University of Hawai'i Press, 2003.

Bryan, David. *Cosmos, Chaos, and the Kosher Mentality.* Sheffield: Sheffield Academic Press, 1995.

Buell, Lawrence. *The Environmental Imagination: Thoreau, Nature Writing, and the Formation of American Culture.* Cambridge: Harvard University Press, 1995.

Butala, Sharon. *Perfection of the Morning: A Woman's Awakening in Nature.* St. Paul: Hungry Mind, 1994.

Callicott, J. Baird. *Earth Insight: A Multicultural Survey of Ecological Ethics from the Mediterranean Basin to the Australian Outback.* Berkeley: University of California Press, 1994.

Callicott, J. Baird, and Roger T. Ames, eds. *Nature in Asian Traditions of Thought: Essays in Environmental Philosophy.* Albany: State University of New York Press, 1989.

Callicott, J. Baird, and Michael P. Nelson. *American Indian Environmental Ethics: An Ojibwa Case Study.* Upper Saddle River, NJ, 2004.

Carrasco, David, ed. *The Imagination of Matter: Religion and Ecology in Mesoamerican Traditions.* Oxford: BAR, 1989.

Carroll, Peter N. *Puritanism and the Wilderness: The Intellectual Significance of the New England Frontier, 1620–1700.* New York: Columbia University Press, 1969.

Carson, Rachel. *Lost Woods: The Discovered Writing of Rachel Carson.* Edited by Linda Lear. Boston: Beacon, 1998.

————. *Silent Spring*. Boston: Houghton Mifflin, 1962.

Chan, Wing-tsit. *A Source Book in Chinese Philosophy*. Princeton: Princeton University Press, 1963.

Chapple, Christopher Key. "Hindu Environmentalism: Traditional and Contemporary Resources." In *Worldviews and Ecology: Religion, Philosophy, and the Environment*. Edited by Mary Evelyn Tucker and John A. Grim. Lewisburg: Bucknell University Press, 1993.

————, ed. *Jainism and Ecology: Non-violence in the Web of Life*. Cambridge: Center for the Study of World Religions, Harvard Divinity School, 2000.

————. "Jainism and Nonviolence." In *Subverting Hatred: The Challenge of Nonviolence in Religious Traditions*. Edited by Daniel L. Smith-Christopher. Cambridge: Boston Research Center for the 21st Century, 1998.

————. *Nonviolence to Animals, Earth, and Self in Asian Traditions*. Albany: State University of New York Press, 1993.

Chapple, Christopher Key, and Mary Evelyn Tucker, eds. 2000. *Hinduism and Ecology: The Intersection of Earth, Sky, and Water*. Cambridge: Center for the Study of World Religions, Harvard Divinity School, 2000.

Chen, Ellen M. "The Meaning of Te in the Tao Te Ching: An Examination of the Concept of Nature in Chinese Taoism." *Philosophy East and West* 23 (1973): 457–70.

Cheng Chung-ying. "On the Environmental Ethics of the Tao and the Ch'i." *Environmental Ethics* 8 (1986): 351–70.

Cherrett, J. M. "Key Concepts: The Results of a Survey of Our Members' Opinions." In *Ecological Concepts: The Contribution of Ecology to an Understanding of the Natural World*. Edited by J. M. Cherrett. London: Blackwell Scientific, 1989.

Christiansen, Drew, and Walter Grazer, eds. *"And God Saw That It Was Good": Catholic Theology and the Environment*. Washington, DC: United States Catholic Conference, 1996.

Chryssavgis, John. *Cosmic Grace, Humble Prayer: The Ecological Vision of the Green Patriarch Bartholomew I*. Grand Rapids: Eerdmans, 2003.

Clark, Stephen R. L. *The Moral Status of Animals*. Oxford: Clarendon, 1977.

Cleary, Edward L., and Timothy J. Steigenga, eds. *Resurgent Voices in Latin America: Indigenous Peoples, Political Mobilization, and Religious Change*. New Brunswick: Rutgers University Press, 2004.

Clunas, Craig. *Fruitful Sites: Garden Culture in Ming Dynasty China*. Durham: Duke University Press, 1996.

————. *Superfluous Things: Material Culture and Social Status in Early Modern China*. Honolulu: University of Hawai'i Press, 2004.

Coates, Peter. *Nature: Western Attitudes since Ancient Times*. Berkeley: University of California Press, 1998.

Cobb, John B., Jr. *Is it Too Late? A Theology of Ecology*. Denton, TX: Environmental Ethics, 1995 (orig. 1972).

Cohen, Jeremy. *"Be Fertile and Increase, Fill the Earth and Master It": The Ancient and Medieval Career of a Biblical Text*. Ithaca: Cornell University Press, 1989.

Cohen, Michael P. *The Pathless Way: John Muir and American Wilderness*. Madison: University of Wisconsin Press, 1984.

Cohen, Noah J. *Tza'ar Ba'ale Hayim: The Prevention of Cruelty to Animals, Its Bases, Development, and Legislation in Hebrew Literature*. 2nd ed. New York: Feldheim, 1976.

Conradi, Ernst, Charity Majiza, Jim Cochrane, Weliel T. Sigabi, Victor Molobi, and David Field. "Seeking Eco-Justice in the South African Context." In *Earth Habitat: Eco-Injustice and the Church's Response.* Edited by Dieter Hessel and Larry Rasmussen. Minneapolis: Fortress, 2001.

Conspirando: Revista de ecofeminismo, spiritualidad y teologia. 1992–.

Cooper, Susan Fenimore. *Rural Hours.* Edited by Rochelle Johnson and Daniel Patterson. Athens: University of Georgia Press, 1998.

Crocker, Jon C. *Vital Souls: Bororo Cosmology, Natural Symbolism, and Shamanism.* Tucson: University of Arizona Press, 1985.

Daly, Herman E., and John B. Cobb Jr. *For the Common Good: Redirecting the Economy toward Community, the Environment, and a Sustainable Future.* 2nd ed. Boston: Beacon, 1994.

Daneel, Marthinus. *African Earthkeepers: Holistic Interfaith Mission.* Maryknoll, NY: Orbis, 2001.

Deane Drummond, Cecelia. *The Ethics of Nature.* Oxford: Blackwell, 2004.

De Barros, Marcelo, and José Luis Caravias. *Teología de la tierra.* Madrid: Paulinas, 1988. Simultaneous publication: *Teologia da terra.* Petrópolis, Brazil: Vozes, 1988.

de Bary, W. Theodore. *The Trouble with Confucianism.* Cambridge: Harvard University Press, 1991.

de Bary, W. Theodore, Irene Bloom, and Richard Lufrano, eds. *Sources of Chinese Tradition.* 2 vols. New York: Columbia University Press, 2000.

Deloria, Vine, Jr. *God Is Red: A Native View of Religion.* Updated ed. Golden, CO: 1972. Reprinted Golden, CO: Fulcrum, 1994.

Deussen, Paul. *Sixty Upanisads of the Veda.* Translated by V. M. Bedekar and G. B. Palsule. Delhi: Motilal Banarsidass, 1980.

Devall, Bill, and George Sessions. *Deep Ecology.* Salt Lake City: Smith, 1985.

DeWitt, Calvin B. "Behemoth and Batrachians in the Eye of God: Responsibility to Other Kinds in Biblical Perspective." In *Christianity and Ecology: Seeking the Well-Being of Earth and Humans.* Edited by Dieter T. Hessel and Rosemary Radford Ruether. Cambridge: Center for the Study of World Religions, Harvard Divinity School, 2000.

———. *Caring for Creation: Responsible Stewardship of God's Handiwork.* Grand Rapids: Baker/Washington, DC: Center for Public Justice, 1998.

———. "Complementarities of Scientific Understanding of Nature with Religious Perspectives of Creation." In *The Good in Nature and Humanity: Connecting Science Religion and Spirituality with the Natural World.* Edited by Stephen R. Kellert and Timothy Farnham. Washington, DC: Island Press, 2002.

———. "A Contemporary Evangelical Perspective." In *The Greening of Faith.* Edited by John E. Carroll, Paul Brockelman, and Mary Westfall. Hanover: University of New Hampshire Press, 1997.

———. *A Sustainable Earth: Religion and Ecology in the Western Hemisphere.* Marcelona, MI: Au Sable Institute, 1987.

Dillard, Annie. *Pilgrim at Tinker Creek.* Toronto: Bantam, 1973.

Dolan, R. J. "Emotion, Cognition, and Behavior." *Science* 298 (2002): 1191–94.

Dombrowski, Daniel. *Hartshorne and the Metaphysics of Animal Rights.* Albany: State University of New York Press, 1988.

Dubos, René. *The Wooing of Earth.* New York: Scribner, 1980.

Dwivedi, O. P. "Classical India." In *A Companion to Environmental Philosophy.* Edited by Dale Jamieson. New York: Blackwell, 2000.

———. "Dharmic Ecology." In *Hinduism and Ecology: The Intersection of Earth, Sky, and Water.* Edited by Christopher K. Chapple and Mary Evelyn Tucker. Cambridge: Center for the Study of World Religions, Harvard Divinity School, 2000.

———. *India's Environmental Policies, Programmes, and Stewardship.* London: Macmillan, 1997.

———. "Our Karma and Dharma to the Environment." Pp. 59–74 in *Environmental Stewardship: History, Theory, and Practice.* Edited by Mary Ann Beavis. Winnipeg: University of Winnipeg, Institute of Urban Studies, 1994.

———. *Vasudhaiv Kutumbakam: A Commentary on Atharvediya Prithivi Sukta.* 2nd ed. Jaipur: Institute for Research and Advanced Studies 1998 (orig. 1995).

———. "Vedic Heritage for Environmental Stewardship." *Worldviews: Environment, Culture, and Religion* 1.1 (April 1997).

Dwivedi, O. P., and B. N. Tiwari. *Environmental Crisis and Hindu Religion.* New Delhi: Gitanjali, 1987.

Dwivedi, O. P., B. N. Tiwari, and R. N. Tripathi. "Hindu Concept of Ecology and the Environmental Crisis." *Indian Journal of Public Administration* 30.1 (**YEAR**): 33–67.[i]

Earth Council. *The Earth Charter: Values and Principles for a Sustainable Future.* San Jose, Costa Rica: Earth Charter International Secretariat, 1998.

Eaton, Heather, and Lois Ann Lorentzen, eds. *Ecofeminism and Globalization: Exploring Culture, Context, and Religion.* Lanham, MD: Rowman & Littlefield, 2003.

Ehrlich, Gretel. *Questions of Heaven.* Boston: Beacon, 1997.

Ehrlich, Paul R. *Human Natures.* Washington, DC: Island Press, 2001.

Ehrlich, Paul R., and Anne H. Ehrlich. *One with Nineveh: Politics, Consumption, and the Human Future.* Washington, DC: Island Press, 2004.

Eiseley, Loren. *All the Strange Hours.* Lincoln: University of Nebraska Press, 2000.

Eisenberg, Evan. *The Ecology of Eden.* New York: Knopf, 1998.

Elder, John. *Imagining the Earth: Poetry and the Vision of Nature.* Urbana: University of Illinois Press, 1985.

Eldredge, Niles. *Life in the Balance: Humanity and the Biodiversity Crisis.* Princeton: Princeton University Press, 1998.

Elon Ari, Naomi Mara Hyman, and Arthur Waskow, eds. *Trees, Earth, and Torah: A Tu B'Shevat Anthology.* Philadelphia: Jewish Publication Society, 1999.

Elvin, Mark. *The Retreat of the Elephants: An Environmental History of China.* New Haven: Yale University Press, 2004.

Elvin, Mark, and Liu Ts'ui-jung, eds. *Sediments of Time: Environment and Society in Chinese History.* Cambridge: Cambridge University Press, 1998.

Emerson, Ralph Waldo. *The Selected Writings of Ralph Waldo Emerson.* New York: Random, 1950.

Engel, J. Ronald, and Joan Gibb Engel, eds. *Ethics of Environment and Development: Global Challenge, International Response.* Tucson: University of Arizona Press, 1990.

Felix, Yehuda. *Nature and Man in the Bible: Chapters in Biblical Ecology.* New York: Soncino, 1981.

Fern, Richard. *Nature, God, and Humanity: Envisioning and Ethics of Nature.* Cambridge: Cambridge University Press, 2002.

Ferro Medina, Alfredo, ed. *A teologia se fez terra: primeiro encontro latino-americano de teologia da terra.* São Leopoldo, Brazil: Sinodal, 1991.

Foltz, Richard C. *Animals in Islamic Tradition and Muslim Cultures*. Oxford: OneWorld, 2005.

———, ed. *Environmentalism in the Muslim World*. Hauppage, NY: Nova Science, 2005.

———, ed. *Worldviews, Religion, and the Environment: A Global Anthology*. Belmont, CA: Wadsworth/Thomson Learning, 2003.

Foltz, Richard C., Frederick M. Denny, and Azizan Baharuddin, eds. *Islam and Ecology: A Bestowed Trust*. Cambridge: Center for the Study of World Religions, Harvard Divinity School, 2003.

Foster, George M., and Barbara Gallatin Anderson. *Medical Anthropology*. Oxford: Pergamon, 1978.

Fox, Alan. "Process Ecology and the Ideal Dao." *Journal of Chinese Philosophy* 32.1 (2005): 47–57.

Fox, Matthew. *The Coming of the Cosmic Christ: The Healing of Mother Earth and the Birth of a Global Renaissance*. San Francisco: Harper & Row, 1988.

———. *Creation Spirituality: Liberating Gifts for the Peoples of the Earth*. San Francisco: Harper, 1991.

———. *Original Blessing*. Santa Fe: Bear, 1983.

Fox, Stephen. *The American Conservation Movement: John Muir and His Legacy*. Madison: University of Wisconsin Press, 1981.

Gardner, Gary. *Invoking the Spirit: Religion and Spirituality in the Quest for a Sustainable World*. Worldwatch Paper 164. Washington, DC: Worldwatch Institute, 2002.

Gatta, John. *Making Nature Sacred: Literature, Religion, and Environment in America from the Puritans to the Present*. New York: Oxford University Press, 2004.

Gebara, Ivone. *Intuiciones ecofeministas: ensayo para repensar el conocimiento y la religión*. Translated by Graciela Pujo. Trotta, 2000.

———. *Longing for Running Water: Ecofeminism and Liberation*. Translated by David Molineaux. Minneapolis: Fortress, 1999.

Gerstenfeld, Manferd. *Judaism, Environmentalism, and the Environment: Mapping and Analysis*. Jerusalem: Jerusalem Institute for Israel Studies/Rubin Mass, 1998.

Gill, Sam D. *Mother Earth: An American Story*. Chicago: University of Chicago Press, 1987.

Giradot, N. J., James Miller, and Liu Xiaogan, eds. *Daoism and Ecology: Ways within a Cosmic Landscape*. Cambridge: Center for the Study of World Religions, Harvard Divinity School, 2001.

Glacken, Clarence. *Traces on the Rhodian Shore: Nature and Culture in Western Thought from Ancient Times to the End of the Eighteenth Century*. Berkeley: University of California Press, 1969.

Gnanadason, Aruna. "Creator God, in Your Grace, Transform the Earth: An Ecofeminist Ethic of Resistance, Prudence, and Care." DMin thesis, San Francisco Theological Seminary, 2004.

Golley, Frank. *A Primer for Environmental Literacy*. New Haven: Yale University Press, 1998.

Gomez-Ibanez, Daniel. "Spiritual Dimensions of the Environmental Crisis." In *A Source Book for the Community of Religions*. Edited by Joel D. Beversluis. Chicago: Council for a Parliament of the World's Religions, 1993.

Good, Byron. *Medicine, Rationality, and Experience: An Anthropological Perspective*. Cambridge: Cambridge University Press, 1994.

Goodenough, Ursula. *The Sacred Depths of Nature.* New York: Oxford University Press, 1998.

Gosling, David L. *Religion and Ecology in India and Southeast Asia.* London: Routledge, 2001.

Gottlieb, Roger. S. *A Greener Faith: Religious Environmentalism and Our Planet's Future.* New York: Oxford University Press, 2006.

———. *Joining Hands: Politics and Religion Together for Social Change.* Cambridge, MA: Westview, 2002.

———, ed. *Liberating Faith: Religious Values for Justice, Peace, and Ecological Wisdom.* Lanham, MD: Rowman & Littlefield, 2003.

———. *A Spirituality of Resistance: Finding a Peaceful Heart and Protecting the Earth.* Lanham, MD: Rowman & Littlefield, 2003.

———, ed. *This Sacred Earth: Religion, Nature, Environment.* 2nd ed. New York: Routledge, 2004.

Gould, Rebecca Kneale. *At Home in Nature: Modern Homesteading and Spiritual Practice in America.* Berkeley: University of California Press, 2005.

Gould, Stephen Jay. "Chance Riches." *Natural History* 89.11 (Nov. 1980): 36–44.

———. *The Structure of Evolutionary Theory.* Cambridge: Harvard University Press, 2002.

———. *Wonderful Life: The Burgess Shale and the Nature of History.* New York: Norton, 1989.

Granberg-Michaelson, Wesley. *A Worldly Spirituality: The Call to Redeem Life on Earth.* San Francisco: Harper & Row, 1984.

Griffith, Ralph T. H. *The Hymns of the Rig Veda.* Edited by J. L. Shastri. Delhi: Motilal Banarasidass, 1973 (orig. 1889).

Grim, John A., ed. *Indigenous Traditions and Ecology: The Interbeing of Cosmology and Community.* Cambridge: Center for the Study of World Religions, Harvard Divinity School, 2001.

———. *The Shaman: Patterns of Religious Healing among the Ojibway Indians.* Norman, OK, 1983.

Grinde, Donald A., and Bruce E. Johansen. *Ecocide of Native America: Environmental Destruction of Indian Lands and People.* Santa Fe, 1995.

Guroian, Vigen. "Ecological Ethics: An Ecclesial Event." In *Ethics after Christendom: Towards an Ecclesial Christian Ethic.* Grand Rapids: Eerdmans, 1994.

Guss, David M. *To Weave and Sing: Art, Symbol, and Narrative in the South American Rainforest.* Berkeley, 1989.

Gustafson, James M. *A Sense of the Divine: The Natural Environment from a Theocentric Perspective.* Cleveland: Pilgrim, 1994.

Gutierrez, Gustavo. *Las casas: In Search of the Poor of Jesus Christ.* Maryknoll, NY: Orbis, 1993.

Habito, Ruben L. F. *Healing Breath: Zen Spirituality for a Wounded Earth.* Maryknoll, NY: Orbis, 1993.

Hallman, D. *Ecotheology: Voices from the South and North.* Maryknoll, NY: Orbis, 1994.

Hallowell, A. Irving. "Ojibwa Ontology, Behavior, and World View." In *Culture and History: Essays in Honor of Paul Radin.* Edited by Stanley Diamond. New York, 1960.

Hardin, Garrett. "The Tragedy of the Commons." *Science* 169 (1968): 1243–48.

Hargrove, Eugene, ed. *Religion and Environmental Crisis.* Athens: University of Georgia Press, 1986.

Hart, John. *Ethics and Technology: Innovation and Transformation in Community Contexts.* Cleveland: Pilgrim, 1997.

———. *Sacramental Commons.* Lanham, MD: Rowman & Littlefield, 2006.

———. *The Spirit of the Earth: A Theology of the Land.* Mahwah, NJ: Paulist Press, 1984.

———. *What Are They Saying about... Environmental Theology?* Mahwah, NJ: Paulist Press, 2004.

Harvey, Graham. *Contemporary Paganism: Listening People, Speaking Earth.* New York: New York University Press, 1997.

Haught, John F. *The Promise of Nature: Ecology and Cosmic Purpose.* Mahwah, NJ: Paulist Press, 1993.

Hedstrom, Ingemar, ed. *Volveran las golondrinas? la reintegracion de la creación desde una perspectiva latinoamericana.* San Jose, Costa Rica: DEI, 1990.

Henderson, John B. *The Development and Decline of Chinese Cosmology.* New York: Columbia University Press, 1984.

Heschel, Abraham Joshua. *God in Search of Man: A Philosophy of Judaism.* Philadelphia: Jewish Publication Society, 1956.

———. *Man Is Not Alone.* New York: Farrar, Straus & Young, 1951.

———. *The Sabbath: The Meaning for Modern Man.* New York: Farrar, Straus & Giroux, 1991 (orig. 1951).

Hessel, Dieter T., and Rosemary Radford Ruether, eds. *Christianity and Ecology: Seeking the Well-Being of Earth and Humans.* 2000.

Hiebert, Theodore. *The Yahwist Landscape: Nature and Religion in Early Israel.* New York: Oxford University Press, 1996.

Hill, Brennan R. *Christian Faith and the Environment: Making Vital Connections.* Maryknoll, NY: Orbis, 1998.

Hinga, Teresia. "The Gikuyu Theology of Land and Environmental Justice." Pp. 172–84 in *Women Healing Earth: Third World Women on Ecology, Feminism, and Religion.* Edited by Rosemary Radford Ruether. Maryknoll, NY: Orbis, 1994.

Hirsch, Samson Raphael. *Eternal Judaism: Selected Essays.* Edited and translated by I. Gurnfeld. 2 vols. London: Soncino, 1959.

Hogan, Linda. *Dwellings: A Spiritual History of the Living World.* New York: Touchstone, 1995.

Holmes, Steven J. *The Young John Muir: An Environmental Biography.* Madison: University of Wisconsin Press, 1999.

Houten, Richard Van. "Nature and Tzu-jan in Early Chinese Philosophical Literature." *Journal of Chinese Philosophy* 15 (1988): 33–49.

Hull, Fritz, ed. *Earth and Spirit: The Spiritual Dimension of the Environmental Crisis.* New York: Continuum, 1993.

Hultkrantz, Ake. "Ecology." In *Encyclopedia of Religion.* Edited by M. Eliade and C. J. Adams. New York: Macmillan, 1987.

Hume, Robert Ernest. *The Thirteen Principal Upanishads.* New York: Oxford University Press, 1977.

Hunt-Badiner, Alan, ed. *Dharma Gaia: A Harvest of Essays in Buddhism and Ecology.* Berkeley: Parallax, 1990.

Hüttermann, Aloys. *The Ecological Message of the Torah: Knowledge, Concepts, and Laws Which Made Survival in a Land of "Milk and Honey" Possible.* Atlanta: Scholars Press, 1999.

Hutton, Ronald. *The Triumph of the Moon: A History of Modern Pagan Witchcraft*. Oxford: Oxford University Press, 2000.

Ignatius IV, Patriarch of Antioch. "Three Sermons on the Environment: Creation, Spirituality, Responsibility." *Sourozh* 38 (1989).

Ikhwan al-Safa. *The Case of the Animals versus Man before the King of the Jinn*. Translated by Lenn Evan Goodman. Boston: Twayne, 1978.

Ingold, Tim. *Perception of the Environment: Essays in Livelihood, Dwelling, and Skill*. London, 2000.

Ip Po-keung. "Taoism and the Foundation of Environmental Ethics." *Environmental Ethics* 5 (1986): 335–43.

Isaacs, Ronald H. *The Jewish Sourcebook on the Environment and Ecology*. Jerusalem, 1998.

Izzi Dien, Mawil Y. *The Environmental Dimensions of Islam*. Cambridge: Lutterworth, 2000.

Jaina Sutras, part 1: *The Akaranga Sutra; the Kalpa Sutra*. Translated by Hermann Jacobi. Reprinted New York: Dover, 1968 (orig. 1884).

Jaini, Padmanabh S. *Gender and Salvation*. Berkeley: University of California Press, 1993.

———. *The Jaina Path of Purification*. Berkeley: University of California Press, 1989.

Jeffers, Robinson. *The Selected Poetry of Robinson Jeffers*. New York: Random, 1938.

Jefferson, Thomas. *Notes on the State of Virginia*. Reprinted New York: Harper & Row, 1964.

Jenkins, Timothy N. "Chinese Traditional Thought and Practice: Lessons for an Ecological Economics Worldview." *Ecological Economics* 16 (2002): 39–52.

John (Zizioulas), Metropolitan of Pergamon. "Man, the Priest of Creation: A Response to the Ecological Problem." In *Living Orthodoxy in the Modern World*. Edited by A. Walker and C. Carras. London: SPCK, 1996.

———. "Preserving God's Creation: Three Lectures on Theology and Ecology." *King's Theological Review* 12 (1989). Reprinted in *Sourozh* 39–41 (1990).

Johnson, Elizabeth A. *Women, Earth, and Creator Spirit*. Mahwah, NJ: Paulist Press, 1993.

Jullien, François. *The Propensity of Things: Towards a History of Efficacy in China*. Translated by Janet Lloyd. New York: Zone, 1995.

Kalland, Arne, ed. *Nature across Cultures: Views of Nature and the Environment in Non-Western Cultures*. Boston: Kluwer, 2003.

Kallistos of Diokleia. *Through the Creation to the Creator*. London: Friends of the Center Papers, 1997.

Kapleau, Philip. *To Cherish All Life: A Buddhist Case for Becoming Vegetarian*. San Francisco: Harper & Row, 1982.

Karlsson, B. G. *Contested Belonging: An Indigenous People's Struggle for Forest and Identity in Sub-Himalayan Bengal*. 2nd ed. Richmond, U.K., 2000.

Kaufman, Gordon D. *In Face of Mystery: A Constructive Theology*. Cambridge: Harvard University Press, 1993.

Kaza, Stephanie, *Hooked! Buddhist Writings on Greed, Desire, and the Urge to Consume*. Boston: Shambhala, 2005.

Kaza, Stephanie, and Kenneth Kraft, eds. *Dharma Rain: Sources of Buddhist Environmentalism*. Boston: Shambala, 2000.

Keller, David R., and Frank B. Golley, eds. *The Philosophy of Ecology: From Science to Synthesis*. Athens: University of Georgia Press, 2000.

Kellert, Stephen, and Timothy Farnham, eds. *The Good in Nature and Humanity: Connecting Science and Spirituality with the Natural World.* Washington, DC: Island Press, 2002.

Khalid, Fazlun, and Joanne O'Brien, eds. *Islam and Ecology.* New York: Cassell, 1992.

Kickingbird, Kirke, and Karen Ducheneaux. *One Hundred Million Acres.* New York, 1973.

Kim, Yung Sik. *The Natural Philosophy of Chu Hsi, 1130–1200.* Philadelphia: American Philosophical Society, 2000.

Kinsley, David. *Religion and Ecology: Ecological Spirituality in a Cross-cultural Perspective.* Englewood Cliffs, NJ: Prentice-Hall, 1995.

Kleinman, Arthur. *Patients and Healers in the Context of Culture: An Exploration of the Borderland between Anthropology, Medicine, and Psychiatry.* Berkeley: University of California Press, 1980.

Knitter, Paul. *One Earth, Many Religions: Multifaith Dialogue and Global Responsibility.* Maryknoll, NY: Orbis, 1995.

Koenig, H. G., M. E. McCullough, and D. B. Larson. *Handbook of Religion and Health.* Oxford: Oxford University Press, 2001.

Kohn, Livia. *Daoism in Chinese Culture.* Cambridge, MA: Three Pines, 2001.

———. *Health and Long Life: The Chinese Way.* Cambridge, MA: Three Pines, 2005.

Krech, Shepard, III. *The Ecological Indian: Myth and History.* New York: Norton, 1999.

Kumar, Satish. *No Destination: An Autobiography.* Totnes: Green Books, 1993.

Kuriyama, Shigehisa. *The Expressiveness of the Body and the Divergence of Greek and Chinese Medicine.* New York: Zone, 1999.

Laduke, Winona. *All Our Relations: Native Struggles for Land and Life.* Boston: South End, 1999.

Lal, S. K. "Pancamahabhutas: Origin and Myths in Vedic Literature." Pp. 5–21 in *Prakrti: An Integral Vision.* Edited by Sampat Narayanan. New Delhi: Indira Gandhi National Centre for the Arts, 1995.

Lane, Belden C. *Landscapes of the Sacred: Geography and Narrative in American Spirituality.* New York: Paulist Press, 1988.

Lee, Charles, ed. *Proceedings of the First National People of Color Environmental Leadership Summit.* New York: United Church of Christ Commission for Racial Justice, 1992.

Le Guin, Ursula. *Always Coming Home.* New York: Harper & Row, 1985.

Leopold, Aldo. *A Sand County Almanac and Sketches Here and There.* New York: Oxford University Press. 1949.

Levy, Zeev, and Nadav Levy. *Ethics, Emotions, and Animals: On the Moral Status of Animals* (Hebrew). Tel Aviv: Sifriat Po'alim/Haifa: University of Haifa Press, 2002.

Li, Guohao, Zhang Mengwen, and Cao Tianqin, eds. *Explorations in the History of Science and Technology in China.* Shanghai: Shanghai Chinese Classics, 1982.

Limouris, Gennadios, ed. *Justice, Peace, and the Integrity of Creation: Insights from Orthodoxy.* Geneva: World Council of Churches, 1990.

Lincoln, Bruce. *Priests, Warriors, and Cattle: A Study in the Ecology of Religions.* Berkeley: University of California Press, 1981.

Linzey, Andrew. *Animal Gospel: Christian Faith as If Animals Mattered.* London: Hodder & Stoughton/Louisville: Westminster John Knox, 1999.

———. *Animal Rites: Liturgies of Animal Care.* London: SCM/Cleveland: Pilgrim, 1999.

———. *Animal Theology.* London: SCM/Chicago: University of Illinois Press, 1994.

———. *Christianity and the Rights of Animals.* London: SPCK/New York: Crossroad, 1987.

Linzey, Andrew, and Paul Barry Clarke, eds. *Animal Rights: A Historical Anthology.* New York: Columbia University Press, 2005.

Linzey, Andrew, and Dan Cohn-Sherbok. *After Noah: Animals and the Liberation of Theology.* London: Mowbray, 1996.

Linzey, Andrew, and Dorothy Yamamoto, eds. *Animals on the Agenda: Questions about Animals for Theology and Ethics.* London: SCM/Chicago: University of Illinois Press, 1998.

Lloyd, G. E. R. *Adversaries and Authorities: Investigations into Ancient Greek and Chinese Science.* Cambridge: Cambridge University Press, 1996.

Lopez, Barry. *Arctic Dreams: Imagination and Desire in a Northern Landscape.* New York: Scribner, 1986.

Macy, Joanna. *Mutual Causality in Buddhism and General Systems Theory.* Albany: State University of New York Press, 1991.

———. *World as Lover, World as Self.* Berkeley: Parallax, 1991.

Maguire, Daniel C. *The Moral Core of Judaism and Christianity: Reclaiming the Revolution.* Philadelphia: Fortress, 1993.

Maguire, Daniel C., and Harold Coward. *Visions of a New Earth: Religious Perspectives on Population, Consumption, and Ecology.* Albany: State University of New York Press, 2000.

Maguire, Daniel C., and Larry L. Rasmussen. *Ethics for a Small Planet: New Horizons on Population, Consumption, and Ecology.* Albany: State University of New York Press, 1998.

Mahabharata. Translated by M. N. Dutta. Delhi: Parimal, 1988.

Manusmriti [The Laws of Manu]. Translated by G. Buhler. Delhi: Motilal Banarsidass, 1975.

Martin, Julia. *Ecological Responsibility: A Dialogue with Buddhism.* Delhi: Tibet House and Sri Satguru, 1997.

Mascall, E. L. *The Christian Universe.* London: Darton, Longman & Todd, 1966.

Masri, Al-Hafiz B. A. *Islamic Concern for Animals.* Petersfield, Hants, England: Athene Trust, 1987.

Matthews, Clifford, Mary Evelyn Tucker, and Philip Hefner, eds. *When Worlds Converge: What Science and Religion Tell Us about the Story of the Universe and Our Place in It.* Chicago: Open Court, 2002.

Matthiessen, Peter. *The Snow Leopard.* New York: Bantam, 1979.

May, Roy H., Jr. *Poor of the Land.* Maryknoll, NY: Orbis, 1991.

McDaniel, Jay. *Of God and Pelicans: A Theology of Reverence for Life.* Louisville: Westminster/John Knox, 1989.

McFague, Sallie. *The Body of God: An Ecological Theology.* Minneapolis: Fortress, 1993.

———. *Models of God: Theology for an Ecological Nuclear Age.* Philadelphia: Fortress, 1987.

———. *Super, Natural Christians: How We Should Love Nature.* Minneapolis: Augsburg Fortress, 1997.

McGrath, Alistair. *The Reenchantment of Nature: The Denial of Religion and the Ecological Crisis.* New York: Doubleday, 2002.

McNeill, J. R. *Something New under the Sun: An Environmental History of the Twentieth Century World.* New York: Norton, 2000.

McPherson, Dennis, and J. Douglas Rabb. *Indian from the Inside: A Study in Ethnometaphysics.* Thunder Bay, ON: Lakehead University, Centre for Northern Studies, 1993.

Merchant, Carolyn. *The Death of Nature: Women, Ecology, and the Scientific Revolution.* San Francisco: Harper & Row, 1980.

Midgley, Mary. *Beast and Man: The Roots of Human Nature.* New York: Routledge, 1979.

Miller, James. "Daoism and Nature." Pp. 393–410 in *Nature across Cultures: Non-Western Views of Nature and Environment.* Edited by Helaine Selin. Dordrecht: Kluwer, 2003.

———. "Daoism and Nature." In *Encyclopedia of Religion and Nature.* Edited by Bron Taylor. London: Continuum, 2004.

———. "Ecology and Religion: Daoism and Ecology." In *Encyclopedia of Religion.* Edited by Lindsay Jones. 2nd ed. New York: Macmillan, 2005.

———. "Envisioning The Daoist Body in the Economy of Cosmic Power." *Daedalus* 130.4 (2001): 265–82.

Mitchell, John Hanson. *Ceremonial Time: Fifteen Thousand Years on One Square Mile.* Readings, MA: Addison-Wesley, 1984.

Momaday, N. Scott. *The Way to Rainy Mountain.* Albuquerque: University of New Mexico Press, 1969.

Muir, John. *My First Summer in the Sierra.* Boston: Houghton Mifflin, 1916.

Nakayama, Shigeru. *Academic and Scientific Traditions in China, Japan, and the West.* Translated by Jerry Dusenbury. Tokyo: University of Tokyo Press, 1984.

Nash, James. *Loving Nature: Ecological Integrity and Christian Responsibility.* Nashville: Abingdon, 1991.

Nash, Roderick Frazier. *Wilderness and the American Mind.* 4th ed. New Haven: Yale University Press, 1967.

Nasr, Seyyed Hossein. *Man and Nature: The Spiritual Crisis of Modern Man.* Dunstable, England: ABC/Chicago: Kazi, 1997.

———. *Religion and the Order of Nature.* New York: Oxford University Press, 1996.

Needham, Joseph, et al. *Science and Civilization in China.* 8 vols. Cambridge: Cambridge University Press, 1954–.

Nelson, Lance, ed. *Purifying the Earthly Body of God.* Albany: State University of New York Press, 1998.

Nelson, Richard K. *The Island Within.* New York: Vintage, 1991.

———. *Make Prayers to the Raven: A Koyukon View of the Northern Forest.* Chicago, 1983.

Northcott, Michael. *The Environment and Christian Ethics.* Cambridge: Cambridge University Press, 1996.

Norton, Bryan. "Biodiversity and Environmental Values: In Search of a Universal Earth Ethic." *Biodiversity and Conservation* 9 (2000): 1029–44.

Novak, David. *Natural Law in Judaism.* Cambridge: Cambridge University Press, 1998.

Oelschlaeger, Max. *Caring for Creation: An Ecumenical Approach to the Environmental Crisis.* New Haven: Yale University Press, 1994.

Olson, Sigurd. *Reflections from the North Country.* Minneapolis: University of Minnesota Press, 1998.

Osborn, Fairfield. *Our Plundered Planet.* Boston: Little, Brown, 1948.

Palmer, Clare. *Environmental Ethics and Process Thinking.* Oxford: Oxford University Press, 1998.

Palmer, Martin. *Faith in Conservation: New Approaches to Religion and the Environment.* Washington, DC: World Bank, 2003.

Paper, Jordan, and Li Chuang Paper. "Chinese Religions, Population, and the Environment." Pp. 173–91 in *Population, Consumption, and the Environment: Religious and*

Secular Responses. Edited by Harold Coward. Albany: State University of New York Press, 1995.

Paulos, Mar Gregorios. *The Human Presence: An Orthodox View of Nature.* Geneva: World Council of Churches, 1978. Reprinted as *The Human Presence: Ecological Spirituality and the Age of the Spirit.* New York, 1987.

Peet, Richard, and Michael Watts. *Liberation Ecologies: Environment, Development, and Social Movements.* New York: Routledge, 1996.

Peterson, Anna. *Being Human: Ethics, Environment, and Our Place in the World.* Berkeley: University of California Press, 2001.

Phelps, Norm. *The Dominion of Love: Animal Rights according to the Bible.* New York: Lantern, 2002.

Pike, Sarah. *New Age and Neopagan Religions in America.* New York: Columbia University, 2004.

Pinches, Charles, and Jay B. McDaniel, eds. *Good News for Animals? Christian Approaches to Animal Well-Being.* Maryknoll, NY: Orbis, 1993.

Porkert, Manfred. *The Theoretical Foundations of Chinese Medicine: Systems of Correspondence.* Cambridge: MIT Press, 1974.

Porkert, Manfred, and Christian Ullmann. *Chinese Medicine.* Translated and adapted by Mark Howson. New York: Morrow, 1982.

Posey, Darrell Addison. *Cultural and Spiritual Values of Biodiversity.* Nairobi, Kenya: United Nations Environmental Program, 1999.

Prescott-Allen, Robert. *The Wellbeing of Nations.* Washington, DC: Island Press/Ottawa: International Development Research Centre, 2001.

Primavesi, Anne. *From Apocalypse to Genesis: Ecology, Feminism. and Christianity.* Minneapolis: Fortress, 1991.

Rachels, James. *Created from Animals: The Moral Implications of Darwinism.* Oxford: Oxford University Press, 1990.

Ramos, Alcida Rita. *Indigenism: Ethnic Politics in Brazil.* Madison: University of Wisconsin Press, 1998.

Rappaport, Roy A. *Ecology, Meaning, and Religion.* Richmond, CA: North Atlantic, 1979.
———. *Pigs for the Ancestors: Ritual in the Ecology of a New Guinea People.* 2nd ed. Prospect Heights, IL, 2000.

Rasmussen, Larry. *Earth Community: Earth Ethics.* Maryknoll, NY: Orbis, 1996.

Reaka-Kudla, Marjorie, Don Wilson, and Edward O. Wilson. *Biodiversity II: Understanding and Protecting Our Biological Resources.* Washington, DC: Joseph Henry, 1997.

Reichel-Dolmatoff, Gerardo. *The Forest Within: The Worldview of the Tukano Amazonian Indians.* Totnes, U.K.: Themis-Green, 1996.

Rexroth, Kenneth. *Complete Poems.* Edited by Sam Hamill and Bradford Morrow. Port Townsend, WA: Copper Canyon, 2003.

Rig Veda. Commentary by Maharishi Dayanand Saraswati. 12 vols. New Delhi: Sarvadeshik Arya Pratinidhi Sabha, 1974.

Risser, Paul G., Jane Lubchenco, and Samuel A. Levin. "Biological Research Priorities: A Sustainable Biosphere." *BioScience* 47 (1991): 625–27.

Rockefeller, Steven, and John Elder, eds. *Spirit and Nature: Why the Environment Is a Religious Issue.* Boston: Beacon, 1991.

Rolston, Holmes, III. "The Bible and Ecology." *Interpretation: Journal of Bible and Theology* 50 (1996): 16–26.

————. *Conserving Natural Value*. New York: Columbia University Press, 1994.

————. "Does Nature Need to Be Redeemed?" *Zygon: Journal of Religion and Science* 29 (1994): 205–29.

————. *Science and Religion: A Critical Survey*. Philadelphia: Temple University Press, 1987.

Roseman, Marina. *Healing Sounds from the Malaysian Rainforest: Temiar Music and Medicine*. Berkeley, 1991.

Roy, Arundhati. *Power Politics*. Boston: South End, 2001.

Rue, Loyal. *Everybody's Story: Wising Up to the Epic of Evolution*. Albany: State University of New York Press, 2000.

Ruether, Rosemary Radford. *Gaia and God: An Ecofeminist Theology of Earth Healing*. San Francisco: Harper, 1992.

————. *Integrating Ecofeminism, Globalization, and World Religions*. Lanham, MD: Rowman & Littlefield, 2005.

————, ed. *Women Healing Earth: Third World Women on Ecology, Feminism, and Religion*. Maryknoll, NY: Orbis, 1994.

Rumi, Jalal al-Din. *Masnavi-yi ma'navi: The Mathnawi of Jalalu'ddin Rumi*. Translated by R. A. Nicholson. London: Gibb Memorial Trust, 1926.

Ryan, P. D. *Buddhism and the Natural World: Toward a Meaningful Myth*. Birmingham: Windhorse, 1998.

Saletore, Bhaskar Anand. *Medieval Jainism*. Bombay: Karnatak, 1938.

Salomonsen, Jone. *Enchanted Feminism: Ritual, Gender, and Divinity among the Reclaiming Witches of San Francisco*. London: Routledge, 2001.

Samuelson, Norbert M. *Judaism and the Doctrine of Creation*. Cambridge: Cambridge University Press, 1994.

Sanders, Scott Russell. *Staying Put: Making a Home in a Restless World*. Boston: Beacon, 1993.

Santmire, Paul. *The Travail of Nature: The Ambiguous Ecological Promise of Christian Theology*. Philadelphia: Fortress, 1985.

Sapontzis, Steve F., ed. *Food for Thought: The Debate over Eating Meat*. Amherst, NY: Prometheus, 2004.

Sardar, Ziauddin, ed. *Touch of Midas: Scientific Values and the Environment in Islam and the West*. Manchester: Manchester University Press, 1984.

Satvalekar, Pandit Shripad Damodar. *Prithivi-Sukta: Atharvaveda Book 12*. Surat: Swadhyay Mandal, 1958.

Schmemann, Alexander. *For the Life of the World: Sacraments and Orthodoxy*. New York: St. Vladimir's Seminary Press, 1973.

Schmithausen, Lambert. *Buddhism and Nature*. Studia philologica buddhica Occasional Paper 7. Tokyo: International Institute for Buddhist Studies, 1991.

Schwartz, Richard H. *Judaism and Global Survival*. New York: Atara, 1987.

Schweitzer, Albert. *Civilization and Ethics*. Translated by C. T. Campion. London: Allen & Unwin, 1923.

Sears, John. *Sacred Places: American Tourist Attractions in the Nineteenth Century*. New York: Oxford University Press, 1989.

Selin, Helaine, ed. *Nature across Cultures: Views of Nature and the Environment in Non-Western Cultures*. Dordrecht: Kluwer, 2003.

Settar, S. *Pursuing Death*. Dharwad: Karnatak University, Institute of Indian Art History, 1990.

Sheldon, Joseph K. *Rediscovery of Creation: A Bibliographical Study of the Church's Response to the Environmental Crisis.* Metuchen, NJ: American Theological Library Association/Scarecrow, 1992.

Shepard, Paul. *Nature and Madness.* San Francisco: Sierra Club, 1982.

Sherrard, Philip. *The Eclipse of Man and Nature: An Enquiry into the Origins and Consequences of Modern Science.* West Stockbridge: Lindisfarne, 1987.

———. *Human Image, World Image.* Ipswich: Golgonooza, 1990.

———. *The Sacred in Life and Art.* Ipswich: Golgonooza, 1990.

Shiva, Vandana. *Biopiracy: The Plunder of Nature and Knowledge.* London: Zed, 1997.

———. *Monocultures of the Mind: Biodiversity, Biotechnology, and the Third World.* London: Zed, 1993.

———. *Stolen Harvest: The Hijacking of the Global Food Supply.* Cambridge, MA: South End, 1999.

———. *The Violence of the Green Revolution: Third World Agriculture, Ecology and Politics.* London: Zed, 1993.

Sideris, Lisa. *Environmental Ethics, Ecological Theology, and Natural Selection.* New York: Columbia University Press, 2003.

Silko, Leslie Marmon. *Ceremony.* New York: Viking Penguin, 1977.

Simkins, Ronald A. *Creator and Creation: Nature in the Worldview of Ancient Israel.* Peabody, MA: Hendrickson, 1994.

Sivaraksa, Sulak, ed. *Santi Pracha Dhamma: Essays in Honor of the Late Puey Ungphakorn.* Bangkok: Santi Pracha Dhamma Institute, 2001.

———. *Seeds of Peace.* Berkeley: Parallax, 1992.

Slovic, Scott. *Seeking Awareness in American Nature Writing: Henry Thoreau, Annie Dillard, Edward Abbey, Wendell Berry, Barry Lopez.* Salt Lake City: University of Utah Press, 1992.

Smith, Huston. "Tao Now." Pp. 62–81 in *Earth Might Be Fair.* Edited by Ian Barbour. Englewood Cliffs, NJ: Prentice-Hall, 1972.

Smith, Nigel J. H. *The Enchanted Amazon Rain Forest: Stories from a Vanishing World.* Gainesville: University Press of Florida, 1996.

Smith, Richard J., and D. W. Y. Kwok, eds. *Cosmology, Ontology, and Human Efficacy: Essays in Chinese Thought.* Honolulu: University of Hawai'i Press, 1993.

Smith, Samantha. *Goddess Earth: Exposing the Pagan Agenda of the Environmental Movement.* Lafayette, LA: Huntington, 1994.

Snyder, Gary. *The Gary Snyder Reader: Prose, Poetry, and Translations, 1952–1998.* Washington, DC: Counterpoint, 1999.

———. *A Place in Space.* Washington, DC: Counterpoint, 1995.

———. *The Practice of the Wild.* San Francisco: North Point, 1990.

Sober, Elliott, and David Sloan Wilson. *Unto Others: The Evolution and Psychology of Unselfish Behavior.* Cambridge: Harvard University Press, 1998.

Social Teaching. Washington, DC: United States Catholic Conference, 1991.

Soloveitchik, Hayim Dov. *The Lonely Man of Faith.* New York: Doubleday, 1992 (orig. *Tradition* 7.2 [1965]).

Soper, David E. *The Geography of Religions.* Englewood Cliffs, NJ: Prentice-Hall, 1967.

Sorrell, Roger D. *St. Francis of Assisi and Nature: Tradition and Innovation in Western Christian Attitudes toward the Environment.* Oxford: Oxford University Press, 1988.

Srimad Bhagavata Mahapurana. Translated by C. L. Goswami and M. A. Shastri. Gorakhpur: Gita, 1982.

Staniloae, Dimitru. "The World as Gift and Sacrament of God's Love." *Sobornost* 5.9 (1969).

Steward, Julian. *Evolution and Ecology: Essays on Social Transformation*. Urbana: University of Illinois Press, 1977.

Strassberg, Richard E. *Inscribed Landscapes: Travel Writing from Imperial China*. Berkeley: University of California Press, 1994.

Sung, Ying-hsing. *T'ien-kung K'ai-wu: Chinese Technology in the Seventeenth Century*. Translated by E-Tu Zen Sun and Shiou-Chuan Sun. University Park: Pennsylvania State University Press, 1966.

Suri, Santi. *Jiva Vicara Prakaranam along with Pathaka Ratnakara's Commentary*. Edited by Muni Ratna-Prabha Vijaya. Translated by Jayant P. Thaker. Madras: Jain Mission Society, 1950.

Suzuki, David, and Peter Knudtson. *Wisdom of the Elders: Honoring Sacred Native Visions of Nature*. New York: Bantam, 1992.

Swimme, Brian, and Thomas Berry. *The Universe Story: From the Primordial Flaring Forth to the Ecozoic Era—A Celebration of the Unfolding of the Cosmos*. San Francisco: HarperCollins, 1992.

Szerszynksi, Bronislaw. *Nature, Technology, and the Sacred*. Malden, MA: Blackwell, 2005.

Takacs, David. *The Idea of Biodiversity: Philosophies of Paradise*. Baltimore: John Hopkins University Press, 1996.

Taylor, Bron Raymond. "Diggers, Wolfs, Ents, Elves, and Expanding Universes: Bricolage, Religion, and Violence from Earth First! and the Earth Liberation Front to the Antiglobalization Resistance." Pp. 26–74 in *The Cultic Milieu: Oppositional Subcultures in an Age of Globalization*. Edited by Jeffrey Kaplan and Heléne Lööw. Lanham, MD: Altamira/Rowman & Littlefield, 2002.

———. "Earth and Nature-Based Spirituality (Part I): From Deep Ecology to Radical Environmentalism." *Religion* 31.2 (2001): 175–93.

———. "Earth and Nature-Based Spirituality (Part II): From Deep Ecology to Scientific Paganism." *Religion* 31.3 (2001): 225–45.

———. "Earthen Spirituality or Cultural Genocide: Radical Environmentalism's Appropriation of Native American Spirituality." *Religion* 17.2 (1997): 183–215.

———, ed. *Ecological Resistance Movements: The Global Emergence of Radical and Popular Environmentalism*. Albany: State University of New York Press, 1995.

———, ed. *The Encyclopedia of Religion and Nature*. London: Continuum, 2005.

———. "Resacralizing Earth: Pagan Environmentalism and the Restoration of Turtle Island." Pp. 97–151 in *American Sacred Space*. Edited by David Chidester and Edward T. Linenthal. Bloomington: Indiana University Press, 1995.

Taylor, Sarah. "Reinhabiting Religion: Green Sisters, Ecological Renewal, and the Biogeography of Religious Landscape." In *This Sacred Earth: Religion, Nature, Environment*. Edited by Roger S. Gottlieb. 2nd ed. New York: Routledge, 2004.

Theokritoff, Elizabeth. "Creation, Incarnation, and Transfiguration: The Material World and Our Understanding of it." *Sobornost/Eastern Churches Review* 11.1–2 (1989).

Thoreau, Henry David. *The Essays of Henry D. Thoreau*. Edited by Lewis Hyde. San Francisco: North Point, 2002.

————. *Walden: A Fully Annotated Edition.* Edited by Jeffrey S. Cramer. New Haven: Yale University Press, 2004.

Tillich, Paul. "Nature, Also, Mourns for a Lost Good." In Tillich's *The Shaking of the Foundations.* New York: Scribner, 1962.

Tirosh-Samuelson, Hava, ed. *Judaism and Ecology: Created World and Revealed Word.* Cambridge: Center for the Study of World Religions, Harvard Divinity School, 2002.

Totman, Conrad. *Early Modern Japan.* Berkeley: University of California Press, 1993.

Tuan, Yi-Fu. *Topophilia: A Study of Environmental Perception, Attitudes, and Values.* Englewood Cliffs, NJ: Prentice-Hall, 1974.

Tucker, Mary Evelyn. *Moral and Spiritual Cultivation in Japanese Neo-Confucianism: The Life and Thought of Kaibara Ekken (1630–1714).* Albany: State University of New York Press, 1989.

————. *Worldly Wonder: Religions Enter Their Ecological Phase.* Chicago: Open Court, 2003.

Tucker, Mary Evelyn, and John Berthrong, eds. *Confucianism and Ecology: The Interrelation of Heaven, Earth, and Humans.* Cambridge: Center for the Study of World Religions, Harvard Divinity School, 1998.

Tucker, Mary Evelyn, and John Grim, eds. *Religion and Ecology: Can the Climate Change?* Special issue of *Daedalus* 130.4. Cambridge, MA: American Academy of Arts and Sciences, 2001.

————, eds. *Worldviews and Ecology: Religion, Philosophy, and the Environment.* Maryknoll, NY: Orbis, 1994.

Tucker, Mary Evelyn, and Duncan Ryuken Williams, eds. *Buddhism and Ecology: The Interconnection of Dharma and Deeds.* Cambridge: Center for the Study of World Religions, Harvard Divinity School, 1997.

Ulanowicz, Robert E. "Ecosystem Dynamics: A Natural Middle." *Theology and Science* 2.2 (2004): 231–53.

Umasvati. *That Which Is (Tattvartha Sutra): A Classic Jain Manual for Understanding the True Nature of Reality.* Translated by Nathmal Tatia. San Francisco: HarperCollins, 1994.

United States Catholic Bishops. *Global Climate Change: A Plea for Dialogue, Prudence, and the Common Good.* Washington, DC: United States Catholic Conference, 2001.

United States Congress. National Environmental Policy Act. 83 Stat. 852. Public Law 91-190. 1969.

Vasileios, Archimandrite. *Ecology and Monasticism.* Montreal: Alexander, 1996.

Vecsey, Christopher, and Robert W. Venables. *American Indian Environments: Ecological Issues in Native American History.* Syracuse, 1980.

Waldau, Paul. *The Specter of Speciesism.* New York: Oxford University Press, 2002.

Walters, Kerry S., and Lisa Portmess, eds. *Religious Vegetarianism from Hesiod to the Dalai Lama.* Albany: State University of New York Press, 2001.

Wang, Aihe. *Cosmology and Political Culture in Early China.* Cambridge: Cambridge University Press, 2000.

Waskow, Arthur. *Down-to-Earth Judaism: Food, Money, Sex, and the Rest of Life.* New York: Morrow, 1995.

————, ed. *Torah of the Earth: Exploring 4000 Years of Ecology in Jewish Thought.* 2 vols. Burlington, VT: Jewish Lights, 2000.

Weaver, Jace, ed. *Defending Mother Earth: Native American Perspectives on Environmental Justice*. Maryknoll, NY: Orbis, 1996.

Webb, Stephen H. *On God and Dogs: A Christian Theology of Compassion for Animals*. New York: Oxford University Press, 1998.

White, Lynn, Jr. "The Historical Roots of Our Ecological Crisis." *Science* 155 (10 March 1967): 1203–7.

Wilbert, Johannes. *Mystic Endowment: Religious Ethnography of the Warao Indians*. Cambridge, MA, 1993.

Wilkinson, Loren, ed. *Earthkeeping in the Nineties: Stewardship and the Renewal of Creation*. Grand Rapids: Eerdmans, 1991.

Williams, Terry Tempest. *Red: Passion and Patience in the Desert*. New York: Pantheon, 2001.

Wilson, David J. *Indigenous South Americans of the Past and Present: An Ecological Perspective*. Boulder, CO: Westview, 1999.

Wilson, David Sloan. *Darwin's Cathedral: Evolution, Religion, and the Nature of Society*. Chicago: University of Chicago Press, 2002.

Wilson, Edward O. *The Future of Life*. New York: Knopf, 2002.

Wilson, Samuel M. *The Indigenous People of the Caribbean*. Gainesville: University Press of Florida, 1997.

Wirzba, Norman. *A Paradise of God: Renewing Religion in an Ecological Age*. Oxford: Oxford University Press, 2003.

World Bank. *Economic Developments in India*. Washington, DC: World Bank, 1995.

World Commission on Environment and Development. *Our Common Future*. Oxford: Oxford University Press, 1987.

Yaffe, Martin D., ed. *Judaism and Environmental Ethics: A Reader*. Lanham, MD: Lexington, 2001.

Yasunaga, Toshinobu. *Ando Shoeki: Social and Ecological Philosopher of Eighteenth-Century Japan*. New York: Weatherhill, 1992.

Zaidi, Iqtidar H. "On the Ethics of Man's Interaction with the Environment: An Islamic Approach." *Environmental Ethics* 3.1 (1981): 35–47.

Zhang Dainian. *Key Concepts in Chinese Philosophy*. Translated and edited by Edmund Ryden. New Haven: Yale University Press, 2002.

INDEX

....

633